Applied Polyoxometalate-based Electrocatalysis

Applied Polyoxometalate-based Electrocatalysis

Edited by Diana M. Fernandes

WILEY-VCH

Editor

Dr. Diana M. Fernandes
LAQV-REQUIMTE
Faculty of Sciences
University of Porto
Rua do Campo Alegre, s/n
Porto 4169-007
Portugal

Cover Image: © Akaradech Pramoonsin/Getty Images, © tommy/Getty Images, © YIUCHEUNG/Shutterstock, © Scharfsinn/Shutterstock

Library of Congress Card No.: applied for

British Library Cataloguing-in-Publication Data
A catalogue record for this book is available from the British Library.

Bibliographic information published by the Deutsche Nationalbibliothek The Deutsche Nationalbibliothek lists this publication in the Deutsche Nationalbibliografie; detailed bibliographic data are available on the Internet at <http://dnb.d-nb.de>.

Print ISBN: 978-3-527-35213-5
ePDF ISBN: 978-3-527-84269-8
ePub ISBN: 978-3-527-84270-4
oBook ISBN: 978-3-527-84271-1

Typesetting Straive, Chennai, India

Printed and bound by CPI Group (UK) Ltd, Croydon, CR0 4YY

C9783527352135_280824

Contents

Part I

Fundamentals

1

Introduction to Polyoxometalates

Daniela Flores and Carlos M. Granadeiro

University of Porto, LAQV/REQUIMTE, Department of Chemistry and Biochemistry, Faculty of Sciences, Rua do Campo Alegre, s/n, Porto 4169-007, Portugal

1.1 Introduction

Polyoxometalates (POMs) represent a captivating and unique class of nanoscale metal–oxide clusters, boasting remarkable structural and chemical versatility, making them a sought-after choice across diverse scientific domains [1–3], namely biomedicine, catalysis, colloid science, electronic or magnetic devices, functional materials, nanotechnology, sensors and surfaces [4, 5]. The exploration of these materials dates back to the 1826 discovery by Berzelius of the ammonium salt of $[PMo_{12}O_{40}]^{3-}$. However, it was not until 1933, when Keggin conducted the first structural determination of the tungsten analogue $[PW_{12}O_{40}]^{3-}$ over a century later, that their structural intricacies began to be unveiled [6]. The scientific community has devoted considerable effort and time to the structural characterization of these intriguing compounds, and so the potential applications of POMs only began to be truly explored after the late 1970s.

POMs are inorganic clusters composed of metallic centres, frequently in their highest oxidation state (M = V, Mo, W, Nb and Ta), named the addenda atoms, connected by bridging or terminal oxygen atoms. Currently, there are a large number of known POMs structures with diverse architectures and distinct compositions. In order to establish correlations between their highly symmetric structures, physical properties and reactivity, POMs are typically divided into two main categories: isopolyanions and heteropolyanions (Figure 1.1) [7].

Isopolyanions, with the general formula $[M_xO_y]^{m-}$, are composed of a metal–oxide framework where the addenda atom is a singular transition-metal ion from group V or VI surrounded by oxygen atoms. Heteropolyanions, represented by the general formula $[X_zM_xO_y]^{n-}$ with $z \leq x$, contain an additional atom (X) named heteroatom. In POM structures, heteroatoms are typically elements from the p-block (X = P, Si, Al, Ga, Ge) but also from the d-block (X = Fe, Co, Ni, Zn), which are located at the centre of a M_xO_y shell [8].

The spatial arrangement of POMs can be visualized as packed arrays of polyhedral MO_x units, typically MO_6 octahedra or, more rarely, MO_5 pyramidal units.

Applied Polyoxometalate-based Electrocatalysis, First Edition. Edited by Diana M. Fernandes.

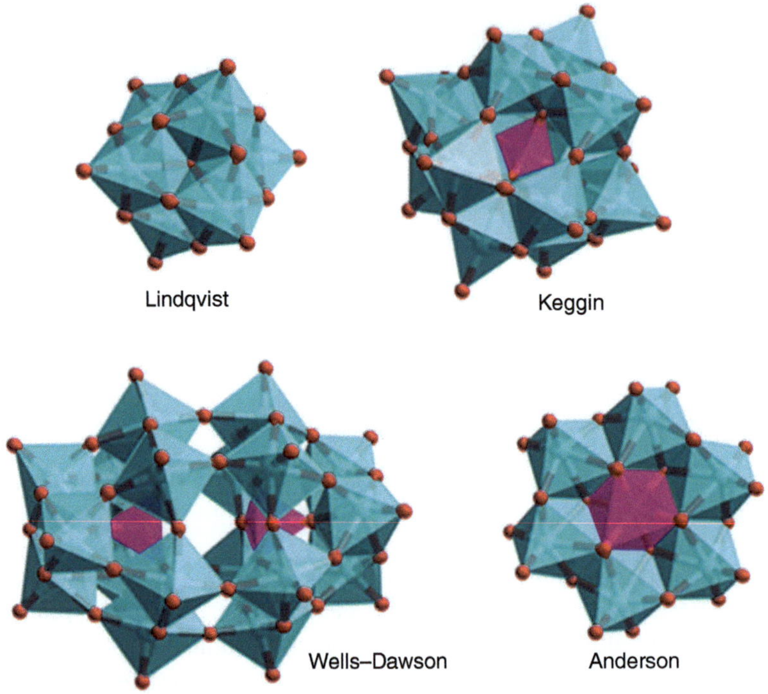

Figure 1.1 Structural representation of the most common POM categories: the Lindqvist isopolyanion ($[M_6O_{19}]^{n-}$), and the heteropolyanions Keggin-($[XM_{12}O_{40}]^{m-}$), Wells–Dawson-($[X_2M_{18}O_{62}]^{w-}$) and Anderson-type ($[XM_6O_{24}]^{n-}$) anions. MO_x: light blue; X: purple; O: red. Source: Granadeiro et al. [4]/with permission of Elsevier.

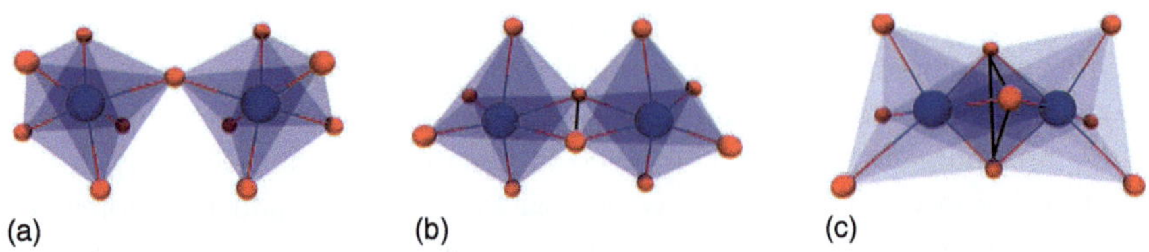

Figure 1.2 Polyhedral and ball-and-stick representation of the different types of MO_6 connectivity in POMs: (a) corner-shared, (b) edge-shared and (c) face-shared octahedra. Source: Bijelic and Rompel [9]/Springer Nature/CC BY 4.0.

These units act as building blocks in the construction of multiple architectures by connecting between themselves and sharing edges, corners or faces (Figure 1.2) [10]. Based on this, the oxygen atoms in a POM structure can be classified according to their position (Figure 1.3). The oxygen atom connected to the heteroatom is denoted as O_a, O_b and O_c represent corner- and edge-sharing atoms, while the unshared (terminal) oxygen atom is denoted as O_d [12].

The vast number of POM structures is further broadened by the use of more than one type of M addenda atoms (mixed-addenda anions) or by structural modification through the removal of one or more MO_x groups (lacunary anions). The formation of lacunary POM anions (Figure 1.4) is achieved by varying the

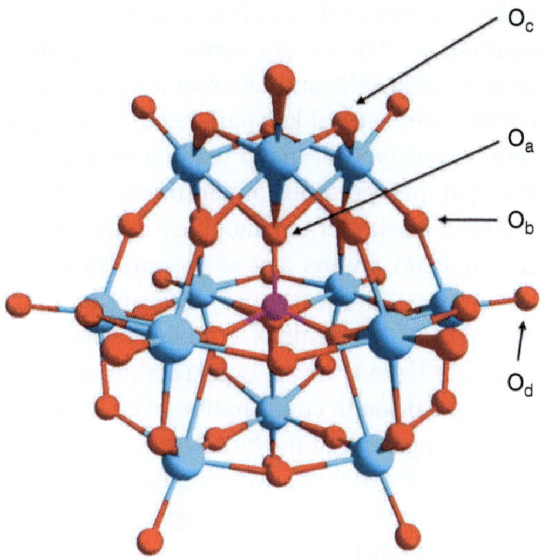

Figure 1.3 Ball-and-stick representation of the Keggin anions with the different type of oxygen atoms: O_a: heteroatom-connected oxygen; O_b: corner-shared oxygen; O_c: edge-shared oxygen; O_d: terminal oxygen. Source: Zheng et al. [11]/The Royal Society of Chemistry/CC BY 3.0.

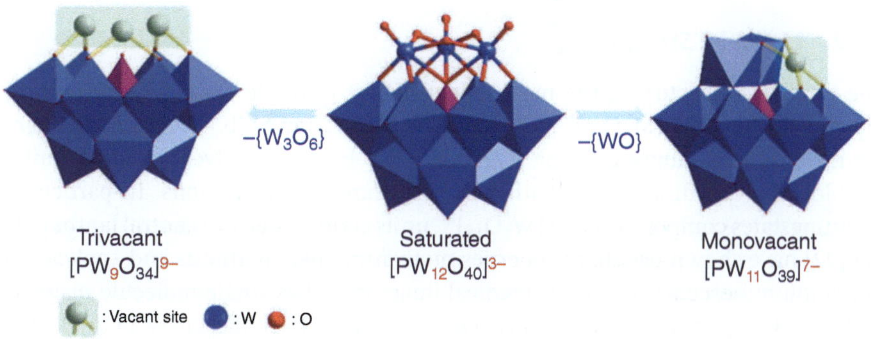

Figure 1.4 Polyhedral and ball-and-stick representations of the formation of mono- and trilacunary (vacant) Keggin-type POMs. Source: Bao et al. [13]/with permission of Elsevier.

experimental conditions, such as temperature, pH or precursors, which are frequently used to achieve enhanced reactivity and superior mechanical properties [14, 15]. The obtained empty lacuna with free oxygen atoms leads the anion to act as a tetradentate or pentadentate ligand capable of coordinating to any electrophilic species, typically lanthanide or transition-metal ions [5]. By doing so, unprecedented structures have been obtained by the connection of two (or more) known anions, including the formation of supramolecular POM structures.

1.2 Polyoxometalate Structures

1.2.1 Synthetic Methodologies

The synthesis of POM structures is relatively straightforward, typically involving an acidic solution containing relevant metal oxide anions. This can be achieved by following two main methodologies: (i) one-pot synthesis and (ii) building block method. In the one-pot synthesis, a condensation reaction takes place between simple metal salts and heteroanions. However, this approach demands the precise control over several key parameters, namely the choice of reducing agent, concentration and type of metal oxide anion, heteroatom type and concentration, ionic strength, pH, presence of additional ligands and reaction temperature [16]. In the building block method, lacunary anions, obtained by removal of MO_x groups from parent POM anions, act as precursors to build bulkier and more complex POM structures [17]. The lacunary derivatives, typically prepared by controlled hydrolysis of parent POM solutions, contain nucleophilic oxygen centres and will act as multidentate ligands in the construction of discrete larger clusters or even polymeric networks through coordination to electrophilic species [18]. The more efficient and elegant approach of the building block method has led to the exponential growth of novel POM structures with increasingly complex arrangements, higher number of functionalities and distinct physico-chemical properties, enlarging the application fields of POMs [19–21].

1.2.2 Lindqvist Structure

The Lindqvist structure is the most representative example of isopolyanions with the general formula $[M_6O_{19}]^{n-}$ with M = W, Mo, Cr and Nb. The monolacunary Lindqvist polyoxotungstate unit $[W_5O_{18}]^{6-}$ has been extensively used as building block by coordination to lanthanide and transition-metal ions. In particular, decatungstates composed of two $[W_5O_{18}]^{6-}$ units coordinated to a central lanthanide ion [22] have shown peculiar properties motivating their synthesis and application in photoluminescence, catalysis, medical imaging and as single-molecule magnets (SMM) [23–27]. In these compounds, with the general formula $[Ln(W_5O_{18})_2]^{9-}$, the lanthanide ion is coordinated to eight oxygen atoms exhibiting a square antiprismatic geometry [28]. The $[Eu(W_5O_{18})_2]^{9-}$ anion is among the most studied POMs due to its exceptionally high quantum yield arising from the highly efficient energy transfer process from the O → W charge transfer band to the lanthanide ion [29].

1.2.3 Keggin Structure

The Keggin structure, with the general formula $[XM_{12}O_{40}]^{n-}$, stands as the most well-known heteropolyanion featuring tetrahedrally coordinated heteroatoms and four trimetallic M_3O_3 groups arranged around a central XO_4 tetrahedron [30]. The Keggin anion exhibits five rotational isomers (α, β, γ, δ and ε) resulting from consecutive 60° rotations of each M_3O_3 unit (Figure 1.5) [7].

Keggin anions, through controlled alkaline hydrolysis, are also able to form lacunary species by removing one or more MO_x groups from the plenary structure. The monolacunary $[XM_{11}O_{39}]^{(n+4)-}$ Keggin anion is able to coordinate to trivalent (or even tetravalent) metallic cations through the available oxygen atoms in the lacuna

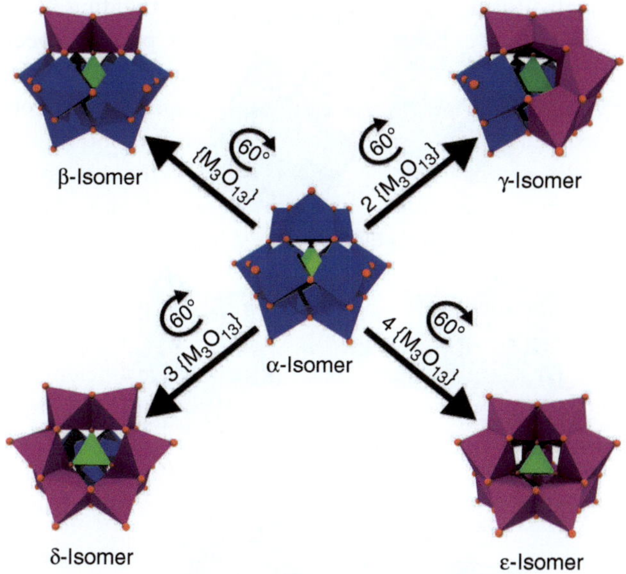

Figure 1.5 Polyhedral representation of the rotational isomers of the Keggin anions. Source: Bijelic and Rompel [9]/Springer Nature/CC BY 4.0.

[4]. These monolacunary anions typically originate as 1 : 1 $[XM_{11}M'(L)O_{39}]^{n-}$ or 1 : 2 $[M'(XM_{11}O_{39})_2]^{n-}$ complexes. The 1 : 1 complexes are mainly obtained when the metallic cation (M′) is a transition-metal or p-group element. For these complexes, L represents a monodentate ligand (generally a water molecule) necessary to complete the octahedral coordination of M′. Bulkier metallic cations (e.g. lanthanide ions) tend to coordinate to two monolacunary units forming a 1 : 2 complex [31]. Nevertheless, a few examples can be found in the literature reporting 1 : 1 complexes with lanthanide ions [17, 19, 32]. The formation of 2 : 2 Keggin-type complexes is even rarer, although some crystalline structures have already been reported, namely $[\{Ln(H_2O)_3(\alpha\text{-}PW_{11}O_{39})\}_2]^{6-}$ with Ln(III) = Nd, Gd (Figure 1.6) [33].

Trilacunary Keggin anions $[XM_9O_{34}]^{(n+6)-}$, obtained from the removal of three MO^{4+} units, are also commonly used in the preparation of POM structures. The removal of the three MO^{4+} units can be from octahedra sharing a corner (A-type) or sharing an edge (B-type) (Figure 1.7).

Several crystalline determinations of POM structures using trilacunary Keggin anions as building blocks have been described, mostly using tungstoarsenate, tungstophosphate, tungstoantimonate and tungstosilicate moieties [34–37].

The major area of application for Keggin-based POMs has been catalysis, mainly oxidative reactions with alcohols, alkenes and aldehydes [38, 39], as well as in the fields of electro- and photocatalysis [40]. Besides, europium-containing Keggin POMs have also shown interesting luminescent features [29], gadolinium-containing Keggin POMs have exhibited promising features as MRI contrast agents [41], while holmium and erbium analogues have been explored as SMMs [25].

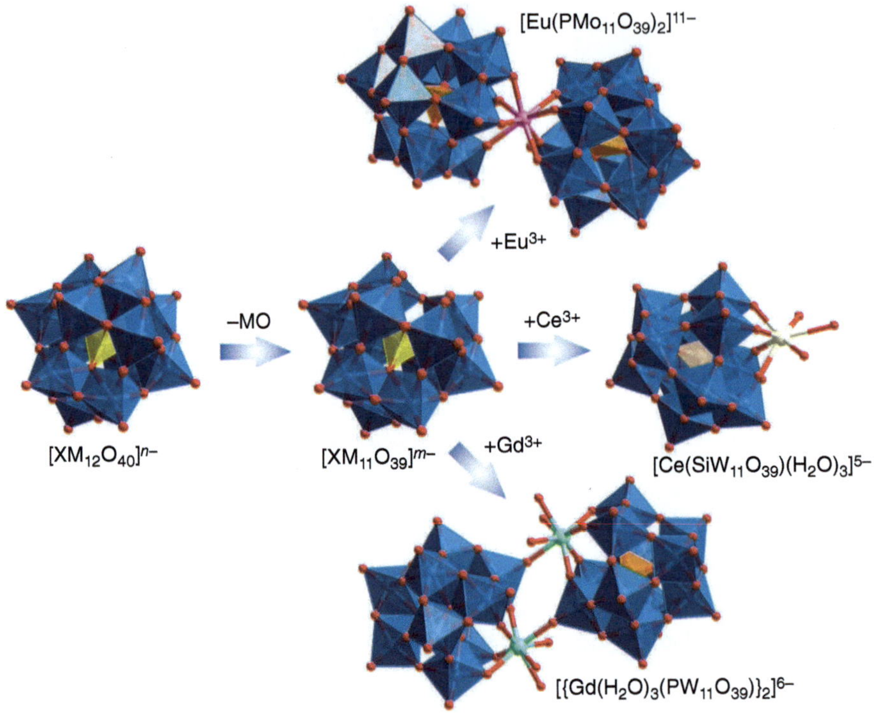

Figure 1.6 Polyhedral representation of the Keggin-type complexes: 1 : 2 $[Eu(PMo_{11}O_{39})_2]^{11-}$, 1 : 1 $[Ce(SiW_{11}O_{39})(H_2O)_3]^{5-}$ and 2 : 2 anionic $[\{Gd(H_2O)_3(\alpha\text{-}PW_{11}O_{39})\}_2]^{6-}$. Source: Granadeiro et al. [21]/with permission of Elsevier.

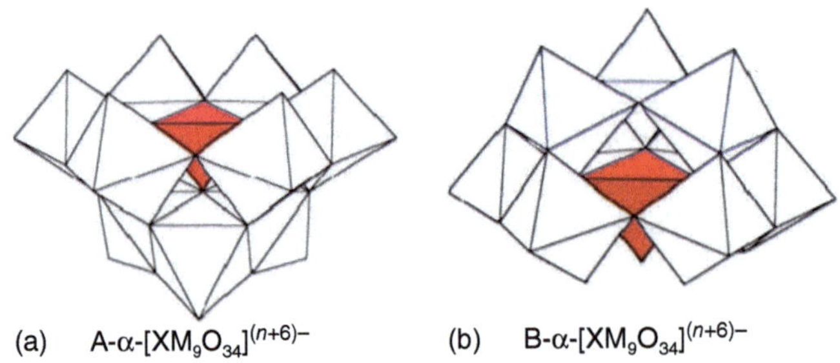

Figure 1.7 Polyhedral representation of the A- and B-type isomers of the Keggin trilacunary structure. (a) A-α-$[XM_9O_{34}]^{(n+6)}$ and (b) B-α-$[XM_9O_{34}]^{(n+6)}$.

1.2.4 Wells–Dawson Structure

The Wells–Dawson structure, with the general formula $[X_2M_{18}O_{62}]^{n-}$, consists of a diamagnetic anion composed of 18 MO_6 octahedra sharing corners and edges with the heteroatoms occupying the tetrahedral positions. Depending on the orientation of the two top M_3O_3 units, the Wells–Dawson structure displays three rotational isomers (α, β or γ) (Figure 1.8) [43]. Wells–Dawson anions can also form lacunary

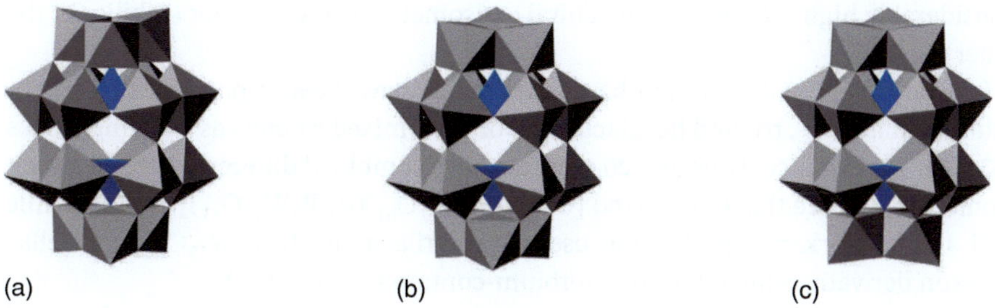

Figure 1.8 Polyhedral representation of the rotational isomers of the Wells–Dawson anions. (a) α-Wells–Dawson, (b) β-Wells–Dawson, and (c) γ-Wells–Dawson. Source: Matsunaga et al. [42]/MDPI/CC BY 4.0.

Figure 1.9 Polyhedral representation of the positional isomers of the α-monolacunary Wells–Dawson structure. Source: Huang et al. [45]/MDPI/CC BY 4.0.

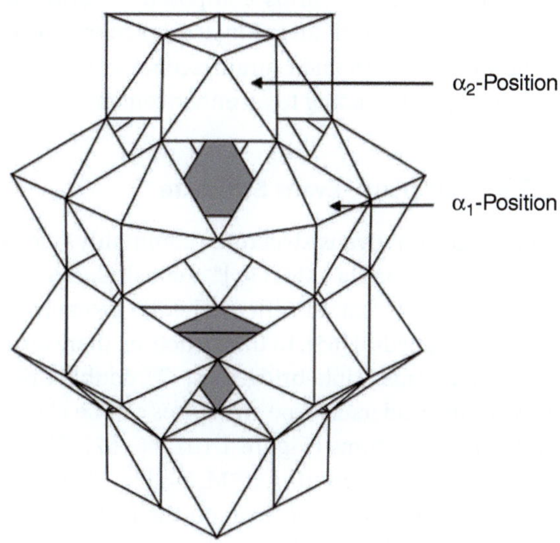

derivatives by removing one or more MO_x groups from the parent structure. In the case of isomers of lacunary species, it is necessary to distinguish between the different isomers by specifying the location of the lacuna [44]. Considering the α-isomer, if the lacuna is located in the equatorial region, the isomer is denoted as α_1, while if the lacuna is located at the top, the isomer is α_2 (Figure 1.9).

Monolacunary Wells–Dawson anions have also been used as building blocks for the preparation of monomeric 1 : 1 or dimeric 1 : 2 or 2 : 2 complexes by coordination to metallic cations [46]. The metallic centre in 2 : 2 dimers can coordinate to an oxygen atom of another unit in a 'cap-to-cap' or 'cap-to-belt' arrangement depending on whether the oxygen atom is located at the top or at the equatorial region, respectively [47, 48].

Numerous Wells–Dawson-based POM structures have been reported to date, consisting of either monomeric or dimeric complexes mainly containing α-isomers due to their higher stability when compared with the other isomers [46, 49, 50]. Among α-isomers, the number of crystalline reports dealing with the α_2-isomer is considerably higher than with the chiral α_1-isomer due to the minor stability of the latter [51, 52].

Several other Wells–Dawson-based structures have been reported using either multilacunary (di-, tri- and hexalacunary) or even mixed isomers as building blocks [53–56]. For instance, Pope and co-workers have combined different Wells–Dawson isomers to prepare the α_1,α_2-mixed $[Ce(\alpha_1\text{-}P_2W_{17}O_{61})(\alpha_2\text{-}P_2W_{17}O_{61})]^{17-}$ [57], while Hill and co-workers reported the use of two trilacunary $[\alpha\text{-}P_2W_{15}O_{56}]^{12-}$ Wells–Dawson derivatives linked by an ytterbium-containing cluster [58].

Wells–Dawson-based POMs have been mainly applied in the fields of catalysis and magnetism [59–61]. Despite the relatively weak emission of lanthanide-containing Wells–Dawson POMs, some reports can be found dealing with their application in photoluminescence [62–64]. Granadeiro et al. proposed the synthesis of organic–inorganic hybrids composed of lanthanide-containing $[\alpha_2\text{-}P_2W_{17}O_{61}]^{10-}$ Wells–Dawson units and 3-hydroxypicolinic acid [62]. In these hybrids, sensitization of the lanthanide luminescence could be more efficiently achieved via the organic ligands leading to an enhancement of the global quantum efficiency.

1.2.5 Anderson–Evans Structure

The Anderson–Evans structure (commonly referred to simply as Anderson), with the general formula $[XM_6O_{24}]^{n-}$, consists in a planar ring arrangement composed of a central heteroatom with octahedral geometry XO_6 surrounded by six MO_6 edge-sharing octahedral [65]. In this structure, there are three distinct coordination modes of oxygen atoms: triple-bridged (μ_3-O), double-bridged (μ_2-O) and terminal (O_d) oxygen atoms. Anderson-type structures can be classified according to the protonation of the oxygen atoms (Figure 1.10) [66, 67]. The non-protonated A-type Anderson with the general formula $[X^{n+}M_6O_{24}]^{(12-n)-}$ and the protonated B-type Anderson in which the protons are located on the μ_3-O atoms around the heteroatom with the formula $[X^{n+}(OH)_6M_6O_{18}]^{(6-n)-}$.

In the Anderson-type structure, despite the addenda atoms being practically limited to W or Mo, nearly all first-row transition-metals have been reported as heteroatoms. Moreover, several other elements, such as metalloids and post-transition-

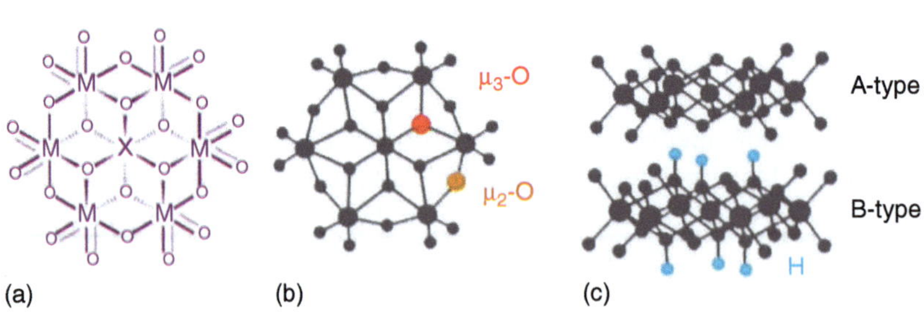

Figure 1.10 Structure of the Anderson-type POM (a), ball-and-stick representation of the Anderson structure highlighting the different coordination of oxygen atoms (b) and the different types (A- or B-) according to the protonation of oxygen atoms (c). Source: Wei et al. [65]/MDPI/CC BY 4.0.

metals (Al, Ga, Sb, Te), have also been used as heteroatoms in Anderson-type structures. Nevertheless, there is a predominance in the number of reported crystalline structures using $[Cr(OH)_6Mo_6O_{18}]^{3-}$, $[TeMo_6O_{24}]^{6-}$, $[AlMo_6(OH)_6O_{18}]^{3-}$ and $[Co(OH)_6Mo_6O_{18}]^{3-}$ as building units [68, 69].

The physico-chemical properties of the Anderson-type POM are mainly governed by the type of heteroatom present in the structure [14, 70]. These anions have been mainly applied in the fields of photochromism, photocatalysis, biomedicine and magnetism [71, 72].

1.2.6 Preyssler Structure

The Preyssler-type POM is a bulky heteropolyoxotungstophosphate with the general formula $[XP_5W_{30}O_{110}]^{14-}$ first described by Preyssler in 1970 [73]. It is composed of five $[PW_6O_{22}]^{9-}$ units linked together in a cyclic arrangement forming an internal cavity occupied by an encrypted cation. In this structure, the presence of water molecules coordinated to the cation displaces it outside the equatorial plane of the framework [74]. Although the original cationic element is sodium, several other crystalline Preyssler structures have been reported using different central cations typically achieved by cationic-exchange procedures. Pope et al. reported a series of novel Preyssler-type structures by replacement of the central sodium cation by trivalent lanthanides as well as tetravalent Ce and U [75]. However, the crystalline structure for these compounds was only determined a few years later for the europium-exchanged $([Eu(OH_2)P_5W_{30}O_{110}])^{14-}$ anion [76]. Kortz and co-workers described the direct synthesis of the silver-analogue $[AgP_5W_{30}O_{110}]^{14-}$ Preyssler structure by solvothermal and microwave-assisted synthesis [77]. Wang and co-workers prepared the first non-phosphate Preyssler-type POM by using tetrahedral $\{SO_4\}$ group in the structure leading to $[XS_5W_{30}O_{110}]^{14-}$ with $X = Na^+$ or K^+ [78] (Figure 1.11).

The central cationic centre strongly influences the physico-chemical properties of the Preyssler-type structure, ultimately determining their fields of application. The europium-containing analogue has been mainly investigated with promising results in luminescent, photo- and electrochromic applications [64, 80–84]. Sodium-

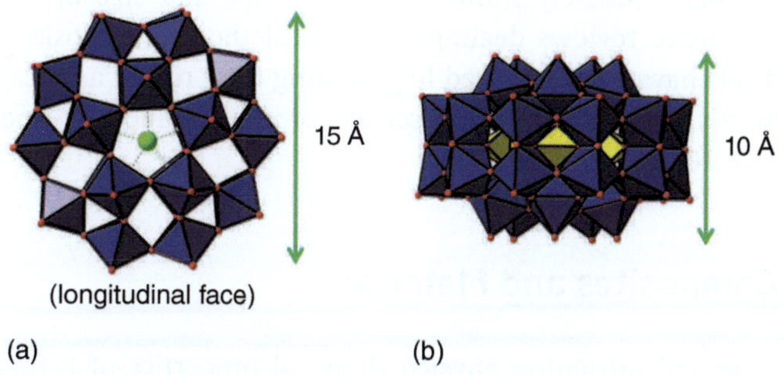

Figure 1.11 Polyhedral representation of the Preyssler-type POM: top view (a) and side view (b). Source: Leclerc et al. [79]/MDPI/CC BY 4.0.

and silver-containing Preyssler POMs have been successfully tested as antibacterial, antitumour and antiviral agents [85–89].

1.2.7 Other POM Structures

Besides the more traditional and well-known POM structures, several other examples of POM compounds have been reported. Dexter and Silverton described the crystal structure of a dodecamolybdocerate anion with the general formula $[Ce^{IV}Mo_{12}O_{42}]^{8-}$ [90]. The central position is occupied by cerium heteroatom surrounded by 12 face-sharing MoO_6 octahedra. The development of novel synthetic methods and the technological advances in the field of crystallography have undoubtedly contributed to the appearance of unprecedented POM structures with high nuclearities and increased complexity. Several authors have described the synthesis of POM-based compounds by combining different types of anions. In these mixed-type POM structures, generally prepared under hydro-/solvothermal synthesis, distinct smaller POM fragments (e.g. Lindqvist, Keggin) are joined together, typically using metallic clusters as linkers, to produce bulkier and more complex POM arrangements [21]. For instance, Yamase et al. reported numerous mixed-POM structures by combining trilacunary Keggin $[B\text{-}\alpha\text{-}SbW_9O_{33}]^{9-}$ and monolacunary Lindqvist $[W_5O_{18}]^{6-}$ units as building blocks [91–94].

The appealing physico-chemical properties and peculiar structural arrangements of high-nuclearity POM clusters have led to the development of supramolecular structures that can reach nanometric dimensions, generally referred as giant POMs [95, 96]. The construction of these nanosized assemblies relies on the use of metal cations, namely lanthanide ions and transition-metal ions, as linkers to assemble the smaller individual POM units [97]. Pope and co-workers have reported the synthesis and crystallographic determination of the wheel-shaped $[As_{12}Ce_{16}(H_2O)_{36}W_{148}O_{524}]^{76-}$ with a diameter of c. 4 nm [15]. This giant cyclic structure contains 148 tungsten atoms and combines $[B\text{-}\alpha\text{-}AsW_9O_{33}]^{9-}$ and $[W_5O_{18}]^{6-}$ units linked together by cerium clusters. Müller et al. have described several wheel-shaped giant POM structures which reach nanometric dimensions [98–100]. Interestingly, in the crystalline structure of the cyclic $[Mo_{120}O_{366}(H_2O)_{48}H_{12}\{Pr(H_2O)_5\}_6]^{6-}$ anion, six $\{Pr(H_2O)_5\}^{3+}$ fragments occupy the positions of $\{Mo\}^{2+}$ units ultimately influencing the shape and size of the ring. Recent and comprehensive reviews dealing with the plethora of nanosized high-nuclearity POM wheels have been reported highlighting their recent applications in the fields of catalysis, magnetism, luminescence, proton conductivity and molecular recognition [101, 102].

1.3 POM-based Composites and Materials

Over the years, the unique and intriguing physico-chemical properties of POMs have raised enormous interest in the scientific community. The structural diversity, coordination flexibility and extensive possibility of elemental compositions in

POMs make them potential candidates in a wide range of application fields. Nevertheless, the direct application of POMs in specific applications can uncover some limitations/drawbacks, such as low surface area, leaching, poor conductivity, mechanical and thermal stabilities and difficulty to remove from reactional media. Several strategies have been employed to overcome such issues by preparing POM-based composites and materials with enhanced processability and reinforced stability. The main strategies to develop POM-based composites and materials are (i) self-assembled hybrids using cationic surfactants, (ii) embedment in thin films/membranes, (iii) immobilization on support materials.

The properties of POMs can be tuned by replacing the original counterions with cationic surfactants resulting in surfactant-encapsulated Polyoxometalates (SEPs). The surfactant molecules surround the POM moieties and create a protective hydrophobic environment which promotes the preservation of the POM properties (Figure 1.12). The use of long alkyl chain surfactants has proved to be a useful strategy to prepare SEPs with enhanced features for applications in catalysis and luminescence. SEPs containing specific surfactants have been reported to behave as heterogeneous catalysts enabling their recovery and consequent recycling from the reactional media [103, 104]. Regarding luminescence, the hydrophobic environment in SEPs avoids the well-known luminescence quenching by water molecules and allows for lanthanide-containing POMs to preserve their emission even in aqueous media [81, 105]. Some of the most widely used surfactants to prepare SEPs include dimethyldioctadecylammonium (DODA), dodecyltrimethylammonium (DDTA), dodecyl(11-methacryloyloxyundecyl)dimethylammonium (DMDA) and di(11-hydroxyundecyl)dimethylammonium bromide (DOHDA).

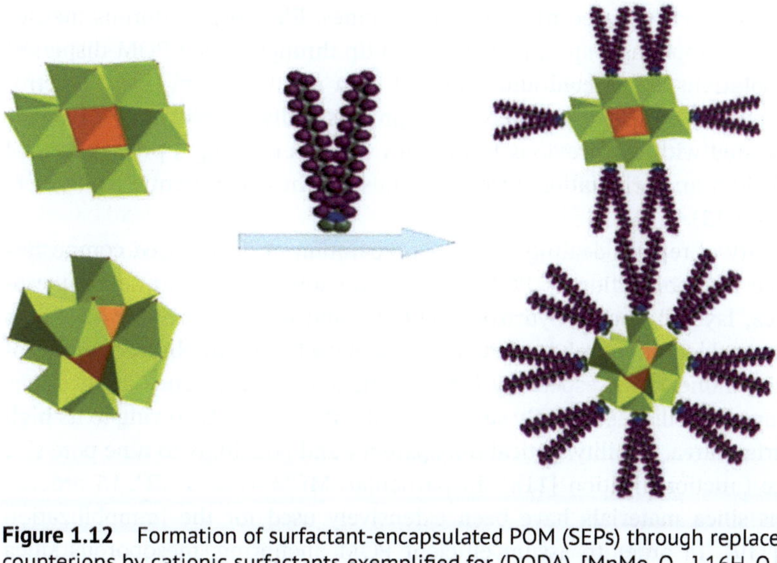

Figure 1.12 Formation of surfactant-encapsulated POM (SEPs) through replacement of counterions by cationic surfactants exemplified for $(DODA)_6[MnMo_9O_{32}]\cdot 16H_2O$ (top) and $(DODA)_8[CeMo_{12}O^{42}]\cdot 9H_2O$ (bottom). Source: Jia et al. [103]/with permission of Elsevier.

Recently, a different methodology to prepare surfactant-assembled POMs has been reported, named 'complex coacervate core micelles' (C3Ms) [106, 107]. These materials, assembled by electrostatic interactions, rely on neutral–cationic block copolymers. In C3Ms, the neutral segments of the copolymer surround the micellar core composed of its cationic segments and the anionic POM units.

The preparation of POM-embedded thin films and membranes is one of the most used approaches to prepare POM-based materials. Numerous stable and ordered films/membranes containing POM moieties have been reported owing to the solubility of POMs in distinct organic solvents [4]. The main techniques for POM-embedded films and membranes are Langmuir–Blodgett (LB), layer-by-layer (LbL), solvent-casting and spin-coating [108, 109]. The LB technique starts with the formation of a cationic surfactant monolayer on the surface of an aqueous dilute POM solution. Through immersion of a solid support, POM-embedded thin films can be obtained with controllable orientation and thickness [110]. The LbL technique involves the alternate immersion of a substrate in a cationic polyelectrolyte solution followed by immersion in anionic POM solutions, resulting in electrostatically assembled multilayered films [111]. POM-based thin films typically use poly(allylamine hydrochloride) (PAH), polyethyleneimine (PEI) and poly(styrenesulphonate) (PSS) as polyelectrolytes. Solvent-casting is a straightforward method for preparing POM-based films and membranes, involving dispersion of POMs in a volatile solvent and direct casting onto a flat surface [112, 113]. Despite being one of the most used methods, it has great limitations regarding homogeneity and thickness of the prepared films. Spin-coating is a highly versatile technique that allows to prepare pure homogeneous POM films with controlled thickness and relies on centrifugal force to spread the coating over a substrate [114]. Recently, electrospinning has proved to be an extremely promising and versatile technique for the production of POM-based films and membranes. Electrospun fibrous meshes are produced by applying a specific voltage to a tip through which POM-dispersed polymeric solutions are ejected and collected on a rotating drum [115]. Electrospinning allows to successfully overcome the processability, stability and scalability issues associated with the previous techniques while permitting a precise control over morphology and orientation by simply adjusting the experimental parameters [116] (Figure 1.13).

The majority of reports dealing with the preparation of POM-based composites relies on the immobilization of POM moieties on solid supports, namely mesoporous silica, layered double hydroxides (LDHs) and metal–organic frameworks (MOFs). A considerable number of POM anions have been immobilized on mesoporous silica frameworks, especially for catalytic and luminescence applications [117]. Mesoporous silica is a highly suitable candidate for support owing to its high internal surface area, stability, optical transparency and possibility to tune pore size and surface functionalization [118]. In particular, MCM-41 and SBA-15 ordered mesoporous silica materials have been extensively used for the immobilization of POMs [119]. In order to assure efficient POM anchoring, mesoporous silica materials are typically functionalized with different organosilanes which allow to minimize the leaching of the POM species [39, 120]. LDHs or anionic clays

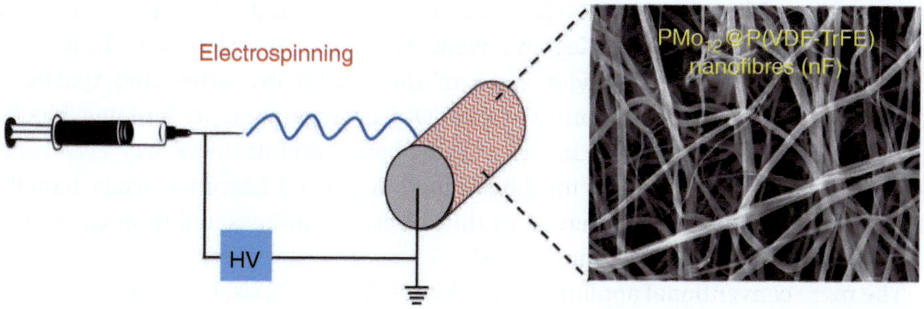

Figure 1.13 Representation of POM-containing electrospun fibre meshes exemplified for the Keggin phosphomolybdate embedded in PVDF polymeric matrix. Source: Reproduced with permission of Mirante et al. [116]/Royal Society of Chemistry.

are lamellar materials composed of positively charged layers intercalated with anionic species [121]. In LDHs, a wide range of negatively charged species can be intercalated by anionic exchange due to the weak interactions between layers. The great interest in LDHs as POM supports is due to their layered structure that allows for a wide range of POM anions with different sizes and spatial orientations to be intercalated by adjustment of the LDH interlayer distance [122]. Most of the reports dealing with POM-intercalated LDHs aim at their application as heterogeneous catalysts [123–125], although some reports can be found in the field of photoluminescence [126, 127]. MOFs are highly porous multidimensional crystalline structures obtained through coordination of metallic centres with organic ligands. The exceptionally high surface area, permanent porosity, post-synthetic functionalization and high thermal stability of MOFs soon attracted the scientific interest in these materials as solid supports able to efficiently accommodate a wide array of active species (including POMs) [128, 129]. The preparation of POM@MOF composites generally follows different synthetic pathways: (i) impregnation of MOF frameworks with POM solutions, (ii) 'bottle-around-the-ship' through the in situ synthesis of MOFs in the presence of POMs and (iii) 'ship-in-a-bottle' by the direct POM synthesis onto the MOF channels and cages [21]. The preparation of POM@MOF composites has proven to be a very efficient strategy for the heterogenization of some of the most catalytically active POM species. The MIL-type MOF structures, especially MIL-100 and MIL-101, have been particularly chosen due to their adequately large pore sizes with accessible channels for POM accommodation as well as their remarkable chemical and thermal robustness [130–134]. A few reports also show the promising potential of POM@MOF composites in optical sensing and electrocatalytic applications [135–138].

1.4 Conclusions

Over the last decades, POM chemistry has gained immense attention from the scientific community on the account of its peculiar physico-chemical properties and the virtually infinite combination of structural arrangements and chemical compositions. This chapter provides a comprehensive summary of the traditional POM structures (Lindqvist-, Keggin-, Wells–Dawson-, Preyssler-type, Anderson–Evans and others) together with some of the related properties and synthesis methodologies. The application of these POM arrangements as building blocks in the construction of more intricate POM-based architectures has also been addressed. The main strategies for the production of POM-based materials, namely surfactant encapsulation, formation of thin films/membranes and immobilization on support materials, were also reviewed.

The more conventional application areas of POMs (catalysis, luminescence, magnetism) have initially triggered much of the research work and interest in POM chemistry. Nevertheless, the technological progress leading to the development of advanced characterization techniques and innovative synthetic strategies allowed to significantly expand the range of applications and highlight the potential of POMs in several emerging areas.

This chapter intends to highlight that, despite the increasing number and type of POM-based composites and materials reported over the last years, there are still plenty of innovative opportunities to be explored in the field of POM chemistry. Considering the unique features of POMs herein revisited, it is expected that, in the near future, increasingly more industrial and technological applications will be developed employing POM-based materials and devices.

References

1 Pope, M.T. and Müller, A. (1991). Polyoxometalate chemistry: an old field with new dimensions in several disciplines. *Angew. Chem. Int. Ed. Engl.* 30: 34–48.

2 Li, F., Xu, L., Wei, Y. et al. (2006). Lanthanide-containing and bridged polyoxometalate chains: syntheses, crystal structures and magnetic properties. *Inorg. Chim. Acta* 359: 3795–3799.

3 Misra, A., Kozma, K., Streb, C., and Nyman, M. (2020). Beyond charge balance: counter-cations in polyoxometalate chemistry. *Angew. Chem. Int. Ed.* 59: 596–612.

4 Granadeiro, C.M., de Castro, B., Balula, S.S., and Cunha-Silva, L. (2013). Lanthanopolyoxometalates: from the structure of polyanions to the design of functional materials. *Polyhedron* 52: 10–24.

5 Das, V., Kaushik, R., and Hussain, F. (2020). Heterometallic 3d-4f polyoxometalates: an emerging field with structural diversity to multiple applications. *Coord. Chem. Rev.* 413: 213271.

6 Keggin, J.F. (1933). Structure of the molecule of 12-phosphotungstic acid. *Nature* 131: 908–909.

7 Coronado, E., Borrás-Almenar, J.J., Müller, A., and Pope, M. (ed.) (2003). *Polyoxometalate Molecular Science*. Dordrecht: Springer.

8 Yurchenko, É.N., Detusheva, L.G., Solov'eva, L.P. et al. (1992). Identification of type heteropolyantions and isopolyanions from the vibrational spectra of their crystalline salts. *J. Struct. Chem.* 33: 373–382.

9 Bijelic, A. and Rompel, A. (2018). Polyoxometalates: more than a phasing tool in protein crystallography. *ChemTexts* 4: 10.

10 Kondinski, A. (2016). On the isomer problem of mixed–addenda and heterogroup–substituted polyoxometalates. Doctoral dissertation. Jacobs University Bremen.

11 Zheng, Z., Zhou, Q., Li, M., and Yin, P. (2019). Poly(ethylene glycol) nanocomposites of sub-nanometer metal oxide clusters for dynamic semi-solid proton conductive electrolytes. *Chem. Sci.* 10: 7333–7339.

12 Braunstein, L.A.O.P. and Raithby, P.R. (ed.) (2008). *Metal Clusters in Chemistry*. Wiley-VCH.

13 Bao, W., Huang, T., Wang, C. et al. (2022). Controlled synthesis of efficient NiWS active phases derived from lacunary polyoxometalate and the application in hydrodesulfurization. *J. Catal.* 413: 374–387.

14 Gouzerh, P. and Proust, A. (1998). Main-group element, organic, and organometallic derivatives of polyoxometalates. *Chem. Rev.* 98: 77–112.

15 Wassermann, K., Dickman, M.H., and Pope, M.T. (1997). Self-assembly of supramolecular polyoxometalates: the compact, water-soluble heteropolytungstate anion $[AsCe(H_2O)_{36}W_{148}O_{524}]^{76-}$. *Angew. Chem. Int. Ed. Engl.* 36: 1445–1448.

16 Long, D.-L., Tsunashima, R., and Cronin, L. (2010). Polyoxometalates: building blocks for functional nanoscale systems. *Angew. Chem. Int. Ed.* 49: 1736–1758.

17 Sadakane, M., Dickman, M.H., and Pope, M.T. (2000). Controlled assembly of polyoxometalate chains from lacunary building blocks and lanthanide-cation linkers. *Angew. Chem. Int. Ed.* 39: 2914–2916.

18 Mirzaei, M., Eshtiagh-Hosseini, H., Alipour, M., and Frontera, A. (2014). Recent developments in the crystal engineering of diverse coordination modes (0–12) for Keggin-type polyoxometalates in hybrid inorganic–organic architectures. *Coord. Chem. Rev.* 275: 1–18.

19 Mialane, P., Lisnard, L., Mallard, A. et al. (2003). Solid-state and solution studies of $\{Ln_n(SiW_{11}O_{39})\}$ polyoxoanions: an example of building block condensation dependent on the nature of the rare earth. *Inorg. Chem.* 42: 2102–2108.

20 Xuan, W., Pow, R., Long, D.-L., and Cronin, L. (2017). Exploring the molecular growth of two gigantic half-closed polyoxometalate clusters $\{Mo_{180}\}$ and $\{Mo_{130}Ce_6\}$. *Angew. Chem. Int. Ed.* 56: 9727–9731.

21 Granadeiro, C.M., Julião, D., Ribeiro, S.O. et al. (2023). Recent advances in lanthanide-coordinated polyoxometalates: from structural overview to functional materials. *Coord. Chem. Rev.* 476: 214914.

22 Peacock, R.D. and Weakley, T.J.R. (1971). Heteropolytungstate complexes of the lanthanide elements. Part I. Preparation and reactions. *J. Chem. Soc. A* 1836–1839.

23 Granadeiro, C.M., Ferreira, R.A.S., Soares-Santos, P.C.R. et al. (2010). Lanthanopolyoxotungstates in silica nanoparticles: multi-wavelength photoluminescent core/shell materials. *J. Mater. Chem.* 20: 3313–3318.

24 Carvalho, R.F.S., Pereira, G.A.L., Rocha, J. et al. (2021). Lanthanopolyoxometalate-silica core/shell nanoparticles as potential MRI contrast agents. *Eur. J. Inorg. Chem.* 2021: 3458–3465.

25 AlDamen, M.A., Cardona-Serra, S., Clemente-Juan, J.M. et al. (2009). Mononuclear lanthanide single molecule magnets based on the polyoxometalates $[Ln(W_5O_{18})_2]^{9-}$ and $[Ln(\beta_2\text{-}SiW_{11}O_{39})_2]^{13-}$($Ln^{III}$ = Tb, Dy, Ho, Er, Tm, and Yb). *Inorg. Chem.* 48: 3467–3479.

26 Fernandes, S., Mirante, F., Castro, B.d. et al. (2022). Lindqvist versus Keggin-type polyoxometalates as catalysts for effective desulfurization of fuels. *Catalysts* 12: 581.

27 Granadeiro, C.M., Ribeiro, S.O., Kaczmarek, A.M. et al. (2016). A novel red emitting material based on polyoxometalate@periodic mesoporous organosilica. *Microporous Mesoporous Mater.* 234: 248–256.

28 Mariichak, O.Y., Ignatyeva, V.V., Baumer, V.N. et al. (2020). Heteropoly decatungstolanthanidates(III) with Peacock–Weakley type anion: synthesis and crystal structure of isostructural salts $Na_9[Ln(W_5O_{18})_2]\cdot35H_2O$ (Ln = Gd, Er). *J. Chem. Crystallogr.* 50: 255–266.

29 Binnemans, K. (2009). Lanthanide-based luminescent hybrid materials. *Chem. Rev.* 109: 4283–4374.

30 Pope, M.T. (2013). Happy birthday Keggin structure! *Eur. J. Inorg. Chem.* 2013: 1561–1561.

31 Müller, A., Peters, F., Pope, M.T., and Gatteschi, D. (1998). Polyoxometalates: very large clusters nanoscale magnets. *Chem. Rev.* 98: 239–272.

32 Zhang, C., Ma, P., Chen, H. et al. (2011). Synthesis, structure, and properties of a 1-D cerium based on monovacant Keggin-type polyoxotungstate. *J. Coord. Chem.* 64: 2178–2185.

33 Niu, J., Wang, K., Chen, H. et al. (2009). Assembly chemistry between lanthanide cations and monovacant Keggin polyoxotungstates: two types of lanthanide substituted phosphotungstates $[\{(\alpha\text{-}PW_{11}O_{39}H)Ln(H_2O)_3\}_2]^{6-}$ and $[\{(\alpha\text{-}PW_{11}O_{39})Ln(H_2O)(\eta_2,\mu\text{-}1,1)\text{-}CH_3COO\}_2]^{10-}$. *Cryst. Growth Des.* 9: 4362–4372.

34 Fukaya, K. and Yamase, T. (2003). Alkali-metal-controlled self-assembly of crown-shaped ring complexes of lanthanide/$[\alpha\text{-}AsW_9O_{33}]^{9-}$: $[K\subset\{Eu(H_2O)_2(\alpha\text{-}AsW_9O_{33})\}_6]^{35-}$ and $[Cs\subset\{Eu(H_2O)_2(\alpha\text{-}AsW_9O_{33})\}_4]^{23-}$. *Angew. Chem. Int. Ed.* 42: 654–658.

35 Hussain, F., Gable, R.W., Speldrich, M. et al. (2009). Polyoxotungstate-encapsulated Gd_6 and Yb_{10} complexes. *Chem. Commun.* 328–330.

36 Khoshnavazi, R., Naseri, E., Tayamon, S., and Ghiasi Moaser, A. (2011). An evidence for conversion of $A\text{-}PW_9O_{34}{}^{9-}$ derivatives to $A\text{-}PW_{10}O_{37}{}^{9-}$ derivatives. Synthesis and characterization of $[(A\text{-}PW_9O_{34})_2(H_2OM)_3CO_3]^{11-}$ (M = Sm(III), Eu(III) and Gd(III)) and $[(PW_{10}Sm_2O_{38})_4(W_3O_8(H_2O)_2(OH)_4)]^{22-}$ (M = Sm(III) and Gd(III)). *Polyhedron* 30: 381–386.

37 Khoshnavazi, R., Bahrami, L., and Davoodi, H. (2012). Decomposition of the lanthanide-containing sandwich-type polyoxometalates: synthesis and characterization of new multi samarium-containing polyoxometalates. *Inorg. Chim. Acta* 382: 158–161.

38 Kholdeeva, O.A., Timofeeva, M.N., Maksimov, G.M. et al. (2005). Aerobic oxidation of formaldehyde mediated by a Ce-containing polyoxometalate under mild conditions. *Inorg. Chem.* 44: 666–672.

39 Ribeiro, S.O., Nogueira, L.S., Gago, S. et al. (2017). Desulfurization process conciliating heterogeneous oxidation and liquid extraction: organic solvent or centrifugation/water? *Appl. Catal. A* 542: 359–367.

40 Wang, K., Zhang, D., Ma, J. et al. (2012). Three-dimensional lanthanide polyoxometalate organic complexes: correlation of structure with properties. *CrystEngComm* 14: 3205–3212.

41 Li, Z., Li, W., Li, X. et al. (2007). The gadolinium complexes with polyoxometalates as potential MRI contrast agents. *Magn. Reson. Imaging* 25: 412–417.

42 Matsunaga, S., Miyamae, E., Inoue, Y., and Nomiya, K. (2016). β,β-Isomer of open-Wells–Dawson polyoxometalate containing a tetra-iron(III) hydroxide cluster: $[\{Fe_4(H_2O)(OH)_5\}(\beta,\beta\text{-}Si_2W_{18}O_{66})]^{9-}$. *Inorganics* 4: 15.

43 Baker, L.C.W. and Glick, D.C. (1998). Present general status of understanding of heteropoly electrolytes and a tracing of some major highlights in the history of their elucidation. *Chem. Rev.* 98: 3–50.

44 Huang, W., Schopfer, M., Zhang, C. et al. (2008). ^{31}P magic angle spinning NMR spectroscopy of paramagnetic rare-earth-substituted Keggin and Wells–Dawson solids. *J. Am. Chem. Soc.* 130: 481–490.

45 Ueda, T., Nishimoto, Y., Saito, R. et al. (2015). Vanadium(V)-substitution reactions of Wells–Dawson-type polyoxometalates: from $[X_2M_{18}O_{62}]^{6-}$ (X = P, As; M = Mo, W) to $[X_2VM_{17}O_{62}]^{7-}$. *Inorganics* 3: 355–369.

46 Luo, Q.-H., Howell, R.C., Dankova, M. et al. (2001). Coordination of rare-earth elements in complexes with monovacant Wells–Dawson polyoxoanions. *Inorg. Chem.* 40: 1894–1901.

47 Sadakane, M., Ostuni, A., and Pope, M.T. (2002). Formation of 1 : 1 and 2 : 2 complexes of Ce(III) with the heteropolytungstate anion $\alpha_2\text{-}[P_2W_{17}O_{61}]^{10-}$, and their interaction with proline. The structure of $[Ce_2(P_2W_{17}O_{61})_2(H_2O)_8]^{14-}$. *J. Chem. Soc. Dalton Trans.* 63–67.

48 Zhang, C., Bensaid, L., McGregor, D. et al. (2006). Influence of the lanthanide ion and solution conditions on formation of lanthanide Wells–Dawson polyoxotungstates. *J. Clust. Sci.* 17: 389–425.

49 López, X., Carbó, J.J., Bo, C., and Poblet, J.M. (2012). Structure, properties and reactivity of polyoxometalates: a theoretical perspective. *Chem. Soc. Rev.* 41: 7537–7571.

50 Iijima, J. and Naruke, H. (2013). Synthesis and structural characterization of $[CeIV(\alpha_2\text{-}P_2W_{17}O_{61})_2]^{16-}$ in the solid state and in aqueous solution. *J. Mol. Struct.* 1040: 33–38.

51 Boglio, C., Lemière, G., Hasenknopf, B. et al. (2006). Lanthanide complexes of the monovacant Dawson polyoxotungstate $[\alpha_1\text{-}P_2W_{17}O_{61}]^{10-}$ as selective and recoverable Lewis acid catalysts. *Angew. Chem. Int. Ed.* 45: 3324–3327.

52 Kortz, U. (2003). Rare-earth substituted polyoxoanions: $[\{La(CH_3COO)(H_2O)_2(\alpha_2\text{-}P_2W_{17}O_{61})\}_2]^{16-}$ and $[\{Nd(H_2O)_3(\alpha_2\text{-}P_2W_{17}O_{61})\}_2]^{14-}$. *J. Clust. Sci.* 14: 205–214.

53 Peng, Q., Li, S., Wang, R. et al. (2017). Lanthanide derivatives of Ta/W mixed-addendum POMs as proton-conducting materials. *Dalton Trans.* 46: 4157–4160.

54 Chen, W.-C., Jiao, C.-Q., Wang, X.-L. et al. (2019). Self-assembly of nanoscale lanthanoid-containing selenotungstates: synthesis, structures, and magnetic studies. *Inorg. Chem.* 58: 12895–12904.

55 Wang, W., Izarova, N.V., van Leusen, J., and Kögerler, P. (2019). Ce^{III}-functionalized polyoxotungstates: discrete vs extended architectures. *Cryst. Growth Des.* 19: 4860–4870.

56 Ostuni, A. and Pope, M.T. (2000). A large heteropolytungstotetracerate(III) based on a new divacant lacunary derivative of the Wells–Dawson tungstophosphate anion. *C. R. Acad. Sci. – Ser. IIC – Chem.* 3: 199–204.

57 Ostuni, A., Bachman, R.E., and Pope, M.T. (2003). Multiple diastereomers of $[M^{n+}(\alpha_m\text{-}P_2W_{17}O_{61})_2]^{(20-n)-}$ ($M = U^{IV}$, Th^{IV}, Ce^{III}; $m = 1, 2$). Syn- and anti-conformations of the polytungstate ligands in $\alpha_1\alpha_1$, $\alpha_1\alpha_2$, and $\alpha_2\alpha_2$ complexes. *J. Clust. Sci.* 14: 431–446.

58 Fang, X., Anderson, T.M., Benelli, C., and Hill, C.L. (2005). Polyoxometalate-supported Y- and Yb^{III}-hydroxo/oxo clusters from carbonate-assisted hydrolysis. *Chem. Eur. J.* 11: 712–718.

59 Zhao, H.-Y., Zhao, J.-W., Yang, B.-F. et al. (2014). A series of organic–inorganic hybrids based on lanthanide-substituted Dawson-type phosphotungstate dimers and copper–en linkers. *CrystEngComm* 16: 2230–2238.

60 Boglio, C., Lenoble, G., Duhayon, C. et al. (2006). Production and reactions of organic-soluble lanthanide complexes of the monolacunary Dawson $[\alpha_1\text{-}P_2W_{17}O_{61}]^{10}$ polyoxotungstate. *Inorg. Chem.* 45: 1389–1398.

61 Zhao, H.-Y., Yang, B.-F., and Yang, G.-Y. (2017). Two new 2D organic–inorganic hybrids assembled by lanthanide-substituted polyoxotungstate dimers and copper–complex linkers. *Inorg. Chem. Commun.* 84: 212–216.

62 Granadeiro, C.M., Ferreira, R.A.S., Soares-Santos, P.C.R. et al. (2009). Lanthanopolyoxometalates as building blocks for multiwavelength photoluminescent organic–inorganic hybrid materials. *Eur. J. Inorg. Chem.* 2009: 5088–5095.

63 Yan, B., Liang, R., Zheng, K. et al. (2021). Multinuclear lanthanide-implanted tetrameric Dawson-type phosphotungstates with switchable luminescence behaviors induced by fast photochromism. *Inorg. Chem.* 60: 8164–8172.

64 Lu, Y., Xu, Y., Li, Y. et al. (2006). New polyoxometalate compounds built up of lacunary Wells–Dawson anions and trivalent lanthanide cations. *Inorg. Chem.* 45: 2055–2060.

65 Wei, Z., Wang, J., Yu, H. et al. (2022). Recent advances of Anderson-type polyoxometalates as catalysts largely for oxidative transformations of organic molecules. *Molecules* 27: 5212.

66 Evans, H.T. Jr., (1948). The crystal structures of ammonium and potassium molybdotellurates. *J. Am. Chem. Soc.* 70: 1291–1292.

67 Pope, M.T. (1983). *Heteropoly and Isopoly Oxometalates*. Berlin, Heidelberg: Springer-Verlag.

68 Drewes, D., Limanski, E.M., and Krebs, B. (2004). A series of novel lanthanide polyoxometalates: condensation of building blocks dependent on the nature of rare earth cations. *Dalton Trans.* 2087–2091.

69 An, H., Lan, Y., Li, Y. et al. (2004). A novel chain-like polymer constructed from heteropolyanions covalently linked by lanthanide cations: $(C_5H_9NO_2)_2$ $[La(H_2O)_7CrMo_6H_6O_{24}]\cdot 11H_2O$ (Proline = $C_5H_9NO_2$). *Inorg. Chem. Commun.* 7: 356–358.

70 Dolbecq, A., Dumas, E., Mayer, C.R., and Mialane, P. (2010). Hybrid organic–inorganic polyoxometalate compounds: from structural diversity to applications. *Chem. Rev.* 110: 6009–6048.

71 Blazevic, A. and Rompel, A. (2016). The Anderson–Evans polyoxometalate: from inorganic building blocks via hybrid organic–inorganic structures to tomorrows "Bio-POM". *Coord. Chem. Rev.* 307: 42–64.

72 Pardo, R., Zayat, M., and Levy, D. (2011). Photochromic organic–inorganic hybrid materials. *Chem. Soc. Rev.* 40: 672–687.

73 Preyssler, C. (1970). Existance of 18-tungsto-3-phosphate. *Bull. Soc. Chim. Fr.* 1: 30.

74 Kim, K.-C., Pope, M.T., Gama, G.J., and Dickman, M.H. (1999). Slow proton exchange in aqueous solution. consequences of protonation and hydration within the central cavity of Preyssler anion derivatives, $[|M(H_2O)|\supset P_5W_{30}O_{110}]n$. *J. Am. Chem. Soc.* 121: 11164–11170.

75 Creaser, I., Heckel, M.C., Neitz, R.J., and Pope, M.T. (1993). Rigid nonlabile polyoxometalate cryptates $[ZP_5W_{30}O_{110}]^{(15-n)-}$ that exhibit unprecedented selectivity for certain lanthanide and other multivalent cations. *Inorg. Chem.* 32: 1573–1578.

76 Dickman, M.H., Gama, G.J., Kim, K.-C., and Pope, M.T. (1996). The structures of Europium(III)- and Uranium(IV) derivatives of $[P^5w^{30}o^{110}]_{15-}$: evidence for "cryptohydration". *J. Clust. Sci.* 7: 567–583.

77 Haider, A., Zarschler, K., Joshi, S.A. et al. (2018). Preyssler–Pope–Jeannin polyanions $[NaP_5W_{30}O_{110}]^{14-}$ and $[AgP_5W_{30}O_{110}]^{14-}$: microwave-assisted synthesis, structure, and biological activity. *Z. Anorg. Allg. Chem.* 644: 752–758.

78 Zhang, Z.-M., Yao, S., Li, Y.-G. et al. (2012). Inorganic crown ethers: sulfate-based Preyssler polyoxometalates. *Chem. Eur. J.* 18: 9184–9188.

79 Leclerc, N., Haouas, M., Falaise, C. et al. (2021). Supramolecular association between γ-cyclodextrin and Preyssler-type polyoxotungstate. *Molecules* 26: 5126.

80 Bu, W., Wu, L., and Tang, A.-C. (2004). Polyoxometalates matrixed into surfactants: synthesis, characterizations, and conformations of surfactants. *J. Colloid Interface Sci.* 269: 472–475.

81 Zhang, T., Liu, S., Kurth, D.G., and Faul, C.F.J. (2009). Organized nanostructured complexes of polyoxometalates and surfactants that exhibit photoluminescence and electrochromism. *Adv. Funct. Mater.* 19: 642–652.

82 Zhao, H.-M., Gan, H.-M., Zhao, L., and Su, Z.-M. (2020). Two 2D coordination polymers based on Preyssler anions: synthesis, crystal structures and photoluminescence properties. *Inorg. Chem. Commun.* 113: 107728.

83 Liu, S., Kurth, D.G., Möhwald, H., and Volkmer, D. (2002). A thin-film electrochromic device based on a polyoxometalate cluster. *Adv. Mater.* 14: 225–228.

84 Zhang, T.R., Lu, R., Liu, X.L. et al. (2003). Photochromic polyoxotungstoeuropate $K_{12}[EuP_5W_{30}O_{110}]$/polyvinylpyrrolidone nanocomposite films. *J. Solid State Chem.* 172: 458–463.

85 Hill, C.L., Weeks, M.S., and Schinazi, R.F. (1990). Anti-HIV-1 activity, toxicity, and stability studies of representative structural families of polyoxometalates. *J. Med. Chem.* 33: 2767–2772.

86 Razavi, S.F., Bamoharram, F.F., Hashemi, T. et al. (2020). Nanolipid-loaded Preyssler polyoxometalate: synthesis, characterization and in vitro inhibitory effects on HepG2 tumor cells. *Toxicol. In Vitro* 68: 104917.

87 Pimpão, C., da Silva, I.V., Mósca, A.F. et al. (2020). The aquaporin-3-inhibiting potential of polyoxotungstates. *Int. J. Mol. Sci.* 21: 2467.

88 Xu, Z., Chen, K., Li, M. et al. (2020). Sustained release of Ag^+ confined inside polyoxometalates for long-lasting bacterial resistance. *Chem. Commun.* 56: 5287–5290.

89 Chen, K., Yu, Q., Liu, Y., and Yin, P. (2021). Bacterial hyperpolarization modulated by polyoxometalates for solutions of antibiotic resistance. *J. Inorg. Biochem.* 220: 111463.

90 Dexter, D.D. and Silverton, J.V. (1968). A new structural type for heteropoly anions. The crystal structure of $(NH_4)_2H_6(CeMo_{12}O_{42})\cdot 12H_2O$. *J. Am. Chem. Soc.* 90: 3589–3590.

91 Yamase, T., Naruke, H., and Sasaki, Y. (1990). Crystallographic characterization of the polyoxotungstate $[Eu_3(H_2O)_3(SbW_9O_{33})(W_5O_{18})_3]^{18-}$ and energy transfer in its crystalline lattices. *J. Chem. Soc. Dalton Trans.* 1687–1696.

92 Naruke, H. and Yamase, T. (1998). Crystal structure of $K_{18.5}H_{1.5}[Ce_3(CO_3)(SbW_9O_{33})(W_5O_{18})_3]\cdot 14H_2O$. *J. Alloys Compd.* 268: 100–106.

93 Naruke, H. and Yamase, T. (2002). Size-dependent population of trivalent rare earth cations (RE^{3+}) in $[(RE)_2(H_2O)_2(SbW_9O_{33})(W_5O_{18})_2]^{15-}$, and structural characterization of a luthetium–polyoxotungstate complex $[Lu_3(H_2O)_4(SbW_9O_{33})_2(W_5O_{18})_2]^{21-}$. *Bull. Chem. Soc. Jpn.* 75: 1275–1282.

94 Naruke, H. and Yamase, T. (2002). Synthesis and structure of $Ln(W_5O_{18})$-capped mixed-ligand polyoxotungstolanthanoate $[Ln(W_5O_{18})\{Ln(H_2O)_2(SbW_9O_{33})(W_5O_{18})\}]^{15-}$(Ln = Sm and Er). *Bull. Chem. Soc. Jpn.* 74: 1289–1294.

95 Wang, D., Jiang, J., Cao, M.-Y. et al. (2022). An unprecedented dumbbell-shaped pentadeca-nuclear W-Er heterometal cluster stabilizing nanoscale hexameric arsenotungstate aggregate and electrochemical sensing properties of its conductive hybrid film-modified electrode. *Nano Res.* 15: 3628–3637.

96 Zhang, J., Lai, R.-D., Wu, Y.-L. et al. (2020). High-dimensional polyoxoniobates constructed from lanthanide-incorporated high-nuclear $\{[Ln(H_2O)_4]_3[Nb_{24}O_{69}(H_2O)_3]_2\}$ secondary building units. *Chem. Asian J.* 15: 1574–1579.

97 Liu, J.-C., Wang, J.-F., Han, Q. et al. (2021). Multicomponent self-assembly of a giant heterometallic polyoxotungstate supercluster with antitumor activity. *Angew. Chem. Int. Ed.* 60: 11153–11157.

98 Müller, A., Krickemeyer, E., Meyer, J. et al. (1995). $[Mo_{154}(NO)_{14}O_{420}(OH)_{28}(H_2O)_{70}]^{(25 \pm 5)-}$: a water-soluble big wheel with more than 700 atoms and a relative molecular mass of about 24000. *Angew. Chem. Int. Ed. Engl.* 34: 2122–2124.

99 Müller, A. and Beugholt, C. (1996). The medium is the message. *Nature* 383: 296–297.

100 Müller, A., Beugholt, C., Bögge, H., and Schmidtmann, M. (2000). Influencing the size of giant rings by manipulating their curvatures: $Na_6[Mo_{120}O_{366}(H_2O)_{48}H_{12}\{Pr(H_2O)_5\}_6]\cdot(\sim 200H_2O)$ with open shell metal centers at the cluster surface. *Inorg. Chem.* 39: 3112–3113.

101 Liu, J.-C., Zhao, J.-W., Streb, C., and Song, Y.-F. (2022). Recent advances on high-nuclear polyoxometalate clusters. *Coord. Chem. Rev.* 471: 214734.

102 Al-Sayed, E. and Rompel, A. (2022). Lanthanides singing the blues: their fascinating role in the assembly of gigantic molybdenum blue wheels. *ACS Nanosci. Au* 2: 179–197.

103 Jia, Y., Zhang, J., Zhang, Z.-M. et al. (2014). Metal-centered polyoxometalates encapsulated by surfactant resulting in the thermotropic liquid crystal materials. *Inorg. Chem. Commun.* 43: 5–9.

104 Ribeiro, S.O., Julião, D., Cunha-Silva, L. et al. (2016). Catalytic oxidative/extractive desulfurization of model and untreated diesel using hybrid based zinc-substituted polyoxometalates. *Fuel* 166: 268–275.

105 Qi, W., Li, H., Wu, L., and Novel, A. (2007). Luminescent, silica-sol–gel hybrid based on surfactant- encapsulated polyoxometalates. *Adv. Mater.* 19: 1983–1987.

106 Zhang, J., Liu, Y., Li, Y. et al. (2012). Hybrid assemblies of Eu-containing polyoxometalates and hydrophilic block copolymers with enhanced emission in aqueous solution. *Angew. Chem. Int. Ed.* 51: 4598–4602.

107 Wei, H., Du, S., Liu, Y. et al. (2014). Tunable, luminescent, and self-healing hybrid hydrogels of polyoxometalates and triblock copolymers based on electrostatic assembly. *Chem. Commun.* 50: 1447–1450.

108 Qi, W. and Wu, L. (2009). Polyoxometalate/polymer hybrid materials: fabrication and properties. *Polym. Int.* 58: 1217–1225.

109 Liu, S. and Tang, Z. (2010). Polyoxometalate-based functional nanostructured films: current progress and future prospects. *Nano Today* 5: 267–281.

110 Clemente-León, M., Coronado, E., Soriano-Portillo, A. et al. (2005). Langmuir–Blodgett films based on inorganic molecular complexes with magnetic or optical properties. *Adv. Colloid Interf. Sci.* 116: 193–203.

111 Liu, S., Kurth, D.G., Bredenkötter, B., and Volkmer, D. (2002). The structure of self-assembled multilayers with polyoxometalate nanoclusters. *J. Am. Chem. Soc.* 124: 12279–12287.

112 Wang, Z., Zhang, R., Ma, Y. et al. (2010). Chemically responsive luminescent switching in transparent flexible self-supporting $[EuW_{10}O_{36}]^{9-}$-agarose nanocomposite thin films. *J. Mater. Chem.* 20: 271–277.

113 Wang, Z., Ma, Y., Zhang, R. et al. (2009). Reversible luminescent switching in a $[Eu(SiW_{10}MoO_{39})_2]^{13-}$-agarose composite film by photosensitive intramolecular energy transfer. *Adv. Mater.* 21: 1737–1741.

114 Wang, B.-H. and Yan, B. (2019). Tunable multi-color luminescence and white emission in lanthanide ion functionalized polyoxometalate-based metal–organic frameworks hybrids and fabricated thin films. *J. Alloys Compd.* 777: 415–422.

115 Denny, M.S., Kalaj, M., Bentz, K.C., and Cohen, S.M. (2018). Multicomponent metal–organic framework membranes for advanced functional composites. *Chem. Sci.* 9: 8842–8849.

116 Mirante, F., Salimian, M., Marques, P.A.A.P. et al. (2022). Polyoxometalate-doped electrospun nanofiber mats as active catalysts for the production of clean fuels under solvent-free systems. *Sustainable Energy Fuels* 6: 4681–4691.

117 Zhou, Y., Chen, G., Long, Z., and Wang, J. (2014). Recent advances in polyoxometalate-based heterogeneous catalytic materials for liquid-phase organic transformations. *RSC Adv.* 4: 42092–42113.

118 Jia, M. and Thiel, W.R. (2002). Oxodiperoxo molybdenum modified mesoporous MCM-41 materials for the catalytic epoxidation of cyclooctene. *Chem. Commun.* 2392–2393.

119 Xiong, J., Zhu, W., Ding, W. et al. (2014). Phosphotungstic acid immobilized on ionic liquid-modified SBA-15: efficient hydrophobic heterogeneous catalyst for oxidative desulfurization in fuel. *Ind. Eng. Chem. Res.* 53: 19895–19904.

120 Choi, M., Heo, W., Kleitz, F., and Ryoo, R. (2003). Facile synthesis of high quality mesoporous SBA-15 with enhanced control of the porous network connectivity and wall thickness. *Chem. Commun.* 1340–1341.

121 Omwoma, S., Chen, W., Tsunashima, R., and Song, Y.-F. (2014). Recent advances on polyoxometalates intercalated layered double hydroxides: from synthetic approaches to functional material applications. *Coord. Chem. Rev.* 258–259: 58–71.

122 Liu, K., Yao, Z., and Song, Y.-F. (2015). Polyoxometalates hosted in layered double hydroxides: highly enhanced catalytic activity and selectivity in sulfoxidation of sulfides. *Ind. Eng. Chem. Res.* 54: 9133–9141.

123 Wang, X., Chen, W., and Song, Y.-F. (2014). Directional self-assembly of exfoliated layered europium hydroxide nanosheets and $Na_9EuW_{10}O_{36}·32H_2O$ for application in desulfurization. *Eur. J. Inorg. Chem.* 2014: 2779–2786.

124 Jia, Y., Zhao, S., and Song, Y.-F. (2014). The application of spontaneous flocculation for the preparation of lanthanide-containing polyoxometalates intercalated layered double hydroxides: highly efficient heterogeneous catalysts for cyanosilylation. *Appl. Catal. A* 487: 172–180.

125 Li, T., Wang, Z., Chen, W. et al. (2017). Rational design of a polyoxometalate intercalated layered double hydroxide: highly efficient catalytic epoxidation of allylic alcohols under mild and solvent-free conditions. *Chem. Eur. J.* 23: 1069–1077.

126 Sousa, F.L., Pillinger, M., Sá Ferreira, R.A. et al. (2006). Luminescent polyoxotungstoeuropate anion-pillared layered double hydroxides. *Eur. J. Inorg. Chem.* 2006: 726–734.

127 Omwoma, S., Lagat, S.C., and Lalah, J.O. (2018). Fine tuning the microenvironment of $[EuW_{10}O_{36}]^{9-}$ anion leads to the large enhancement of the red light luminescence. *J. Lumin.* 196: 294–301.

128 Zhu, Q.-L. and Xu, Q. (2014). Metal–organic framework composites. *Chem. Soc. Rev.* 43: 5468–5512.

129 Piscopo, C.G., Granadeiro, C.M., Balula, S.S., and Bošković, D. (2020). Metal-organic framework-based catalysts for oxidative desulfurization. *ChemCatChem* 12: 4721–4731.

130 Granadeiro, C.M., Silva, P., Saini, V.K. et al. (2013). Novel heterogeneous catalysts based on lanthanopolyoxometalates supported on MIL-101(Cr). *Catal. Today* 218–219: 35–42.

131 Ribeiro, S., Granadeiro, C.M., Silva, P. et al. (2013). An efficient oxidative desulfurization process using terbium-polyoxometalate@MIL-101(Cr). *Cat. Sci. Technol.* 3: 2404–2414.

132 Granadeiro, C.M., Nogueira, L.S., Julião, D. et al. (2016). Influence of a porous MOF support on the catalytic performance of Eu-polyoxometalate based materials: desulfurization of a model diesel. *Cat. Sci. Technol.* 6: 1515–1522.

133 Juan-Alcañiz, J., Ramos-Fernandez, E.V., Lafont, U. et al. (2010). Building MOF bottles around phosphotungstic acid ships: one-pot synthesis of bi-functional polyoxometalate-MIL-101 catalysts. *J. Catal.* 269: 229–241.

134 Férey, G., Mellot-Draznieks, C., Serre, C. et al. (2005). A chromium terephthalate-based solid with unusually large pore volumes and surface area. *Science* 309: 2040–2042.

135 Salomon, W., Dolbecq, A., Roch-Marchal, C. et al. (2018). A multifunctional dual-luminescent polyoxometalate@metal-organic framework EuW_{10}@UiO-67 composite as chemical probe and temperature sensor. *Front. Chem.* 6: 425.

136 Viravaux, C., Oms, O., Dolbecq, A. et al. (2021). Temperature sensors based on europium polyoxometalate and mesoporous terbium metal–organic framework. *J. Mater. Chem. C* 9: 8323–8328.

137 Abdelkader-Fernández, V.K., Fernandes, D.M., Balula, S.S. et al. (2020). Advanced framework-modified POM@ZIF-67 nanocomposites as enhanced oxygen evolution reaction electrocatalysts. *J. Mater. Chem. A* 8: 13509–13521.

138 Abdelkader-Fernández, V.K., Fernandes, D.M., Balula, S.S. et al. (2022). Unveiling the structural transformations of the $PW_{11}Co$@ZIF-67 nanocomposite induced by thermal treatment. *Dalton Trans.* 51: 17844–17857.

2

Design and Strategies to Enhance the Electrochemical Properties of POM Nanomaterials for Electrocatalysis

Adedayo Oghenenerhuvwu Dada, Kieran Dylan Jones, Darren Anthony Walsh, and Graham Neil Newton

University of Nottingham, GSK Carbon Neutral Laboratory for Sustainable Chemistry, School of Chemistry, Jubilee Campus, 6 Triumph Road, Lenton, Nottingham NG7 2TU, UK

2.1 Introduction

Polyoxometalates (POMs) are a diverse class of anionic metal oxide clusters, composed of early transition metals in their highest oxidation states connected by bridging and terminal oxo ligands. The discovery of the first POM species is credited to Berzelius in 1826 when he investigated molybdenum compounds [1, 2]. In his paper, he observed yellow (and reduced blue) compounds formed when combining ammonium molybdate with excess phosphoric acid. This ultimately led to the first detailed crystallographic characterization of $[PMo_{12}O_{40}]^{3-}$ by Keggin in 1933 [3]. Known for their structural diversity, rich and reversible redox properties, photoactivity and excellent thermal and chemical stability, POMs have garnered attention as ideal 'building blocks' for functional materials [4] with a myriad of applications including photo- and electrocatalysis [5], energy storage [6], battery technologies [7], magnetism [8], medicine [9] and sensing [10].

Their appeal as catalysts lies both in their stability under harsh conditions and the degree to which their electronic and redox properties can be modified through controllable changes to their structure and composition. There are various approaches to enhance or modify the electronic characteristics of POMs, either by altering their environments or directly manipulating their structure, through, for example, hybridization with organic molecular fragments, supramolecular assembly and immobilization. This flexibility renders them useful functional building blocks in a range of fields.

The purpose of this review is not simply to present the best electrocatalysts within the field, but to discuss the modifications of POMs used to produce tailored electrocatalysts. We aim to provide an overview of the recent developments in the designs and strategies on POM nanomaterials to enhance their chemical properties for future applications The focus will be on the multifunctionality of POMs and how each level of manipulation is necessary to produce hierarchical functional molecular-cluster-based materials. Additionally, new areas and

Applied Polyoxometalate-based Electrocatalysis, First Edition. Edited by Diana M. Fernandes.

possibilities for POM electrochemistry will be explored. Specifically, attention will be given to predominant organofunctionalization, supramolecular and immobilization strategies that can be used in the production of redox-active hybrid POM materials.

2.1.1 Structure Bonding and Formation

POMs come in all shapes and sizes, but typically consist of the same basic building units. The smallest unit of a POM is the $[MO_x]$ polyhedron (where x = 4–7, typically 6) and comprises a central metal atom (addenda atom) coordinated to 4–7 oxo-groups (formally O^{2-} anions) [2, 11] (Figure 2.1a). To form discrete clusters, addenda atoms must possess certain properties, such as the ability to change coordination number from 4 to 6 upon polymerization in solution, high positive charges and the ability to form double bonds with unshared (terminal) oxygen atoms through $p\pi$–$d\pi$ orbital interactions. Transition metals from groups V and VI (V, Nb, Ta, Mo, W) in their highest oxidation states typically obey these rules when in a d^0 or d^1 configuration [12].

The oxo-groups located at each apex of the polyhedron serve as the structurally versatile components (Figure 2.1b). These ligands act as bridging atoms between metals, allowing for polycondensation and preventing the formation of infinite metal oxide structures through the formation of stable terminal M=O oxo-groups. The terminal oxo-groups play a crucial role in the stability of many POM clusters as they have a strong *trans*-effect, which weakens the bond to the bridging oxygen atom located at the opposite apex of the octahedron that shares its orbital. This

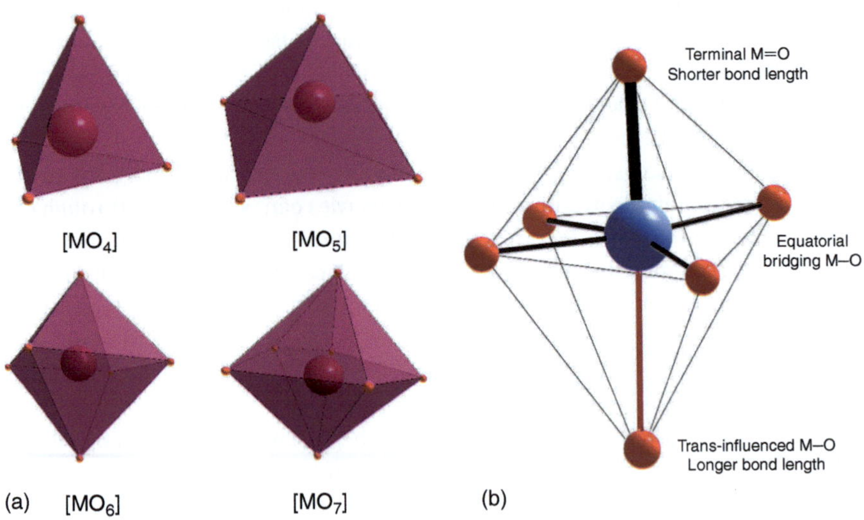

Figure 2.1 (a) Common polyhedral units within POM structures. (b) Visual representation of common addenda in POM structures including an $[MO_6]$ octahedral unit comprising one terminal oxygen atom and five bridging oxygen atoms that are shared with other metal or hetero-atom addenda.

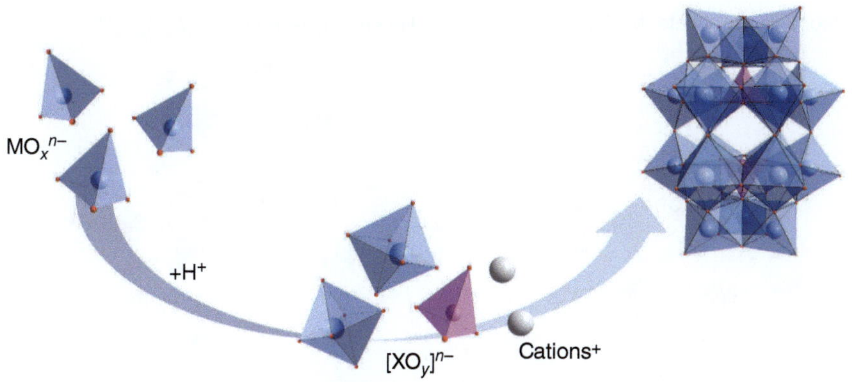

Figure 2.2 Polycondensation reaction of $[MO_x]$ polyhedra into POMs.

results in a displacement of the metal centre towards the terminal oxo-group, which causes the structure to deviate from the idealized octahedral geometry [13].

POMs are formed through acid driven condensation reactions between metal oxide fragments (Figure 2.2). The unstable intermediates reacting to form kinetically stable species (as shown in Scheme 2.1) [14]. This procedure is highly dependent on various experimental factors such as the pH, temperature, solvent, countercation and the ratio of addenda atom to heteroatom.

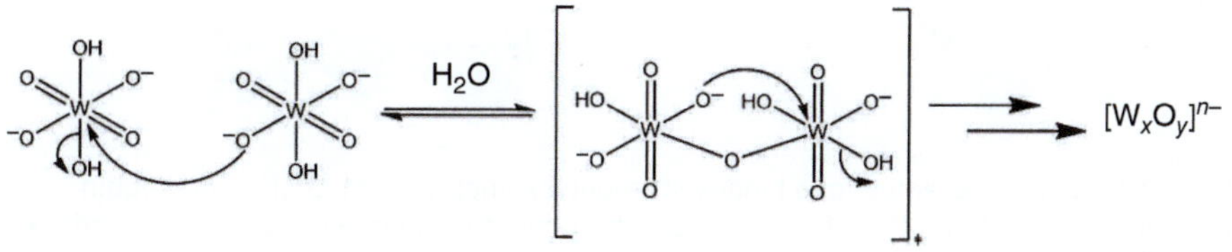

Scheme 2.1 Condensation reaction between $[MO_6]$ octahedra forming polynuclear M_xO_y units.

The $[MO_6]$ polyhedral units can bind in various ways, depending on whether they are connected through one, two or three oxo-groups; these connections can be considered corner-sharing, edge-sharing or face-sharing (Figure 2.3). Intact POM structures can exhibit a combination of these polyhedral binding modes, with each

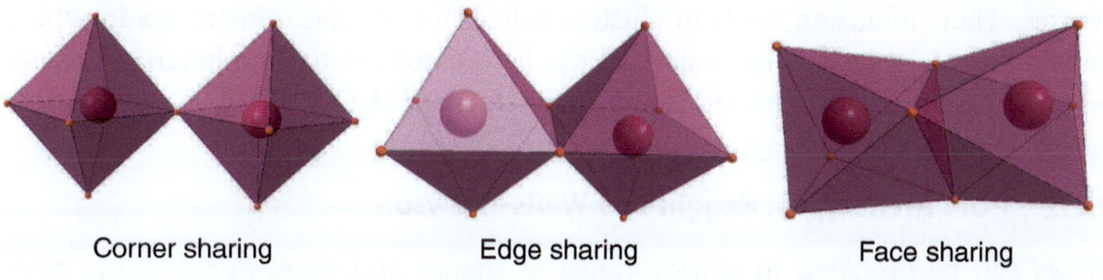

Figure 2.3 Connectivity modes of $[MO_x]$ polyhedra after condensation based on sharing one (corner), two (edge) or three (face) oxygen atoms between two metal centres.

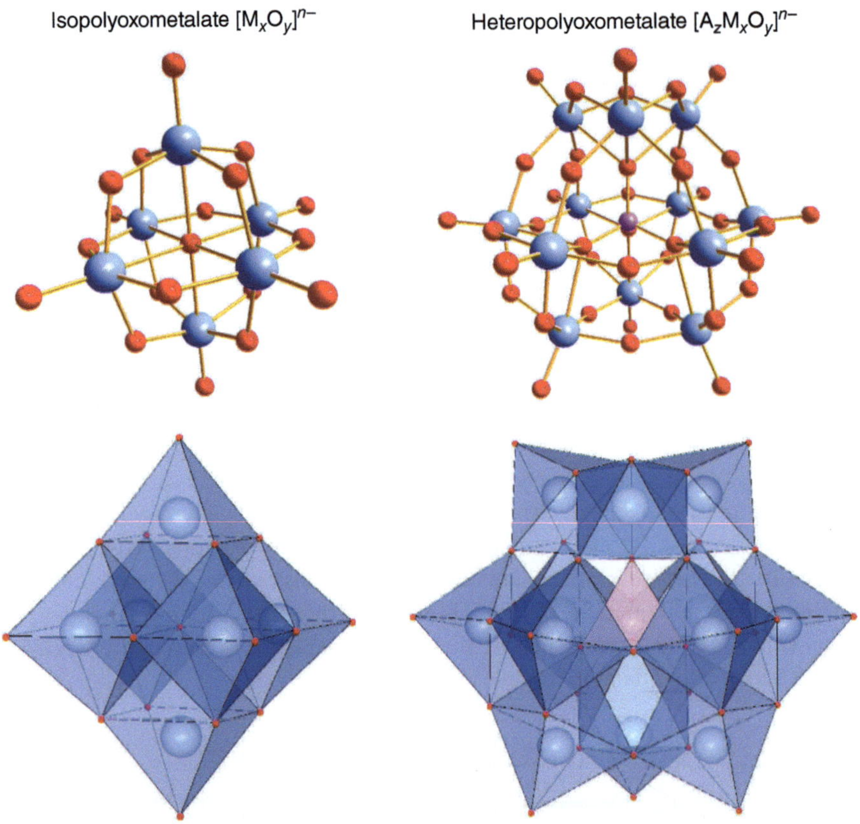

Figure 2.4 Generic example of a Lindqvist isopolyoxometalate $[M_6O_{19}]^{2-}$ and Keggin heteropolyoxometalate $[XM_{12}O_{40}]^{n-}$ where X is a non-addenda heteroatom. Colour code: W (blue), P (pink), O (red).

polyhedron containing either one or two terminal oxo-groups [15]. The number of terminal oxo atoms that each metal centre possesses has a significant impact on the electronic properties of POMs, which we will explore further in this chapter.

Heteroatoms can reside within the core of POMs, typically in a tetrahedral binding site. They play a crucial role in the formation of the structure and are often referred to as 'templating' atoms (Figure 2.4). These atoms can be either transition metals or main group elements, and they must be able to bond with at least three oxygen centres. Their inclusion leads to clusters called heteropolyoxometalates in which metal-oxo polyhedra assemble around the heteroatom. When no heteroatoms are present, the clusters are termed isopolyoxometalates $[M_xO_y]^{n-}$ [16].

2.1.2 POM Archetypes: Keggin and Wells–Dawson

POMs can be prepared in a wide range of shapes and sizes. Some of the most versatile and well-defined archetypes include Lindqvist, Anderson, Keggin and Wells–Dawson clusters (Figure 2.5). In this chapter, we will mainly focus on

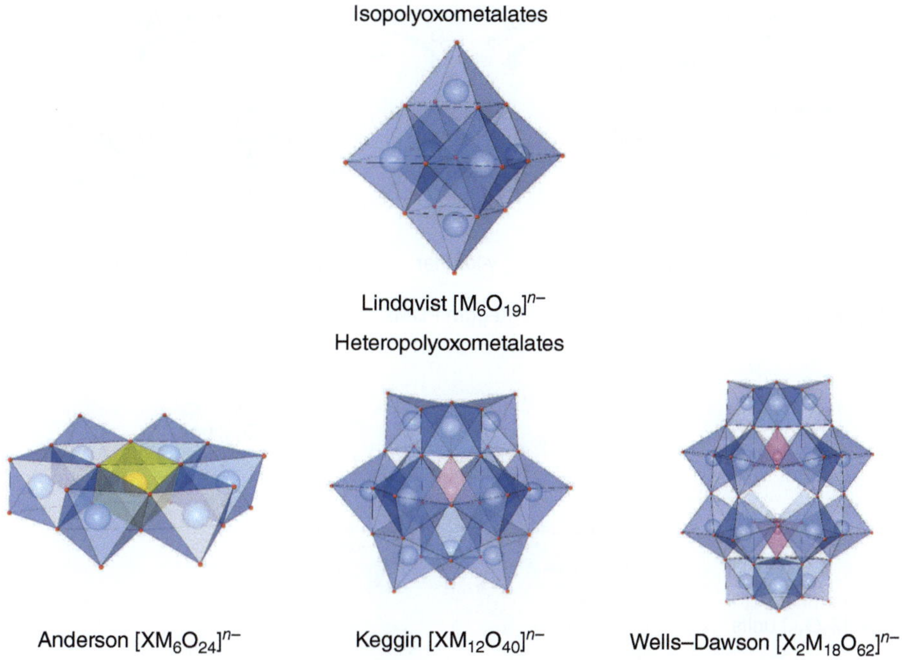

Figure 2.5 Examples of well-defined iso- and hetero-polyoxometalate clusters.

the Keggin and Wells–Dawson structures [17]. These architectures lead to stable, redox-active POMs that lend themselves to application in electrochemical systems [18].

The Keggin ($[XM_{12}O_{40}]^{n-}$; X = P, Si...) and Wells–Dawson ($[X_2M_{18}O_{62}]^{n-}$) heteropolyanion clusters have been extensively researched for their structural, photochemical and redox properties [3, 17, 19]. The Keggin structure was first identified in 1933 using the α-phosphotungstate anion [20]. The structure features a central tetrahedral heteroanion, $[XO_4]^{n-}$, surrounded by 12 octahedral $[MO_6]$ polyhedra. These polyhedra are arranged in four triads, each containing three grouped octahedra, $[M_3O_{13}]$. The Keggin structure has five known isomers, α, β, γ, δ and ε, that are defined by 60° rotations of the $[M_3O_{13}]$ triads (Figure 2.6) [21]. Experimental and theoretical studies on the systems indicate differing stability across the series [22].

The Wells–Dawson cluster was first discovered and fully characterized in 1953 [23]. It is closely related to the Keggin structure and can be viewed as two fused $[XM_9]$ fragments (common to both Keggin and Wells–Dawson clusters). The structure contains two templating heteroatoms that are typically tetrahedral. Each centre is connected to one $[M_3O_{13}]$ unit and three edge-sharing bimetallic units, $[M_2O_{10}]$. The Dawson cluster has two trimetallic units that form the 'caps' of the cluster, and six of bimetallic units that serve as the 'belt' of the cluster (Figure 2.7). Similar to the Keggin structure, the Wells–Dawson structure can also be isolated in a range of rotational isomers (α, α^*, β, β^*, γ and γ^*) [21].

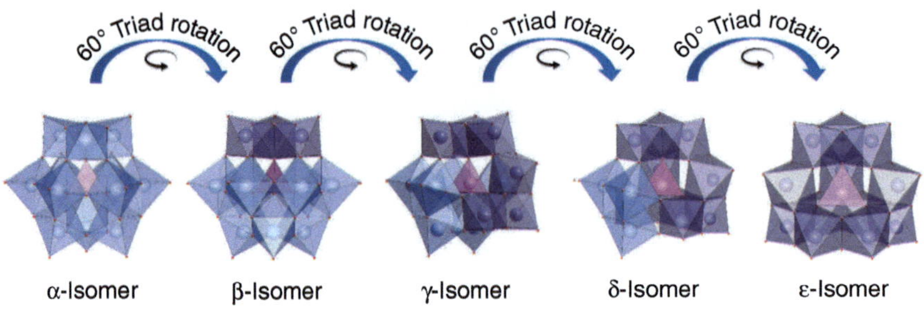

Figure 2.6 α, β, γ, δ, and ε isomers of Keggin POMs via rotation of $[M_3O_{13}]$ triads.

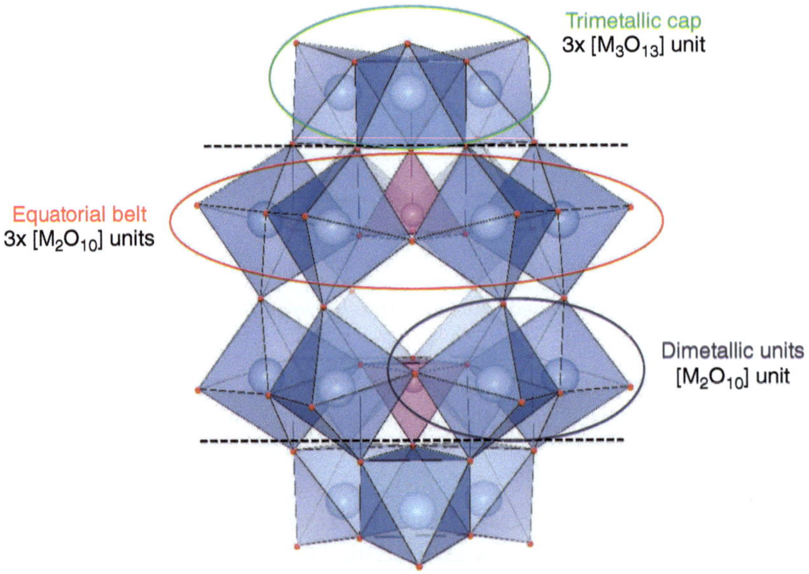

Figure 2.7 Anatomy of Wells–Dawson structure.

Under certain pH conditions, POM clusters, such as Keggin and Dawson structures, can undergo partial hydrolysis that results in a 'lacunary' species. This means that one or more $[MO_x]$ units are removed, which exposes nucleophilic oxygen vacancies in the lacuna (unfilled space or gap) and increases the cluster's overall charge. Typically, lacunary tungstates can be isolated, however, lacunary polyoxomolybdates and polyoxovanadates are relatively unstable [24]. Lacunary species are highly desirable because the exposed oxo-ligands provide opportunities for coordination with transition metals [25] or organo-functionalization with organic groups that are attached to specific *p*-block-based oxophilic linkers [26].

POMs can be modified to develop various nanostructures, due to their uniform and tunable architectures, allowing incorporation into composite, polymer and film materials [27]. They have been found to interact with biomolecules [28] and exhibit

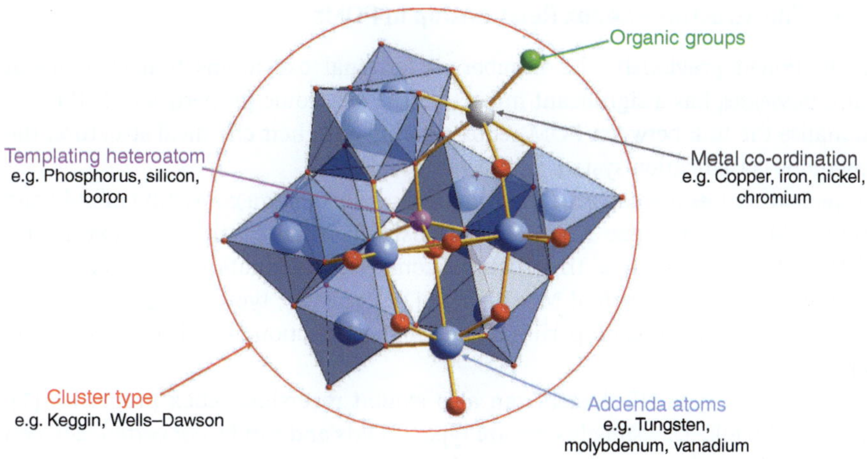

Figure 2.8 Design possibilities for POM modification. Source: Herrmann et al. [6]/Royal Society of Chemistry/CC BY 3.0.

biological activity across numerous studies, making them useful in emerging pharmaceuticals and biotechnology [29]. The choice of addenda metal ions, templating heteroatoms, and coordinated guest in the lacunary site offer a range of handles through which the properties of POMs can be tuned (Figure 2.8) [30].

2.1.3 Factors Influencing the Catalytic Role of POMs

Due to their wide range of properties including composition, structure, size and charge, POMs have attracted considerable and emerging interest for their diverse applications. They are akin to 'soluble metal oxides' and possess high solubility across several polar solvents that is typically determined by their shape, size and charge, and the nature of their charge-balancing counter-cations.

Catalysis is arguably the most studied application of POMs [31]. They have been widely used as acid catalysts in various reactions particularly esterification and hydrolysis [32]. Additionally, their electron-rich surface can be used to attach electrophiles or abstract protons, especially when using highly charged lacunary structures [33].

In recent years, there has been a surge in interest towards using POMs to address environmental issues [34, 35]. POMs are now gaining attention as electrocatalysts to assist the efficiency of green energy vectors, including hydrogen production, water oxidation and CO_2 reduction [34, 36, 37]. The rich electrochemistry of POMs comes from their highly oxidized transition-metal ion framework, which gives them a unique capacity to store and transfer multiple electrons. Many POMs can undergo rapid multi-electron redox processes while maintaining their structural stability, particularly the Dawson and Keggin clusters. In addition to the redox-active metals in the POM framework, the electrochemistry of POMs is influenced by their charge density, geometry and the nature of the counter-cations.

2.1.4 The Structure–Redox Relationship in POMs

As mentioned previously, the number of terminal oxo atoms that each metal centre possesses has a significant impact on the electronic properties of POMs. To rationalise the link between POM redox stability and their chemical structure, the Lipscombe classification system was developed [1].

Type I structures are constructed from addenda octahedra consisting of only one terminal M=O bond, type II structures consist of metal centres with two terminal M=O bonds, while type III structures comprise combinations of addenda centres with one or two terminal M=O bonds [1]. Typically, type I structures possess rich and reversible redox properties related to the reduction of their metal addenda atoms.

Furthermore, type III clusters can also exhibit reversible redox properties [1]. Keggin and Wells–Dawson clusters are type I POMs and can be converted between multiple stable redox states accompanied by minimal alterations to their molecular structure. This chapter will focus on type I and III systems. Type II POMs are not generally stable under strongly reducing conditions [38].

The correlation between POM structures and their capacity to accept electrons was first described by Pope et al. [16]. POMs with the same addenda atoms can accept electrons in their lowest unoccupied molecular orbital (LUMO), the relative energy of which is determined by their size-to-charge ratio (q/m, where q is the overall negative charge and m is the number of metal ions). POMs that are larger in size or have a low overall charge can be easily reduced. However, when comparing POMs of the same structure with mixed addenda atoms, the LUMO energies of the POMs decrease as the electronegativity of the metal addenda increases [39]. Additionally, the nature of the templating heteroatom, as well as the presence of different addenda atoms, impacts the overall charge of the POM and therefore its capacity to accept electrons (Figure 2.9) [40]. This nuanced understanding is essential for tailoring POMs with specific redox properties for applications in catalysis and energy storage.

2.2 Design Approaches via Organofunctionalization

Hybrid materials that combine molecular entities with contrasting chemistries offer unique opportunities based on the synergistic interplay between the components [41]. Recently, the field of organic–inorganic hybrid chemistry has garnered attention for its potential to create new functional molecular materials [42].

POM hybridization is achieved through coupling of inorganic and organic species to preassembled clusters to impart additional functionality and further tailor their intra- and intermolecular properties. This chapter presents a concise overview of the synthetic strategies employed to merge organic components with POMs to produce organic–inorganic hybrid compounds with new electrochemical properties. There are three key approaches to the functionalization of POMs, as depicted in Figure 2.10.

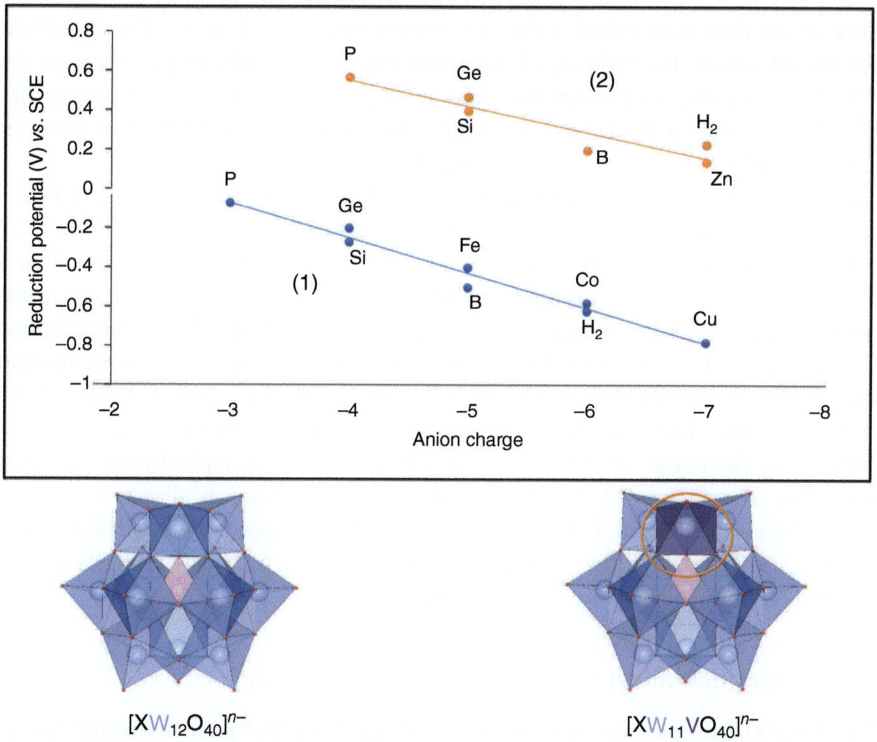

Figure 2.9 Linear dependence between the charge of the templating heteroatom and its surface potential for (1) $[XW_{12}O_{40}]^{n-}$ and (2) $[XW_{11}VO_{40}]^{n-}$. Source: Altenau et al. [40]/with permission of American Chemical Society.

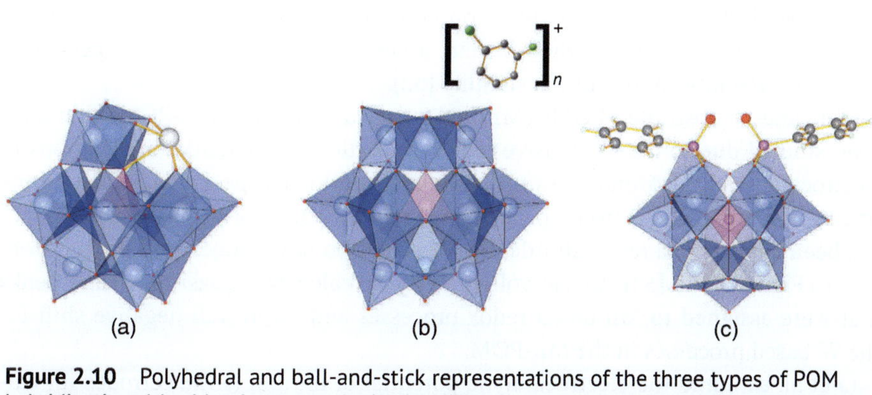

Figure 2.10 Polyhedral and ball-and-stick representations of the three types of POM hybridization: (a) addenda atom substitution, (b) non-covalent association and (c) covalent grafting.

Firstly, the coordination of transition metal ions, lanthanides or main group elements in the POM lacunary site, lanthanides or main-group elements. Secondly, the non-covalent association of cationic organic or inorganic molecules with POM complexes. Lastly, the binding of organofunctionalized main group atoms at the POM lacunary site. The latter two approaches give rise to two distinct categories of organic–inorganic hybrid POMs, determined by the type of interaction between the POM and its associated organic components.

Class I hybrids are based on the non-covalent interactions between anionic POMs and functional counterions. Intricate cationic species, including organometallic complexes, photosensitizers and ionic liquids, have been employed in class I hybrids [43–45]. Despite the opportunities offered by the systems, these hybrids can be affected by issues of instability in different solvent conditions (polarity, ionic strength and pH), something that can hinder their application in electrochemical studies [46]. In class II hybrids, the organic components are connected to the POM through covalent bonds. Although the synthesis of class II hybrids is typically more demanding than that of their class I analogues, they are typically more robust than class I systems [47]. Notably, covalent functionalization allows for precise modifications of molecular structure enabling effective approaches to design novel functional materials. In the context of this chapter, our primary emphasis will be on class II organic–inorganic hybrid POMs.

2.2.1 Transition-metal-substituted POMs (TMS-POMs)

TMS-POMs involve the incorporation of transition metals or *f*-block metals into the POM's structure. This can be achieved through two main methods: one-pot or stepwise synthesis, where the POM is formed in the presence of the cationic metal, or post-synthetic addition of the cation to solutions of lacunary POMs [48].

Extensive reviews of TMS-POMs have summarized progress in the incorporation of transition metals into POMs and their applications in magnetism, catalysis and electrocatalysis [49–51]. Recent experimental and theoretical reports have noted mixed-metal POMs display complex configurational isomerism and differences in their properties compared to their plenary archetypes in aspects such as speciation, solid-state assembly and solution stability [52].

Manganese-substituted POMs (Mn-POMs) are a family of TMS-POMs that have been studied due to their extensive redox properties and potential applications in electrocatalysis [25]. Metal substitution has proven to be especially valuable in the fine-tuning of POM electronic properties. The Mn-POM, $[SiW_{11}O_{39}Mn^{II}(H_2O)]^{6-}$, has been applied as a redox shuttle for the evolution of hydrogen and oxygen from water (Figure 2.11) [53]. Cyclic voltammetry revealed two quasi-reversible peaks that were assigned to Mn-based redox processes and an overall negative shift for the W-based processes in the Mn-POM.

Li et al. explored the electrocatalytic potential of indium (In) electrodes for carbon dioxide reduction reactions (CO_2RR), using aqueous solutions enriched with $SiW_{11}Mn$ POMs as electrolytes [54]. The outcomes demonstrated that the In electrode, when immersed in the electrolyte, effectively facilitated the electrocatalytic

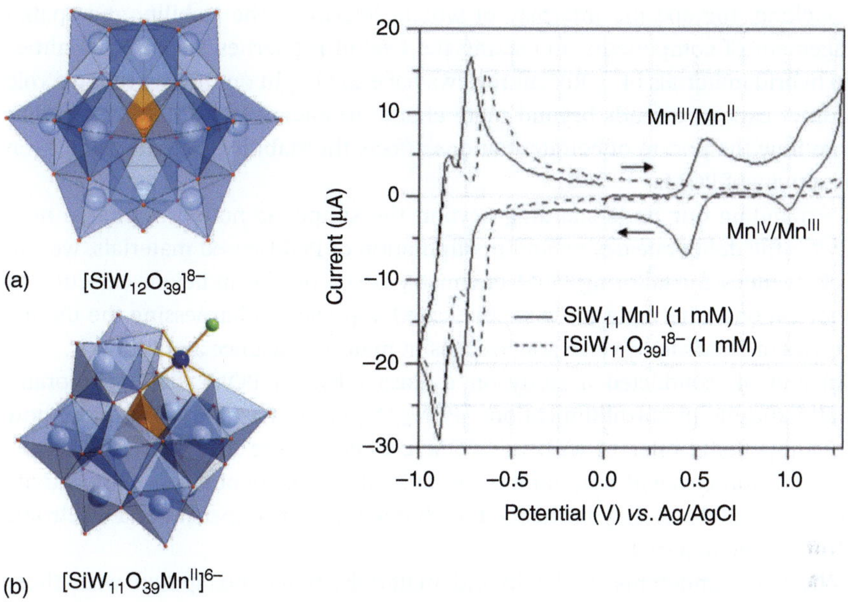

Figure 2.11 Diagrams showing the structures and cyclic voltammograms of (a) $[SiW_{12}O_{39}]^{8-}$ and (b) $[SiW_{11}O_{39}Mn^{II}]^{6-}$. Source: Reproduced with permission from Tsuji et al. [53], © 2016 Wiley Publishers.

reduction of CO_2 to acetic acid at potentials of −0.55 V *vs.* RHE. Surface analysis of the In electrode, conducted through X-ray photoelectron spectroscopy and in situ infrared (IR) spectroscopy, revealed a notable decrease in oxide species concentration on the electrode surface, accompanied by the formation of W^{6+} species. In situ IR spectroscopy further detected In-CO_2* intermediates, illustrating the activation of CO_2 at the In electrode.

The efficient electron transfer capability significantly enhances the selectivity to produce acetic acid, showcasing the potential of this electrocatalytic system. Notably, this research reveals that the reduction of tungsten in the Mn-POM further amplifies the selectivity of the In electrode in catalysing the reduction of CO_2 to formic acid or acetic acid. This enhancement is attributed to the electron transfer pathway established in the electrolyte. The system remains stable for up to 12 hours. The research introduces a novel strategy for augmenting the product selectivity of electrocatalytic CO_2 reduction reactions through the careful design of electrolyte solutions. These findings offer invaluable insights for advancing the development of electrolytes that are not only efficient but also highly selective in the electrochemical reduction of CO_2.

2.2.2 Class I Hybrid POMs

Non-covalent interactions between POMs and their counter-cations form the basis of class I hybrid POMs. The interactions between components can be based on a combination of hydrogen bonding, van der Waals forces, ionic interactions and π–π stacking, the specific interplay of which determine the stability and spatial arrangement of components, impacting the overall properties and functionalities of the hybrid materials [43]. Recent reviews have shed light on the fact that the role of counter-cations extends beyond mere charge balancing [55, 56]. These studies explore how the nature of counter-cations affects the stability, solubility and even the reactivity of POMs.

By expanding our understanding beyond the simplistic notion of charge neutrality for the deliberate design and manipulation of POM-based materials, we can unlock avenues for tailoring POM chemistry based on the judicious selection of counter-cations, offering a more sophisticated approach to harnessing the unique properties of POMs for diverse applications in materials science and catalysis.

Kibler et al. conducted a study on a class I hybrid POM that incorporates benzothiadiazole-imidazolium cations $[P_2W_{18}O_{62}][BTD\text{-}4,7\text{-}ImH]_3$ to explore how these components interact with the POM cluster (Figure 2.12) [57]. The study found that non-covalent associations are crucial to the cooperative photoactivity of the hybrid POM as they influence the charge transfer properties and electronic structure of the material.

Cyclic voltammograms of the hybrid materials, when juxtaposed with those of its precursors, revealed a subtle negative shift in the peak potential for each

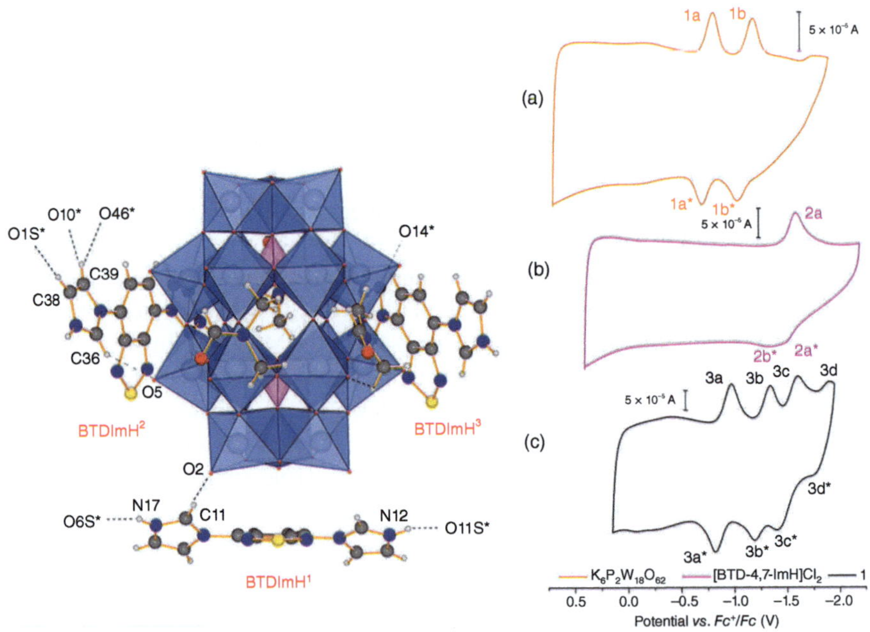

Figure 2.12 Diagrams showing $[BTD\text{-}4,7\text{-}ImH]_3[P_2W_{18}O_{62}]$ and solid-state cyclic voltammograms of (a) $K_6P_2W_{18}O_{62}$ (orange), (b) $[BTD\text{-}4,7\text{-}ImH]Cl_2$ (pink), and (c) $[BTD\text{-}4,7\text{-}ImH]_3[P_2W_{18}O_{62}]$ (black) deposited as inks onto glassy carbon working electrodes and performed in anhydrous acetonitrile (10 ml) with 0.1 M $TBAPF_6$ as supporting electrolyte and reported vs. $Fc^+|Fc$ as an internal reference. Source: Kibler et al. [57]/Frontiers/CC BY 4.0.

electrochemical process. Notably, an additional POM redox process emerged in the voltammogram of the hybrid system, attributed to enhanced charge compensation facilitated by the close packing of the organic BTD cations in the hybrid structure.

The tunability of redox properties, achieved through controlled non-covalent associations, further underscores their significance in tailoring POMs for specific electrochemical applications. Overall, harnessing non-covalent interactions presents a versatile strategy to design advanced POM-based materials with optimized electrochemical functionality, promising advancements in fields such as energy storage and catalysis.

2.2.3 Class II Hybrid POMs

Class II hybrid systems involve the direct covalent attachment of organic molecules to the POM. The methodologies for achieving this are highly specific, contingent upon both the structure of the POM and the identity of the addenda atoms present on the POM. These hybridization strategies have been comprehensively reviewed and grouped based on several fundamental motifs (Figure 2.13) [26].

Comprehensive reviews have extensively explored the diverse range of synthetic methods for covalently organofunctionalizing POMs [50, 58, 59]. In class II hybrid POMs, organic groups are frequently covalently linked to the POM structure through *p*-block elements, such as arsenic, silicon, phosphorus, which sit in the lacunary site [30, 60]. Anchoring groups exert electronic effects that can influence the energy of the POM molecular orbitals. These electronic effects can be further modified by considering the electronic effects inflicted by the organic moiety itself [60, 61].

Kibler et al. produced an in-depth study on the effects of organoarsenic groups on the redox chemistry of hybrid POM clusters [60]. Within this study, they developed a new approach which enabled modulation of redox-active materials to produce a

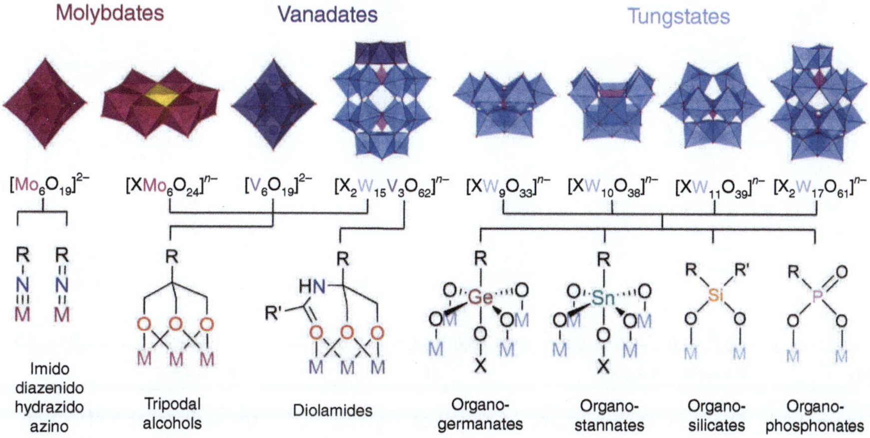

Figure 2.13 Covalent organo-functionalization routes of a broad range of POMs and lacunary POMs. Source: Kibler and Newton [26]/with permission of Elsevier.

new family of organofunctionalized Wells–Dawson polyoxotungstates $[P_2W_{17}O_{61}(AsOC_6H_5)_2]^{6-}$. Compared to its plenary POM $[P_2W_{18}O_{62}]^{6-}$, the organoarsenic hybrids displayed a positive shift in the initial reduction potential (Figure 2.14a). These hybrids were further expanded by modulating the aryl groups on the arsenic linkers $[P_2W_{17}O_{61}(AsOC_6H_4R)_2]^{6-}$, where R = NH_2 or NO_2, showcasing how the electronic structure of the aromatic motif can alter the redox properties of the hybrid system (Figure 2.14b).

Functionalization of redox-rich Keggin and Dawson clusters through organophosphonate chemistry garners significant attention as the electronic properties of hybrid systems can be modified through meticulous design of the organophosphonate group [62]. Primary electrochemical investigations conducted by Boujtita et al. on an organophosphonate-functionalized Dawson cluster $[P_2W_{17}O_{61}\{(POC_6H_5)_2\}]$ revealed that the first two reversible redox processes exhibited a positive shift relative to the unmodified 'plenary' Dawson cluster $[P_2W_{18}O_{62}]$ ($\{P_2W_{18}\}$). This shift was more pronounced than observed in the phenylsiloxane analogue, where the silyl group led to a slight negative shift in POM redox potentials [63]. This observation revealed the overall electron withdrawing effects induced by the phosphonate group on the cluster. Subsequently, numerous studies exploring functionalization via phosphonate linkers have emerged including the incorporation of porphyrin moieties [64].

Further investigations by Cameron and Fujimoto and co-workers elucidated that organophosphonate hybrid Wells–Dawson POMs offer a remarkable degree of tunability in their orbital energies, achieved by altering the electron-withdrawing properties of the aryl organophosphonate ligands (Figure 2.15). These studies unveiled a relationship of electron density within the HOMO and LUMO levels, showcasing delocalization across both the POM and the ligand's π* orbitals. This phenomenon highlights the presence of inductive and resonance effects within these hybrid structures [66].

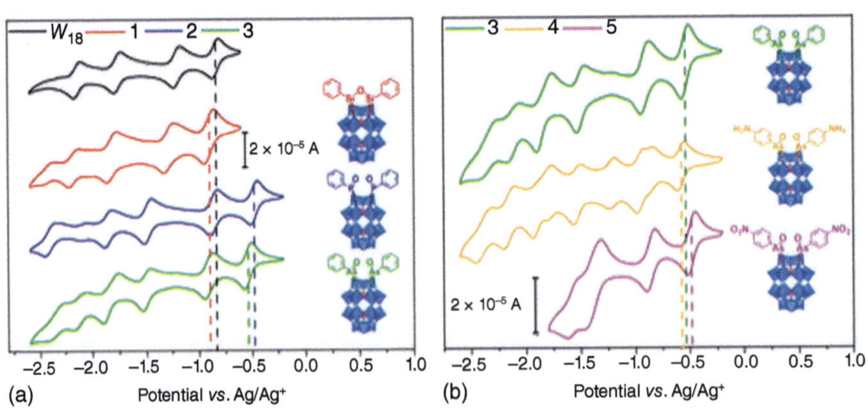

Figure 2.14 Cyclic voltammograms of polyoxotungstate POMs showing (a) the effect of different linkers between the organic group and plenary POM and (b) relationship between the As-hybrid POMs and different aryl groups. Source: Kibler et al. [60]/American Chemical Society/CC BY 4.0.

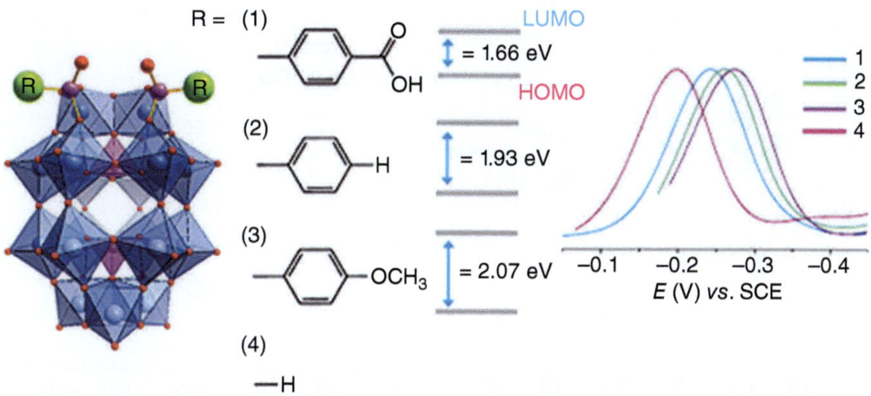

Figure 2.15 From left: Structural representation of organophosphonate hybrid POMs (1)–(4) showing the different substituent R-groups, the respective estimated HOMO–LUMO energy gaps based on time-dependent density functional theory calculations, and differential pulse voltammograms highlighting the first reduction wave of each species (current plotted on an arbitrary scale to aid comparison). Source: Fujimoto et al. [65]/with permission of American Chemical Society.

Furthermore, these investigations revealed a shift in the redox potentials of these hybrid POMs, which manifested as a positive displacement relative to the parent cluster $[P_2W_{18}O_{62}]^{6-}$. The more electron-withdrawing the substituent group, the greater the positive shift in the first redox potential. This was coupled with an increase in the visible-light sensitivity of the compounds [65].

Cameron et al. investigated the post-functionalization of an organophosphonate hybrid Dawson POM by introducing various transition metals coordinated with neighbouring P=O sites, resulting in the formation of novel dimeric hybrid POM moieties (Figure 2.16) [61]. These coordination properties of bis-organophosphonyl POM hybrids had previously only been observed in trilacunary Keggin derivatives. In this study, it was found that metal functionalization had an impact on the

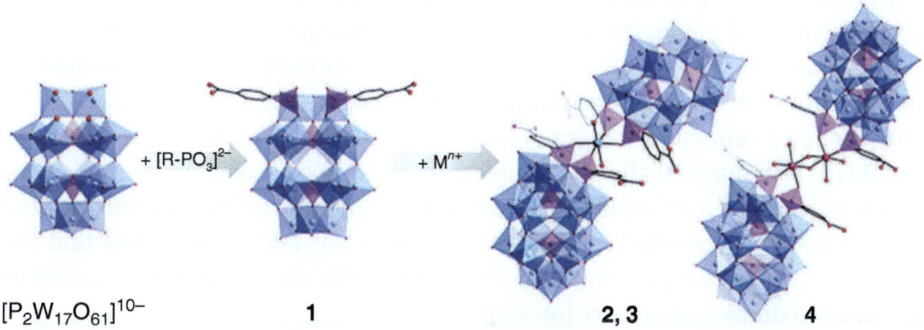

Figure 2.16 Schematic displaying the hybridization and post-functionalization of a Wells–Dawson POM. Colour code: blue polyhedra = $\{WO_6\}$, purple polyhedra = $\{PO_4\}$ or $\{PO_3C\}$, green sphere = Zn or Co, orange spheres = Mn, red spheres = O. Cations, protons and solvent molecules have been removed for clarity. Source: Reproduced with permission from Cameron et al. [61]. © 2018 Royal Society of Chemistry.

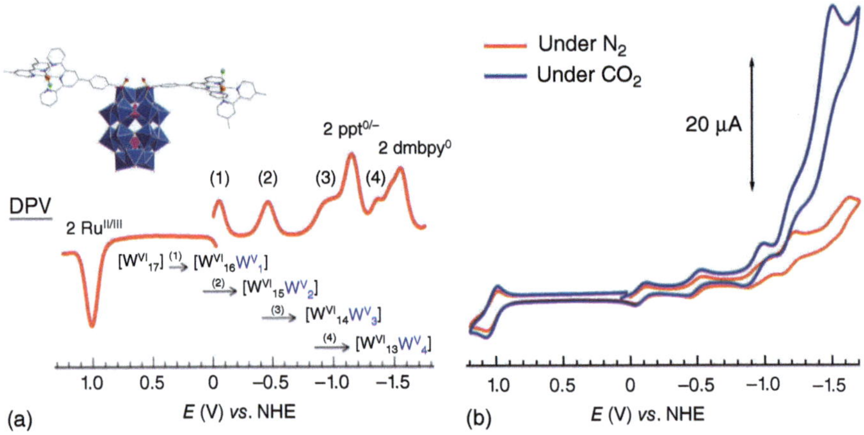

Figure 2.17 Single crystal X-ray structure of $(^n Bu_4N)_3H[(Ru(dmbpy)(ppt)Cl)_2P_2W_{17}O_{57}]$ and the reduction events at the POM body via differential pulse voltammetry (a) and cyclic voltammograms (b) of $(^n Bu_4N)_3H[(Ru(dmbpy)(ppt)Cl)_2P_2W_{17}O_{57}]$ under N_2 (red) or CO_2 (blue). Source: Kuramochi et al. [68]/Royal Society of Chemistry/CC BY 3.0.

electronic structure and photoactivity of the Dawson cluster. The binding of metal centres led to positive shifts in the POM redox processes. This discovery is important because it differs from what was seen in previous examples of transition metal-coordinating hybrids that showed poor electronic communication between the metal and POM, leaving the redox properties relatively unchanged [67].

Kuramochi et al. investigated the use of a POM-connected transition-metal complex as an electrocatalyst for reductive conversion of CO_2 [68]. A bis-Ru-polypyridyl-substituted hybrid POM, $(^n Bu_4N)_3H[(Ru(dmbpy)(ppt)Cl)_2P_2W_{17}O_{57}]$ was produced with solution-state electrochemical properties allowing selective conversion of CO_2 to H_2/CO or formic acid. Differential pulse voltammetry of the hybrid POM revealed multiple waves corresponding to POM-based $W^{VI} \rightarrow W^{V}$ redox processes and the reduction of the polypyridyl ligands to their radical anions (Figure 2.17a). The reduced hybrid POM serves as an active electrocatalyst for CO_2 conversion, as the polypyridyl radical anions are rapidly quenched in the presence of CO_2. Ru-bound Cl centres are lost, allowing CO_2 coordination at the ruthenium centre followed by rapid electron transfer from the radical anion ligands to the CO_2 (Figure 2.17b).

Other related ruthenium catalysts can similarly drive CO_2 electroreduction, however, this hybrid material allows selective isolation of formic acid under acidic conditions which cannot be replicated without direct hybridization of the derivatized mononuclear Ru^{2+} tethered group. The novelty of this approach implies that the tethering of electrocatalytic centres to POMs may offer significant opportunities in CO_2 electroreduction chemistry [68, 69].

2.2.4 Asymmetric Systems

Typically, hybridization of POM clusters with organosilyl, phosphoryl or arsenyl moieties produces symmetrically bi-functionalized hybrid materials; two identical organic groups bound to the lacunary site. Recently, approaches have emerged through which two different organic groups can be attached at the same lacunary site, forming so-called asymmetric hybrid POMs. These systems give the synthetic chemist increased control over the physiochemical properties of the hybrid system and will lead to new molecular nanomaterials with tailored modular electrochemical functionalities [70, 71]. Whilst the synthesis and purification of asymmetrically functionalized hybrid POMs remains a formidable challenge, this section presents key examples that have been reported in recent years.

There are very few reported examples of asymmetric hybrid POMs [72]. The vast majority of these are based on the Anderson-Evans cluster type, a Lipscombe type II cluster, which shows limited photo and redox activity compared to many other POMs [73]. The electronic properties of hybrid structures are usually dominated by the transition-metal heteroatom or organic group, and the poor conjugation across the triol functionalities limits their potential as electrocatalysts [26]. Despite this, a range of interesting cluster types have been prepared [73].

Thus, it is valuable to develop asymmetric functionalization strategies targeting clusters, such as the Wells–Dawson, which have the capacity for multi-electron redox processes and highly tunable organic hybrid structures. These approaches can unlock new pathways for isolating more advanced multifunctional hybrid systems, which can potentially be used as redox-, photo- and catalytically active nanomaterials. Hampson et al. reported the concept of using asymmetric POMs as a highly tunable platform for designing nanomaterials and successfully isolated the first stable asymmetric Wells–Dawson hybrid POM $[P_2W_{17}O_{61}(POC_{21}H_{14}N_3)(PO_2C_{24}H_{41})]^{6-}$ through a one-pot acid-catalysed condensation approach (Figure 2.18) [74].

Through this study, a significant breakthrough in the field of nanomaterials was discussed as the hybrid material displayed amphiphilic properties which

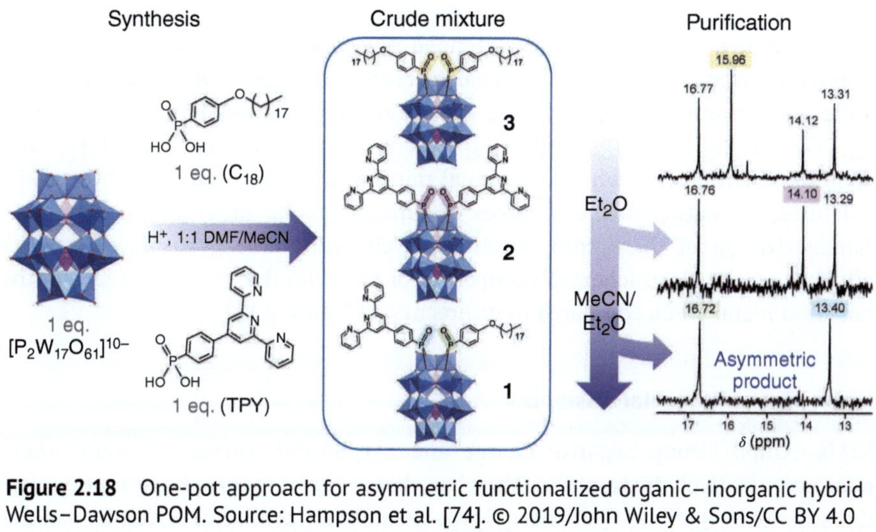

Figure 2.18 One-pot approach for asymmetric functionalized organic–inorganic hybrid Wells–Dawson POM. Source: Hampson et al. [74].

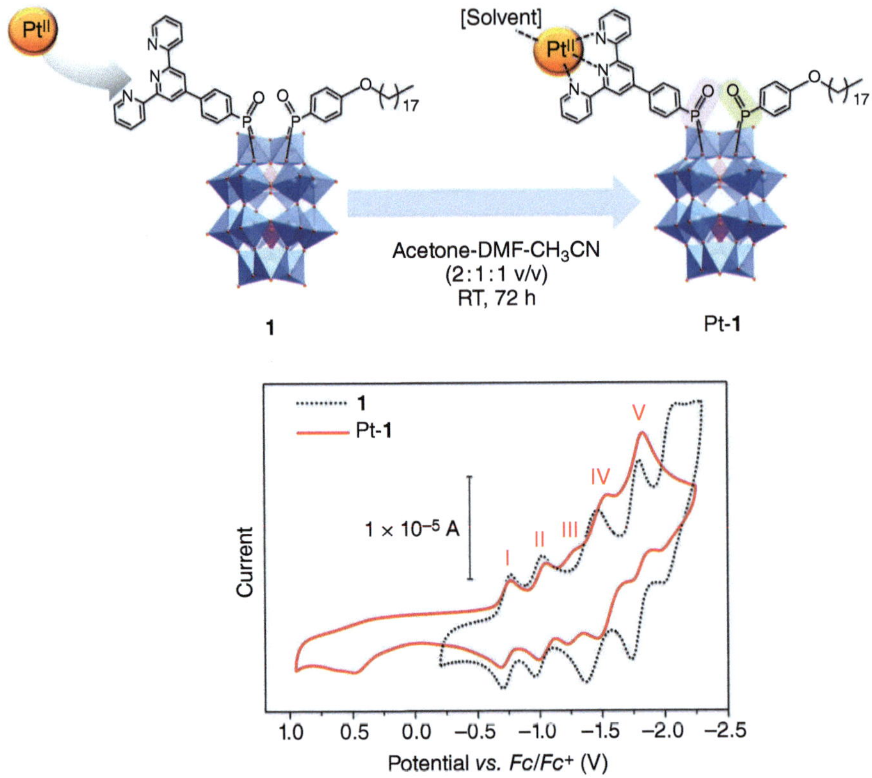

Figure 2.19 One-pot approach for Pt^{2+}-coordinated asymmetric functionalized hybrid Wells–Dawson POM and their respective cyclic voltammograms. Source: Hampson et al. [75]/Royal Society of Chemistry/CC BY 3.0.

enabled spontaneous self-assembly into redox-active micellar structures in aqueous conditions.

Subsequently, Hampson et al. extended this approach to facilitate solvent-dependent assembly into micellar nanostructures and transition-metal binding to produce platinum(II) complexes [75]. Results showed that the hybrid POM $[P_2W_{17}O_{61}(POC_{21}H_{14}N_3Pt(solv.))(PO_2C_{24}H_{41})]$ retained its micellar structure and amphiphilic properties, whilst individual micelles coalesced into small worm-like structures, indicating further degrees of supramolecular aggregation. Electronic characterization of these compounds via cyclic voltammetry displayed a weak influence on the electrochemical properties of the POM due to the presence of the transition metal when compared to its precursor (Figure 2.19).

2.2.5 Supramolecular Assembly

POMs exhibit strong negative charge and oxygen-rich surfaces, making them inherently relatively hydrophilic. They have been shown to form large supramolecular structures, such as "blackberry"-type assemblies, primarily driven by van der

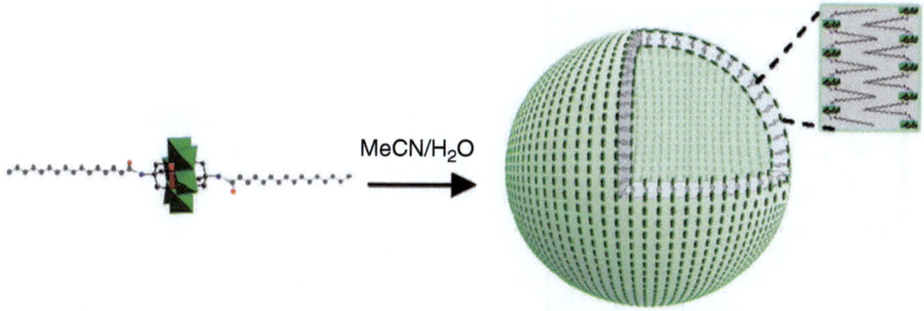

Figure 2.20 Schematic of the Mn-Anderson C_{16} POM and its vesicle formation in MeCN/H_2O. Source: Zhang et al. [79]/with permission of American Chemical Society.

Waals forces [76]. Examples of POM clusters that exhibit this property include large 'Molybdenum Blues' and mixed metal tungstates and vanadates, however, the assembly of these structures is slow, taking days to weeks [76–78]. To speed up this process, POMs can be organo-functionalized, which introduces a hydrophobic element that transforms the POMs into amphiphilic compounds encouraging rapid self-assembly via solvophobic interactions.

In this section, we will explore a range of unique examples of class II hybrid POM amphiphiles, their supramolecular assemblies and their functionalities.

Cronin and Liu et al. synthesized an unprecedented amphiphilic molecule $(^nBu_4N)_3[MnMo_6O_{18}\{(OCH_2)_3CNHCO(CH_2)_{14}CH_3\}_2]$ (Mn-Anderson-C_{16}) by grafting two aliphatic C_{16} chains to an Anderson POM via an amide bond. As one of the first examples of a class II hybrid–POM amphiphile, the molecule exhibited solvent-dependent supramolecular assembly (Figure 2.20) [79]. When mixed with MeCN/H_2O, the hybrid was observed to assemble into vesicles the interaction between the hydrophobic chains in a hydrophobic layer. DLS analysis revealed that large assemblies grew in a solution containing 40% acetonitrile, and after 57 days, the assemblies reached growth equilibrium at 115 nm hydrodynamic radii. Subsequent research from this group found that by changing the solvent composition to MeCN/toluene, reverse micelles could be achieved [80].

By exploring the self-assembly mechanisms, the study highlights the significant impact and potential utility of POMs in shaping nanoscale organic–inorganic systems. The controlled self-assembly of hybrid-POMs from discrete 0D to extended 3D assemblies continues to develop towards applications in areas such as nanotechnology and materials science [81].

Kastner et al. explored the behaviour of an alkylthiol-functionalised redox-active POM amphiphile $H_6[P_2W_{17}O_{61}(H_{27}C_{17}O_2PS)_2]$ which exhibited solvent-dependent self-assembly to form regular micellar assemblies (in water) and exist as molecular species in dimethylformamide (DMF) (Figure 2.21a) [82]. Cyclic voltammograms of the compound under the different conditions showed significantly different electrochemical properties, depending on the nature of the supramolecular assembly (Figure 2.21b).

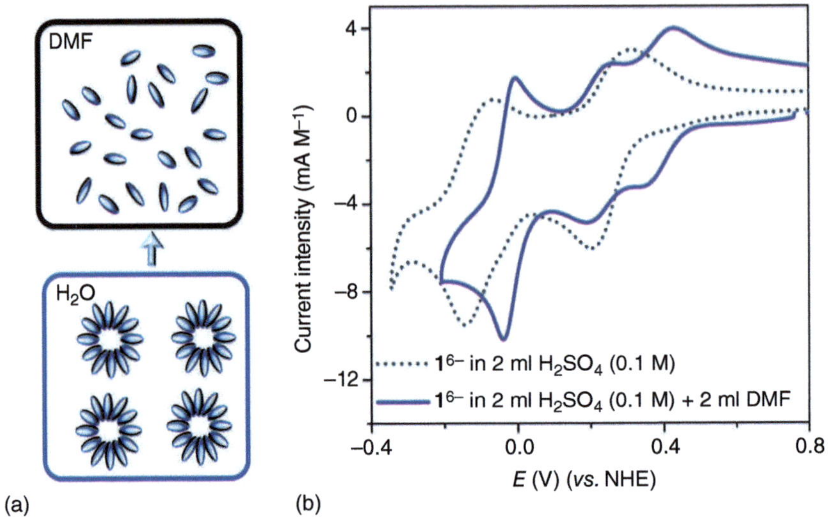

Figure 2.21 Schematic of the solvent-dependent behaviour of the alkylthiol-functionalised amphiphilic POM (a) and its respective cyclic voltammogram (b). Source: Reproduced with permission from Kastner et al. [82], © 2017 Royal Society of Chemistry.

The transformation from the molecular to the supramolecular regime and the accompanying change in electrochemical properties offers opportunities for environment-specific applications in catalysis, photo-catalysis and advanced switchable materials. This breakthrough suggests that future research could focus on developing the self-assembly and solvent-dependent disassembly of the system, with the goal of developing redox-tunable soft materials as stable, water-soluble supramolecular capsules for use in organic solvent-free catalytic systems.

Subsequently, Amin et al. delved into the spontaneous self-assembly of amphiphilic hybrid POMs on various carbon surfaces to investigate their electrochemical behaviour, stabilities and morphology [83]. In this study, they observed monodisperse micelles of their hybrid species $K_6[P_2W_{17}O_{61}(POC_6H_4OC_{20}H_{41})_2]$ ($\{W_{17}C_{20}\}$) and substrate-dependent hierarchical assembly on different carbons (Figure 2.22).

The surface-immobilised supramolecular assemblies displayed distinct multi-redox behaviour over hundreds of electrochemical cycles. This approach to nanostructure electrode fabrication may ultimately offer interesting opportunities in electrocatalysis and energy storage systems.

2.2.6 Immobilization Techniques

In the field of materials science, POMs are increasingly studied on surfaces or as encapsulated redox-active species in host-guest systems [50, 84, 85]. Surface immobilization involves attaching POMs to solid supports or interfaces, which

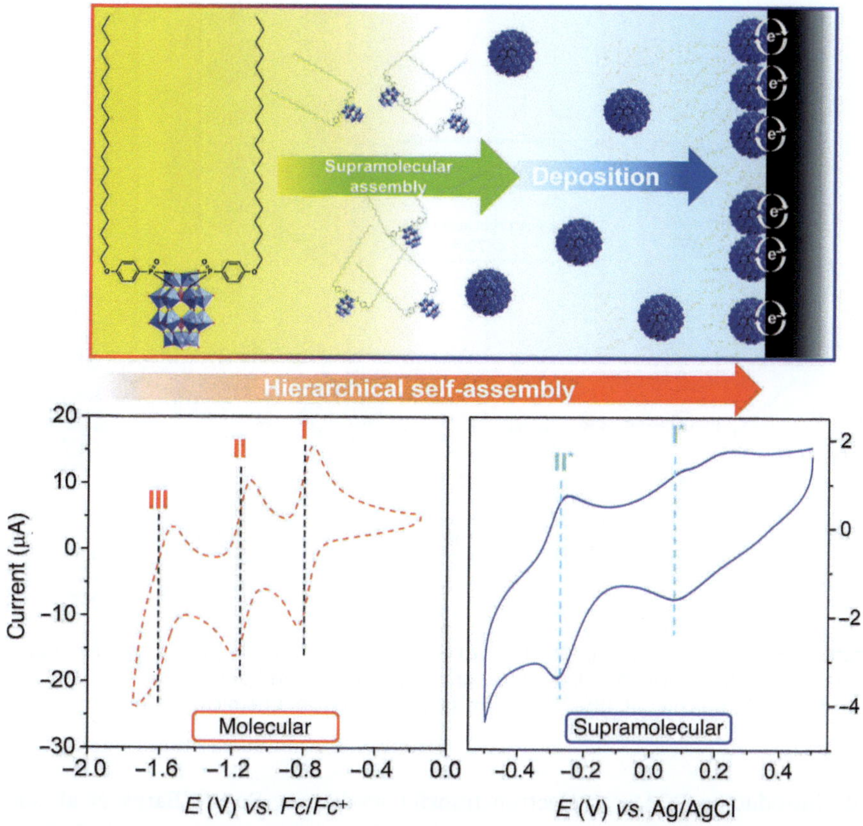

Figure 2.22 Schematic of the hierarchical supramolecular assembly of redox-active $\{W_{17}C_{20}\}$ in aqueous solutions and their respective cyclic voltammograms in DMF (red) and aqueous H_2SO_4 (blue). Source: Amin et al. [83]/Royal Society of Chemistry/CC BY 3.0.

improves their stability, catalytic efficiency and recyclability. Nanoencapsulation, on the other hand, provides a protective environment for POMs, preventing them from degrading or aggregating and allowing precise control over their release and interactions. Both techniques offer excellent control over the properties and functionalities of hybrid POMs, making them ideal for use in a range of applications with electrocatalysis.

2.2.6.1 Surface Immobilization

Gam Derouich et al. explored the covalent bonding of Keggin and Dawson hybrids to surfaces such as gold and carbon electrodes (Figure 2.23) [86]. Recently, they extended this to attachment to silicon/silicon oxide and indium tin oxide substrates using different functional groups, including diazonium and carboxylic acid groups [87]. Research on the electronic transport properties of these POM-modified materials has highlighted the significance of both the hybrid POM molecular structure and the influence of covalent anchoring towards tailoring their electronic characteristics and photoelectrochemical performance [87–89].

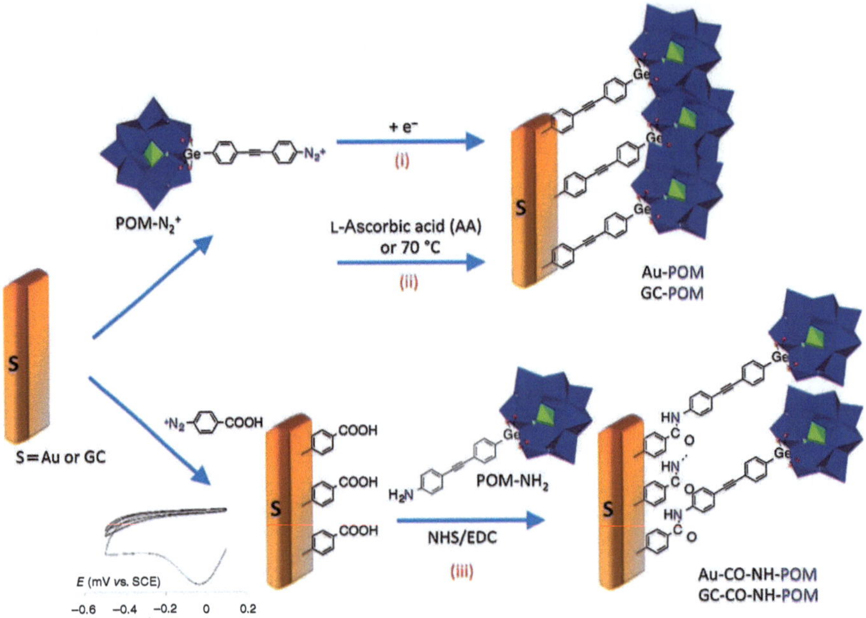

Figure 2.23 Covalent grafting strategies to immobilize POMs onto glassy carbon and gold substrates via electrografting (i), chemical grafting (ii) and peptidic coupling (iii). Source: Gam Derouich et al. [86]/with permission of American Chemical Society.

By introducing POMs as electron transfer mediators, Ben M'Barek et al. successfully mitigated charge recombination, leading to improved overall efficiency in dye-sensitized photocathodes [89]. POMs act as efficient electron shuttles, facilitating the smooth transport of electrons and minimizing their loss through undesirable pathways. The significance of this breakthrough lies in its potential to boost the efficiency of light-to-electricity conversion, a critical aspect in the development of sustainable and renewable energy sources.

Direct immobilization onto the SWCNTs addresses the POM desorption and leaching challenges from prior electrostatic anchored POM/CNT composites. Ji et al. demonstrated the formation of a covalent linkage between hybrid POMs $[N(^nBu)_4]_3[MnMo_6O_{18}\{(OCH_2)_3CNH_2\}_2]$ and single-walled carbon nanotubes (SWCNTs). The POMs are immobilised via an amide bond between the organic moieties of a hybrid Mn-Anderson POM and carboxylate-functionalized SWCNTs (Figure 2.24) [90].

When used as anode material for lithium-ion batteries, the composite material exhibited improved electrochemical performance compared to pure SWCNTs or pure Anderson clusters with enhanced conductivity, enabling faster electron transport and improved cycling stability. This improvement was attributed to the covalent linkage of the POM to SWCNTs due to the short distance between the POM and SWCNT inducing a less dense π–π packing of the nanotubes and preventing desorption.

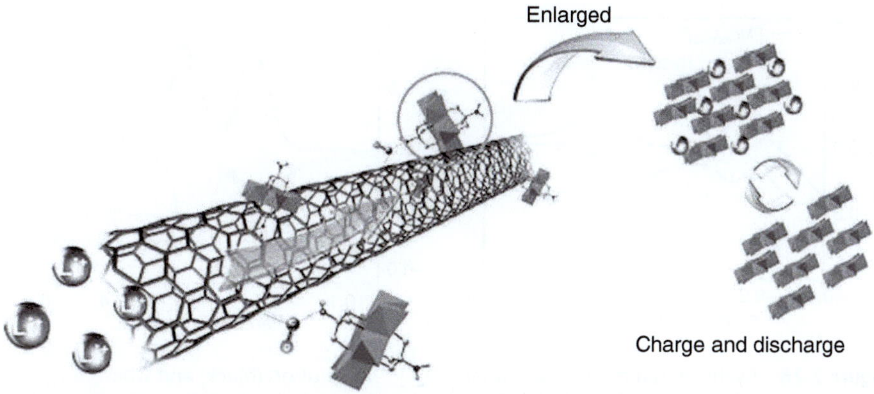

Figure 2.24 Schematic illustrating hybrid Anderson POMs covalently linked to SWCNTs for use as an anode material in lithium-ion batteries. Source: Reproduced with permission from Ji et al. [90]. © 2015 Wiley Publishers.

2.2.6.2 Nanoencapsulation

Jordan et al. demonstrated the spontaneous encapsulation of POMs ($[PW_{12}O_{40}]^{3-}$ and $[P_2W_{18}O_{62}]^{6-}$) within the cavities of SWCNTs, driven by an electron transfer process [91]. In this process, SWCNTs transfer electrons to POMs, leading to the formation of reduced POMs and cationic nanotubes. The irreversible encapsulation of POMs is then partially driven by electrostatic forces between the components (Figure 2.25).

Subsequently, Jordan et al. successfully applied this mechanism to electronically distinct polyoxomolybdate clusters $[PMo_{12}O_{40}]^{3-}$ to explore the effects of encapsulation on the redox chemistry of POM clusters [92]. Cyclic voltammograms revealed three reversible redox processes with small positive shifts in comparison to the solution phase molecular analogue (Figure 2.26a) and excellent redox stability

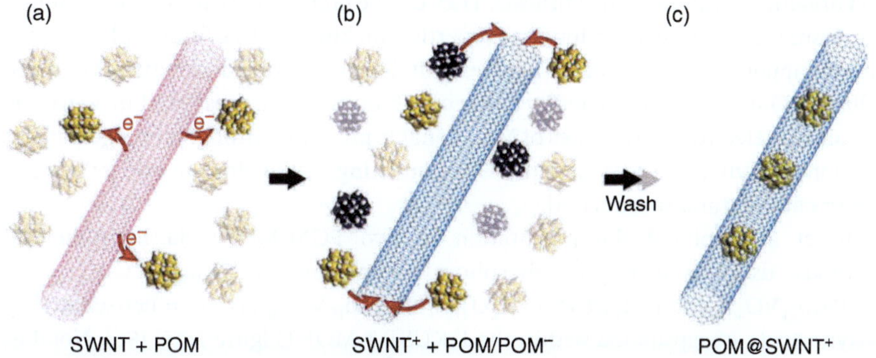

Figure 2.25 Diagram showing: (a) electron transfer from SWNTs (pink – neutral, blue – cationic) to POMs (yellow – native state, dark blue – reduced state); (b) Coulombic force-driven encapsulation of POM anions into the cationic SWNTs; (c) nonencapsulated POM anions and all potassium cations are washed away using water. Source: Reproduced with permission from Jordan et al. [91], © 2019 Wiley Publishers.

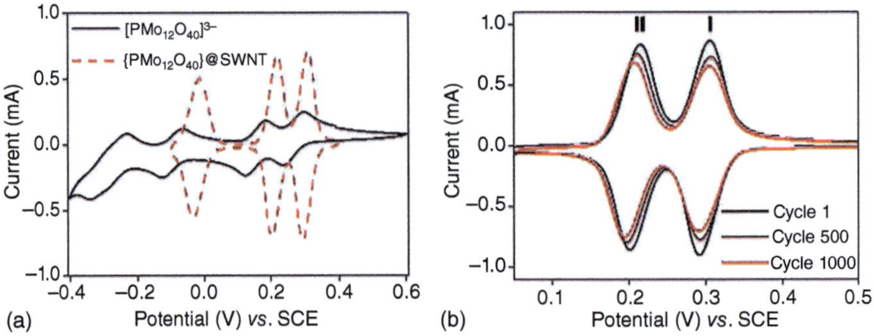

Figure 2.26 Cyclic voltammograms of $[PMo_{12}O_{40}]^{3-}$ in solution (black) and when encapsulated within the walls of SWCNTs (red) (a) and repeated potential cycling over 1000 cycles (b). Source: Reproduced with permission from Jordan et al. [92]/John Wiley & Sons/CC BY 4.0.

(Figure 2.26b). The encapsulated POMs retain up to 86% of their charge capacity even after undergoing 1000 redox cycles (Figure 2.26b).

The encapsulation of redox-active POMs within SWCNTs is a significant advancement in developing new design principles for enhancing their electrochemical properties. The coating can provide significant protection for the POMs, allowing them even to retain their redox chemistry under solvent conditions in which they would usually decompose. There have already been many reports of POM clusters with additional heterometals or organic groups, making them theoretically suitable for incorporation into SWCNTs [93]. The strategies presented in this study offer new opportunities to explore the compatibility of different redox molecules and nanotubes, which vary in size and chemistry.

Metal–organic frameworks (MOFs) are a class of porous materials known for their diverse network topologies which can be tailored by selecting specific metal ions and organic ligands in their synthesis. They can be prepared to exhibit semiconducting, insulating and even conducting properties, making MOFs a flexible platform for various applications. The porous nature of MOFs serves as an ideal matrix for immobilizing POMs, enhancing stability and accessibility to active sites [94]. Furthermore, the tunable electronic properties of MOFs enable precise control over charge transfer phenomena when interacting with POMs, enabling applications in energy storage, electrocatalysis and sensors [95].

Du et al. explored the preparation of two POM-MOFs via hydrothermal methods using Keggin-type phosphovanadomolybdates, $[Cu_{12}(BTC)_8(H_2O)_{12}][H_4PMo_{11}VO_{40}]$ and $[Cu_{12}(BTC)_8(H_2O)_{12}][H_5PMo_{10}V_2O_{40}]$ (BTC = benzenetricarboxylic acid), encapsulated within the HKUST-1 MOF (Figure 2.27) [96]. Notably, the systems displayed catalytic activity for benzene hydroxylation and oxidative desulfurization.

The discovery of distinct, high-nuclearity cage-like water clusters within POM-MOFs is likely to be an important area of research in the future application of POMs in electrocatalysis.

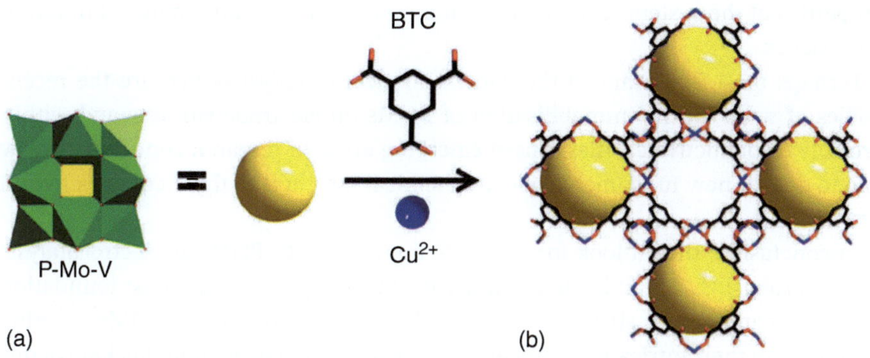

Figure 2.27 Simplified schematic of Keggin phosphovanadomolybdate (a); constructed into a $Cu_3(BTC)_2$ MOF (b). Source: Reproduced with permission from Du et al. [96]. © 2020 American Chemical Society.

2.3 Conclusion

Renowned for their redox activity, stability and tunable electronic structures, POMs have emerged as promising candidates for a wide array of electrocatalytic processes. Interest in POMs within the field of electrocatalysis stems from their unique combination of attributes, including high catalytic activity, stability and tunable properties. Their redox versatility allows for tailoring electrocatalytic reactions, and their stability ensures prolonged functionality in diverse applications.

Electrocatalysts play a pivotal role in addressing diverse challenges across various disciplines, influencing energy storage, clean energy production, pollution reduction, environmental remediation, electrochemical sensing, carbon capture, fuel cells, electrochemical synthesis, nitrogen reduction and biomedical applications. Their impact extends across these fields, enhancing efficiency and sustainability, and driving technological progress with a focus on clean energy, pollution mitigation and biomedical innovations.

In this contribution, we have set out to present some of the emerging strategies that can be deployed when considering how to design bespoke POM electrocatalysts. The primary tools involve the basic building units of the clusters themselves – the addenda atoms and templating groups that give the chemist significant control over both the structure and the redox properties of POM clusters. When moving beyond this, we can consider the functionalisation of POMs, either through substitution of addenda atoms with transition metals or through organofunctionalisation strategies. It seems likely that the new levels of control over composition, electronic structure and redox properties that can be accessed following these approaches will lead to significant breakthroughs in the area.

Running alongside the approaches that allow us to control the molecular properties are the opportunities we have to tune the supramolecular chemistry of POMs. The recent work highlighting the translation of POM redox properties across the size scales, from molecular systems to nanoscale soft materials, all the way to extend micron-scale solid materials illustrate further how electrochemical properties of the systems can be tuned and optimised through a range of different approaches.

Perhaps most important in the context of electrocatalyst design are the recent bodies of work on the immobilisation of POMs on electrode surfaces and within protective conductive cavities. These emerging areas of research continue to grow and to deliver new fundamental electrochemical insight into these complex hybrid systems.

In conclusion, the outlook for synthetic approaches to POMs as electrocatalysts is very promising. The existing literature in the area provides a robust foundation for understanding the structure-activity relationships intrinsic to POMs, offering insights into their intricate electrocatalytic mechanism. Leveraging this knowledge sets the stage for future synthetic endeavours, enabling the systematic design of POM-based electrocatalysts. By strategically aligning POM design, researchers can pursue a rationale-driven approach to address current challenges in electrocatalysis. As this exploration continues to expand, the potential for tailored and efficient electrocatalysts continues to grow, propelling the field towards novel solutions for sustainable energy conversion and storage.

References

1 Pope, M.T. (2003). *Polyoxometalate Molecular Science* (ed. J.J. Borrás-Almenar, E. Coronado, A. Müller, and M. Pope), 3–31. Netherlands, Dordrecht: Springer.

2 Berzelius, J.J. (1826). Contribution to more detailed knowledge of molybdenum. *Ann. Phys.* 82: 369–392.

3 Keggin, J.F. and Bragg, W.L. (1997). Containing papers of a mathematical and physical character. *Proc. R. Soc. Lond. Ser. A* 144: 75–100.

4 Ji, Y., Huang, L., Hu, J. et al. (2015). Polyoxometalate-functionalized nanocarbon materials for energy conversion, energy storage and sensor systems. *Energy Environ. Sci.* 8: 776–789.

5 Walsh, J.J., Bond, A.M., Forster, R.J., and Keyes, T.E. (2016). Hybrid polyoxometalate materials for photo (electro-) chemical applications. *Coord. Chem. Rev.* 306: 217–234.

6 Herrmann, S., Ritchie, C., and Streb, C. (2015). Polyoxometalate–conductive polymer composites for energy conversion, energy storage and nanostructured sensors. *Dalton Trans.* 44: 7092–7104.

7 Li, Q., Zhang, L., Dai, J. et al. (2018). A hybrid fuel cell for water purification and simultaneously electricity generation. *J. Chem. Eng.* 351: 441–461.

8 Clemente-Juan, J.M., Coronado, E., and Gaita-Ariño, A. (2012). Magnetic polyoxometalates: from molecular magnetism to molecular spintronics and quantum computing. *Chem. Soc. Rev.* 41: 7464–7478.

9 Bijelic, A., Aureliano, M., and Rompel, A. (2018). The antibacterial activity of polyoxometalates: structures, antibiotic effects and future perspectives. *Chem. Commun.* 54: 1153–1169.

10 Lin, C.G., Hu, J., and Song, Y.F. (2017). Chapter seven—polyoxometalate-functionalized nanocarbon materials for energy conversion, energy storage, and sensor systems. *Adv. Inorg. Chem.* 69: 181–212.

11 Hill, C.L. and Prosser-McCartha, C.M. (1995). The first combinatorially prepared and evaluated inorganic catalysts. Polyoxometalates for the aerobic oxidation of the mustard analog tetrahydrothiophene (THT). *Coord. Chem. Rev.* 143: 407–455.

12 Winkler, J.R. and H. B. (2012). Gray. In: *Molecular Electronic Structures of Transition Metal Complexes I* (ed. D.M.P. Mingos, P. Day, and J.P. Dahl), 17–28. Springer Berlin Heidelberg.

13 Margerum, D.W. (1979). *Science* 1968 (160): 873–873.

14 Herrmann, S. (2015). *New Synthetic Routes to Polyoxometalate Containing Ionic Liquids*, 1–17. Wiesbaden: Springer Fachmedien Wiesbaden.

15 Lipscomb, W.N. (1965). Synthesis and reactions of hexafluoro-1,2-diisocyancyclobutane chromium complexes. *Inorg. Chem.* 4: 132–134.

16 Pope, M.T. (1983). *Heteropoly and Isopoly Oxometalates*. Springer.

17 Long, D.-L., Tsunashima, R., and Cronin, L. (2010). Polyoxometalates: building blocks for functional nanoscale systems. *Angew. Chem. Int. Ed* 49: 1736–1758.

18 Hill, C.L. (1998). Introduction: polyoxometalates multicomponent molecular vehicles to probe fundamental issues and practical problems. *Chem. Rev.* 98: 1–2.

19 Kondinski, A. and Parac-Vogt, T.N. (2018). Keggin structure, where are you going? *Front. Chem.* 6: 346.

20 Keggin, J.F. (1933). Structure of the molecule of 12-phosphotungstic acid. *Nature* 131: 908–909.

21 Baker, L.C.W. and Figgis, J.S. (1970). New fundamental type of inorganic complex: hybrid between heteropoly and conventional coordination complexes. Possibilities for geometrical isomerisms in 11-, 12-, 17-, and 18-heteropoly derivatives. *J. Am. Chem. Soc.* 92: 3794–3797.

22 Zhang, F.-Q., Zhang, X.-M., Wu, H.-S. et al. (2007). Inversed stability order in Keggin polyoxothiometalate isomers: a DFT study of 12-electron reduced α, β, γ, δ, and ε $[(MoO_4)\ Mo_{12}O_{12}S_{12}(OH)_{12}]^{2-}$ anions. *J. Phys. Chem. A* 111: 1683–1687.

23 Dawson, B. (1953). The structure of the 9 (18)-heteropoly anion in potassium 9(18)-tungstophosphate, $K_6(P_2W_{18}O_{62})$. $14H_2O$. *Acta Crystallogr.* 6: 113–126.

24 Hashikawa, A., Fujimoto, M., Hayashi, Y., and Miyasaka, H. (2011). Isolation of a stable lacunary Dawson-type polyoxomolybdate cluster. *Chem. Commun.* 47: 12361.

25 Patel, A. and Sadasivan, R. (2021). Modified Mn substituted POMs: synthetic strategies, structural diversity to applications. *Prog. Mater. Sci.* 118: 100759.

26 Kibler, A.J. and Newton, G.N. (2018). Tuning the electronic structure of organic–inorganic hybrid polyoxometalates: the crucial role of the covalent linkage. *Polyhedron* 154: 1–20.

27 Song, Y.-F. and Tsunashima, R. (2012). Recent advances on polyoxometalate-based molecular and composite materials. *Chem. Soc. Rev.* 41: 7384.

28 Zhang, B., Zhao, M., Qi, Y. et al. (2019). The intrinsic enzyme activities of the classic polyoxometalates. *Sci. Rep.* 9: 14832.

29 Gao, N., Sun, H., Dong, K. et al. (2014). Transition-metal-substituted polyoxometalate derivatives as functional anti-amyloid agents for Alzheimer's disease. *Nat. Commun.* 5: 3422.

30 Steffler, F., De Lima, G.F., and Duarte, H.A. (2020). The effect of the heteroatom (X= P, As, Si and Ge) on the geometrical and electronic properties of α-Keggin polyoxometalates (M= Mo, W and Nb)–A DFT investigation. *J. Mol. Struct.* 1213: 128159.

31 Wang, S.-S. and Yang, G.-Y. (2015). Recent advances in polyoxometalate-catalyzed reactions. *Chem. Rev.* 115: 4893–4962.

32 Timofeeva, M.N. (2003). Acid catalysis by heteropoly acids. *Appl. Catal. A Gen.* 256: 19–35.

33 Kamata, K. and Sugahara, K. (2017). Base catalysis by mono-and polyoxometalates. *Catalysts* 7: 345.

34 Zhang, Y., Liu, J., Li, S.-L. et al. (2019). Polyoxometalate-based materials for sustainable and clean energy conversion and storage. *EnergyChem* 1: 100021.

35 Omwoma, S., Gore, C.T., Ji, Y. et al. (2015). Environmentally benign polyoxometalate materials. *Coord. Chem. Rev.* 286: 17–29.

36 Zeb, Z., Huang, Y., Chen, L. et al. (2023). Comprehensive overview of polyoxometalates for electrocatalytic hydrogen evolution reaction. *Coord. Chem. Rev.* 482: 215058.

37 Wang, Z.-H., Wang, X.-F., Tan, Z., and Song, X.-Z. (2021). Polyoxometalate/metal–organic framework hybrids and their derivatives for hydrogen and oxygen evolution electrocatalysis. *Mater. Today Energy* 19: 100618.

38 Gumerova, N.I. and Rompel, A. (2018). Polyoxometalates in solution: speciation under spotlight. *Nat. Rev. Chem.* 2: 0122.

39 López, X., Bo, C., and Poblet, J.M. (2002). Electronic properties of polyoxometalates: electron and proton affinity of mixed-addenda Keggin and Wells–Dawson anions. *J. Am. Chem. Soc.* 124: 12574–12582.

40 Altenau, J.J., Pope, M.T., Prados, R.A., and So, H. (1975). Models for heteropoly blues. Degrees of valence trapping in vanadium(IV)-and molybdenum(V)-substituted Keggin anions. *Inorg. Chem.* 14: 417–421.

41 Kickelbick, G. (2014). Hybrid materials-past, present and future, doi:https://doi.org/10.2478/hyma-2014-0001.

42 Sanchez, C., Julián, B., Belleville, P., and Popall, M. (2005). Applications of hybrid organic–inorganic nanocomposites. *J. Mater. Chem.* 15: 3559.

43 Hammami, I., Dhifallah, F., Ouari, K. et al. (2019). A new assembly of a Wells-Dawson polyoxometalate-based iron(III) coordination complex hybrid: synthesis, crystal structure, thermal and electrochemical properties. *Polyhedron* 160: 63–67.

44 Schaming, D., Costa-Coquelard, C., Sorgues, S. et al. (2010). Photocatalytic reduction of Ag_2SO_4 by electrostatic complexes formed by tetracationic zinc porphyrins and tetracobalt Dawson-derived sandwich polyanion. *Appl. Catal. A Gen.* 373: 160–167.

45 Martinetto, Y., Pégot, B., Roch-Marchal, C. et al. (2020). Designing functional polyoxometalate-based ionic liquid crystals and ionic liquids. *Eur. J. Inorg. Chem.* 2019: 228–247.

46 Kirchhoff, B., Rau, S., and Streb, C. (2016). Detecting and preventing the formation of photosensitizer-catalyst colloids in homogeneous light-driven water oxidation. *Eur. J. Inorg. Chem.* 1425–1429.

47 Ren, Y., Wang, M., Chen, X. et al. (2015). Heterogeneous catalysis of polyoxometalate based organic–inorganic hybrids. *Materials* 8: 1545–1567.

48 Glass, E.N., Fielden, J., Huang, Z. et al. (2016). Transition metal substitution effects on metal-to-polyoxometalate charge transfer. *Inorg. Chem.* 55: 4308–4319.

49 Zheng, S.-T. and Yang, G.-Y. (2012). Recent advances in paramagnetic-TM-substituted polyoxometalates (TM= Mn, Fe, Co, Ni, Cu). *Chem. Soc. Rev.* 41: 7623.

50 Proust, A., Matt, B., Villanneau, R. et al. (2012). Functionalization and post-functionalization: a step towards polyoxometalate-based materials. *Chem. Soc. Rev.* 41: 7605.

51 Sun, M., Zhang, J., Putaj, P. et al. (2014). Catalytic oxidation of light alkanes (C1–C4) by heteropoly compounds. *Chem. Rev.* 114: 981–1019.

52 Simms, C., Kondinski, A., and Parac-Vogt, T.N. (2020). Metal-addenda substitution in plenary polyoxometalates and in their modular transition metal analogues. *Eur. J. Inorg. Chem.* 2559–2572.

53 Tsuji, K., Tomita, O., Higashi, M., and Abe, R. (2016). Manganese-substituted polyoxometalate as an effective shuttle redox mediator in Z-scheme water splitting under visible light. *ChemSusChem* 9: 2201–2208.

54 Li, C., Zha, B., and Li, J. (2020). A SiW11Mn-assisted indium electrocatalyst for carbon dioxide reduction into formate and acetate. *J. CO2 Util.* 38: 299–305.

55 Misra, A., Kozma, K., Streb, C., and Nyman, M. (2020). Recent advances in confining polyoxometalates and the applications. *Angew. Chem. Int. Ed.* 59: 596–612.

56 Guo, L., He, L., Zhuang, Q. et al. (2023). Recent advances in confining polyoxometalates and the applications. *Small* 19: 2207315.

57 Kibler, A.J., Souza, V.S., Fernandes, J.A. et al. (2021). A cooperative photoactive class-I hybrid polyoxometalate with benzothiadiazole–imidazolium cations. *Front. Chem.* 8: 2296–2646.

58 Dolbecq, A., Dumas, E., Mayer, C.R., and Mialane, P. (2010). Hybrid organic–inorganic polyoxometalate compounds: from structural diversity to applications. *Chem. Rev.* 110: 6009–6048.

59 Anyushin, A.V., Kondinski, A., and Parac-Vogt, T.N. (2020). Hybrid polyoxometalates as post-functionalization platforms: from fundamentals to emerging applications. *Chem. Soc. Rev.* 49: 382–432.

60 Kibler, A.J., Tsang, N., Winslow, M. et al. (2023). Electronic structure and photoactivity of organoarsenic hybrid polyoxometalates. *Inorg. Chem.* 62: 3585–3591.

61 Cameron, J.M., Fujimoto, S., Wei, R.-J. et al. (2018). Post-functionalization of a photoactive hybrid polyoxotungstate. *Dalton Trans.* 47: 10590–10594.

62 Amin, S.S., Jones, K.D., Kibler, A.J. et al. (2023). Diphosphoryl-functionalized polyoxometalates: structurally and electronically tunable hybrid molecular materials. *Angew. Chem.* 62: e202302446.

63 Boujtita, M., Boixel, J., Blart, E. et al. (2008). Redox properties of hybrid Dawson type polyoxometalates disubstituted with organo-silyl or organo-phosphoryl moieties. *Polyhedron* 27: 688–692.

64 Zhang, S.-X., Liu, L., Li, F.-B., and Du, Z.-L. (2016). Synthesis, characterization and photoelectric properties of Dawson-type polyoxometalate covalently linked to zinc(II) porphyrin. *Inorg. Chim. Acta* 453: 179–185.

65 Fujimoto, S., Cameron, J.M., Wei, R.-J. et al. (2017). A simple approach to the visible-light photoactivation of molecular metal oxides. *Inorg. Chem.* 56: 12169–12177.

66 Cameron, J.M., Fujimoto, S., Kastner, K. et al. (2016). Orbital engineering: photoactivation of an organofunctionalized polyoxotungstate. *Chem. A Eur. J.* 23: 47–50.

67 Parrot, A., Izzet, G., Chamoreau, L.-M. et al. (2013). Photochromic properties of polyoxotungstates with grafted spiropyran molecules. *Inorg. Chem.* 52: 11156–11163.

68 Kuramochi, S., Cameron, J.M., Fukui, T. et al. (2023). Selective electrochemical CO_2 conversion with a hybrid polyoxometalate. *Chem. Commun.* 59: 10801–10804.

69 Roy, S., Sharma, B., Pécaut, J. et al. (2017). Molecular cobalt complexes with pendant amines for selective electrocatalytic reduction of carbon dioxide to formic acid. *J. Am. Chem. Soc.* 139: 3685–3696.

70 Song, Y.-F., Long, D.-L., Kelly, S.E., and Cronin, L. (2008). Sorting the assemblies of unsymmetrically covalently functionalized Mn-Anderson polyoxometalate clusters with mass spectrometry. *Inorg. Chem.* 47: 9137–9139.

71 Macdonell, A., Johnson, N.A.B., Surman, A.J., and Cronin, L. (2015). Configurable nanosized metal oxide oligomers via precise "click" coupling control of hybrid polyoxometalates. *J. Am. Chem. Soc.* 137: 5662–5665.

72 Song, Y.-F., McMillan, N., Long, D.-L. et al. (2009). Micropatterned surfaces with covalently grafted unsymmetrical polyoxometalate-hybrid clusters lead to selective cell adhesion. *J. Am. Chem. Soc.* 131: 1340–1341.

73 Zhuang, Q., Sun, Z., Lin, C.-G. et al. (2023). Latest progress of asymmetrically functionalized Anderson-type polyoxometalates. *Inorg. Chem. Front.* 10: 1695–1711.

74 Hampson, E., Cameron, J.M., Amin, S. et al. (2019). Asymmetric hybrid polyoxometalates: a platform for multifunctional redox-active nanomaterials. *Angew. Chem. Int. Ed.* 58: 18281–18285.

75 Hampson, E., Cameron, J.M., Watts, J.A., and Newton, G.N. (2020). Transition metal decorated soft nanomaterials through modular self-assembly of an asymmetric hybrid polyoxometalate. *Chem. Commun.* 56: 8237–8240.

76 Müller, A. and Serain, C. (2000). Soluble molybdenum blues "des pudels kern". *Acc. Chem. Res.* 33: 2–10.

77 Liu, G., Liu, T., Mal, S.S., and Kortz, U. (2006). Wheel-shaped polyoxotungstate $[Cu_{20}Cl(OH)_{24}(H_2O)_{12}(P_8W_{48}O_{184})]_{25}$-macroanions form supramolecular "blackberry" structure in aqueous solution. *J. Am. Chem. Soc.* 128: 10103–10110.

78 Kistler, M.L., Patel, K.G., and Liu, T. (2009). Accurately tuning the charge on giant polyoxometalate type keplerates through stoichiometric interaction with cationic surfactants. *Langmuir* 25: 7328–7334.

79 Zhang, J., Song, Y.-F., Cronin, L., and Liu, T. (2008). Self-assembly of organic–inorganic hybrid amphiphilic surfactants with large polyoxometalates as polar head groups. *J. Am. Chem. Soc.* 130: 14408–14409.

80 Zhang, J., Song, Y., Cronin, L., and Liu, T. (2010). Reverse-vesicle formation of organic–inorganic polyoxometalate-containing hybrid surfactants with tunable sizes. *Chem. A Eur. J.* 16: 11320–11324.

81 Chakrabarty, R., Mukherjee, P.S., and Stang, P.J. (2011). Supramolecular coordination: self-assembly of finite two-and three-dimensional ensembles. *Chem. Rev.* 111: 6810–6918.

82 Kastner, K., Kibler, A.J., Karjalainen, E. et al. (2017). Redox-active organic–inorganic hybrid polyoxometalate micelles. *J. Mater. Chem. A Mater.* 5: 11577–11581.

83 Amin, S.S., Cameron, J.M., Cousins, R.B. et al. (2022). Redox-active hierarchical assemblies of hybrid polyoxometalate nanostructures at carbon surfaces. *Inorg. Chem. Front.* 9: 1777–1784.

84 Dufaud, V. and Lefebvre, F. (2010). Inorganic hybrid materials with encapsulated polyoxometalates. *Materials* 3: 682–703.

85 Cherevan, A.S., Nandan, S.P., Roger, I. et al. (2020). Polyoxometalates on functional substrates: concepts, synergies, and future perspectives. *Adv. Sci.* 7: 1903511.

86 Gam Derouich, S., Rinfray, C., Izzet, G. et al. (2014). Control of the grafting of hybrid polyoxometalates on metal and carbon surfaces: toward submonolayers. *Langmuir* 30: 2287–2296.

87 Laurans, M., Trinh, K., Dalla Francesca, K. et al. (2020). Covalent grafting of polyoxometalate hybrids onto flat silicon/silicon oxide: insights from POMs layers on oxides. *ACS Appl. Mater. Interfaces* 12: 48109–48123.

88 Laurans, M., Dalla Francesca, K., Volatron, F. et al. (2018). Molecular signature of polyoxometalates in electron transport of silicon-based molecular junctions. *Nanoscale* 10: 17156–17165.

89 Ben M'Barek, Y., Rosser, T., Sum, J. et al. (2020). Dye-sensitized photocathodes: boosting photoelectrochemical performances with polyoxometalate electron transfer mediators. *ACS Appl. Energy Mater.* 3: 163–169.

90 Ji, Y., Hu, J., Huang, L. et al. (2015). Covalent attachment of Anderson-type polyoxometalates to single-walled carbon nanotubes gives enhanced performance electrodes for lithium ion batteries. *Chem. Eur. J.* 21: 6469–6474.

91 Jordan, J.W., Lowe, G.A., McSweeney, R.L. et al. (2019). Host–guest hybrid redox materials self-assembled from polyoxometalates and single-walled carbon nanotubes. *Adv. Mater.* 31: 1904182.

92 Jordan, J.W., Cameron, J.M., Lowe, G.A. et al. (2022). Stabilization of polyoxometalate charge carriers via redox-driven nanoconfinement in single-walled carbon nanotubes. *Angew. Chem. Int. Ed.* 61: e202115619. Angew. Chem. 2022, 134, e202115619.

93 Fabre, B., Falaise, C., and Cadot, E. (2022). Polyoxometalates-functionalized electrodes for (photo) electrocatalytic applications: recent advances and prospects. *ACS Catal.* 12: 12055–12091.

94 Zhang, S., Ou, F., Ning, S., and Cheng, P. (2021). Polyoxometalate-based metal–organic frameworks for heterogeneous catalysis. *Inorg. Chem. Front.* 8: 1865–1899.

95 Zhang, Y., Liu, Y., Wang, D. et al. (2023). State-of-the-art advances in the syntheses, structures, and applications of polyoxometalate-based metal–organic frameworks. *Polyoxometalates* 2: 9140017.

96 Du, Z.-Y., Yu, Y.-Z., Hong, Y.-L. et al. (2020). Polyoxometalate-based metal–organic frameworks with unique high-nuclearity water clusters. *ACS Appl. Mater. Interfaces* 12: 57174–57181.

Part II

Polyoxometalates for Oxidative Electrocatalysis

3

POM-based Electrocatalysts for L-Cysteine and NADH Oxidation

Israël Mbomekallé, Anne-Lucie Teillout, and Pedro de Oliveira

Equipe d'Electrochimie et de Photo-électrochimie (EPEC), Institut de Chimie Physique (ICP), UMR 8000 CNRS, Faculté des Sciences d'Orsay, Université Paris-Saclay, Orsay Cedex 91405, France

3.1 Introduction

The interest in profiting from the interactions of polyoxometalates (POMs) with molecules of biological relevance in order to develop analytical methods, namely in the biomedical field, has been growing in the last years.

Since POMs are molecular oxides constituted of transition metals and the latter may exist in several oxidation states, POMs possess a rich and varied redox chemistry and are particularly suited to exchange electrons with biomolecules. In addition, many of the redox processes in which they are involved are susceptible to be probed and/or triggered by electrochemistry, so electrochemical techniques are often used to study POMs.

This chapter deals specifically with the oxidation by POM-based electrocatalysts of two relevant biochemical species, the amino acid L-cysteine (Cys) [1, 2] and the reduced co-enzyme nicotinamide adenine dinucleotide (NADH) [3] (Figure 3.1), as well as its phosphorylated counterpart (NADPH).

3.2 The Electrocatalytic Oxidation of L-cysteine (Cys)

3.2.1 V-containing POMs as Electrocatalysts in Homogeneous Phase

V-containing POMs belonging to the Wells–Dawson family have been obvious choices as electrocatalysts for redox reactions involving biological substrates, because they are stable in large pH ranges (often including physiological pH values), their V centres are stable in both the oxidation states +V and +IV in the presence of air and the redox potential values for the interconversion between these two oxidation states covers several hundreds of millivolts, depending on the elemental composition of the POM [4]. Symmetrical (α_1- and α_2-$[X_2VW_{17}O_{62}]^{7-\text{ or }8-}$ (X = As or P)) and dissymmetrical α_2-$[H_4XVW_{17}O_{62}]^{8-\text{ or }9-}$ (X = P or As) species (Figure 3.2), the latter exhibiting lower redox potentials than the former, were

Applied Polyoxometalate-based Electrocatalysis, First Edition. Edited by Diana M. Fernandes.

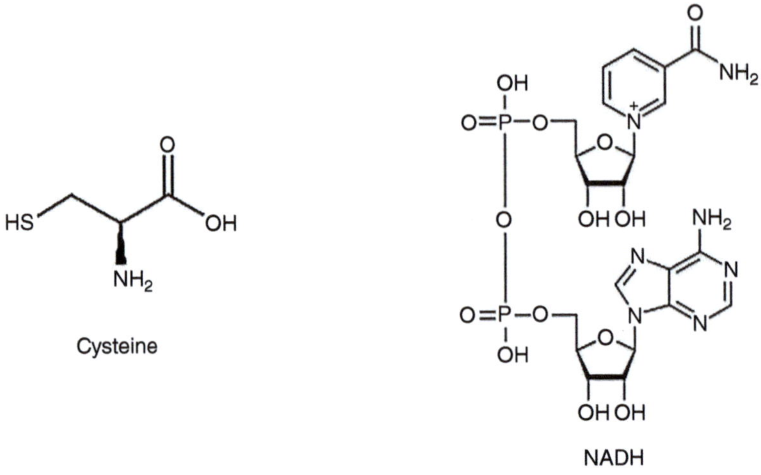

Figure 3.1 Chemical formulae of cysteine and NADH.

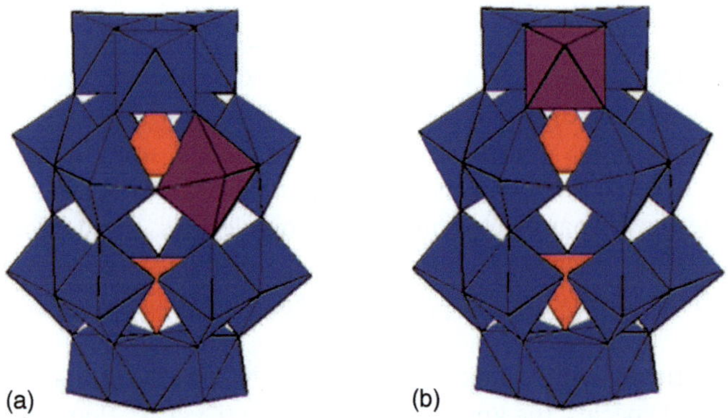

Figure 3.2 Polyhedral representations of the α_1 (a) and the α_2 (b) symmetrical isomers of the Wells–Dawson structure α-$[X_2M_{18}O_{62}]^{n-}$. The dissymmetrical counterparts possess a single red tetrahedron inside the scaffold. Source: Dika Manga et al. [5]/MDPI/CC BY-4.0.

employed in the homogeneous electrocatalysis of Cys. Ideally, electrocatalysts of oxidation reactions should exchange electrons at the lowest possible redox potentials, and smaller than the oxidation potential of the substrate of interest alone, in order to minimize both the energy input and the probability of interference from other substrates susceptible of being oxidized at higher potentials. The cyclic voltammetric response of the V centres of some of the mentioned POMs is shown in Figure 3.3 and allowed to select the ones having the lowest formal potentials for the electrocatalysis study: α_2-H_4XVW_{17} (X = P or As) and α_2-P_2VW_{17} (the formal redox potentials, $E^{0\prime}$ *vs.* SCE, for α_2-H_4PVW_{17} and α_2-P_2VW_{17} being 0.291 V and 0.415 V, respectively) [6, 7].

The stoichiometry of the reaction between Cys and α_2-$P_2V^VW_{17}$ was deduced to be 1 : 1 by UV-visible spectrophotometry. Stopped flow experiments revealed the existence of a single isosbestic point (Figure 3.4 shows an example concerning

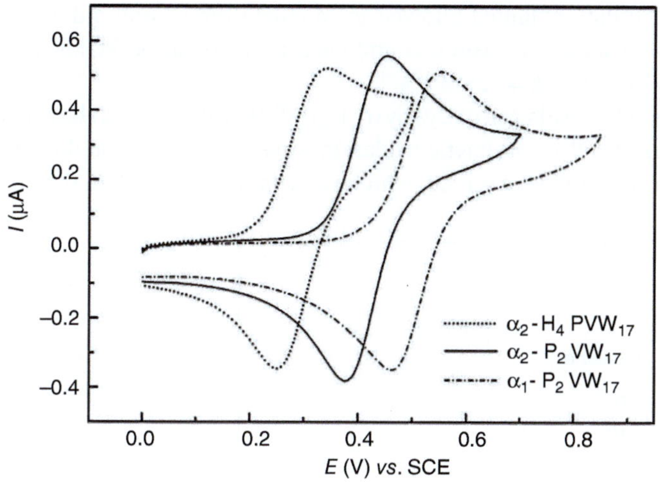

Figure 3.3 Cyclic voltammograms (CVs) restricted to the V centres for 2×10^{-4} M α_1-P_2VW_{17}, α_2-P_2VW_{17} and α_2-H_4PVW_{17}, recorded in a pH 7.00 (0.4 M NaH_2PO_4/NaOH) buffer solution, at a scan rate of $10\,mV\,s^{-1}$, with a glassy carbon working electrode. Source: Keita et al. [4]/with permission of Springer Nature.

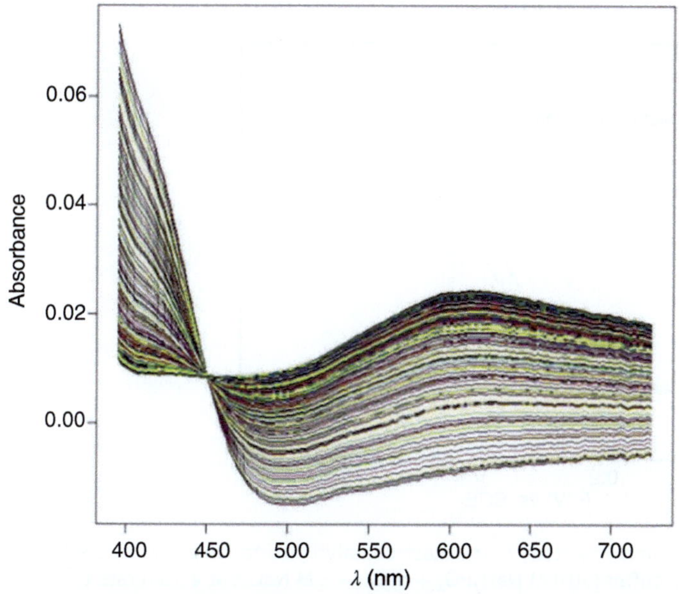

Figure 3.4 Sets of UV–visible spectra of a mixture of Cys and α_2-H_4PVW_{17} as a function of time. The concentrations of the solutions before mixing were 0.1 mM for the POM and 1 mM for Cys. The experiments were run in a pH 7.00 (0.4 M NaH_2PO_4/NaOH) buffer solution at room temperature. Source: Keita et al. [4]/with permission of Springer Nature.

α_2-$H_4PV^VW_{17}$), indicative of the transformation of the oxidized V^V centre into its reduced form, V^{IV}. The data obtained allowed to calculate the second-order rate constant for the redox reaction between Cys and each of the oxidized POMs, and the value for α_2-$H_4PV^VW_{17}$ was $k = 925\ M^{-1}\ s^{-1}$.

The electrocatalysis of the oxidation of Cys by α_2-$H_4PV^{IV}W_{17}$ at pH = 7 is shown in Figure 3.5. An irreversible, electrocatalytic oxidation wave at the potential domain of the oxidation of the V centre is observed, which is far more intense than the one

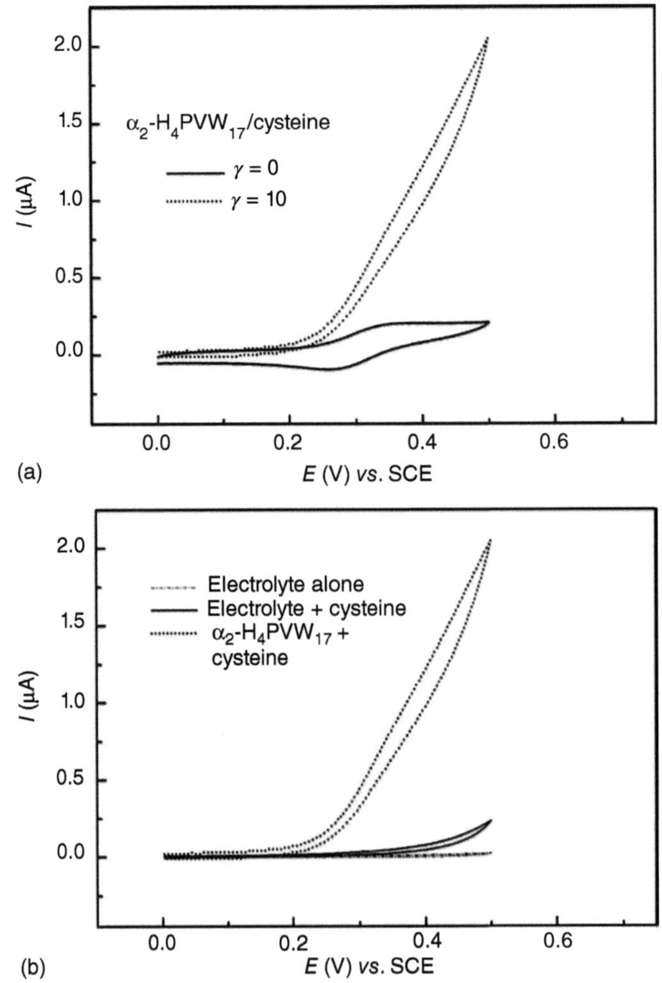

Figure 3.5 Cyclic voltammetry study of the electrocatalytic oxidation of Cys by 2×10^{-4} M α_2-$H_4PV^{IV}W_{17}$, in pH 7 buffer (50 mM NaH_2PO_4 + NaOH + 1 M NaCl), at a scan rate of 2 mV s^{-1}, with a glassy carbon working electrode. The excess parameter was defined as $\gamma = C^0(\text{Cys})/C^0(\text{POM})$. The cyclic voltammograms are restricted to the potential domain of the V centre. (a) The excess parameter values were $\gamma = 0$ and $\gamma = 10$; (b) Comparison of the catalytic process (at $\gamma = 10$) by α_2-$H_4PV^{IV}W_{17}$ with the background current of the supporting electrolyte and with the direct oxidation of Cys on the glassy carbon electrode in the absence of the POM. Source: Keita et al. [4]/with permission of Springer Nature.

recorded in the absence of the POM. Interestingly, the value of the second-order rate constant for this process determined by double potential step chronocoulometry in another study, $k = 1000\ M^{-1}\ s^{-1}$ [8], is in good agreement with that inferred from the stopped flow experiments previously mentioned ($k = 925\ M^{-1}\ s^{-1}$). In this very same study, multi-vanadium POMs were tested, namely $[P_2V_2W_{16}O_{62}]^{8-}$ ($P_2V_2W_{16}$), $[P_2MoV_2W_{15}O_{62}]^{8-}$ ($P_2MoV_2W_{15}$) and $[P_2V_3W_{15}O_{62}]^{9-}$ ($P_2V_3W_{15}$). The results show that the presence of an increasing number of V centres in the POM scaffold renders the electrocatalytic oxidation of Cys more efficient.

A more recent study added supplementary electrochemical data on V-containing, Wells–Dawson-type POMs susceptible to electro-oxidize Cys and other thiols. The α_1 and α_2 isomers of the species $[X_2V^{IV}W_{17}O_{62}]^{8-}$, with X = As or P (Figure 3.2), were studied by cyclic voltammetry, in particular the effect of the pH on the behaviour of both the W and the V centres (Figure 3.6). Both the symmetry of the isomer and the nature of the hetero-element influence the outcome, and some waves split or merge as the pH evolves.

Clearly, the response of the W centres is pH dependent, meaning that the electron transfer is coupled to the uptake or the release of protons. This is not the case for the V centres, whose waves are close to being superimposable, irrespective of the pH of the medium. The pH may, therefore, be modulated in order to optimize

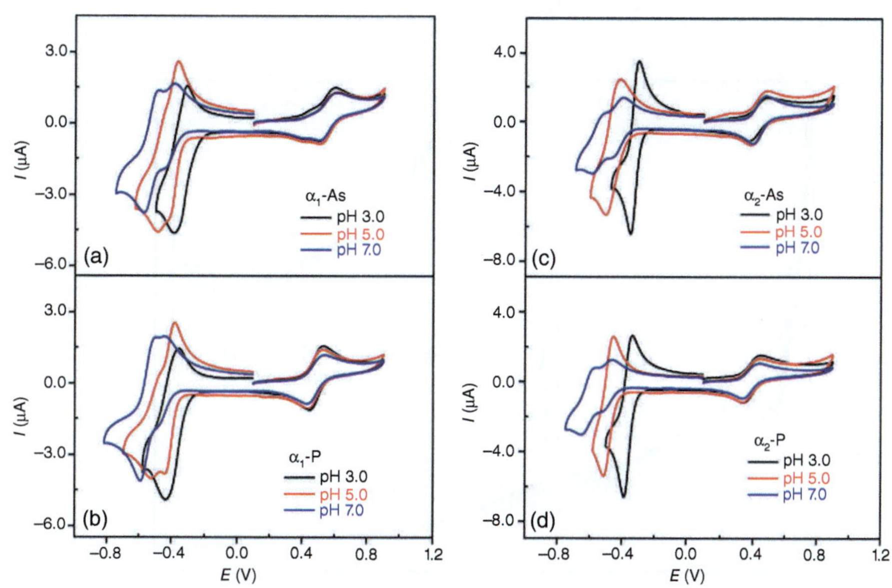

Figure 3.6 Cyclic voltammograms of (a) α_1-$[As_2V^{IV}W_{17}O_{62}]^{8-}$ (α_1-As); (b) α_1-$[P_2V^{IV}W_{17}O_{62}]^{8-}$ (α_1-P); (c) α_2-$[As_2V^{IV}W_{17}O_{62}]^{8-}$ (α_2-As) and (d) α_2-$[P_2V^{IV}W_{17}O_{62}]^{8-}$ (α_2-P) recorded in 0.2 M Na_2SO_4/H_2SO_4 at pH = 3.0 (black), 0.4 M CH_3CO_2Na/CH_3CO_2H at pH = 5.0 (red) and 0.2 M Na_2HPO_4/H_3PO_4 at pH = 7.0 (blue) at a scan rate of 10 mV s^{-1}, with a glassy carbon working electrode and a SCE reference electrode. Source: Dika Manga et al. [5]/MDPI/CC BY-4.0.

the electrocatalytic oxidation of a particular thiol on the V centres, which allows to mimic oxidative stress situations, namely at physiological pH values [5].

As previously stated, the V centres of POMs undergo oxidation processes at potential values lower than those required for the oxidation of Cys, which renders them particularly suited for the electrocatalysis of the latter. It is not a surprise that V-containing POMs are often chosen as electrocatalysts for the oxidation of Cys. The group of Rajasekaran did a first study on the kinetics of the electron transfer reactions between Cys (and thioglycolic acid [TGA]) and the Keggin-type species $[PV^{V}W_{11}O_{40}]^{4-}$ (called HPA for 'heteropolyanion' in their articles) in homogeneous aqueous acidic medium. They concluded that the products of the reactions were cystine (and dithioglycolic acid) and $[PV^{IV}W_{11}O_{40}]^{5-}$, the reduced form of the POM being called heteropolyblue (HPB) due to its intense blue colour. The latter is easily identified with the naked eye and also by UV–vis spectrophotometry. A suitable reaction mechanism was proposed based on the previous formation of an intermediate Cys-POM (or TGA-POM) complex which undergoes facile (even if slow), outer sphere, one-electron transfer steps between the substrate (Cys or TGA) and the POM. The stoichiometry of the reaction was 1 : 1 and its global order 3, the partial orders being 2 for the substrate and 1 for the POM [9].

The same group did a second study using Cys as the substrate and a divanadium derivative of the Keggin-type POM as the catalyst, $[PV^{V}V^{V}W_{10}O_{40}]^{5-}$. Again, cystine and the HPB $[PV^{IV}V^{IV}W_{10}O_{40}]^{7-}$ were the reaction products, the presence of the latter being monitored by UV–vis spectrophotometry (Figure 3.7).

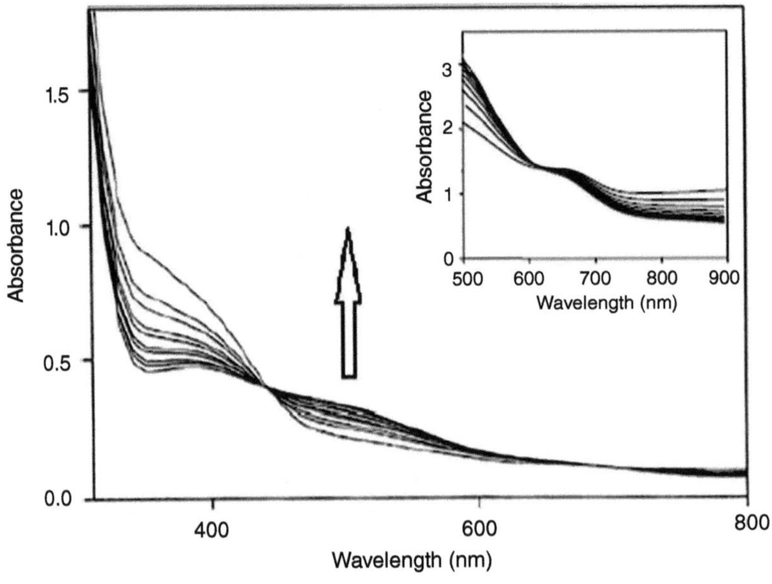

Figure 3.7 Time-dependent UV–visible spectrum for the reduction of $[PV^{V}V^{V}W_{10}O_{40}]^{5-}$ by Cys at 25 °C ([POM] = 5×10^{-4} M, [Cys] = 0.02 M, pH = 4.58). Inset: spectra recorded with [POM] = 5×10^{-3} M and [Cys] = 0.05 M. Source: Sami et al. [10]/with permission of Springer Nature.

The proposed reaction mechanism does not include the previous formation of a Cys-POM complex this time, but consists of a sequence of one-electron transfer steps involving either the thiol or the deprotonated thiol forms of Cys and the POM, until the final products are reached (Scheme 3.1). The overall process speeds up upon increasing the pH, which suggests that deprotonated Cys is probably the species involved in electron transfer with the POM. The stoichiometry of the reaction Cys:POM was 2 : 1 (Figure 3.8) and its global order 2, the partial orders being 1 for the substrate and 1 for the POM [10].

$$[PV^{V}V^{V}W_{10}O_{40}]^{5-} + HSCH_2CH\,(NH_3^+)\,COO^- \xrightarrow{\text{slow}} [PV^{IV}V^{V}W_{10}O_{40}]^{6-} + H^+ + (^{\bullet}SCH_2CH\,(NH_3^+)\,COO^-) \qquad \text{thiol form}$$

$$[PV^{V}V^{V}W_{10}O_{40}]^{5-} + {}^{-}SCH_2CH\,(NH_3^+)\,COO^- \xrightarrow{\text{slow}} [PV^{IV}V^{V}W_{10}O_{40}]^{6-} + (^{\bullet}SCH_2CH\,(NH_3^+)\,COO^-) \qquad \text{thiolate form}$$

$$2\,^{\bullet}SCH_2CH\,(NH_3^+)\,COO^- \xrightarrow{\text{fast}} (SCH_2CH\,(NH_3^+)\,COO^-)_2$$

$$^{\bullet}SCH_2CH\,(NH_3^+)\,COO^- + HSCH_2CH\,(NH_3^+)\,COO^- \overset{\text{fast}}{\rightleftharpoons} (^{\bullet}SCH_2CH\,(NH_3^+)\,COO^-, HSCH_2CH\,(NH_3^+)\,COO^-)$$

$$(^{\bullet}SCH_2CH\,(NH_3^+)\,COO^-,\ HSCH_2CH\,(NH_3^+)\,COO^-) + [PV^{IV}V^{V}W_{10}O_{40}]^{6-} \xrightarrow{\text{fast}} (SCH_2CH\,(NH_3^+)\,COO^-)_2 + [PV^{IV}V^{IV}W_{10}O_{40}]^{7-} + H^+$$

Scheme 3.1 Proposed reaction mechanism of the oxidation of Cys by a divanadium POM.

3.2.2 Ce-containing POMs as Electrocatalysts in Homogeneous Phase

Apart from V, other redox centres have been used for the homogeneous electrocatalytic oxidation of Cys, one of them being Ce (Co and Ru are mentioned below in examples of heterogeneous electrocatalysis). The group of Parac-Vogt have been making several important contributions to unravel the interactions between POMs and molecules of biological relevance [11], and one of them concerns the oxidation of amino acids and peptides with Ce^{IV}-containing POMs belonging to the Keggin (K) and the Wells–Dawson (WD) families (Figure 3.9). The latter have been shown to behave as artificial proteases which selectively hydrolyse peptide bonds, the Ce^{IV} being reduced to Ce^{III} in the process [13–15].

The nature of the possible reductants of the Ce centre remained elusive until a study with amino acid residues revealed that Cys was the most efficient species, followed by tryptophan (Trp), tyrosine (Tyr), histidine (His) and phenylalanine (Phe), which are all aromatic. Cys has been shown to be oxidized to cystine, namely by ^{1}H NMR, but no oxidation products were identified for Trp, Tyr, His and Phe, which may

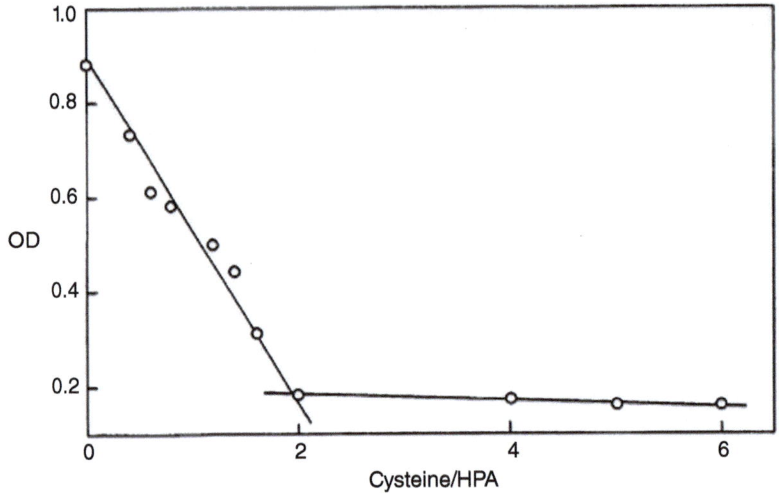

Figure 3.8 Spectrophotometric titration of Cys with $[PV^{V}V^{V}W_{10}O_{40}]^{5-}$. $[POM(=HPA)] = 2.5 \times 10^{-4}$ M and pH = 4.58, at 25 °C. Source: Sami et al. [10]/with permission of Springer Nature.

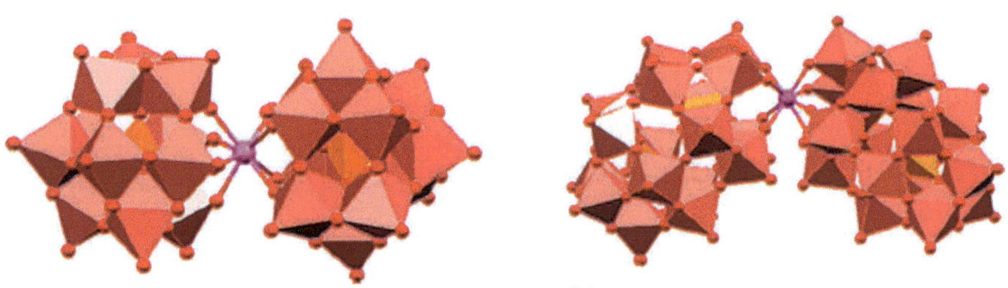

Figure 3.9 Structures of (a) $[Ce^{IV}(PW_{11}O_{39})_2]^{10-}$ ($Ce^{IV}K$) and (b) $[Ce^{IV}(\alpha_2\text{-}P_2W_{17}O_{61})_2]^{16-}$ ($Ce^{IV}WD$). The $Ce^{IV/III}$ metal is represented by a purple ball. W groups are represented by brown octahedra, oxygen is represented by red balls and the internal PO_4 groups are represented by gold tetrahedra. Source: Abdelhameed et al. [12]/with permission of American Chemical Society.

be explained by the regeneration of the reduced amino acid residues by the solvent in a mechanism involving radicals. The proposed reaction mechanism (Scheme 3.2) for Cys is the following:

$$\text{Cys} - \text{SH} \rightleftharpoons \text{Cys} - \text{S}^- + \text{H}^+ \tag{3.1}$$

$$\text{Cys} - \text{SH} + \text{Ce}^{\text{IV}}\text{K} \rightleftharpoons \text{Cys} - \text{S}^\bullet + \text{Ce}^{\text{III}}\text{K} \tag{3.2}$$

$$2\text{Cys} - \text{S}^\bullet \rightleftharpoons \text{Cys} - \text{S} - \text{S} - \text{Cys} \tag{3.3}$$

Scheme 3.2 Proposed reaction mechanism of the oxidation of Cys by a Keggin Ce-POM.

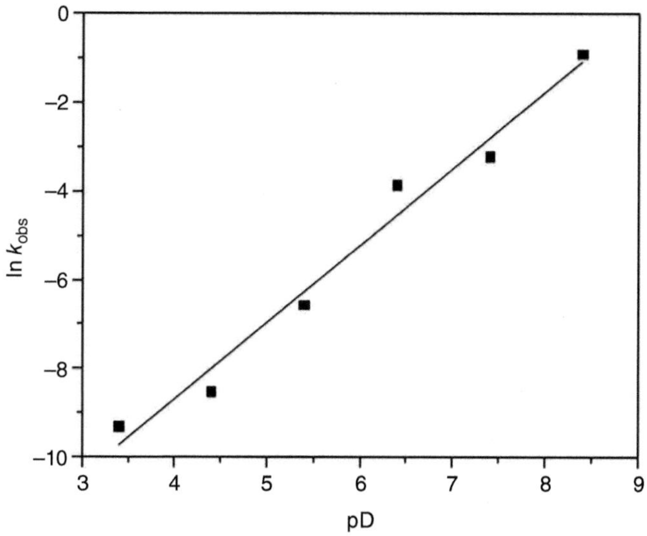

Figure 3.10 Effect of pD on the rate constant of Ce^{IV}K reduction in the presence of Cys. Source: Abdelhameed et al. [12]/with permission of American Chemical Society.

It is supported by the fact that the observed rate constant, in deuterated water, for the reduction by Cys of Ce^{IV} to Ce^{III} in the POM increases with the pD (Figure 3.10), suggesting that the thiolate species takes part in the sequence, since its concentration increases with pD. The cysteinyl radicals formed by oxidation of Cys in step (2) [16], which is expected to be the rate-limiting step of the mechanism [17], combine in pairs to give cystine in step (3.3).

The molecular structure of the POM in which the Ce centre is inserted has a strong influence on the redox reaction with Cys. Ce^{IV}K turned out to be a more efficient oxidant than Ce^{IV}WD, the reaction rate being one order of magnitude higher for the former, in agreement with the higher redox potential of Ce^{IV}K in comparison to Ce^{IV}WD. Interestingly, when these POMs were tested with peptides containing Cys, Trp and Tyr, the formation of the disulphide bridge was detected, but not the hydrolysis of the peptide bonds, implicating that the redox reactions are much faster than the hydrolysis processes [12].

3.2.3 POM-containing Hybrids as Electrocatalysts in Heterogeneous Phase: Carbon Paste Electrodes

POMs may take part in systems having Cys as one of the substrates without being responsible for its electrocatalytic oxidation. Examples come from the group of Shams, who used organic–inorganic hybrid compounds composed of the *N*,*N*′-bis(salicylidene)-1,2-phenylenediaminocobalt(III) (CoSal) complex and of a Keggin-type species in order to modify carbon paste electrodes (CPE) useful for electrocatalytic applications. In a first study, CoSal and $H_4SiW_{12}O_{40}$ (SiW_{12}) yielded a compound with a stoichiometry of 4 : 1, $(CoSal)_4SiW_{12}$, which was incorporated in

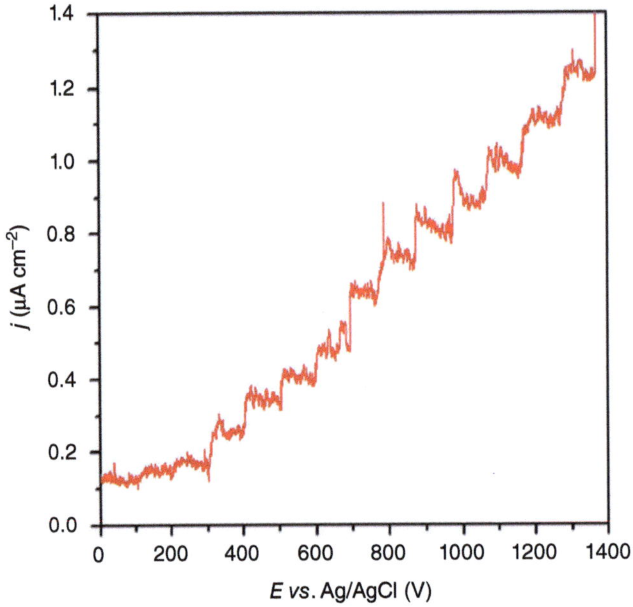

Figure 3.11 Chronoamperogram for successive additions of Cys in 0.1 M Britton–Robinson buffer (pH = 5.0) recorded with a $(CoSal)_4SiW_{12}$ modified CPE at +0.3 V *vs.* Ag/AgCl. Source: Kakhki and Shams [18]/with permission of Elsevier.

a CPE. Briefly, each CPE was prepared by mixing convenient amounts of graphite powder and paraffin oil (typically in a proportion of 70 : 30 in mass), together with an appropriate mass of the catalyst, in a mortar with a pestle. A portion of the composite mixture was packed in a hollow tube from one side and put in contact with a copper wire inserted from the other side, which served as an external electric contact. The carbon paste surface was smoothed on a piece of weighing paper. The presence of the POM in the CPE was confirmed by cyclic voltammetry in 1 M H_2SO_4 (pH = 0.5), and the peak potentials of the W-centre waves shifted towards more negative electric potentials upon increasing the pH, as expected. The modified CPE worked as a bi-functional electrochemical sensor, exhibiting electrocatalytic activity towards both the reduction of IO_3^- and the oxidation of Cys. The former took place on the W centres of the POM, whereas the latter occurred on the Co centre of CoSal. The detection of Cys was carried out by chronoamperometry at $E = 0.3$ V *vs.* AgCl/Ag in 0.1 M Britton–Robinson buffer (pH = 5.0) (Figure 3.11), and a linear dependence of the current density on the concentration of Cys was observed in the range 0.01–1.0 μM, with a sensitivity of 0.16 $\mu A\ \mu M^{-1}\ cm^{-2}$ and a detection limit of 9.67 nM [18].

In a second study, the modified CPE was prepared upon mixing appropriate amounts of carbon paste, CoSal and $(n\text{-butyl})_4SiW_{12}O_{40}$ (SiW_{12}), and tested by cyclic and differential pulse voltammetry, namely as a function of the pH (from 2.0 to 10.0) in Britton–Robinson buffer. An overall reaction mechanism was suggested, which includes an electron transfer step implicating the POM moiety (reaction (3.7)), which would imply an indirect role played by the POM in the oxidation of Cys on the Co centre of CoSal (Scheme 3.3).

$$[Co^{III}Sal]^+ + CysSH + H_2O \longrightarrow [CysS^- - Co^{III}Sal^+] + H_3O^+ \quad (3.4)$$

$$[CysS^- - Co^{III}Sal^+] \longrightarrow CysS^{\bullet}_{ads} + Co^{II}Sal \quad (3.5)$$

$$2CysS^{\bullet}_{ads} \longrightarrow CysS - SCys \quad (3.6)$$

$$POM_{(ox)} + Co^{II}Sal \longrightarrow POM_{(red)} + [Co^{III}Sal]^+ \quad (3.7)$$

$$POM_{(red)} \longrightarrow POM_{(ox)} + e^- \quad (3.8)$$

Scheme 3.3 Proposed reaction mechanism of the electro-oxidation of Cys on a CPE modified with CoSal and the POM SiW_{12}.

The most intense signal for Cys oxidation was obtained at pH = 5.0. This could be explained by the fact that the formation of $[CyS^--Co^{III}Sal^+]$ in reaction (3.4) is favoured as the pH increases from 2.0 to 5.0, but beyond pH 5.0 the Co centres are co-ordinated by OH^- ions, which renders them inactive. A differential pulse voltammetry calibration revealed two linear ranges, one from 0.01 to 1.0 μM (with a detection limit of 4.9 nM) and the other from 10 to 300 μM (Figure 3.12), the sensitivities being 1.57 (corresponding to a normalized current density of 50.1 $\mu A\ \mu M^{-1}\ cm^{-2}$) and 0.07 $\mu A\ \mu M^{-1}$, respectively. The loss in sensitivity at higher Cys concentrations was attributed to a partial fouling of the electrocatalytic sites of the modified CPE. This sensor turned out to be reliable for the assessment of the concentration of Cys in solutions obtained with effervescent tablets, as well as in more complex matrix samples, like human blood serum and urine [19].

The oxidation of Cys has also been performed on CPEs modified with inorganic–organic complexes containing POMs, namely V-containing ones, synthesized using hydrothermal methods. An example is $[Cu(phen)_2]_2PVW_{11}O_{40}$ (phen = 1,10-phenanthroline), which was tested in acetate buffer at pH = 3.84, the electro-oxidation taking place on the V centres. These modified CPEs were polyvalent in the sense that they exhibited also electrocatalytic activity towards the reduction of IO_3^-, NO_2^- and H_2O_2 [20].

3.2.4 POM-containing Hybrids as Electrocatalysts in Heterogeneous Phase: Layer-by-layer modified Electrodes

Another strategy consists of depositing films containing the electrocatalyst on the surface of the working electrode through a layer-by-layer (LBL) process. The species in alternate layers have opposite charges, and the electrostatic interactions between them are expected to hold them together. An example is the fabrication of multilayer films composed of tris(1,10-phenanthroline) ruthenium(II) ($[Ru(phen)_3]^{2+}$) and the POM $[P_2W_{18}O_{62}]^{6-}$ (P_2W_{18}) on ITO electrodes. The fabrication of the

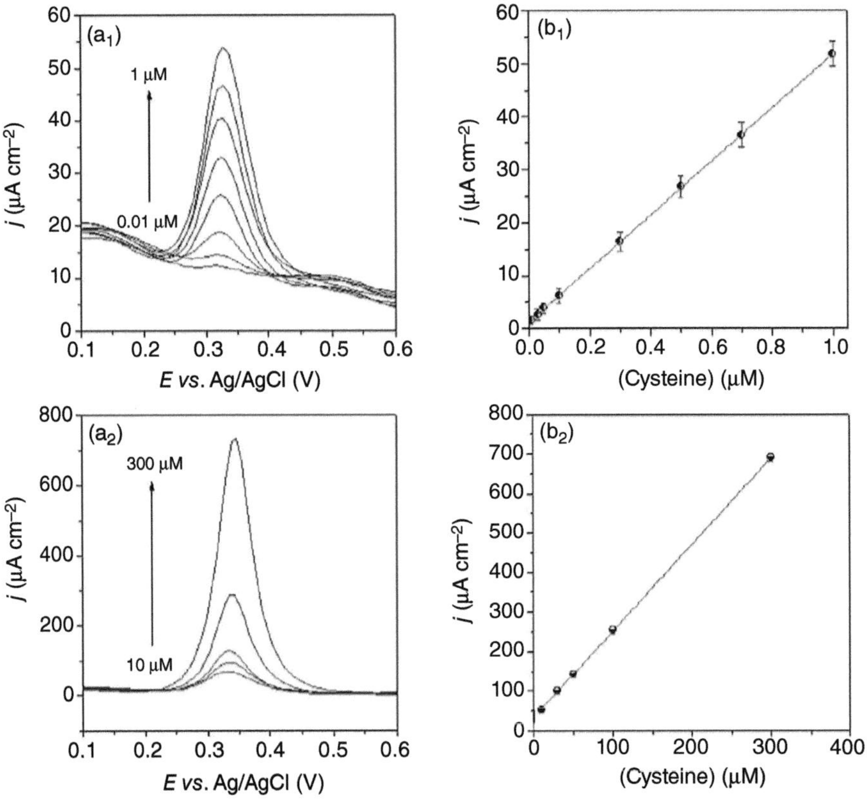

Figure 3.12 Differential pulse voltammograms recorded with a SiW_{12}/CoSal CPE in 0.1 M Britton–Robinson buffer solution pH = 5.0 containing (a_1) 0.01–1.0 μM and (a_2) 10–300 μM of Cys with corresponding calibration carves (b_1) and (b_2), respectively. Source: Kakhki et al. [19]/with permission of John Wiley & Sons.

device includes a previous treatment of the ITO electrode in order to increase the hydrophilic character of its surface, followed by dipping in a 5 mM solution of poly(ethylenimine) (PEI) for 20 minutes, rinsing with deionized water and drying under nitrogen. The obtained PEI-coated ITO is then alternately immersed for 20 minutes in 5 mM solutions of P_2W_{18} and of $[Ru(phen)_3]^{2+}$, with rinsing with water and drying with nitrogen steps after each immersion, and giving $\{PEI/[P_2W_{18}/[Ru(phen)_3]^{2+}]_n\}$ (n = 1–9) films. The progressive accumulation of layers was followed by UV spectra (Figure 3.13), revealing a rather regular and reproducible growth of the film, and the presence of the chemical elements constituting both P_2W_{18} and $[Ru(phen)_3]^{2+}$ in the film was confirmed by XPS.

Cyclic and differential pulse voltammetry experiments carried out in 0.5 M Na_2SO_4/H_2SO_4 buffer, pH = 0.93, revealed the presence of the electro-active centres of the species constituting the films. On the one hand, a study on the reduction side exhibited a linear dependence of the peak currents of the W centres on the scan rate, confirming that P_2W_{18} (and the film as a whole) was immobilized on the electrode surface (Figure 3.14). The electrocatalytic reduction of IO_3^- was

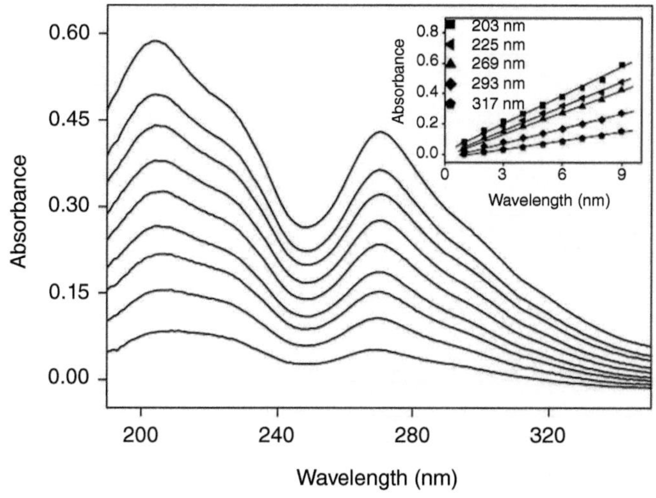

Figure 3.13 UV spectra of $\{PEI/[P_2W_{18}/[Ru(phen)_3]_n\}$ multilayer film ($n = 1-9$) deposited on quartz substrate (on both sides). Inset: plots of the absorbance values at 203, 225, 269, 293 and 317 nm. Source: Dong et al. [21]/with permission of Springer Nature.

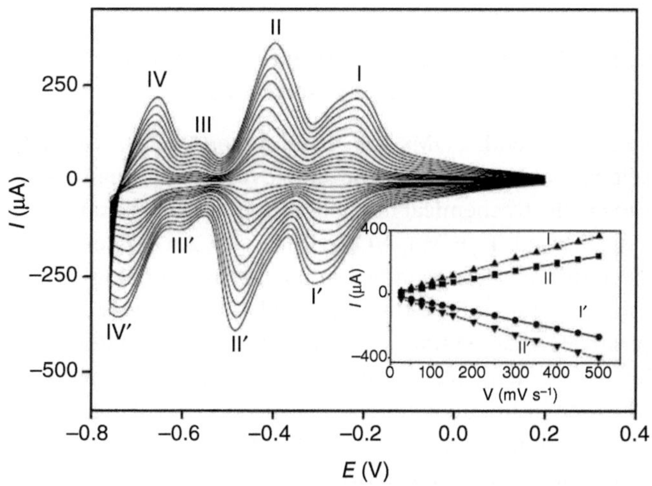

Figure 3.14 Cyclic voltammograms of $\{PEI/[P_2W_{18}/[Ru(phen)_3]_7/P_2W_{18}\}$ multilayer film in 0.5 M Na_2SO_4/H_2SO_4 buffer solution (pH = 0.93) at different scan rates (ranging from 25 to 500 mV s^{-1}). The inset shows a linear dependence of the currents of I–I′ and II–II' with the scan rates. Source: Dong et al. [21]/with permission of Springer Nature.

observed on the W centre waves, the measured currents being proportional to the concentration of the substrate. Similarly, on the positive side of the potential axis, Cys was electrocatalytically oxidized on the Ru centre, again with a proportionality between the current and the concentration of Cys (Figure 3.15). This device may therefore be used as a bi-functional electrochemical sensor both for IO_3^- and Cys [21].

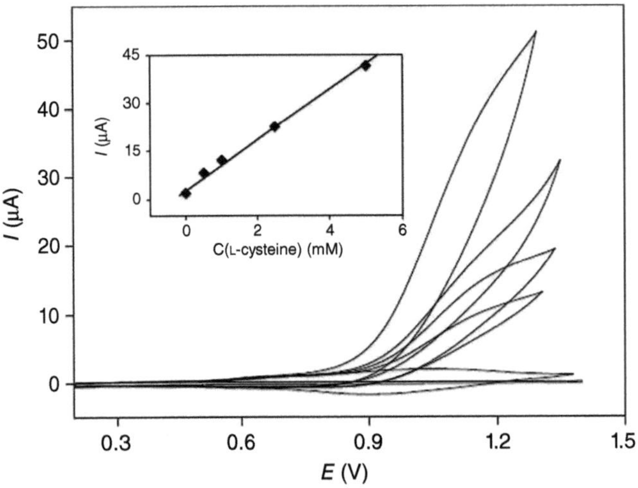

Figure 3.15 Cyclic voltammograms of {PEI/[P_2W_{18}/[Ru(phen)$_3$]$_7$/P_2W_{18}} multilayer film in 0.5 M Na_2SO_4/H_2SO_4 buffer solution (pH = 0.93) containing Cys at various concentrations: 0.0, 0.5, 1.0, 2.5 and 5.0 mM, respectively, and a bare ITO in the same solution containing 5.0 mM Cys. The inset shows the relationship between the catalytic current and concentration of Cys. Source: Dong et al. [21]/with permission of Springer Nature.

3.2.5 POM-containing Hybrids as Electrocatalysts in Heterogeneous Phase: Layer-by-layer and Nanoparticle-modified Electrodes

More elaborate films, containing both POMs and metal nanoparticles, were fabricated with the LBL self-assembly technique and employed in the quantitative detection of Cys thanks to an electrochemical method. In the following example, a V-containing POM was again used: $H_7P_2Mo_{17}VO_{62}$ ($P_2Mo_{17}V$). An ITO electrode previously rendered hydrophilic was first modified upon alternate dipping for five minutes in 1.5 M NaCl solutions containing either 1.0 mg ml^{-1} of negatively charged PEI or 1.0 mg ml^{-1} of positively charged poly(sodium-4-styrenesulphonate) (PSS), followed by rinsing with water. Then, Au-Ag bimetallic core–shell nanoparticles (Au@Ag NPs) were synthesized in situ upon the successive reduction of convenient 1 mM gold and silver salt solutions with 0.1 M $NaBH_4$ and 0.1 M ascorbic acid, respectively, the cycle with silver being done twice. Finally, the obtained film, denoted {[PEI/PSS]$_2$/Au@2Ag}, was completed with a series of alternate layers of PEI and $P_2Mo_{17}V$, which was achieved by successive dipping in 2 mM solutions of either PEI or $P_2Mo_{17}V$, followed by rinsing with water and drying with nitrogen after each immersion. The final film was called {[PEI/PSS]$_2$/Au@2Ag/[PEI/$P_2Mo_{17}V$]$_n$}, n (=1–6) being the number of bilayers. Figure 3.16 shows a schematic illustration of the different steps of the fabrication of the composite film.

The progressive fabrication of the composite film was monitored by UV–visible spectrophotometry, and the presence of the chemical elements of the different components was confirmed by XPS. Cyclic voltammetry experiments revealed the presence of redox waves attributed to the W and V centres of the POM, and again the dependence of the peak currents on the scan rate was coherent with

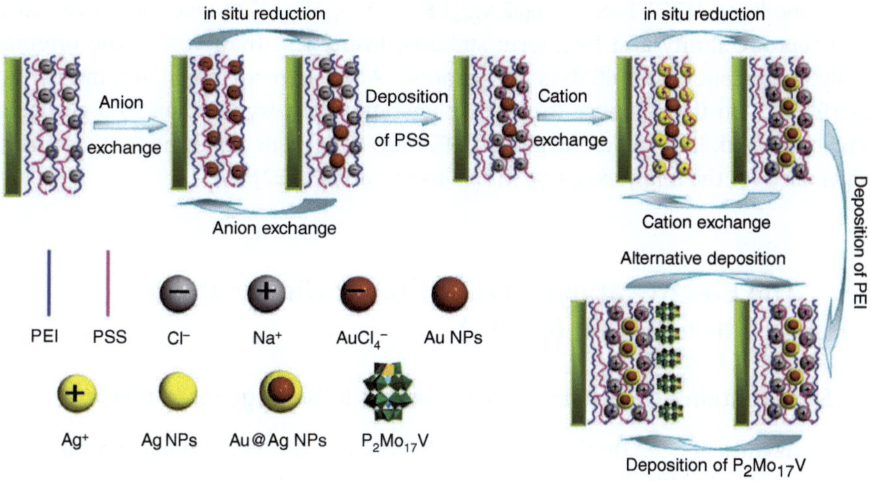

Figure 3.16 Schematic illustration of the different steps of the fabrication of the $\{[PEI/PSS]_2/Au@2Ag/[PEI/P_2Mo_{17}V]_2\}$ composite film. Source: Zhang et al. [22]/with permission of Elsevier.

the electro-active species being immobilized on the electrode surface. The device exhibited an enhanced electrocatalytic activity towards the oxidation of Cys in 0.2 M $NaCH_3COO/CH_3COOH$ buffer, pH = 5.0, probably due to the synergy between Au and Ag in the nanoparticles, the measured current being proportional to the concentration of Cys (Figure 3.17). The system showed a broad linear range of 2.5×10^{-8}–7.625×10^{-6} M, a low detection limit of 2.76×10^{-8} M and a high sensitivity of 1.7946 μA μM^{-1}.

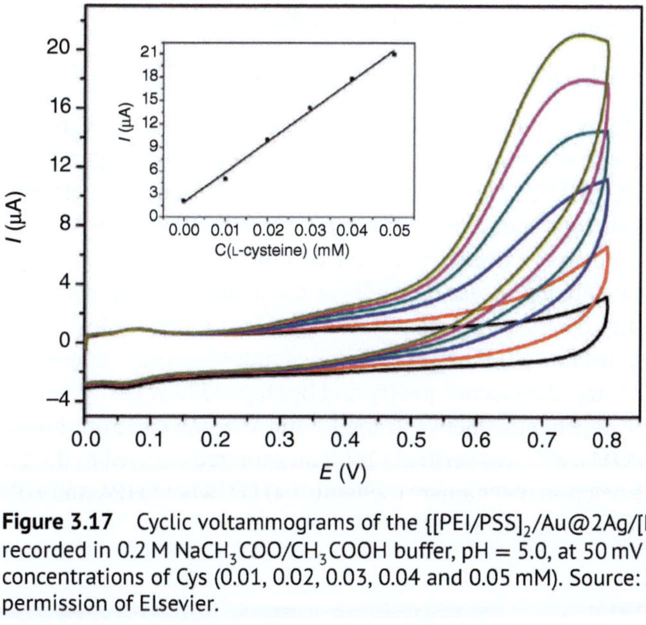

Figure 3.17 Cyclic voltammograms of the $\{[PEI/PSS]_2/Au@2Ag/[PEI/P_2Mo_{17}V]_2\}$ films recorded in 0.2 M $NaCH_3COO/CH_3COOH$ buffer, pH = 5.0, at 50 mV s^{-1} with different concentrations of Cys (0.01, 0.02, 0.03, 0.04 and 0.05 mM). Source: Zhang et al. [22]/with permission of Elsevier.

The modified $\{[PEI/PSS]_2/Au@2Ag/[PEI/P_2Mo_{17}V]_n\}$ ITO electrode revealed a good reproducibility and long-term stability, losing less than 10% of the original catalytic efficiency after 60 days of operation. Also, there are small and mainly no interference on the response from other analytes like L-tryptophan, L(+)-glutamic acid, citric acid, lactic acid and glucose. Finally, the sensor performed satisfactorily when used for the analysis of Cys in real milk samples [22].

3.3 The Electrocatalytic Oxidation of Nicotinamide Adenine Dinucleotide (NADH)

3.3.1 V-containing POMs as Electrocatalysts in Homogeneous Phase

Many enzymes depend on the $NAD(P)H/NAD(P)^+$ redox couples to carry out their biochemical roles [3]. Since they are involved in electron transfer, they may be probed by electrochemical methods or integrated in electrochemical devices used in analytical applications. When performing quantitative studies, care should be taken to avoid aqueous media having a pH below 6, where NADH is known to hydrate rapidly (a few minutes), leading to the depletion of the active form of the coenzyme [23]. Another consequence of this restriction is the quest for POM species stable at pH higher than 6 so that they may be employed together with NADH in convenient experimental conditions. Nadjo and Keita undertook several studies on the electrocatalytic oxidation of NADH with different V-containing POMs verifying the need to be stable at pH > 6 [6, 24–28]. By and large, UV–visible spectrophotometry and coulometry experiments led to the conclusion that the stoichiometry of the reaction between NADH and each POM was 1 : 2, implying that the chosen POMs behave as one-electron oxidants according to the following overall equation concerning an example of a mono-vanadate, tungstic Wells–Dawson POM (Figure 3.2):

$$2\alpha_2 - [P_2V^VW_{17}O_{62}]^{7-} + NADH \rightleftharpoons 2\alpha_2 - [P_2V^{IV}W_{17}O_{62}]^{8-} + NAD^+ + H^+$$

Figure 3.18 shows the example of the electrocatalytic oxidation of NADH by α_1-$[P_2VW_{17}O_{62}]^{7-}$ with both reactants in homogeneous phase, the excess parameter being defined as $\gamma = C^0(NADH)/C^0(POM)$. The diffusion-controlled, one-electron reversible voltammogram corresponding to the interconversion between V^{IV} and V^V when $\gamma = 0$ evolves into an electrocatalytic wave as γ increases up to 2.2.

The efficiency of the catalytic process depends on the formal redox potential of the POM, suggesting that the system probably behaves as a purely redox catalysis one. Reaction rate constants (k) measured under pseudo-first-order conditions both by double potential step chronocoulometry and by stopped flow gave coherent results [7, 25]. A linear dependence between $\log(k)$ and the one-electron, formal redox potential of the POMs, $E^{0\prime}$, was verified [28]. The gathered data led to the following proposal for the reaction mechanism (Scheme 3.4) [7], where HPA and HPB are the oxidized and the reduced forms of the POM, respectively:

$$(\mathrm{HPB} - \mathrm{e}^- \rightleftharpoons \mathrm{HPA}) \tag{3.9}$$

$$\mathrm{NADH} + \mathrm{HPA} \underset{k_{-1}}{\overset{k_1}{\rightleftharpoons}} \mathrm{NADH}^{\bullet +} + \mathrm{HPB} \tag{3.10}$$

$$\mathrm{NADH}^{\bullet +} + \mathrm{B} \underset{k_{-2}}{\overset{k_2}{\rightleftharpoons}} \mathrm{NAD}^{\bullet} + \mathrm{BH}^+ \tag{3.11}$$

$$\mathrm{NAD}^{\bullet} + \mathrm{HPA} \underset{k_{-3}}{\overset{k_3}{\rightleftharpoons}} \mathrm{NAD}^+ + \mathrm{HPB} \tag{3.12}$$

$$\mathrm{NADH} + \mathrm{HPA} \underset{k_{\mathrm{diff}}}{\overset{k_{\mathrm{diff}}}{\rightleftharpoons}} (\mathrm{NADH}, \mathrm{HPA}) \underset{k_{-1\mathrm{act}}}{\overset{k_{1\mathrm{act}}}{\rightleftharpoons}} (\mathrm{NADH}^{\bullet +}, \mathrm{HPB}) \underset{k_{\mathrm{diff}}}{\overset{k_{\mathrm{diff}}}{\rightleftharpoons}} \mathrm{NADH}^{\bullet +} + \mathrm{HPB} \tag{3.13}$$

Scheme 3.4 Proposed reaction mechanism of the electro-oxidation of NADH by a POM (HPA and HPB are the oxidised and the reduced forms of the POM).

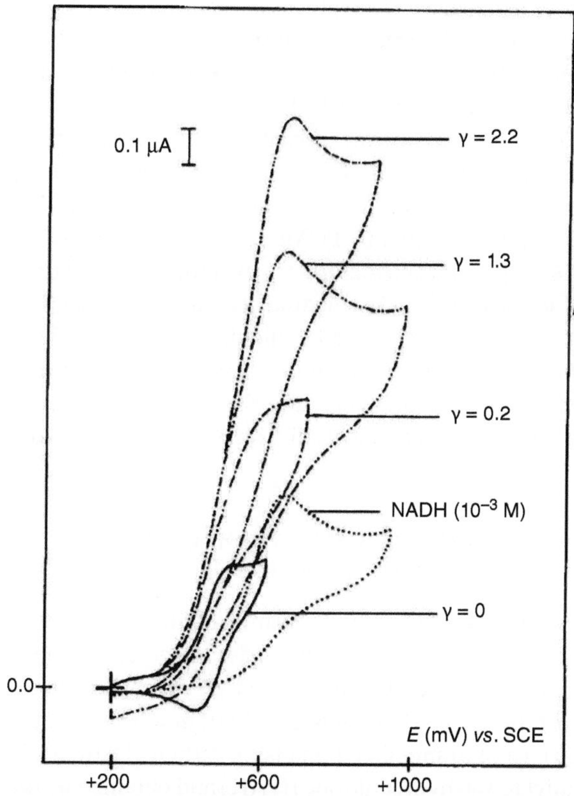

Figure 3.18 Cyclic voltammograms recorded in phosphate buffer, pH = 7, for the system α_1-$[P_2VW_{17}O_{62}]^{8-}$/NADH, at a scan rate of 5 mV s^{-1}, with a 1 mm diameter Pt disk working electrode. The concentrations were [POM] = 10^{-3} M and [NADH] = $\gamma \times 10^{-3}$ M, except for the one containing just 10^{-3} M NADH and no POM. Source: Keita and Nadjo [7]/with permission of Elsevier.

Equation (3.10) corresponds to the first single-electron transfer from NADH to the POM, Eq. (3.11) is the deprotonation of the $NADH^{\cdot+}$ radical, with the proton being transferred to a base B and Eq. (3.12) the second electron transfer from the same substrate molecule (under the radical form $NAD^{\cdot}$) to another POM species. No hydride transfer is considered, and independent measurements of the rate constants revealed that the rate-limiting step is the single-electron transfer between NADH and HPA (Eq. (3.10)). In order to account for the gathered data, the process corresponding to Eq. (3.10) was split into the steps shown in Eq. (3.13), where (NADH, HPA) and ($NADH^{\cdot+}$, HPB) correspond to the species being present within reacting distance, k_{diff} is the diffusion-limited rate constant, and k_{1act} and k_{-1act} are the activation-controlled rate constants.

The linear dependence between log(k) and $E^{0\prime}$ mentioned above proves that there is no rate limitation as far as the electron transfer is concerned, the conclusion being that the rate-limiting step is the separation of the ($NADH^{\cdot+}$, HPB) pair by diffusion. The formal redox potential of the $NADH^{\cdot+}$/NADH couple was estimated at $E^0 = +0.80$ V, at pH = 7 [28]. The overall outcome is coherent with a purely redox catalytic process (as opposed to a chemical catalytic process) [7].

NADPH was also tested according to the same procedures and exhibited similar responses. The rate constant for the reaction between 1,2-$[P_2V_2W_{16}O_{62}]^{8-}$ and NADPH was estimated at $k = 3.5 \times 10^3\ M^{-1}\ s^{-1}$, at pH 8 [6].

The electrocatalytic oxidation of NADH by the mixed-valence 14-vanadogermanate $[GeV^{V}{}_{12}V^{IV}{}_2O_{40}]^{8-}$ was reported in 2006 [29]. The novelty in this study was the fact that this species is an active two-electron oxidation catalyst for NADH at pH 8, contrary to previous results where the POMs used were involved in one-electron transfer systems [6, 27]. A chronocoulometry experiment carried out between −0.030 and +0.400 V *vs.* SCE in a pH 8 medium containing 2×10^{-4} M $[GeV^{V}{}_{12}V^{IV}{}_2O_{40}]^{8-}$ and 2×10^{-3} M NADH allowed to calculate a relatively high rate constant value of $k = 2.80\times10^3\ M^{-1}\ s^{-1}$. The POM in question is stable in aqueous media up to pH = 8, which makes it an interesting choice to be tested as an electrocatalyst for the oxidation of relevant biological substrates in conditions close to the physiological ones.

3.3.2 POM-containing Hybrids as Electrocatalysts in Heterogeneous Phase: Layer-by-layer and Precipitate-deposition-modified Electrodes

Similar to what was described previously for Cys, the deposition of films on a surface has been applied to the fabrication of POM-modified electrodes aimed at the detection of NADH. A couple of articles authored by Ammam et al. cover the subject [30, 31]. In the first one, POMs of the Wells–Dawson family, both mono-substituted and sandwich-type (Figure 3.19), were immobilized on glassy carbon electrodes in multilayer films made of poly(allylamine hydrochloride) (PAH) and obtained by the LBL procedure.

Briefly, glassy carbon electrodes were rendered positively charged upon treatment with 3-aminopropyltriethoxysilane (ATS) under nitrogen atmosphere for 10 minutes, then rinsed and dried. The presence of amine groups in ATS makes the electrode positively charged. Then, they were dipped in 1 M CH_3COOLi/CH_3COOH

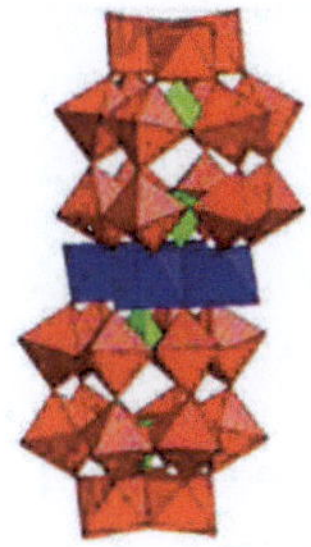

$[M_4(H_2O)_2(As_2W_{15}O_{56})_2]^{y-}$; M = Fe^{III}, Mn^{II} or Co^{II}; y = 12 or 16

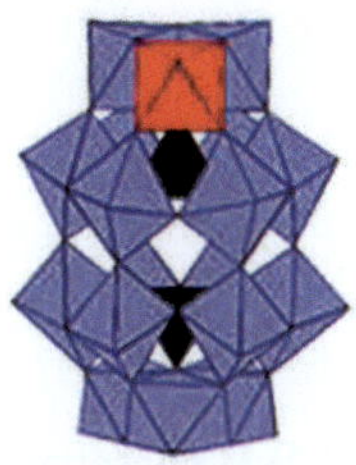

α_2-$[P_2Fe^{III}(H_2O)W_{17}O_{61}]^{7-}$

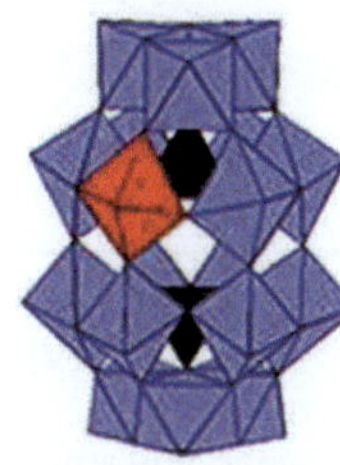

α_1-$[P_2V^{IV}W_{17}O_{62}]^{8-}$

Figure 3.19 Polyhedral representations of the POMs immobilized on glassy carbon electrodes with PAH by LBL: $[Fe_4^{III}(H_2O)_2(As_2W_{15}O_{56})_2]^{12-}$, $[Mn_4^{II}(H_2O)_2(As_2W_{15}O_{56})_2]^{16-}$, $[Co_4^{II}(H_2O)_2(As_2W_{15}O_{56})_2]^{16-}$, α_2-$[P_2Fe^{III}(H_2O)W_{17}O_{61}]^{7-}$, and α_1-$[P_2V^{IV}W_{17}O_{62}]^{8-}$.

buffer, pH = 5, containing 10^{-3} M poly(sodium 4-styrenesulphonate) (PSS) for 10 minutes. The latter adds a negatively charged layer on top of the ATS one and is composed of aromatic rings which are expected to facilitate electron transfer between the components of the film. Then, the electrodes partially modified with ATS/PSS were rinsed, dried and dipped in another acetate buffer solution, pH = 5, containing 10^{-3} M PAH for 10 minutes in order to create a positively charged surface capable of electrostatically immobilizing POM anions. The final step consists of an immersion in 5×10^{-4} M POM solution in acetate buffer, pH = 5, for 10 minutes, followed by rinsing and drying. The last two steps involving the PAH and the POM were repeated several times in order to have multiple layers. The modified electrodes were characterized by UV–visible spectrophotometry and cyclic voltammetry, which revealed that the POMs had lost the majority of the substituting metal centres and had become lacunary tungstic scaffolds. A possible explanation of this outcome is the existence of a repulsive interaction between the both positively charged PAH and the substituting metal cations in the POMs. Nevertheless, traces of the latter have been detected, namely in cyclic voltammetry at low scan rates, which may explain that the modified glassy carbon electrodes exhibited electrocatalytic activity towards the reduction of O_2 and H_2O_2, and the oxidation of NADH.

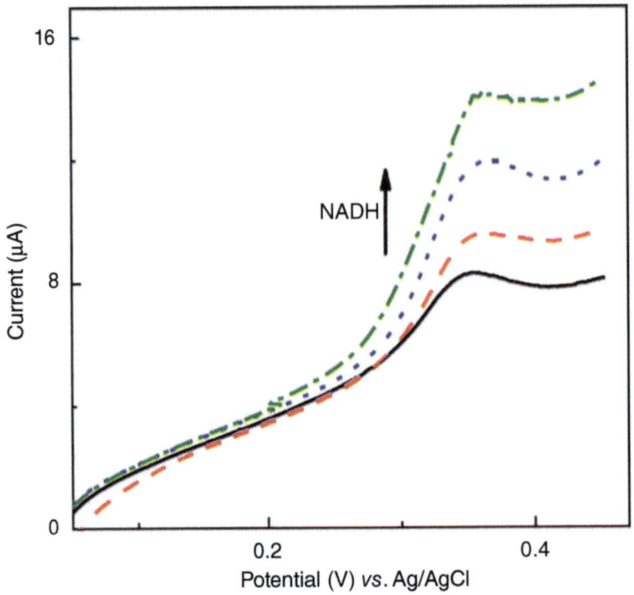

Figure 3.20 Linear sweep voltammograms recorded at 20 mV s^{-1} in 0.25 M phosphate buffer, pH = 7.4, with a glassy carbon electrode modified with $[BMIM]_6P_2Mo_{18}O_{62}$ in the absence of NADH (solid black line) and after successive additions of 0.1 mM NADH. Source: Ammam and Easton [31]/with permission of Springer Nature.

The second article by Ammam et al. concerns the modification of a glassy carbon electrode with an insoluble compound obtained upon precipitation of the POM $[P_2Mo_{18}O_{62}]^{6-}$ with the cation of the ionic liquid 1-butyl-3-methylimidazolium tetrafluoroborate ($[BMIM]BF_4$). The hybrid compound $[BMIM]_6P_2Mo_{18}O_{62}$ was immobilized on the electrode surface by depositing 5 μL of a previously sonicated suspension containing 50 mg of the precipitate, 0.3 ml of 2-propanol and 5 μL of Nafion. The modified electrode was dried overnight at 80 °C and used afterwards.

Electrodes modified with $[BMIM]_6P_2Mo_{18}O_{62}$ were employed in linear sweep voltammetry experiments and exhibited an efficient electrocatalytic activity towards the oxidation of NADH (Figure 3.20), particularly when compared with the response obtained either with the soluble POM in the presence of NADH, or with NADH alone, for which the overpotential was at least 200 mV higher.

The integration of the area below the solid black line in Figure 3.23 led to an estimation of approximately 2 electrons being transferred per molecule of NADH. The following reaction mechanism was proposed (Scheme 3.5):

$$[BMIM]_6P_2Mo_{18}O_{62} \rightleftharpoons ([BMIM]_6P_2Mo_{18}O_{62})^+ + H^+ + 2e^- \quad (3.14)$$

$$NADH + ([BMIM]_6P_2Mo_{18}O_{62})^+ \rightleftharpoons NAD^+ + [BMIM]_6P_2Mo_{18}O_{62} \quad (3.15)$$

Scheme 3.5 Proposed reaction mechanism of the electro-oxidation of NADH by the insoluble salt $[BMIM]_6P_2Mo_{18}O_{62}$.

In the first step (Eq. (3.14)), $[BMIM]_6P_2Mo_{18}O_{62}$ undergoes a two-electron oxidation under the influence of the electrode and loses a proton, yielding the charged species $([BMIM]_6P_2Mo_{18}O_{62})^+$. The latter oxidizes in turn NADH (Eq. (3.15)), upon abstraction of two electrons and a proton. The overall equilibrated equation corresponds to the electrocatalytic oxidation of NADH to NAD^+:

$$NADH \rightleftharpoons NAD^+ + H^+ + 2e^-$$

Complementary data gathered from amperometry led to a calculated sensitivity close to $2\,\mu A\,mM^{-1}\,mm^{-2}$ and revealed the existence of a linear response up to 9 mM and of a detection limit below 0.5 μM (Figure 3.21). The enhanced currents may be due to a synergy between the POM moieties and the imidazolium rings of BMIM, the latter playing the role of mediators that facilitate electron transfer from NADH to the electrode. Stability tests revealed that the modified electrodes kept more than 93% of their performance after two weeks of storage.

A similar study used V-containing, Keggin-type POMs and PAH in order to fabricate modified glassy carbon electrodes with the purpose of developing sensors for the detection of NADH [32]. Briefly, a buffer solution made of 5 mg of positively charged PAH in 10 ml 0.2 M PBS, pH = 7.5, was subject to successive cycling between −0.5 and + 0.5 V *vs.* Ag/AgCl at $50\,mV\,s^{-1}$ using a glassy carbon electrode until the voltammograms were superimposable. The PAH-modified electrode was immersed for 20 minutes in a solution containing $3.0\,mg\,ml^{-1}$ of the POM $[PMo_{11}VO_{40}]^{4-}$ ($PMo_{11}V$), rinsed and dried, yielding the composite film ($PAH/PMo_{11}V$). The same procedure was repeated several times in order to obtain multilayer films $(PAH/PMo_{11}V)_n$. The growth of the composite films was followed by cyclic voltammetry, where the current magnitude of the waves attributed to the V and the Mo centres increased with the number of layers in the film (Figure 3.22).

Figure 3.23 shows the response of an electrode with a four-layer film, $(PAH/PMo_{11}V)_4$, in the presence of an increasing concentration of NADH. The peak current magnitude of the oxidation wave at +0.6 V *vs.* Ag/AgCl increases with [NADH].

Complementary data gathered from amperometry revealed a low detection limit of 1.9×10^{-8} M, and a wide range with two successive linear zones with different slopes, one from 3.2×10^{-7} to 8.2×10^{-5} M, and another from 1.1×10^{-5} to 1.01×10^{-3} M. The stability of the response was assessed over several days and revealed a reliable sensor, the magnitude of the measured currents being kept at around 90% after 60 days of regular use. The selectivity of the device towards the detection of NADH was studied by amperometry at an applied potential of +0.6 V *vs.* Ag/AgCl in the presence of other substrates which might be oxidized in the same potential range (Figure 3.24). The latter include glucose, fructose, sucrose, folate, uric acid, ascorbic acid and dopamine. Clearly, the sensor is rather selective for NADH and was successfully tested for the detection of NADH in human serum samples.

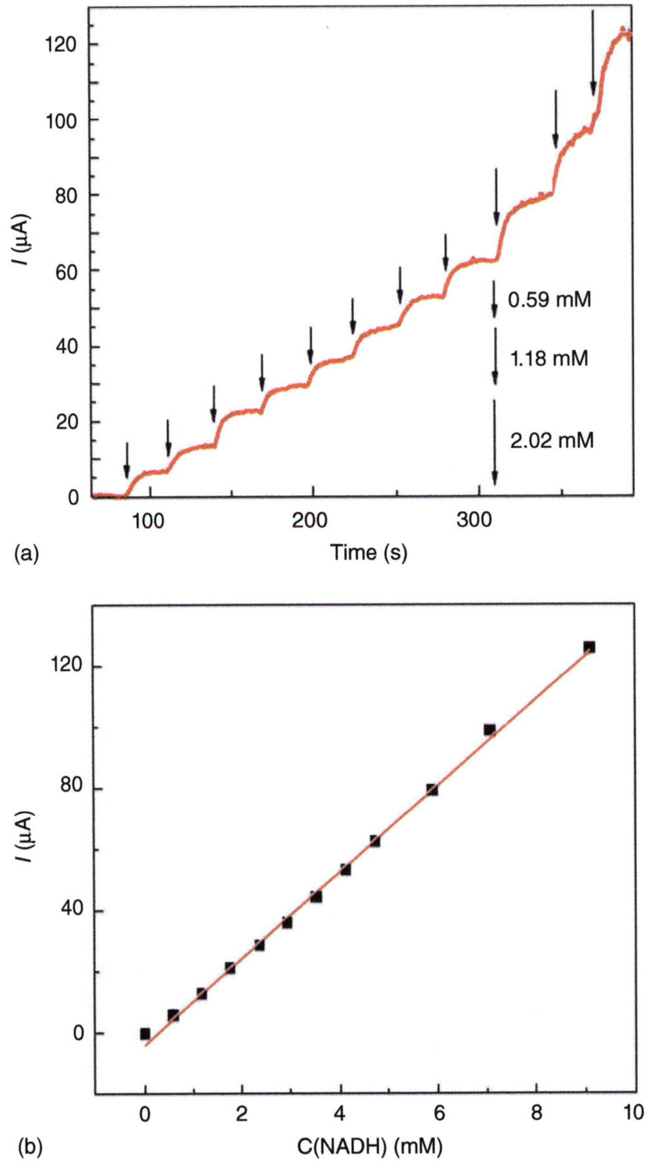

Figure 3.21 (a) Amperometric response obtained in 0.25 M phosphate buffer, pH = 7.4, under constant hydrodynamic agitation with a glassy carbon electrode modified with $[BMIM]_6P_2Mo_{18}O_{62}$ polarized at 0.45 V *vs.* Ag/AgCl to successive additions of 0.59, 1.18 and 2.02 mM NADH. (b) Dependence of the steady-state current on the concentration of NADH. Source: Ammam and Easton [31]/with permission of Springer Nature.

3.3.3 POM-containing Hybrids as Electrocatalysts in Heterogeneous Phase: Layer-by-layer and Nanoparticle-modified Electrodes

Other reports on the employment of POMs in more complex systems aimed at either carrying out the electro-oxidation of NADH or profiting from its behaviour as a reductant may be found in the literature. An example concerns the hydrothermal

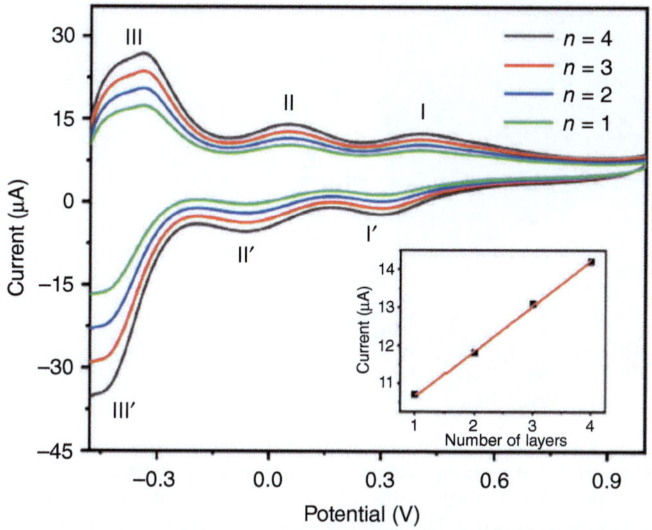

Figure 3.22 Cyclic voltammograms obtained in PBS buffer, pH = 7.5, at a scan rate of 100 mV s^{-1} with a glassy carbon electrode modified with a $(PAH/PMo_{11}V)_n$ multilayer composite film ($n = 1$–4). The reference electrode was Ag/AgCl. Source: Chu et al. [32]/with permission of Elsevier.

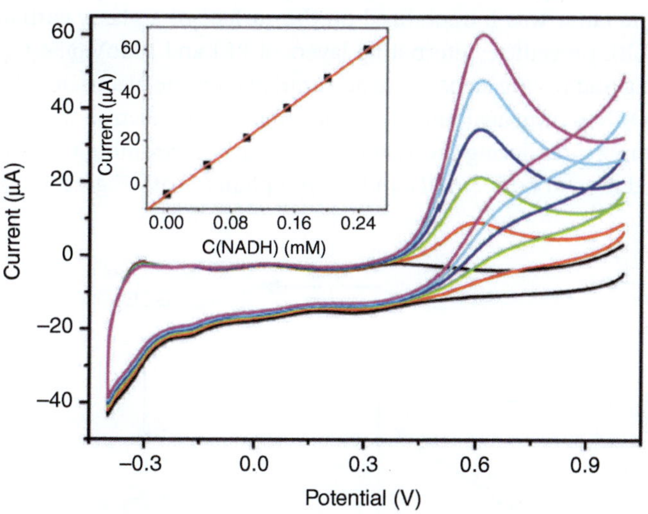

Figure 3.23 Cyclic voltammograms obtained in PBS buffer, pH = 7.5, at a scan rate of 100 mV s^{-1} with a glassy carbon electrode modified with a $(PAH/PMo_{11}V)_4$ composite film, with increasing concentrations of NADH. The inset shows the dependence of the catalytic current on the NADH concentration. The reference electrode was Ag/AgCl. Source: Chu et al. [32]/with permission of Elsevier.

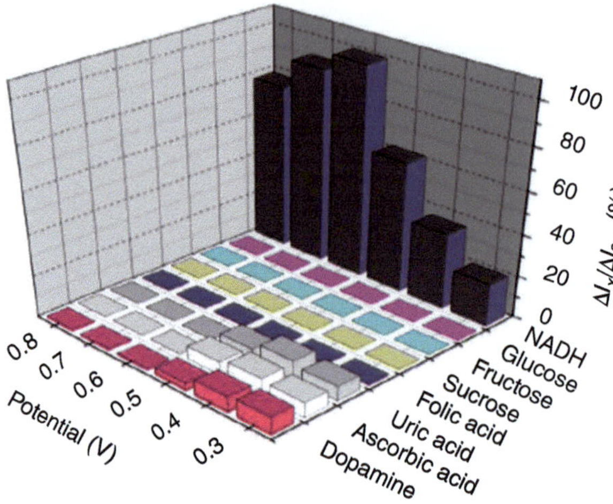

Figure 3.24 The selectivity profile of a glassy carbon electrode modified with a $(PAH/PMo_{11}V)_4$ composite film obtained in PBS buffer, pH = 7.5. The currents were measured at different potentials *vs.* Ag/AgCl. The samples contained NADH (0.25 mM), glucose (0.5 mM), fructose (0.5 mM), sucrose (0.5 mM), folic acid (0.25 mM), uric acid (0.5 mM), ascorbic acid (0.5 mM) and dopamine (0.25 mM). The maximum catalytic current corresponds to the oxidation of NADH at +0.60 V *vs.* Ag/AgCl. Source: Chu et al. [32]/with permission of Elsevier.

synthesis of hybrids composed of metal oxide nanoparticles (NPs) and POMs, namely Fe_3O_4 and $[PMo_9V_3O_{40}]^{6-}$ (PMoV), respectively, which yielded the hybrid PMoV@Fe_3O_4 [33]. The latter was immobilized on the surface of a glassy carbon electrode following a LBL procedure: alternating layers of PEI and PMoV@Fe_3O_4, with subsequent rinsing and drying steps, were accumulated on the electrode. The POMs prevent the Fe_3O_4 NPs from agglomerating and enhance the electron transfer capabilities of the ensemble, rendering the system particularly performing towards the electrocatalytic oxidation of both NADH and L-tryptophan (L-trp) (Figure 3.25).

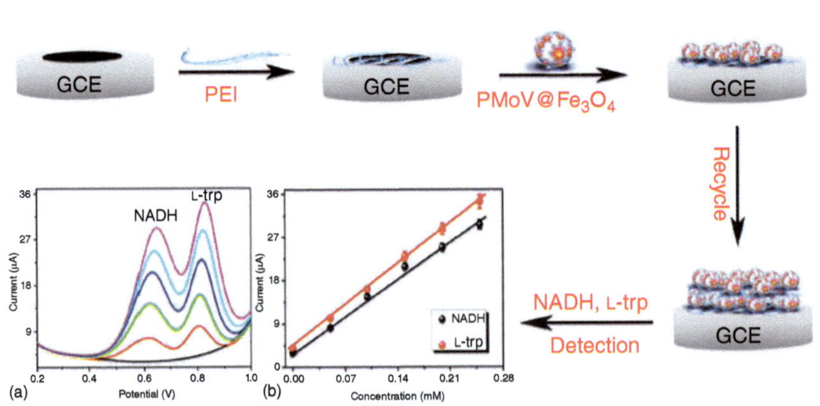

Figure 3.25 Schematic illustration of the fabrication of glassy carbon electrodes modified with PMoV@Fe_3O_4 hybrids. (a) Linear sweep curves in the presence of increasing concentrations of NADH and L-trp. (b) Dependence of the oxidation peak currents on the concentration of the substrates. Source: Zhu et al. [33]/with permission of Elsevier.

The device exhibited a low detection limit, a wide linear range and good selectivity and stability towards the two substrates.

3.3.4 POM-containing Hybrids as Electrocatalysts in Heterogeneous Phase: Precipitate-deposition-modified Electrodes and Electro-generated Chemiluminescence

In another work, ruthenium(II) tris(bipyridine) ($[Ru(bpy)_3]^{2+}$), known for its luminescent properties, was precipitated with the POM $[PMo_{12}O_{40}]^{3-}$ and an aliquot of a suspension of the resulting hybrid $[Ru(bpy)_3]^{2+}/[PMo_{12}O_{40}]^{3-}$ was deposited on the surface of an ITO electrode. The modified ITO-electrode constitutes a solid-state electro-generated chemiluminescence (ECL) sensor for the detection of NADH [34]. Indeed, an enhanced ECL signal was obtained when the POM was included in the device, as compared to the employment of Nafion instead (Figure 3.26). The sensor showed a wide linear range from 2.5×10^{-7} to 5.0×10^{-3} M, a detection limit of 1.67×10^{-8} M and good stability, retaining 90% of the response after two weeks under operation.

3.3.5 POMs in Artificial Reductase Systems for Oxidation Catalysis

Finally, the reducing properties of NADH were used with artificial reductase systems associating PEI and flavin mononucleotide (FMN) with the purpose of carrying out the aerobic oxidation of thioanisole derivatives [35]. This sulphoxidation activity was demonstrated both with polyoxotungstates and polyoxomolybdates, each category following a different catalytic pathway. The former uses H_2O_2 as an intermediate previously formed by the reduction of O_2 via the NADH/FMN system (Figure 3.27), whereas the latter takes the electrons directly from the NADH/FMN system in order to do the reductive activation of O_2 for the sulphoxidation reaction to ensue (Figure 3.28). In both cases, the sulphoxidation reaction takes place, with H_2O as the second final product. Even if this study does not concern electrocatalysis as such, its results are very promising in the context of the application of POMs in oxidation catalysis.

3.4 Conclusion

POMs have been used on their own, or associated with other compounds in mixed hybrids, as electrocatalysts for the oxidation of both Cys and NADH. The fact that they exist under different structures and allow for the metal centres to be substituted by other electro-active elements results in a very rich and varied electrochemical behaviour which renders them particularly versatile redox catalysts. The choice of the POMs employed in the oxidation of Cys and NADH is often guided by the potential domain in which their metal centres respond; indeed, an efficient electrocatalysis is expected to require small or negligible overpotentials, meaning that there is a quest for POMs containing metal centres whose oxidation peak potentials

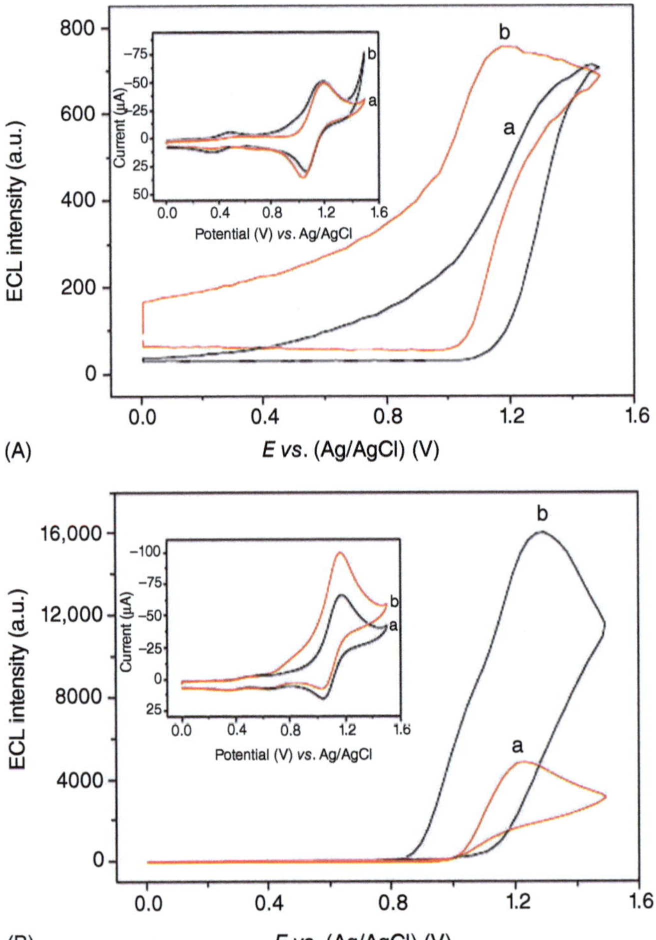

Figure 3.26 ECL intensity–potential curves recorded with ITO electrodes modified with $[Ru(bpy)_3]^{2+}$/Nafion (curve a) and $[Ru(bpy)_3]^{2+}/[PMo_{12}O_{40}]^{3-}$ (curve b) in the absence of (A) and in the presence of (B) 0.1 mM NADH at a scan rate of 500 mV s^{-1}. Inset: the corresponding cyclic voltammograms obtained at a scan rate of 100 mV s^{-1}. Source: Li et al. [34]/with permission of Elsevier.

are a few tens of millivolts smaller than those observed for the two substrates alone. Therefore, it is not a surprise that the majority of the substituted POMs mentioned in this chapter have metals such as V and Ce, other than the W and Mo centres which constitute the scaffold. Several approaches have been presented in order to render visible the electrocatalytic effect, always resorting to an appropriate working electrode which can exchange electrons with the POM: (i) the use of the POM and the substrate in a homogeneous phase; (ii) the inclusion of the POM in the electrode itself, as in the case of CPE; (iii) the immobilization of the POM on the electrode surface, either in multi-layer films, as in the case of the LBL procedure, or in the form of a deposit of an insoluble precipitate. In most of the cases, the system could be studied by several electrochemical techniques, the outcome confirmed the existence

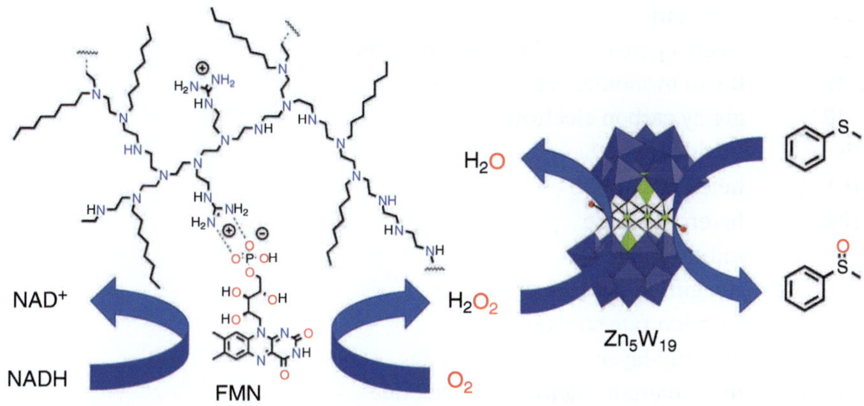

Figure 3.27 Reaction scheme for the oxidation of thioanisole by the tungstic POM $[(ZnW_9O_{34})_2WZn_3(H_2O)_2]^{12-}$ (Zn_5W_{19}) and FMN-PEI in the presence of NADH. Source: Naim et al. [35]/with permission of Royal Society of Chemistry.

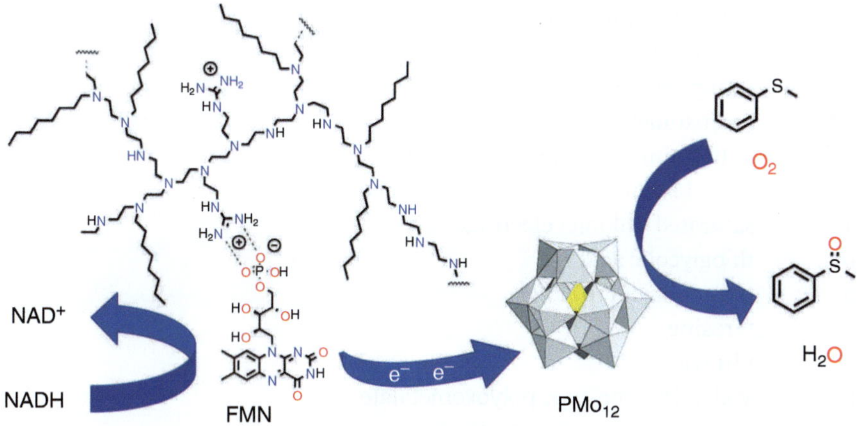

Figure 3.28 Reaction scheme for the oxidation of thioanisole by the molybdic POM $[PMo_{12}O_{40}]^{3-}$ (PMo_{12}) and FMN-PEI in the presence of NADH. Source: Naim et al. [35]/with permission of Royal Society of Chemistry.

of oxidative electrocatalysis and the gathered data revealed that the devices were promising for both the quantitative and qualitative analyses of samples containing either Cys or NADH. Results from future studies involving the electrocatalytic oxidation of these two substrates by POMs, also within artificial and bio-inspired systems, are awaited with interest.

List of Abbreviations

ATS	3-aminopropyltriethoxysilane
BMIM	1-butyl-3-methylimidazolium
CPE	carbon paste electrode

Cys	L-cysteine
ECL	electro-generated chemiluminescence
FMN	flavin mononucleotide
GCE	glassy carbon electrode
His	histidine
HPA	heteropolyanion
HPB	heteropolyblue
ITO	indium tin oxide
K	Keggin-type polyoxometalate
k	reaction rate constant
LBL	layer-by-layer
NADH	nicotinamide adenine dinucleotide
NADPH	nicotinamide adenine dinucleotide phosphate
NMR	nuclear magnetic resonance
NP	nanoparticle
PAH	poly(allylamine hydrochloride)
PEI	poly(ethylenimine)
Phe	phenylalanine
phen	1,10-phenanthroline
POM	polyoxometalate
PSS	poly(sodium-4-styrenesulphonate)
Sal	*N*,*N*′-bis(salicylidene)-1,2-phenylenediamine
SCE	saturated Calomel electrode
TGA	thioglycolic acid
Trp	tryptophan (Trp)
Tyr	tyrosine
UV	ultraviolet
WD	Wells–Dawson-type polyoxometalate
XPS	X-ray photo-electron spectroscopy

References

1 Garrido Ruiz, D., Sandoval-Perez, A., Rangarajan, A.V. et al. (2022). Cysteine oxidation in proteins: structure, biophysics, and simulation. *Biochemistry* 61: 2165–2176.

2 Fra, A., Yoboue, E.D., and Sitia, R. (2017). Cysteines as redox molecular switches and targets of disease. *Front. Mol. Neurosci.* 10: 167.

3 Sellés Vidal, L., Kelly, C.L., Mordaka, P.M., and Heap, J.T. (2018). Review of NAD(P)H-dependent oxidoreductases: properties, engineering and application. *Biochim. Biophys. Acta Proteins Proteom.* 1866: 327–347.

4 Keita, B., Mbomekalle, I.M., de Oliveira, P. et al. (2006). Polyoxometallates and green chemistry: molecules and nanostructured materials for energy conversion and the environment. *J. Cluster Sci.* 17: 221–233.

5 Dika Manga, J., Teillout, A.-L., Riviere, E. et al. (2021). Electrochemical, electrocatalytic, and magnetic properties of vanadium-containing polyoxometalates. *Magnetochemistry* 7: 157.

6 Keita, B., Mbomekalle, I.M., Nadjo, L. et al. (2005). Vanadium-substituted Dawson-type polyoxometalates as versatile electrocatalysts. *C. R. Chim.* 8: 1057–1066.

7 Keita, B. and Nadjo, L. (2007). Polyoxometalate-based homogeneous catalysis of electrode reactions: recent achievements. *J. Mol. Catal. A Chem.* 262: 190–215.

8 Keita, B., Contant, R., Mialane, P. et al. (2006). Multi-vanadium substituted polyoxometalates as efficient electrocatalysts for the oxidation of l-cysteine at low potential on glassy carbon electrodes. *Electrochem. Commun.* 8: 767–772.

9 Sami, P., Venkateshwari, K., Mariselvi, N. et al. (2009). Studies on electron transfer reactions: reduction of heteropoly 11-tungstophosphovanadate(V) by l-cysteine and thioglycolic acid in aqueous acid medium. *Transit. Met. Chem.* 34: 733–737.

10 Sami, P., Venkateshwari, K., Mariselvi, N. et al. (2010). Studies on electron transfer reactions: reduction of heteropoly 10-tungstodivanadophosphate by l-cysteine in aqueous acid medium. *Transit. Met. Chem.* 35: 137–142.

11 Van Rompuy, L.S. and Parac-Vogt, T.N. (2019). Interactions between polyoxometalates and biological systems: from drug design to artificial enzymes. *Curr. Opin. Biotechnol.* 58: 92–99.

12 Abdelhameed, S.A.M., Vandebroek, L., de Azambuja, F., and Parac-Vogt, T.N. (2020). Redox activity of Ce(IV)-substituted polyoxometalates toward amino acids and peptides. *Inorg. Chem.* 59: 10569–10577.

13 Stroobants, K., Moelants, E., Ly, H.G.T. et al. (2013). Polyoxometalates as a novel class of artificial proteases: selective hydrolysis of lysozyme under physiological pH and temperature promoted by a cerium(IV) Keggin-type polyoxometalate. *Chem. Eur. J.* 19: 2848–2858.

14 Moons, J., Van Rompuy, L.S., Rodriguez, A. et al. (2019). Hydrolysis of transferrin promoted by a cerium(IV)-Keggin polyoxometalate. *Polyhedron* 170: 570–575.

15 Quanten, T., De Mayaer, T., Shestakova, P., and Parac-Vogt, T.N. (2018). Selectivity and reactivity of ZrIV and CeIV substituted Keggin type polyoxometalates toward cytochrome c in surfactant solutions. *Front. Chem.* 6: 372.

16 Buzuk, M., Brinić, S., Vladislavić, N. et al. (2016). Kinetics of the oxidation of L-cysteine by trans-and cis-cobalt(III) and iron(III) complexes. *Monatsh. Chem.* 147: 359–367.

17 Abdel-Halim, H.M., Abu-Surrah, A.S., and Baker, H.M. (2006). Kinetics of the oxidation of L-cysteine by trans-and cis-cobalt(III) and iron(III) complexes. *Z. Naturforsch., B: J. Chem. Sci.* 61: 1346–1350.

18 Kakhki, S. and Shams, E. (2013). A new bifunctional electrochemical sensor for oxidation of cysteine and reduction of iodate. *J. Electroanal. Chem.* 704: 249–254.

19 Kakhki, S., Shams, E., and Barsan, M.M. (2013). Electrocatalytic oxidation of cysteine at a CoSalophen/n-(butyl)$_4$SiW$_{12}$O$_{40}$ carbon paste electrode. *Electroanalysis* 25: 2100–2108.

20 Li, C., Cao, R., O'Halloran, K.P. et al. (2008). Preparation, characterization and bifunctional electrocatalysis of an inorganic–organic complex with a vanadium-substituted polyoxometalate. *Electrochim. Acta* 54: 484–489.

21 Dong, T., Chen, F., Du, J., and Hu, C. (2010). A multilayer film for bifunctional electrocatalyst and electrochemical sensor based on polyoxometalate and tris (1, 10-phenanthroline) ruthenium. *J. Clust. Sci.* 21: 779–787.

22 Zhang, L., Ning, L., Zhang, Z. et al. (2015). Fabrication and electrochemical determination of l-cysteine of a composite film based on V-substituted polyoxometalates and Au@ 2Ag core–shell nanoparticles. *Sens. Actuators B Chem.* 221: 28–36.

23 Acheson, S.A., Kirkman, H.N., and Wolfenden, R. (1988). Equilibrium of 5, 6-hydration of NADH and mechanism of ATP-dependent dehydration. *Biochemistry* 27: 7371–7375.

24 Keita, B., Lu, Y.W., Nadjo, L. et al. (1999). Electrochemical and catalytic behaviour of Dawson-type complexes derived from [(1), 2, 3-P2Mo2W15O61] 10−and first transition metal ions. *J. Electroanal. Chem.* 477: 146–157.

25 Keita, B., Essaadi, K., Nadjo, L., and Desmadril, M. (1995). Rate-limiting one-electron transfer in the oxidation of NADH by polyoxometalates. *Chem. Phys. Lett.* 237: 411–418.

26 Essaadi, K., Keita, B., Nadjo, L., and Contant, R. (1994). Oxidation of NADH by oxometalates. *J. Electroanal. Chem.* 367: 275–278.

27 Keita, B., Essaadi, K., Nadjo, L. et al. (1996). Oxidation kinetics of NADH by heteropolyanions. *J. Electroanal. Chem.* 404: 271–279.

28 Keita, B., Essaadi, K., Belhouari, A. et al. (1998). Electrochemical characterization of new molecules of polyoxometallates: application to electrocatalysis. *C. R. Acad. Sci. Paris, Ser. IIc* 1: 343–350.

29 Bi, L.-H., Kortz, U., Dickman, M.H. et al. (2006). Polyoxometallate Chemistry. *J. Clust. Sci.* 17: 143–165.

30 Ammam, M., Keita, B., Nadjo, L. et al. (2010). Attempts to immobilize catalytically active substituted-heteropolytungstates in multilayer film of charged polyelectrolyte poly(allylamine hydrochloride). *J. Electroanal. Chem.* 645: 65–73.

31 Ammam, M. and Easton, E.B. (2013). Hybrid material based on 1-butyl-3-methylimidazolium tetrafluoroborate ionic liquid and Dawson polyoxoanion $[P_2Mo_{18}O_{62}]^{6-}$. Part II: further characterizations and catalytic oxidation of the NADH coenzyme. *J. Solid State Electrochem.* 17: 137–143.

32 Chu, M., Bai, Z., Zhu, D. et al. (2022). A β-nicotinamide adenine dinucleotide electrochemical sensor based on polyoxometalate built by the combination of electrodeposition and self-assembly. *J. Electroanal. Chem.* 907: 116083.

33 Zhu, D., Bai, Z., Ma, H. et al. (2020). High performance simultaneous detection of β-nicotinamide adenine dinucleotide and l-tryptophan in human serum based on a novel nanocomposite of ferroferric oxide-functionalized polyoxometalates. *Sens. Actuators B Chem.* 309: 127787.

34 Li, Y., Yang, X., Yang, F. et al. (2012). Effective immobilization of Ru (bpy) 32+ by functional composite phosphomolybdic acid anion on an electrode surface for solid-state electrochemiluminescene to sensitive determination of NADH. *Electrochim. Acta* 66: 188–192.

35 Naim, A., Chevalier, Y., Bouzidi, Y. et al. (2020). Aerobic oxidation catalyzed by polyoxometalates associated to an artificial reductase at room temperature and in water. *Inorg. Chem. Front.* 7: 2362–2369.

4

POM-based Electrocatalysts for Pharmaceutical Molecules Oxidation

Diana M. Fernandes

Universidade do Porto, Faculdade de Ciências, LAQV-REQUIMTE, Departamento de Química e Bioquímica, Rua do Campo Alegre, s/n, Porto 4169-007, Portugal

4.1 Introduction

Polyoxometalates (POMs) are a unique and fascinating class of metal–oxide clusters which have been used in numerous applications over the past decades. This is in part due to their high variability in terms of composition, shape, size and physical and chemical properties. Apart from these, POMs have the ability to undergo fast and reversible multi-valence reductions/oxidations, leading to the formation of mixed-valence species, which makes them excellent candidates for electrocatalytic applications.

POMs have the general formula $[X_xM_yO_z]^{n-}$, where X is the heteroatom (usually P, As, Si, Ge) and M is the addenda atom (V, Nb, Ta, Mo, W). The most common structures are the Keggin-type ($[XM_{12}O_{40}]^{n-}$) and the Wells–Dawson-type ($[X_2M_{18}O_{62}]^{n-}$). These anions can also be modified by substituting one or more addenda atoms with other metals, usually transition metals, leading to the metal-substituted POMs. Beside the above-mentioned capability to carry out reversible multi-valence electron transfer reactions, these clusters have high proton mobility, strong Brønsted acidity, high solubility in numerous solvents and are resistant to oxidative degradation and hydrolysis in solution.

The Keggin- and Dawson-type anions, their mixed-valence derivatives, as well as their transition-metal-substituted derivatives are the most studied POMs in aqueous media [1–3]. The heteropolyanions undergo several rapid one- and two-electron reversible reductions to produce the so-called 'heteropoly blue' and further irreversible multielectron reductions with concomitant decomposition. The electrons are accepted by the addenda atoms of the heteropolyanions, and if these addenda atoms are all identical, the electrons are delocalized on the addenda ion oxide framework at room temperature by rapid electron hopping [1].

This chapter aims to provide an overview of the enormous variety of efficient and powerful applications of free POMs and POMs-based nanocomposites used as electrode modifiers focusing on those applied to oxidative electrocatalysis of pharmaceutical molecules like dopamine, uric acid, acetaminophen and so on. The general

Applied Polyoxometalate-based Electrocatalysis, First Edition. Edited by Diana M. Fernandes.

methodologies used to immobilize them on electrodes are briefly described, and the influence of the immobilization strategy on their electrocatalytic performance is discussed.

4.2 Preparation Methods of POM-based Films and (Nano)composites

The major difficulty of using POMs as functional materials for various practical applications stems from their properties, like high water solubility and low specific surface area. Immobilizing or dispersing POMs on various supports or entrapping them into various matrices, such as conducting polymers, carbon materials, nanoparticles and metal–organic frameworks (MOFs), is expected to increase the number of active centres and should improve their electrochemical properties and enhance their electrocatalytic performances.

One of the most used strategies to prepared POM-based composites is the POM entrapment in a polymer matrix. The reversible redox chemistry and conductivity demonstrated by conductive organic polymers combined with their relatively low cost and good processability makes these materials excellent candidates for POM integration. Numerous conducting polymers have been used to prepare POM-based hybrids, but the most common are polyaniline (PANI) [4–6], polypyrrole (PPy) [7, 8], polythiophene (PT) and their derivatives [4, 9]. The most common strategies for POMs immobilization in polymeric matrices follow one of two approaches (Figure 4.1): (i) the two step one, where a polymer film is deposited on a substrate through spin coating or electropolymerization. The polymer film is then dipped in a POM solution allowing for diffusing and incorporation of the POM into the polymer matrix [10]; (ii) the one step one which involves the chemical or electrochemical oxidation of a monomer molecule to form a polymer film in the presence of a POM solution. The high ionic conductivity, strong oxidizing power, and acidic character of POMs provide the ideal conditions for the polymerization of monomers like

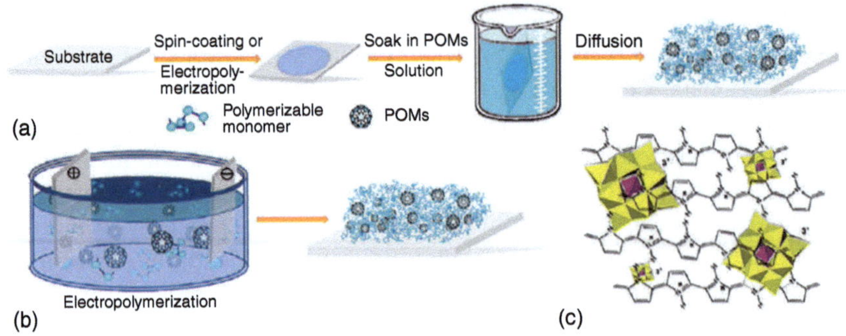

Figure 4.1 Scheme of POM–polymer matrix preparation via two-step (a) and one-step (b) methods. (c) Schematic diagram of the structure of the hybrid polypyrrole/$PMo_{12}O_{40}$. Source: Liu and Tang [10]/with permission of Elsevier.

pyrrole, thiophene and aniline [5]. In most cases the electrolyte used for monomer electrochemical polymerization is a solution containing the POM. The result is a nanostructured hybrid material with POM molecules trapped in the polymer matrix.

The above-described methods can also be adapted to prepare a wide variety of new POM-based polymer hybrids. For example, Fernandes et al. [9] prepared three new hybrids based on Keggin-type $[SiW_{12}O_{40}]^{4-}$, $[SiW_{11}O_{39}]^{8-}$, $[SiW_{11}Fe^{III}(H_2O)O_{39}]^{5-}$ or $[SiW_{11}Co^{II}(H_2O)O_{39}]^{6-}$ and conducting polymer poly(3,4-ethylenedioxythiophene) (PEDOT). The group of Gomez-Romero [6] have also used a combination of a chemical–electrochemical method and a direct electrochemical polymerization procedure to prepare PANI-POM hybrid materials.

Another method to prepare POM-based hybrid films/composites is the layer-by-layer (LbL) self-assembly method initiated by Decher et al. [11] which is based upon the electrostatic attractions between oppositely charged moieties and has proven to be a promising method for the fabrication of ultrathin multilayer films. This method is widely used due to its numerous advantages [10–13]. First, it is a low-cost, simple, environmentally friendly, wet-bench technique that allows the fabrication of very simple, mostly one-dimensionally ordered nanocomposites. Second, it provides extensive control of structure, order and functionality. Third, the materials in LbL deposition can be chosen from a wide range of materials, and the technique can be applied to solvent-accessible surfaces of almost any type and any shape, and the more exotic ones being microcapsules, colloids or biological cells. Finally, this method allows multi-material assembly of several compounds without special chemical modifications, thus giving access to multilayer films with tailored functionality and fabrication of surface-based devices.

A variety of charged materials have been employed for the construction of LbL films, including polycations, such as poly(ethylenimine) (PEI) [14–16], poly(allylamine hydrochloride) (PAH) [17] and poly(diallyldimethylammonium chloride) (PDDA) [18], biopolymers, like chitosan [19], transition-metal complexes [20] and to a much lower extent dye molecules [21–23].

The parent Keggin silicotungstate $[\alpha\text{-}SiW_{12}O_{40}]^{4-}$ has been used as the anion with chitosan [24] or polyaniline [25] as the counteraction. Lacunary Keggin-type $[SiW_{11}O_{39}]^{8-}$ and other related polyoxometalates like $[SiW_{11}Fe^{III}(H_2O)O_{39}]^{5-}$, $[SiW_{11}Co^{II}(H_2O)O_{39}]^{6-}$, $[PW_{11}Fe^{III}(H_2O)O_{39}]^{4-}$ and $[Ln(SiW_{11}O_{39})_2]^{13-}$ (Ln $(SiW_{11})_2$), Ln^{3+} = Eu, Tb, Dy have also been immobilized on glassy carbon electrodes using PEI as the counteraction (Figure 4.2) [14–16]. Dawson-type $[P_2W_{17}VO_{62}]^{7-}$ ($P_2W_{17}V$) have also been used to prepare LbL films with chitosan [26] and PEI [27]. It is important to mention that the use of polymeric cations is not mandatory for growing films with POM anions. Some examples can be found where films were prepared using the heteropolyacids $H_3PW_{12}O_{40}$, $H_3PMo_{12}O_{40}$ and $H_4SiMo_{12}O_{40}$ and methyl viologen or the cationic meso-tetra(4-*N*-methylpyridyl porphyrin) [28]. Dawson-type $[P_2Mo_{18}O_{62}]^{6-}$, $[P_2W_{17}FeO_{61}]^{7-}$ and $[P_2W_{17}CrO_{61}]^{7-}$ have also been used to prepare multilayer films with different metal cations ($NH_3^+(CH_2)_xNH_3^+$, Hg_2^{2+} or Ag^+) [29, 30].

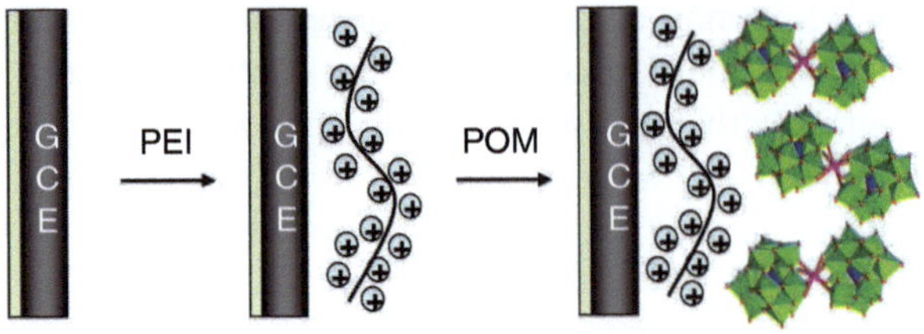

Figure 4.2 Schematic diagram of the formation of the multilayer structure via alternate adsorption of cationic PEI and anionic POM on a substrate. Source: Fernandes et al. [14]/with permission of Elsevier.

Hybrid (nano)composites based on POMs and carbon materials have also been widely reported, thus broadening the range of applications for these materials [31–33]. Simple POM-based compounds have weak electronic conductivity and low specific surface area, which significantly limits the electrocatalytic activity, so favourable conductive support materials are required to counterbalance these defects. Carbon materials have high conductivity and stability, low cost and toxicity, and more importantly, strong affinity for POMs [34]. A variety of carbon materials such as graphene (GF) [34, 35], graphene oxide (GO) [36], reduced graphene oxide (rGO) [37], carbon nanotubes (CNTs) [33–35, 38, 39], carbon black [40], ordered mesoporous carbon (OMC) [41, 42] and activated carbons (AC) [43] have been widely used as conductive support materials to prepare conductive POM-based composites.

Their preparation can be achieved by covalent or non-covalent functionalization routes. The latter can be achieved using a range of intermolecular interactions between POMs and nanocarbons, such as electrostatic interactions, π–π interactions and layer-by-layer assembly, while covalent attachment of POMs requires access to stable organo-functionalized POMs [31]. Electro-assisted and photo-assisted assembly have also been used to fabricate POM/nanocarbon composite materials [44–46].

Awaga et al. [47, 48] developed POM@carbon hybrids where the attachment of the cluster to the single-walled carbon nanotubes (SWCNT) was achieved by electrostatic interactions, as the n-Bu_4N^+ counter ions interacted with the SWCNT and created positive surface charges which were used to bind the anionic PMo_{12}. Similar work was developed by Fernandes et al. [34] where several POMs (PMo_{12}, PMo_{11}, $PMo_{11}V$ and $PMo_{10}V_2$) where immobilized on SWCNT or graphene flakes. The group of Jiang et al. [49] focused on the functionalization of CNT with cationic, protonate chitosan that was used to bind the Keggin anions PMo_{12} and PW_{12} (Figure 4.3a). In 2013, Song et al. [50] studied new modes of chemically attaching redox-active POMs to SWCNT. To this end, the authors developed a novel method to modify SWCNT with organo-functionalized $[SiW_{11}O_{39}]^{7-}$ clusters. Pyrene (Py) moieties were covalently attached to the cluster by silanol linkages, and the functionalized clusters were bound to the SWCNT through π–π stacking, allowing the assembly of Py-SiW_{11}/SWCNT/nanocomposite materials. Bonchio's group developed POM@nanocarbon materials with the

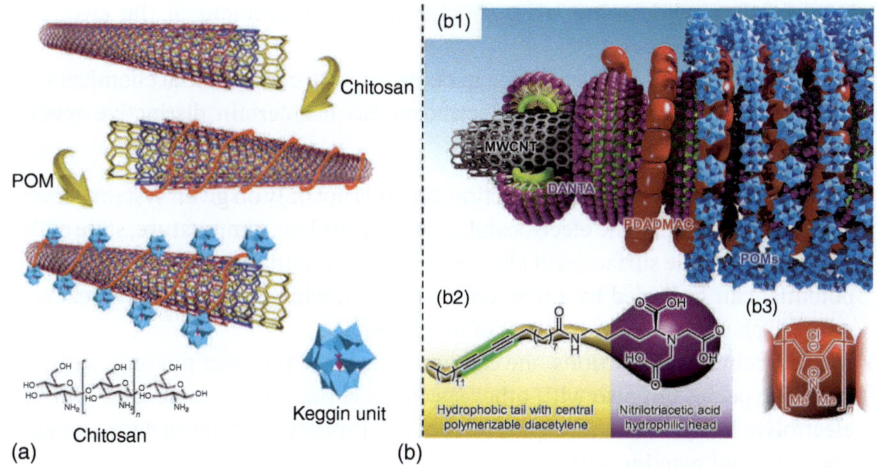

Figure 4.3 (a) Schematic diagram of the functionalization of CNTs with chitosan and Keggin-type POMs (PMo_{12} and PW_{12}). The combination of the cationic chitosan (schematically shown as a red line) and the anionic Keggin units is achieved through electrostatic interactions. Source: Ji et al. [31]/Royal Society of Chemistry/CC BY 4.0. (b) Overview of the [POM]CNT assembly (1), structure of DANTA (2) and PDDA (3). Source: Jawale et al. [39]/with permission of Elsevier.

ultimate goal to used them as water oxidation catalysts (WOC). Examples are the $(Ru_4(\mu\text{-}O)_4(\mu\text{-}OH)_2(H_2O)_4[\gamma\text{-}SiW_{10}O_{36}]_2)^{10-}$ (Ru_4POM) with few layer graphene (FLG) [51] or with SWCNT [33]. Kulesza's group [52–55] have also been actively working in the field of carbon nanotubes modified with Keggin- and Dawson-type POMs, namely $[PMo_{12}O_{40}]^{3-}$ and $[P_2W_{17}VO_{62}]^{8-}$ for different applications. More recently Mbomekallé et al. [39] have also prepared carbon nanotube–polyoxometalate nanohybrids resorting to the LbL approach and diacetylene-nitrilotriacetate (DANTA) amphiphiles bearing a polar nitrilotriacetic head group and a diacetylene lipophilic chain (Figure 4.3b).

Another very attractive type of POM-based nanocomposites are the POMOFs, in which the properties of the active support (MOF) can be enhanced by the incorporated guest (POM). These have been widely used in different fields of research ranging from heterogeneous catalysis to energy-related reactions [56]. Introducing POM into MOF through covalent or van der Waals effect can increase specific surface area, expose more active sites and enhance efficiently the POMs solubility.

4.3 POM-based Electrocatalysis

4.3.1 Electrocatalysis

The science of electrocatalysis provides the connection between the rates of electrochemical reactions and the bulk and surface properties of the electrodes on which these reactions proceed. In electrocatalysis, the major subject is redox reactions occurring on 'inert', non-consumable electrodes and involving substances dissolved in the electrolyte while there is no stoichiometric involvement of the electrode material [57].

Electrocatalytic reactions have much in common with conventional chemical heterogeneous catalytic reactions, but electrocatalysis has certain distinctive special features:

(1) The rate of an electrochemical reaction depends not only on given system parameters (composition of the electrocatalyst and electrolyte, temperature, state of the catalytic electrode surface) but also on electrode potential. In a given system, the potential can be varied by a few tenths of a volt, while as a result, the reaction rate will change by several orders of magnitude.
(2) In electrochemical reactions, the catalyst surface is in contact not only with the reacting species but also with other species (i.e. the solvent molecules and the electrolyte ions), which in turn influence the properties of the surface and give rise to special reaction features.
(3) In electrochemical systems, not only the reactant species but also electrons must be supplied to and/or withdrawn from the catalyst particles, since the electrons are directly involved in all electrochemical reactions [57].

Although strictly speaking, electrocatalysis applies to the dependence of the electrode reaction rate on the nature of the electrode material, in the following section this term will be used in a broader sense. The catalyst may be attached to the electrode surface as well as homogeneously dissolved in the electrolyte solution. The high redox activity of POMs makes them ideal for the catalytic transfer of electrons to or from a substrate. This is particularly appealing for the development of multi-electron transfer reactions which are typically hampered by high overpotentials leading to low overall conversion efficiencies. POMs are ideal candidates to address this challenge as they can be chemically modified with highly redox-active metals (e.g. Co, Mn, Ru). Different polyoxometalates have been used as oxidative electrocatalysts, and several examples can be found in the literature concerning the electrochemical oxidation of different biomolecules. In this section, we will focus on the electrochemical oxidation of dopamine, ascorbic acid, uric acid and other similar biomolecules.

4.3.2 Dopamine Oxidation

Dopamine (DA) is one of the most important catecholamine neurotransmitters since it is involved in many functions of the human central nervous system, including motor control, reward or reinforcement. Its determination is of utmost importance since abnormal levels can cause a variety of medical and behavioural problems. High dopamine levels indicate cardiotoxicity leading to rapid heart rates, hypertension and heart failure, while low levels are implicated as a major cause of several neurological diseases such as Parkinson's disease, schizophrenia, Alzheimer's disease, stress and depression [58].

Various methods have been developed to measure dopamine levels, and electrochemical one seems to be the most viable due to its robustness, selectivity, sensitivity and the possibility to achieve real-time measurements. Still, electrochemical detection is not always easy due to the presence of electroactive interfering species with similar redox potentials in real biological samples like ascorbic acid (AA) or uric acid (UA). These compounds have electrochemical activity at different solid electrodes; however, in general, at bare electrodes the oxidations are irreversible and involve undesirable high overvoltage [59]. In addition, the direct redox reactions of DA, AA and UA take place at very similar potential at bare electrodes, which results in rather poor selectivity. Consequently, modification of the surface of electrodes with suitable electrocatalysts should improve both the sensitivity and the selectivity of detection of these molecules.

Different POM-based nanocomposites and/or films have been applied for the detection of DA. A promising strategy involved the combination of POMs with metal organic frameworks (MOFs). A hybrid $PMo_{10}V_2$@MIL-101 [59] was reported by Fernandes et al., where the vanadium atoms acted as the electroactive centres. The electrocatalytic oxidation of DA was performed at $PMo_{10}V_2$@MIL-101 pyrolytic graphite (PG)-modified electrode in pH 2.5 H_2SO_4/Na_2SO_4 buffer solution containing different DA concentrations. Results showed that the oxidation peak currents for vanadium peak ($E_p \approx 0.510$ V *vs.* Ag/AgCl) increased with increasing DA concentration, suggesting that in this case the first vanadium centre acts as the active site for DA electro-oxidation. It is important to note that at bare PG electrode dopamine is oxidized at ≈0.600 V. Therefore, the overpotential was lowered to approximately 0.90 V and a catalytic efficiency (CAT) of 1465% was obtained. The CAT which is defined by the following equation: CAT = $100\% \times [i_p(\text{POM, substrate}) - i_p(\text{POM})]/i_p(\text{POM})$, where i_p (POM) and i_p (POM, substrate) are the current intensities of POMs in the absence and presence of the substrates, respectively.

Later, Liu et al. [60] used the same MOF, MIL-101(Fe), to prepare a PMo_{12}@MIL-100(Fe)@PVP composite. In this work, the POM was the parent Keggin-type $PMo_{12}O_{40}$, and the authors also used polyvinylpyrrolidone (PVP). According to the authors, the synergistic effect between the three active components (POM, MOF and PVP) allowed a higher electrical conductivity and larger specific surface area which resulted in better electrocatalytic oxidation performance of DA and UA simultaneously with a spacing of 0.146 V between the two peak positions. A limit of detection (LOD) of 0.586 μmol dm^{-3} in 0.1 M phosphate buffer solution (PBS, pH = 6.0) was achieved with linear ranges (LR) of 1–247 μmol dm^{-3}.

In 2020, Kung and co-workers [61] reported a novel heterogeneous electrocatalyst, $V_{10}O_{28}$@NU-902, based on the post-synthetical immobilization of redox-active vanadium-based polyoxometalate, $V_{10}O_{28}$, into a water-stable zirconium-based metal–organic framework, NU-902 (Figure 4.4a). The $V_{10}O_{28}$@NU-902 thin film presented a significantly larger and linear increase in current density with increasing concentration of DA at around +0.6 V, indicating that the thin films are effective for the electrocatalysis of DA oxidation (Figure 4.4b). The authors also suggested that DA was oxidized to dopamine-*o*-quinone (DOQ). Amperometric DA detection showed a sensitivity of 85 mA mM^{-1} cm^{-2} and LOD of 2.1 mmol dm^{-3} in a linear range from 25 to 400 mmol dm^{-3}.

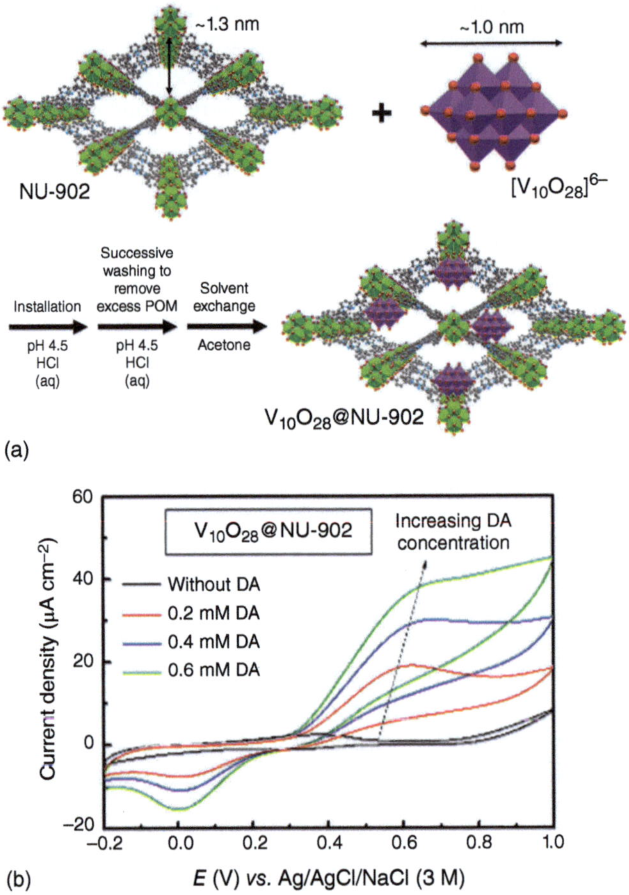

Figure 4.4 (a) Schematic representation for the installation of decavanadate into a Zr-MOF, NU-902. Green: Zr; red: O; purple: V; grey: C; blue: N. For clarity hydrogen atoms are not shown; (b) CV curves of the $V_{10}O_{28}$@NU-902 thin film, measured in the presence of various concentrations of DA. All CV curves were measured at a scan rate of $10\,mV\,s^{-1}$ with $0.05\,mol\,dm^{-3}$ MOPS/NaOH aqueous solutions (pH = 4.5) as the electrolytes. Source: Ho et al. [61]/with permission of Royal Society of Chemistry.

A basket-type 3D polyoxometalate metal–organic framework (POMOF) assembly was also applied as an electrocatalyst for DA oxidation by Zhao et al. [62]. The $P_6Mo_{18}O_{73}$ POM was successfully grafted on a lock-like metal–organic chain to generate a multi-porous coordination polymer, $[(Na(H_2O)(H_2btb))(Cu_4{}^{I}(H_2O)(pz)_5Cl)\text{-}(H_2Sr\subset P_6Mo_2{}^{V}Mo_{16}{}^{VI}O_{73})]\cdot 3H_2O$ (where pz = pyrazine and btb = 1,4-bis(1,2,4-triazole) butane). Electrocatalytic tests showed that the first oxidation peak at 0.4 V, corresponding to a two-electron molybdenum process (Mo^{VI}/Mo^{V}), increased sharply with the addition of DA and that the increased amplitude was much larger than that of the reduction peak current which suggested that DA was oxidated by the first redox pair in the $P_6Mo_{18}O_{73}$ poly(oxoanion) clusters. A CAT of 191.48% was achieved for DA concentration of 0.7 mM. In addition, a low LOD of 0.19 μM was obtained (signal-to-noise ratio (S/N = 3)) and a higher selectivity in the linear range from 0.56 μmol dm^{-3} to 0.24 mmol dm^{-3}. The authors attributed the excellent performance to the POMOF unique crystal structure. The $P_6Mo_{18}O_{73}$ bridge lock-like metal–organic chain formed an inorganic–organic network with multiple active centres, which enhanced the ion/electron delivery efficiency of the redox process. Also, the regular pores formed between the POM clusters and the metal–organic units provided both the required passageway for ion/electron transport and increased the active sites per unit area.

Hybrid materials based on POMOF and carbon materials have also been as DA electrocatalysts successfully. Fu and co-workers [63] reported a novel electrocatalyst material through the integration of a POMOF and rGO sheets via a simple one-pot approach. The as-prepared POMOF/rGO/GCE electrode (GCE = glassy carbon electrode) presented an excellent electrocatalytic activity towards DA oxidation, in pH 7.0 (0.1 mol dm^{-3} PBS solution), owing to matching functionality and structure between the POMOF and rGO. A low detection limit of 80.4×10^{-9} mol dm^{-3} (S/N = 3) was obtained in a linear range between 1×10^{-6} to 2×10^{-4} mol dm^{-3}. These results were only possible due to the promoted electron transfer, as well as abundant active sites, open channels and aromatic planes in the hybrid material. Additionally, its unique structure allowed the adsorption of DA molecules through π–π interaction with graphene as well as 1,10-phenanthroline groups of POMOF that led to a good selectivity and anti-interference performance.

An unprecedented POMOF with eight-fold meso-helical feature, $[Ag_5(trz)_4]_2[PMo_{12}O_{40}]$ abbreviated as PAZ (where trz = 1,2,4-triazol), was synthesized and applied to prepare a nanocomposite for posterior use as electrode material for dopamine oxidation [64]. The nano-PAZ (NPAZ) was combined with carboxyl-functionalized SWCNTs (SWCNTs-COOH) to improve the conductivity of PAZ and the binding ability with SWCNT. The resulting NPAZ@SWCNTs-COOH composite showed a very low detection limit of 8.6 nmol dm^{-3} (S/N = 3) over a linear range from 0.05 to 100 μmol dm^{-3}. The superior electrocatalytic performance of NPAZ@SWCNTs-COOH was attributed to different aspects (Figure 4.5): (i) the

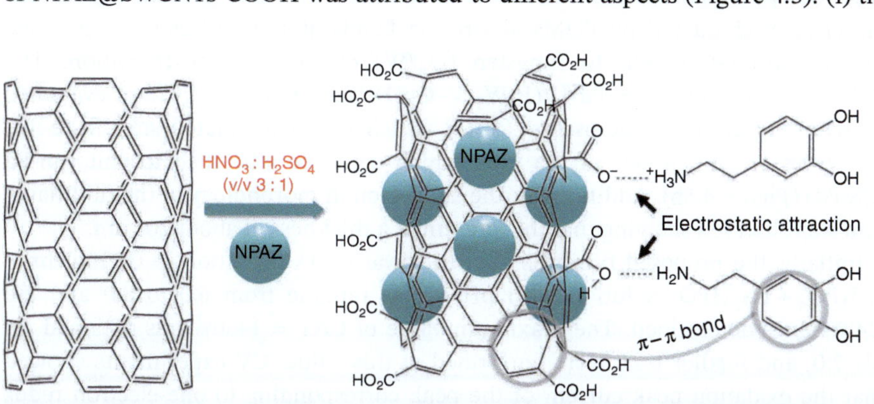

Figure 4.5 Structure of hydrogen-bonding interactions between of DA and carboxylic groups of functionalized SWCNTs. Source: Zhou et al. [64]/with permission of Springer Nature.

—COOH as functional group and anchor points improved the conductivity and the direct intermolecular charge transfer between SWCNTs-COOH and DA, (ii) the —COOH group also played additional role in the electrooxidation of DA, since DA exists as cations resulting in the stronger interaction between DA and —COOH and (iii) more defect sites in SWCNTs-COOH can pinpoint NPAZ resulting in more active centres and higher electrical conductivity leading to higher synergy effect.

The same POM ($PMo_{12}O_{40}$) was used to prepare a POMOF crystalline compound with special catenane structure, PMo_{12}[6]catenane. This was also prepared in the presence of rGO giving rise to the PMo_{12}[6]catenane/rGO. Both were used to modify GCE and evaluated towards DA oxidation. Although the first presented good results with a wide linear detection range from 1 to 90 μmol dm^{-3} and a limit of detection of 0.890 μmol dm^{-3}, the electrocatalytic behaviour was enhanced by the presence of rGO with a LR from 1 to 44 μmol dm^{-3} and a LOD of 0.065 μmol dm^{-3} (S/N = 3). The excellent performance of this composite was assigned to different factors: (1) the unique catenane structure which provides excellent stability and an electron transmission pathway; (2) the good electrochemical reversibility, fast electron transfer ability and stable redox forms of PMo_{12}[6]catenane and (3) the introduction of rGO leads to a composite with large surface area, good conductivity and abundant active sites.

As referred to in the previous section, another very popular method for POMs immobilization onto electrode surfaces is through the preparation of composite films using the LbL strategy. Wang and co-workers [65] reported the preparation of composite films by alternative deposition of phosphomolybdic acid and poly(ethyleneimine) on an indium tin oxide (ITO) glass substrate. The resulting film, PMo_{12}/PEI/ITO, showed in 0.1 mol L^{-1} acetate buffer (pH 5.0) a LOD of 0.2 μg ml^{-1} over a linear range between 5.70×10^{-4} and 3.80×10^{-2} mg ml^{-1}.

A Dawson-type vanadium tungstophosphate, α_2-$K_7P_2W_{17}VO_{62}\cdot 18H_2O$ ($P_2W_{17}V$), and CuO and Pd nanoparticles were also used for the construction of an electrochemical sensor by the LbL self-assembly method [27]. UV–vis spectroscopy (Figure 4.6a) was used to monitor the film growth due to the presence of characteristic POMs absorption bands at 199 and 292 nm, assigned to the terminal oxygen to tungsten (O_d/W) charge transfer transition. The PEI/[($P_2W_{17}V$–CuO)/(CS–Pd)]$_7$/($P_2W_{17}V$–CuO) films morphology was evaluated by scanning electron microscopy (SEM) which revealed isolated snowflake-like structures on the surface of the film, which could be attributed to chitosan-Pd (CS-Pd) (Figure 4.6b). Additionally, the cross-section morphology of the multilayer film was observed showing that this presented a thickness of about 200 nm.

Initially, the proposed film was applied towards DA oxidation in 0.2 mol dm^{-3} $KH_2PO_4 + Na_2HPO_4$ solution with pH values ranging from 6.0 to 8.0 and the CAT values determined. The maximum value of CAT = 1450% was obtained for pH 7.0, and further tests were performed at this value. CV experiments showed that the oxidation peak current of the peak corresponding to one-electron redox process (V^{V}/V^{IV}) increases with the increase of DA concentration. Amperometric response exhibited a sensitivity of 0.92 mA mM^{-1}, a low detection limit of 0.045 mmol dm^{-3} (S/N = 3), a wide linear range of 0.25–217 mmol dm^{-3} and fast

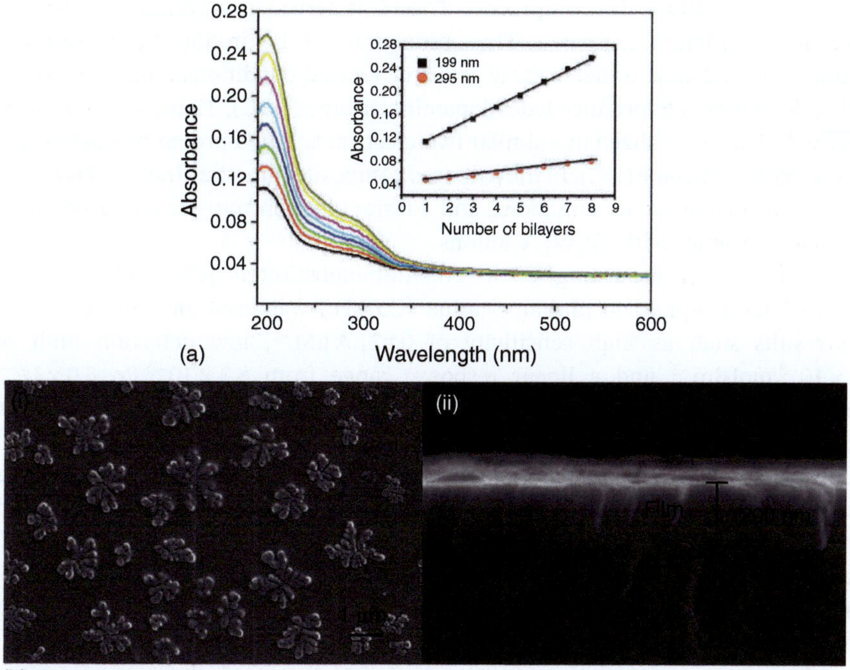

Figure 4.6 (a) UV–vis spectra of the PEI/[(P_2W_{17}V–CuO)/(CS–Pd)]$_n$ film with $n = 1–8$ (from bottom to top) for the quartz substrate (on both sides). The inset: plots of the absorbance values at 199 and 295 nm against bilayer number n. (b)The surface (i) and cross-section (ii) SEM images of the composite film PEI/[(P_2W_{17}V–CuO)/(CS–Pd)]$_7$/(P_2W_{17}V–CuO) on an ITO-coated glass. Source: Adapted from Zhang et al. [27]/Royal Society of Chemistry.

response (<2 seconds). The good performance of the proposed film was attributed to the synergistic effects among the three active species. On the one hand, CuO and CS–Pd nanoparticles with high surface-to-volume ratio could increase immobilized amount of POM with inherent outstanding redox activity on the surface of the modified electrode to provide more active sites for DA molecules to adsorb and react. On the other hand, highly conductive CuO and Pd nanoparticles could decrease the charge transfer resistance at the surface of electrode and accelerate electron transfer rate facilitating the electron transfer between DA and the electrode surface.

The same POM (P_2W_{17}V) and cationic chitosan were also used for the construction of LbL films [26]. The composite film-modified electrode presented excellent electrochemical electrocatalytic activity towards DA, in 0.05 M PBS (pH = 7.0), with a linear response between 1.25×10^{-8} M and 3.04×10^{-4} mol dm^{-3}, a low detection limit of 0.18 μmol dm^{-3} (S/N = 3), a high sensitivity of 0.23 μA μM^{-1} and fast response time (2.2 s). The CAT values for the (PEI/[P_2W_{17}V/CS]$_n$/P_2W_{17}V) composite films with n = 4, 6, 8, 10 and 12 were determined, for a DA concentration of 0.05 mmol dm^{-3}, achieving 192.61%, 532.48%, 316.63%, 385.64% and 277.37%, respectively. According to the authors, DA is transformed into dopamine ortho-quinone (DOQ) through a two-proton and two-electron process. The POM served as a proton and electron reservoir to mediate this process. The amine group of the produced DOQ exists in its unprotonated state under weakly acidic or neutral conditions and undergoes a cyclization process to produce leucodopaminechrome (LDC). Then, at less positive potentials, LDC is oxidized in a similar two-electron and two-proton process to produce dopaminechrome (DC). In the prepared composite film, the kinetics of the electrochemical oxidation are enhanced, and a faster electron transfer rate is obtained due to the presence of the $P_2W_{17}V$ anions.

Composite film consisting of a vanadium-substituted $[PMo_{11}VO_{40}]^{4-}$ and cobalt(II) tetrasulphonate phthalocyanine (CoTsPc) was used and showed excellent results such as high sensitivity of $0.82\,\mu A\,\mu M^{-1}$, low detection limit of $1.3\times10^{-8}\,mol\,dm^{-3}$ and a linear response range from 8.3×10^{-8} to $5.0\times10^{-4}\,mol\,dm^{-3}$ [66]. The film exhibited enhanced electrocatalytic performance for the oxidation of dopamine, which was related to the V^{V}/V^{IV} and Co^{II}/Co^{III} couples. The authors also observed that for $[PEI/PMo_{11}V/PEI/CoTsPc]_4$ the catalytic efficiency was 482.3%. Another example is the composite film based on the tri-vanadium-substituted phosphomolybdate $[PMo_9V_3O_{40}]^{6-}$ (PMo_9V_3) decorated by platinum nanoparticles (Pt) [67]. The films showed high sensitivity of $1.2\,\mu M\,\mu A^{-1}$ dopamine, fast response time (<1 s), low detection limit ($\sim1.3\times10^{-7}\,mol\,dm^{-3}$) and a linear range from 4.2×10^{-7} to $1.3\times10^{-3}\,mol\,dm^{-3}$ towards amperometric sensing of dopamine, as well as no interference from the common interfering species such as ascorbic acid, glucose, uric acid and l-cysteine.

A Dawson-type $[P_2W_{16}V_2O_{62}]^{8-}$ decorated by Au–Pd alloy nanoparticles (Au–Pd) was also used to fabricated LbL film onto an ITO electrode [68]. The composite film was employed for sensitive and simultaneous determination of dopamine and ascorbic acid at biological pH (pH 7.0). Linear analytical curves were obtained in the ranges from 2.1×10^{-6} to $2.06\times10^{-3}\,mol\,dm^{-3}$ and 1.2×10^{-6} to $1.61\times10^{-3}\,mol\,dm^{-3}$ for dopamine and ascorbic acid by differential pulse voltammetry (DPV) methods, respectively. The low detection limits for dopamine and ascorbic acid were 8.3×10^{-7} and $4.3\times10^{-7}\,mol\,dm^{-3}$, and no interference was observed from the common interfering species such as glucose, uric acid, L-cysteine, CH_3OH, CH_3CH_2OH and H_2O_2. Additionally, the CAT values obtained were 288.6% and 345.19% for DA and AA, respectively.

Hybrid (nano)composites based on POMs and carbon materials have also been widely reported. Fernandes et al. [34] prepared a $PMo_{11}V$@GF/GCE, and the square wave voltammetry (SWV) experiments showed that this modified electrode was able to separate the peaks corresponding to DA and AA that at bare GCE and $PMo_{11}V$-modified electrodes appeared as one peak (Figure 4.7a). SWV was further used to determine DA in a solution containing a constant AA concentration (Figure 4.7b) and detection limit of $8.8\times10^{-7}\,mol\,dm^{-3}$ was obtained with a $LR = 2\times10^{-6}$ to $3\times10^{-4}\,mol\,dm^{-3}$.

The latest example, from 2023, of a two-component hybrid reports the synthesis of carbon black/phosphotungstic acid composites by doping PW_{12} on carbon black nanoparticles by impregnation method [40]. Differential pulse voltammetry was used to determine DA, and the calibration plot showed linearity over a

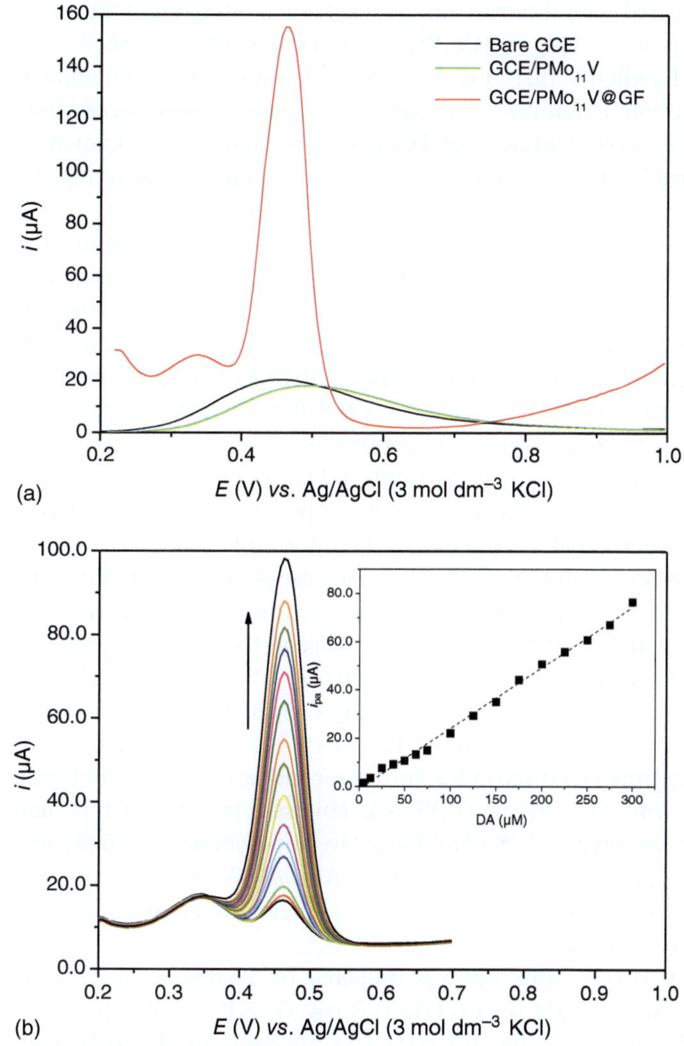

Figure 4.7 (a) SWV responses of 5 mM AA and 0.05 mM DA in pH 2.5 buffer solution at bare GCE (black), GCE/$PMo_{11}V$ (green) and GCE/$PMo_{11}V$@GF (red). (b) SWV responses in the presence of 0.01 mM AA for concentrations of DA from 0.002 to 0.3 mM at GCE/$PMo_{11}V$@GF in pH 2.5 buffer solution; inset shows i_{pa} *vs.* DA concentration at 0.463 V. Source: Fernandes and Freire [34]/with permission of John Wiley & Sons.

concentration of DA between 10 μmol dm^{-3} and 600 μmol dm^{-3} with a LOD of 0.1 μmol dm^{-3}.

Wang and co-workers used a vanadium-substituted POM ($PMo_{10}V_2$) to prepare a hybrid composite with graphene oxide and PDDA by self-assembly and electrodeposition methods [69]. The resulting $PMo_{10}V_2$/GO/PDDA/GCE electrode showed a LOD of 0.18 μmol dm^{-3} over a linear range of 10 μmol dm^{-3} to 5.0×10^4 μmol dm^{-3}. The good behaviour was assigned to the addition of GO which provided favourable sites for electron transfer to probes and to the high amounts of POM on the modified electrode.

Zhao et al. [70] prepared a novel cerium-encapsulated selenotungstate comprising three POM building units $Na_{16}H_6([Ce_3W_4O_{10}(H_2O)_9\text{-}(CH_3COO)_3]_2(Se_2W_7O_{30})(B\text{-}\alpha\text{-}SeW_9O_{33})_4)\cdot(C_5H_8NBO_3)\cdot 119H_2O$ and combined it with carboxyl-functionalized multiwalled carbon nanotube. The composite material was successfully applied in the simultaneous detection of DA and paracetamol (also known as acetaminophen) with a LOD of 0.053 μmol dm^{-3} in the linear range of concentration of 4–100 μmol dm^{-3}.

Ma group have also reported two works on POMs with carbon materials and metal nanoparticles [71, 72]. In the first, an ITO electrode was modified with PMo_9V_3, multiwalled carbon nanotubes and Pd–Pt alloy nanoparticles (Pd@Pt) LbL self-assembly method [71]. The authors studied the effect of different loadings of Pd@Pt (10 : 1, 5 : 1, 2 : 1, 1 : 1 for variational ratio of PMo_9V_3 to Pd@Pt) and layer numbers in $(PMo_9V_3\text{-Pd@Pt/CNTs})_n$/ITO ($n$ = 4, 5, 6, 7, 8, 9, 10). The decrease of the ratio of POM to Pd@Pt leads to a significant current increase with 1 : 1 presenting the highest current response. Regarding the number of layers, there was an increase of response up to n = 6 and after this the current drops. As the number of layers increases, the loading amount of all components becomes larger, which resulted in the increase of CAT; however, this also leads to an increase of film thickness turning the path of the electron exchange from the surface of the composite film to the electrode surface longer, which ultimately led to a decrease in the electron tunnelling probability and electron transfer rate. The optimized $(PMo_9V_3\text{-Pd@Pt/CNTs})_6$/ITO electrode was applied towards DA oxidation achieving a sensitivity of 0.690 μA μM^{-1}, a LOD of 1.25×10^{-8} mol dm^{-3} over a wide LR (2.50×10^{-8} to 1.80×10^{-4} mol dm^{-3}) and a short response time (<1.0 seconds). In the second, a composite film with Dawson-type $\alpha\text{-}K_7P_2VW_{17}O_{62}\cdot 18H_2O$ ($P_2W_{17}V$), chitosan modified CNTs and AuCo nanoparticles was prepared by combining layer-by-layer self-assembly and electrodeposition [72]. The PEI/$P_2W_{17}V$/CNTs-CS/AuCo/$P_2W_{17}V$ composite film showed a detection limit of 0.15 μmol dm^{-3} and a linear detection range from 1.25×10^{-6} to 2.81×10^{-4} mol dm^{-3}.

A PVIM–Co_5POM/NCNT composite was prepared by Thakur et al. [73], by combining the sandwich POM $[WCo_3(H_2O)_2(CoW_9O_{34})_2]^{12-}$ with polyionic liquid poly(vinylimidazolium) cation [$PVIM^+$] and nitrogen-doped carbon nanotubes (N-CNTs). This novel composite demonstrated a superior selectivity and sensitivity with the lowest OD of 0.0005 μmol dm^{-3} and a wide linear range of 0.0005–600 μmol dm^{-3} even in the presence of high concentrations of AA (500 μmol dm^{-3}). Shafiei and Darzi also combined an ionic liquid (IL) with POM ($PW_{12}O_{40}$), reduced graphene oxide and Ag nanoparticles to modify a carbon paste electrode (CPE) [74]. The Ag@POM@rGO-IL/CPE showed good electrocatalytic response for the determination of DA with a LOD of 0.017 μmol dm^{-3} and linear range of 0.05–115.04 μmol dm^{-3} in 0.1 M phosphate buffer solution (pH 7.0).

A 3D porous electrode material, RGO-CD-MWCNT-POM nanocomposite, was synthesized by assembly of graphene oxide, β-cyclodextrin (CD) and multiwall carbon nanotube (MWCNT), with subsequent reduction of GO in the assistance of polyoxometalate ($PW_{12}O_{40}$) [75]. Due to the different binding properties of CD towards different analytes, the proposed electrode material was used for the simultaneous determination of DA, AA and uric acid (UA). The combination of the advantages of each component led to a composite with excellent electrochemical sensing performances. Under optimized conditions, the LOD obtained were 0.04, 0.84 and 0.05 μmol dm^{-3} for DA, AA and UA, respectively, with linear ranges of 0.5--300, 5--2000 and 1–400 μmol dm^{-3}.

4.3.3 Ascorbic Acid Oxidation

Ascorbic acid, also known as vitamin C, is a hexonic sugar acid with two dissociable protons (pK_a 4.1 and 11.6), is an important biomolecule present in mammalian brain and is a vital component in the human diet. It is considered critical for scavenging endogenously generated free radicals, serving as one of the most important neuroprotectors for various oxidative stress-associated brain dysfunctions such as cerebral ischemia injury, epilepsy and Parkinson's disease. AA is also considered as sort of neuromodulator and cofactor of enzymes involved in both the synthesis of neurotransmitters and the modulation of glutamate-mediated neurotransmission. Given that it can be found in many fruits, vegetables and beverages, its determination is essential for several industrial companies especially those related to pharmaceuticals, clinical studies and food studies. AA has been used for the prevention and treatment of common cold, mental illness, infertility, cancer and AIDS [76].

Although several POM-based composites/films have been explored as electrocatalysts for AA oxidation, the use of POMs alone is rare. In fact, only two works can be found in the literature concerning the AA oxidation by POMs. One refers to a phosphomolybdovanadate ($PMo_{10}V_2$), and it is used to compare with the results obtained by the same POM immobilized at a metal organic framework [59], while the other reports on the synthesis and application of a tri-vanadium-capped Keggin phosphomolybdate $[Na_3PMo_{12}V_3O_{43}]\cdot 4H_2O$ [77]. The latter presented a CAT of approximately 160% for a AA concentration of 2.5 mmol dm^{-3}, while the former showed a decrease in the overpotential of approximately 0.150 V *vs.* Ag/AgCl and a CAT = 734%.

In contrast, several POM-based composites have been used as electrocatalysts for AA oxidation. Fernandes et al. prepared two composite materials based on the 3D structure of MIL-101(Cr) and two POMs, $[SiW_{11}Fe^{III}(H_2O)O_{39}]^{5-}$ [78] and $[PMo_{10}V_2]^{5-}$ [59], and immobilized them at pyrolytic graphite electrodes. Both composite materials proved to be excellent electrocatalysts towards AA oxidation, with a significant decrease in the overpotential (0.087 and 0.149 V) and CAT values of 833.0% and 1009%, respectively. The linear range for the catalytic current extends up to an AA concentration of 3×10^{-4} mol dm^{-3}, with a detection limit of 3.07×10^{-5} mol dm^{-3} for the $SiW_{11}Fe$@MIL-101 and 9.0×10^{-4} mol dm^{-3}, with a detection limit of 2.75×10^{-6} mol dm^{-3} for the $PMo_{10}V_2$@MIL-101.

Zhou group have also prepared different POMOFs and applied them as AA oxidation electrocatalysts [79–81]. One of the works used different metal–organic units which were introduced into the (PMo_{12}) POM system to yield three porous coordination polymers with distinct characteristics, $(Cu(pra)_2)[(Cu\text{-}(pra)_2)_3(PMo_{11}{}^{VI}Mo^{V}O_{40})]$ (1), $[(Ag_5(pz)_6(H_2O)_{0.5}Cl)(PMo_{11}{}^{VI}Mo^{V}O_{40})]$ (2) and $[(Cu_3(bpz)_5(H_2O))$

$(PMo_{12}O_{40})]$ (3) (pra = pyrazole; pz = pyrazine; bpz = benzopyrazine), by an in situ hydrothermal method [79]. The three POMOFs presented values of CAT between ~55% and 110% with compound (2) presenting the best results. This was further used as AA sensor achieving a sensitivity and detection limit of 244.38 μA mM^{-1} cm^{-2} and 0.08 μmol dm^{-3}, respectively, with a LR of AA concentration of 0.25–192.8 mol dm^{-3}. In the other two, Wells–Dawson-type POMs were used ($As_2W_{18}O_{62}$ [80] and $P_2W_{18}O_{62}$ [81]) with rigid short ligands (bim = 2,2′-biimidazole; pz = pyrazine) and flexible ligands (bip = 1,5-bis (imidazole-1-yl)pentane; bipy = 2,2′-bipy), respectively. All compounds prepared presented good electrocatalytic performance towards AA oxidation.

A Lindquist-type POM $[Co(Mo_6O_{19})]$ has also been used together with N-donor ligand 9,10-bis(imidazol-1-ylmethyl)anthracene (9,10-BIMA) to prepare a POMOF [82]. The prepared material exhibited good electrocatalytic performance in the AA concentration range of 10–100 μmol dm^{-3} with a LOD of 2.5 μmol dm^{-3}. In this case the redox process of Co^{III}/Co^{II} was the active centre for AA oxidation.

Many other examples of organic–inorganic hybrid compounds are reported such as $[Zn(phen)_2(H_2O)]_2[VW_{12}O_{40}]$, proposed by Ma's group [83], where the vanadium is the active centre for AA oxidation; the $[BMIM]_6P_2Mo_{18}O_{62}$ [84] hybrid based on Wells–Dawson $[P_2Mo_{18}O_{62}]^{6-}$ and a room temperature ionic liquid 1-butyl-3-methylimidazolium tetrafluoroborate ($[BMIM][BF_4]$) that showed a sensitivity of ~63 nA/μM AA, fast response time (<9 s), low detection limit ($<0.1\times10^{-6}$ mol dm^{-3}) and high selectivity towards endogenous interferences such as uric acid, acetaminophen and dopamine; the $H[Cd(L)(CrMo_6(OH)_6O_{18})]\cdot2H_2O$ and $H[Co(L)(CrMo_6(OH)_6O_{18})(H_2O)_2]\cdot3H_2O$ [85] (L = *N*,*N*′-bis(3-pyridinecarboxamide)-1,4 cyclohexane) with LOD of 0.018 mmol dm^{-3} and 0.041 mmol dm^{-3}, respectively, and the $[Cu_2(Htib)_2(\alpha\text{-}SiW_{12}O_{40})]$, $[Cu_2(Htib)_2(H_2tib)_2(H_2O)_4(\alpha\text{-}HBW_{12}O_{40})_2]\cdot8H_2O$ and $[Cu_2(H_2tib)_4(H_2O)_2(\alpha\text{-}HP_2W_{18}O_{62})_2]\cdot14H_2O$ prepared by Zhou et al [86]. The latter were employed to fabricate bulk-modified carbon paste electrode, and the results showed that the Keggin-type α-SiW_{12} hybrid exhibits better performance towards electrocatalytic oxidation of AA than that for Dawson-type hybrids.

The Peng group has also studied the effect of different POM structures on the AA electrocatalytic performance of organic–inorganic hybrid compounds [87–92]. These studies included isopolyanions and heteropolyanions and within the latter the Keggin and Wells–Dawson types. In one of the first works, they observed that the $[Mo_3O_{10}]^{2-}$-based organic–inorganic hybrid exhibited better electrocatalytic performances than $[Mo_8O_{26}]^{4-}$-type hybrids towards AA oxidation [87]. In another work, results showed that $[V_2O_6]^{2-}$ hybrid outperformed β-$[Mo_8O_{26}]^{4-}$ using the same Ag-bimpy segments (bimpy = 3,5-bis(1-imidazoly) pyridine) [88]. Cu–bimpy complexes were also prepared using $[\alpha\text{-}BW_{12}O_{40}]^{5-}$, $[\alpha\text{-}SiW_{12}O_{40}]^{4-}$, $[\alpha\text{-}PMo_{12}O_{40}]^{3-}$ and $[\alpha\text{-}HPW_{12}O_{40}]^{3-}$ [91]. Results showed that all presented electrocatalytic activity for the oxidation of AA in the following order: $[\alpha\text{-}HPW_{12}O_{40}]^{3-} > [\alpha\text{-}BW_{12}O_{40}]^{5-} > [\alpha\text{-}SiW_{12}O_{40}]^{4-} > [\alpha\text{-}PMo_{12}O_{40}]^{3-}$. In the following work, the authors explored the use of Keggin-type $[\alpha\text{-}PW_{12}O_{40}]^{3-}$, bi-capped Keggin $[HPMo^{VI}{}_8V^{IV}{}_4O_{40}(V^{IV}O)_2]$ and Wells–Dawson-type $[\alpha\text{-}P_2Mo_{18}O_{62}]^{6-}$ [92].

The results showed that the PW_{12}–Keggin-type hybrid exhibits better performance than bi-Keggin-type and Wells–Dawson hybrids.

Several POMs have also been immobilized on carbon materials with the ultimate goal to serve as efficient electrocatalysts. In 2013, aniline containing suspended silicotungstic acid and carbon nanotubes was electropolymerized on the surface of a GCE in a single step, and the SiW_{12}-CNTs/GCE showed a detection limit of 5.1×10^{-7} mol dm^{-3} (LR: 1×10^{-6} to 9×10^{-3} mol dm^{-3}) [93]. A modified electrode with PMo_{12} and graphene was also applied to the AA oxidation, and it showed a decrease in overpotential of ~0.210 mV and a LOD = 5×10^{-7} mol dm^{-3} (LR: 1×10^{-6} to 8×10^{-3} mol dm^{-3}) [94] which are very similar results with the ones obtained for SiW_{12}-CNTs/GCE.

In 2015, Fernandes et al [34]. immobilized several phosphomolybdates (PMo_{12}, PMo_{11}, $PMo_{11}V$ and $PMo_{10}V_2$) on single-walled carbon nanotubes or graphene flakes. Initial electrochemical tests showed that all the peaks were much better resolved and had higher current intensities in both types of nanocomposite-modified electrodes compared with the analogous POM-modified electrodes (Figure 4.8), suggesting faster electron transfer kinetics, which is associated with the exceptional electronic properties of SWCNTs and graphene layers. Regarding electrocatalytic performance, except for compounds with PMo_{11}, all the other presented excellent electrocatalytic activities towards AA oxidation. The authors observed that for the composites containing $PMo_{11}V$ and $PMo_{10}V_2$ there was an increase of vanadium peak current with increasing AA concentration suggesting that the vanadium centres acted as the active site for AA electrooxidation. For the PMo_{12}-containing composites the active site for AA was the Mo centre at more positive potentials. The CAT values obtained for PMo_{12}@SWCNT, $PMo_{11}V$@SWCNT, $PMo_{10}V_2$@SWCNT, PMo_{12}@GF, $PMo_{11}V$@GF and $PMo_{10}V_2$@GF were 209.0, 88.3, 105.1, 93.0, 666.3 and 273.7%, respectively, with LR of 5–500, 5–450, 5–925, 5–850, 5–800 and 250–925 μmol dm^{-3}.

Muqaddas et al. [95] have also prepared an electrochemical sensor based on Wells–Dawson type cobalt-based polyoxometalate (Co-POM, $K_8[CoP_2W_{17}O_{61}.H_2O]\cdot 16H_2O$) and fibre-shaped flexible CNTs fibre (Co-POM/CNTFs). The modified electrode showed fast response time (5 seconds), high sensitivity of 300 μA mM^{-1} cm^{-2}, low LOD of 0.10 μmol dm^{-3} and broad linear range of up 10 mmol dm^{-3}. The efficient heterogenization of Co-POM catalyst over CNTs fibre provided a synergistic effect, enhancing the electron transport at the electrode/electrolyte interface.

Layer-by-layer films were also used where depending on the POM employed the active centre could be either tungsten, molybdenum or vanadium. Examples are the (PEI/RuSiW$_{10}$)$_n$/ITO ($Rb_8K_2[(Ru_4O_4(OH)_2(H_2O)_4)(\gamma\text{-}SiW_{10}O_{36})_2]\cdot 25H_2O$) with LOD = 8×10^{-5} mol dm^{-3} [96], the [PR/$PMo_{11}VO_{40}$]$_n$/GCE (PR-pararosaniline) with LOD = 1.2×10^{-6} mol dm^{-3} and CAT = 3879% [22] and the ($SiNi(H_2O)W_{11}O_{39}$)/cysteamine/gold electrode (LOD = 14.6×10^{-3} mol dm^{-3}) [97].

Wang and co-workers [65] reported on the preparation of composite films by alternative deposition of phosphomolybdic acid and poly(ethyleneimine) on an ITO glass substrate. The resulting film, PMo_{12}/PEI/ITO, showed in 0.1 mol L^{-1} acetate buffer (pH 5.0) a LOD of 0.43 μg ml^{-1} over a linear range between 8.80×10^{-4}

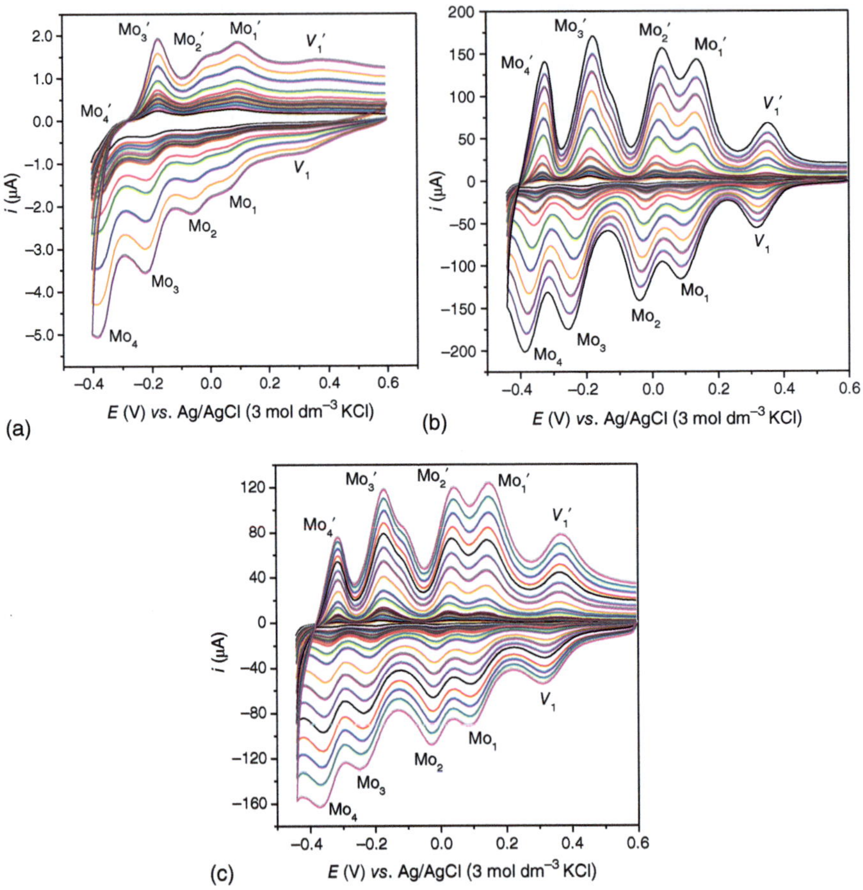

Figure 4.8 Cyclic voltammograms in pH 2.5 H_2SO_4/Na_2SO_4 buffer solution at different scan rates from 0.03 to 0.8 V s^{-1} of (a) $PMo_{11}V$, (b) $PMo_{11}V$@SWCNT and (c) $PMo_{11}V$@GF. Source: Fernandes and Freire [34]/with permission of John Wiley & Sons.

and 4.30×10^{-2} mg ml^{-1}. Wang et al. [69] prepared a hybrid composite based on vanadium-substituted POM ($PMo_{10}V_2$), graphene oxide and PDDA by self-assembly and electrodeposition methods [69]. The resulting $PMo_{10}V_2$/GO/PDDA/GCE electrode showed a LOD of 0.03 μmol dm^{-3} over a linear range of 10 μmol dm^{-3} to 1.0×10^4 μmol dm^{-3}.

In 2020, Ma and co-workers [98] prepared a composite film based on Keggin-type vanadium-substituted POM (PMo_9V_3) decorated by graphene and tris(2.2-bipyridine) ruthenium(II) [$Ru(bpy)_3^{2+}$] using the LbL self-assembly method. The films presented good sensing performance to AA in 0.2 mol dm^{-3} PBS (pH = 6.5), with a linear range from 7.5×10^{-7} to 2.07×10^{-4} mol dm^{-3}, a LOD = 1.0×10^{-7} mol dm^{-3} and sensitivity of 0.24 μM $μA^{-1}$. In the following year, the same authors used a similar approach to construct films based on $H_7P_2Mo_{17}VO_{62}$ ($P_2Mo_{17}V$), [$Ru(bpy)_3^{2+}$] and chitosan-palladium (CS-Pd) [99]. The effect of the number of layers (n = 2, 4, 6, 8, 10) on the $(P_2Mo_{17}V/Ru(bpy)_3/CS\text{-}Pd)_n$

performance was evaluated. The authors observed that electrocatalytic current increased with the increase of n reaching a maximum at $n = 6$. For n higher than 6 there was a decrease in the current, which can be justified by the increase of film thickness hindering the electron transfer on the sensor surface. The optimized film displayed excellent sensing performance towards AA, in 0.2 M PBS (pH = 7.0), with a fast response time (<2 s), a wide linear range of 0.125–118 μmol dm^{-3} and a low LOD of 0.1 μmol dm^{-3} (S/N = 3). Different aspects were pointed out by the authors as the reason for such a good performance. First, when pH is higher than 4.1, most AA molecules are negatively charged due to the protons lost, and positively charged CS-Pd nanoparticles facilitate their adsorption via electrostatic bonding enhancing electrocatalytic activity. Second, CS-Pd nanoparticles with high surface-to-volume ratio increased the amount of immobilized POM. Third, the presence of highly conductive Pd nanoparticles might decrease the charge transfer resistance and accelerate the electron transfer rate. Fourth, $Ru(bpy)_3$ acts as a mediator to form an electrical network between $P_2Mo_{17}V$ and CS-Pd, greatly promoting electron transfer from the modified electrode to AA molecules during the electrocatalytic reaction. Fifth, POMs anions have reversible redox activity, which can proceed with fast, reversible electron transfer.

4.3.4 Other Molecules

POM-based composites and films have also been explored as electrocatalysts for the oxidation of other biomolecules like uric acid and acetaminophen (APAP) known commonly as paracetamol. UA is the primary end product of purine metabolism [100]. It has been shown that extreme abnormalities of UA levels are symptoms of several diseases, such as gout and hyperuricemia; therefore, their determination is important.

Ma group [101] developed composite films based on reduced graphene oxide functionalized with PDDA (PDDA-rGO) and anionic Au nanoparticles mixed with polyoxometalates clusters $K_8P_2W_{16}V_2O_{62}$ ($P_2W_{16}V_2$-Au) using the LbL method. The resulting PEI/[$P_2W_{16}V_2$-Au/PDDA-rGO]$_8$ composite film showed a CAT of 400.2% for a concentration of UA = 0.05 mmol dm^{-3}, a LOD of 8.0×10^{-8} mol dm^{-3} (S/N = 3) in 0.2 M PBS (pH = 7.0) and a linear range from 2.5×10^{-7} to 1.5×10^{-4} mol dm^{-3}. The rGO and Au nanoparticles with large surface-to-volume ratio and high electro-conductivity allowed the immobilization of higher amounts of POM clusters with high electrocatalytic activity for UA oxidation. Additionally, the π–π interactions of the rGO layers, and the electrical network formed between these and the $P_2W_{16}V_2$-Au improved electron transfer between electrode and UA molecules. The same group reported, in 2020, a PEI/$P_2W_{17}V$/CNTs-CS/AuCo/$P_2W_{17}V$ composite film using a similar approach [72]. In this case, chitosan-modified CNTs and AuCo nanoparticles were combined with Dawson-type α-$K_7P_2VW_{17}O_{62}\cdot 18H_2O$ ($P_2W_{17}V$). The film showed a LR = 0.75×10^{-6} to 1.66×10^{-4} mol dm^{-3}, LOD = 0.25 μmol dm^{-3} (S/N = 3) in 0.2 M pH = 7.4 PBS.

At the same time, Feng and co-workers [102] fabricated a composite film with rGO, gold nanoparticles (Au NPs) and $P_2W_{18}O_{62}$ by combining electrodeposition and electrochemical reduction in situ processes. The reduced POM was used as reducing agents and bridging molecules for rGO and Au nanoparticles and the encapsulating molecules for the Au NPs. The film with only one monolayer showed impressive results with LR from 0.5 to 450 μmol dm^{-3}, a detection limit of 0.15 μmol dm^{-3} and a sensitivity of 0.487 μA $μM^{-1}$.

Instead of films, Sha group [103] prepared POM-based organic–inorganic hybrids with electrocatalytic activity for UA. In one work, the authors prepared a novel $(H_2bix)_3(P_2W_{18}O_{62})\cdot H_2O$ ($bixP_2W_{18}$) with bix = 1,4-bis(imidazol-1-ylmethyl) benzene. The $bixP_2W_{18}$-based sensor showed a LOD of 5.85×10^{-7} mol dm^{-3} (S/N = 3) in 0.1 mol dm^{-3} PBS (pH 3) and a LR = 2.5×10^{-7} to 6.99×10^{-4} mol dm^{-3}. In the other, a $H_{10}[Cu(bix)_2(P_2W_{18}O_{62})_2](bix)_6\cdot16H_2O$ hybrid was prepared with a LOD of 4.97×10^{-7} mol dm^{-3} (S/N = 3) and a LR = 2.5×10^{-7} to 6.79×10^{-4} mol dm^{-3} [104].

Liu et al. [60] combined a MOF, MIL-101(Fe) with parent Keggin-type PMo_{12} and PVP to prepare a PMo_{12}@MIL-100(Fe)@PVP composite. The resulting material presented electrocatalytic activity towards UA oxidation with a LOD of 0.372 μmol dm^{-3} in 0.1 M PBS (pH = 6.0) with linear range of 5–406 μmol dm^{-3}.

Fernandes et al. [105] prepared a new nanocomposite by immobilizing $PMo_{11}V$ into nitrogen-doped carbon nanotubes (CNT-N). The $PMo_{11}V$@CNT-N nanocomposite was evaluated with success, as an electrochemical sensor, in the detection of acetaminophen in the presence of tryptophan. Acetaminophen is the most widely used analgesic and antipyretic drug in the world, usually used for pain relief and fever. When used appropriately, the side effects are rare, but overdoses of APAP can cause liver and kidney damage and possibly mortality. So, its determination alone and/or in mixtures is important to control not only the quality of pharmaceutical formulations but also their amount in biological fluids (blood and urine). The authors observed that anodic peak current of APAP increases linearly with the increasing concentration of APAP from 1.5×10^{-6} to 3.9×10^{-4} mol dm^{-3} in the presence of tryptophan, leading to a LOD of 1.0×10^{-6} mol dm^{-3}. A ternary nanocomposite of ferrocene-based reduced graphene oxide (Fc-rGO) with PMo_{12} (Fc-rGO/PMo_{12}, rGFP-n) was also applied for the determination of APAP [106]. A LOD of 13.27 nmol dm^{-3} (S/N = 3) and a sensitivity of 36.81 μA $μM^{-1}$ cm^{-2} were achieved with a LR from 1×10^{-6} to 1×10^{-3} mol dm^{-3}. Shafiei and Darzi used a similar POM (PW_{12}) together with rGO and Ag nanoparticles to modify a carbon paste electrode [74]. The Ag@POM@rGO-IL/CPE showed good electrocatalytic response for the determination of APAP with a LOD of 0.037 μmol dm^{-3} and linear range of 0.1–137.9 μmol dm^{-3} in 0.1 mol dm^{-3} PBS (pH 7.0).

4.4 Conclusions

After a long period of progress, POM-based (nano)composites have become a very important class of promising materials in electrochemical sensor field due to their exceptional electrochemical performances. A huge variety of modified electrodes with POM-based composites and films have been explored in the oxidation of different molecules such as dopamine, ascorbic acid and uric acid, among others. In general, Keggin and Dawson-type POMs are the most studied ones with the W-, Mo- and V-centres of the POM moieties being responsible for the electrocatalytic activities in sensor applications.

The construction of POMs-based composites with conductive functional materials, like carbon materials, conducting polymers and metal nanoparticles, is highly advantageous as it enhances the conductivity of the composite material and allows the immobilization of more POM active sites leading to better performances and higher sensibility. Still, the immobilization of POMs on carbon materials can only improve the electrical conductivity at the macroscopic level. Another problem when dispersing/immobilizing POMs on conductive matrixes is the control of uniformity. Their combination with metal nanoparticles also has some difficulties related to the fine control of morphology and size of resulting material. Another commonly used technology to enhance both stability and repeatability of measurements is the LbL self-assembly method. It prevents the detachment of the material from the electrodes and provides an easier path to the electrons. The combination with MOFs has also led to very promising sensors, but the major drawback is their weak conductivity. The low stability of POMs and some POM-based (nano)composites in mediums with high pH values is still a major drawback. Consequently, new strategies to improve this subject are still needed in order to broaden the range of efficient POM-based electrocatalysts.

Despite great advances, there is still margin to develop innovative research namely regarding the understanding of the intrinsic activity of POMs in order to fully know the mechanisms involved and also the composition–structure–function relationships. Developing new synthetic strategies for POM-based composites is also an important key point as it will enable the appearance of novel materials with enhanced electrocatalytic properties.

Acknowledgments

This work received support and help from FCT/MCTES (LA/P/0008/2020 DOI 10.54499/LA/P/0008/2020, UIDP/50006/2020 DOI 10.54499/UIDP/50006/2020 and UIDB/50006/2020 DOI 10.54499/UIDB/50006/2020), through national funds. D.M. Fernandes also thanks FCT for funding through the Individual Call to Scientific Employment Stimulus 2021.00771.CEECIND/CP1662/CT0007.

List of Abbreviations

[BMIM][BF_4]	1-butyl-3-methylimidazolium tetrafluoroborate
[$Ru(bpy)_3$]	tris-(2,2′-bipyridine) ruthenium(II)
9,10-BIMA	9,10-bis(imidazol-1-ylmethyl)anthracene
Ag/AgCl	Ag/AgCl reference electrode
Au NPs	gold nanoparticles

bim	2,2′-biimidazole
bimpy	3,5-bis(1-imidazoly) pyridine
bip	1,5-bis(imidazole-1-yl)pentane
bipy	2,2′-bipyridine
bix	1,4-bis(imidazol-1-ylmethyl)benzene
CAT	catalytic efficiency
CD	β-cyclodextrin
CNT	carbon nanotubes
CoTsPc	cobalt(II) tetrasulphonate phthalocyanine
CPE	carbon paste electrode
CS	chitosan
CV	cyclic voltammetry
DANTA	diacetylene-nitrilotriacetate
DC	dopaminechrome
DEMS	differential electrochemical mass spectrometry
DOQ	dopamine ortho-quinone
DPV	differential pulse voltammetry
Fc	ferrocene
FLG	few layer graphene
GCE	glassy carbon electrode
GF	graphene
GO	graphene oxide
IL	ionic liquid
ITO	indium titanium oxide
LbL	layer-by-layer
LDC	leucodopaminechrome
LOD	limit of detection
LR	linear range
MOF	metal organic framework
MWCNT	multi-walled carbon nanotubes
NPs	nanoparticles
OMC	ordered mesoporous carbon
ox-SWCNTs	oxidized single-walled carbon nanotubes
PANI	polyaniline
pbpy	5-phenyl-2-(4-pyridinyl) pyridine
PBS	phosphate buffer solution
PDDA	poly(diallyldimethylammonium chloride)
PEDOT	poly(3,4-ethylenedioxythiophene)
PEI	polyethylenimine
P-ERGO	phosphorus-doped electrochemically reduced graphene oxide
PG	pyrolytic graphite
POMOF	POM–organic frameworks
POMs	polyoxometalates
PPy	polypyrrole
PVP	poly(4-vinylpyridine)

pz	pyrazine
rGO	reduced graphene oxide
SWCNT	single-walled carbon nanotubes
SWCNT-COOH	carboxyl functionalized SWCNT
SWV	square wave voltammetry
TBA	tetrabutylammonium
trz	1,2,4-triazol
WOC	water oxidation catalysts

References

1 Sadakane, M. and Steckhan, E. (1998). Electrochemical properties of polyoxometalates as electrocatalysts. *Chem. Rev.* 98 (1): 219–237.

2 Baker, L.C.W. and Glick, D.C. (1998). Present general status of understanding of heteropoly electrolytes and a tracing of some major highlights in the history of their elucidation. *Chem. Rev.* 98 (1): 3–49.

3 Fernandes, D.M., Simões, S.M.N., Carapuça, H.M. et al. (2008). Functionalisation of glassy carbon electrodes with deposited tetrabutylammonium microcrystalline salts of lacunary and metal-substituted alpha-Keggin-polyoxosilicotungstates. *Electrochim. Acta* 53 (22): 6580–6588.

4 Vaillant, J., Lira-Cantu, M., Cuentas-Gallegos, K. et al. (2006). Chemical synthesis of hybrid materials based on PAni and PEDOT with polyoxometalates for electrochemical supercapacitors. *Prog. Solid State Chem.* 34 (2-4): 147–159.

5 Gomez-Romero, P., Cuentas-Gallegos, K., Lira-Cantú, M. et al. (2005). Hybrid nanocomposite materials for energy storage and conversion applications. *J. Mater. Sci.* 40 (6): 1423–1428.

6 Gomez-Romero, P., Chojak, M., Cuentas-Gallegos, K. et al. (2003). Hybrid organic-inorganic nanocomposite materials for application in solid state electrochemical supercapacitors. *Electrochem. Commun.* 5 (2): 149–153.

7 Balamurugan, A. and Chen, S.M. (2007). Silicomolybdate doped polypyrrole film modified glassy carbon electrode for electrocatalytic reduction of Cr(VI). *J. Solid State Electrochem.* 11 (12): 1679–1687.

8 Wang, X.L., Zhang, H., Wang, E. et al. (2004). Phosphomolybdate-polypyrrole composite bulk-modified carbon paste electrode for a hydrogen peroxide amperometric sensor. *Mater. Lett.* 58 (10): 1661–1664.

9 Fernandes, D.M., Brett, C.M.A., and Cavaleiro, A.M.V. (2011). Preparation and electrochemical properties of modified electrodes with Keggin-type silicotungstates and PEDOT. *J. Electroanal. Chem.* 660 (1): 50–56.

10 Liu, S.Q. and Tang, Z.Y. (2010). Polyoxometalate-based functional nanostructured films: current progress and future prospects. *Nano Today* 5 (4): 267–281.

11 Decher, G. (1997). Fuzzy nanoassemblies: toward layered polymeric multicomposites. *Science* 277 (5330): 1232–1237.

12 Dubas, S.T. and Schlenoff, J.B. (1999). Factors controlling the growth of polyelectrolyte multilayers. *Macromolecules* 32 (24): 8153–8160.

13 Bertrand, P., Jonas, A., Laschewsky, A. et al. (2000). Ultrathin polymer coatings by complexation of polyelectrolytes at interfaces: suitable materials, structure and properties. *Macromol. Rapid Commun.* 21 (7): 319–348.

14 Fernandes, D.M., Julião, D., Pereira, C. et al. (2012). Hybrid layer-by-layer films based on lanthanide-bridged silicotungstates and poly(ethylenimine). *Colloids Surf. A: Physicochem. Eng. Asp.* 415: 302–309.

15 Fernandes, D.M., Carapuça, H.M., Brett, C.M.A. et al. (2010). Electrochemical behaviour of self-assembly multilayer films based on iron-substituted alpha-Keggin polyoxotungstates. *Thin Solid Films* 518 (21): 5881–5888.

16 Fernandes, D.M., Brett, C.M.A., and Cavaleiro, A.M.V. (2011). Layer-by-layer self-assembly and electrocatalytic properties of poly(ethylenimine)-silicotungstate multilayer composite films. *J. Solid State Electrochem.* 15 (4): 811–819.

17 Caruso, F., Kurth, D.G., Volkmer, D. et al. (1998). Ultrathin molybdenum polyoxometalate-polyelectrolyte multilayer films. *Langmuir* 14 (13): 3462–3465.

18 Wang, B., Vyas, R.N., and Shaik, S. (2007). Preparation parameter development for layer-by-layer assembly of keggin-type polyoxometalates. *Langmuir* 23 (22): 11120–11126.

19 Fernandes, D.M. and Freire, C. (2014). Hybrid photochromic multilayer films based on chitosan and europium phosphomolybdate. *J. Appl. Electrochem.* 44 (5): 655–665.

20 Fernandes, D.M., Vos, J.G., and Freire, C. (2014). Europium phosphomolybdate and osmium metallopolymer multi-functional LbL films: redox and electrocatalytic properties. *J. Colloid Interface Sci.* 420: 127–135.

21 Novais, H.C., Fernandes, D.M., and Freire, C. (2015). Hybrid methyl green/cobalt-polyoxotungstate nanostructured films: self-assembly, electrochemical and electrocatalytic properties. *Appl. Surf. Sci.* 347: 40–47.

22 Fernandes, D.M., Teixeira, A., and Freire, C. (2015). Multielectrocatalysis by Layer-by-Layer Films Based on Pararosaniline and Vanadium-Substituted Phosphomolybdate. *Langmuir* 31 (5): 1855–1865.

23 Gao, S.Y., Cao, R., and Yang, C.P. (2008). Dye-polyoxometalate composite films: self-assembly, thermal and photochemical properties. *J. Colloid Interface Sci.* 324 (1-2): 156–166.

24 Feng, Y.H., Han, Z., Peng, J. et al. (2006). Fabrication and characterization of multilayer films based on Keggin-type polyoxometalate and chitosan. *Mater. Lett.* 60 (13-14): 1588–1593.

25 Wang, Y.H., Guo, C., Chen, Y. et al. (2003). Self-assembled multilayer films based on a Keggin-type polyoxometalate and polyaniline. *J. Colloid Interface Sci.* 264 (1): 176–183.

26 Wang, X.M., Zhao, Y., Zhang, D. et al. (2019). Dawson-type vanadium-substituted tungstophosphate-modified ITO electrode: preparation, characterization and electrochemical determination of dopamine. *Int. J. Electrochem. Sci.* 14 (4): 3595–3609.

27 Zhang, L., Ning, L., Li, S. et al. (2016). Selective electrochemical detection of dopamine in the presence of uric acid and ascorbic acid based on a composite film modified electrode. *RSC Adv.* 6 (71): 66468–66476.

28 Martel, D. and Gross, M. (2007). Electrochemical study of multilayer films built on glassy carbon electrode with polyoxometalate anions and two multi-charged molecular cationic species. *J. Solid State Electrochem.* 11 (3): 421–429.

29 Kuhn, A., Mano, N., and Vidal, C. (1999). Polyoxometalate modified electrodes: from a monolayer to multilayer structures. *J. Electroanal. Chem.* 462 (2): 187–194.

30 Kuhn, A. and Anson, F.C. (1996). Adsorption of monolayers of P2Mo18O626- and deposition of multiple layers of Os(bpy)(3)(2+)-P2Mo18O626- on electrode surfaces. *Langmuir* 12 (22): 5481–5488.

31 Ji, Y.C., Huang, L., Hu, J. et al. (2015). Polyoxometalate-functionalized nanocarbon materials for energy conversion, energy storage and sensor systems. *Energ. Environ. Sci.* 8 (3): 776–789.

32 Song, Y.F. and Tsunashima, R. (2012). Recent advances on polyoxometalate-based molecular and composite materials. *Chem. Soc. Rev.* 41 (22): 7384–7402.

33 Toma, F.M., Sartorel, A., Iurlo, M. et al. (2010). Efficient water oxidation at carbon nanotube-polyoxometalate electrocatalytic interfaces. *Nat. Chem.* 2 (10): 826–831.

34 Fernandes, D.M. and Freire, C. (2015). Carbon nanomaterial-phosphomolybdate composites for oxidative electrocatalysis. *ChemElectroChem* 2 (2): 269–279.

35 Fernandes, D.M., Novais, H.C., Bacsa, R. et al. (2018). Polyoxotungstate@carbon nanocomposites as oxygen reduction reaction (ORR) electrocatalysts. *Langmuir* 34 (22): 6376–6387.

36 Gao, J., Yang, T., Wang, X. et al. (2020). Spherical phosphomolybdic acid immobilized on graphene oxide nanosheets as an efficient electrochemical sensor for detection of diphenylamine. *Microchem. J.* 158.

37 Suma, B.P., Adarakatti, P.S., Kempahanumakkagari, S.K. et al. (2019). A new polyoxometalate/rGO/Pani composite modified electrode for electrochemical sensing of nitrite and its application to food and environmental samples. *Mater. Chem. Phys.* 229: 269–278.

38 Zhang, D., Ma, H., Chen, Y. et al. (2013). Amperometric detection of nitrite based on Dawson-type vanodotungstophosphate and carbon nanotubes. *Anal. Chim. Acta* 792: 35–44.

39 Jawale, D.V., Fossard, F., Miserque, F. et al. (2022). Carbon nanotube-polyoxometalate nanohybrids as efficient electrocatalysts for the hydrogen evolution reaction. *Carbon* 188: 523–532.

40 Liu, X., Zhong, C., Ji, J. et al. (2023). Polyoxometalate/carbon black modified glassy carbon electrode for the detection of dopamine. *Electroanalysis* 12.

41 Zhou, M., Guo, L., Lin, F. et al. (2007). Electrochemistry and electrocatalysis of polyoxometalate-ordered mesoporous carbon modified. *Anal. Chim. Acta* 587 (1): 124–131.

42 Zhang, Y.F., Bo, X., Nsabimana, A. et al. (2015). Green and facile synthesis of an Au nanoparticles@polyoxometalate/ordered mesoporous carbon tri-component nanocomposite and its electrochemical applications. *Biosens. Bioelectron.* 66: 191–197.

43 Ruiz, V., Suarez-Guevara, J., and Gomez-Romero, P. (2012). Hybrid electrodes based on polyoxometalate-carbon materials for electrochemical supercapacitors. *Electrochem. Commun.* 24: 35–38.

44 Li, H., Pang, S., Feng, X. et al. (2010). Polyoxometalate assisted photoreduction of graphene oxide and its nanocomposite formation. *Chem. Commun.* 46 (34): 6243–6245.

45 Wang, S., Li, H., Li, S. et al. (2013). Electrochemical-reduction-assisted assembly of a polyoxometalate/graphene nanocomposite and its enhanced lithium-storage performance. *Chem. A Eur. J.* 19 (33): 10895–10902.

46 Li, H.L., Pang, S., Wu, S. et al. (2011). Layer-by-layer assembly and uv photoreduction of graphene-polyoxometalate composite films for electronics. *J. Am. Chem. Soc.* 133 (24): 9423–9429.

47 Wang, H., Kawasaki, N., Yokoyama, T. et al. (2012). Molecular cluster batteries of nano-hybrid materials between Keggin POMs and SWNTs. *Dalton Trans.* 41 (33): 9863–9866.

48 Kawasaki, N., Wang, H., Nakanishi, R. et al. (2011). Nanohybridization of polyoxometalate clusters and single-wall carbon nanotubes: applications in molecular cluster batteries. *Angew. Chem. Int. Ed.* 50 (15): 3471–3474.

49 Cui, Z.M., Li, C.M., and Jiang, S.P. (2011). PtRu catalysts supported on heteropolyacid and chitosan functionalized carbon nanotubes for methanol oxidation reaction of fuel cells. *Phys. Chem. Chem. Phys.* 13 (36): 16349–16357.

50 Ma, D., Liang, L., Chen, W. et al. (2013). Covalently tethered polyoxometalate-pyrene hybrids for noncovalent sidewall functionalization of single-walled carbon nanotubes as high-performance anode material. *Adv. Funct. Mater.* 23 (48): 6100–6105.

51 Hadad, C., Ke, X., Carraro, M. et al. (2014). Positive graphene by chemical design: tuning supramolecular strategies for functional surfaces. *Chem. Commun.* 50 (7): 885–887.

52 Kulesza, P.J., Skunik, M., Baranowska, B. et al. (2006). Fabrication of network films of conducting polymer-linked polyoxometallate-stabilized carbon nanostructures. *Electrochim. Acta* 51 (11): 2373–2379.

53 Cui, Z.M., Kulesza, P.J., Li, C.M. et al. (2011). Pd nanoparticles supported on HPMo-PDDA-MWCNT and their activity for formic acid oxidation reaction of fuel cells. *Int. J. Hydrogen Energy* 36 (14): 8508–8517.

54 Skunik, M. and Kulesza, P.J. (2009). Phosphomolybdate-modified multi-walled carbon nanotubes as effective mediating systems for electrocatalytic reduction of bromate. *Anal. Chim. Acta* 631 (2): 153–160.

55 Sosnowska, M., Goral-Kurbiel, M., Skunik-Nuckowska, M. et al. (2013). Hybrid materials utilizing polyelectrolyte-derivatized carbon nanotubes and vanadium-mixed addenda heteropolytungstate for efficient electrochemical charging and electrocatalysis. *J. Solid State Electrochem.* 17 (6): 1631–1640.

56 Freire, C., Fernandes, D.M., Nunes, M. et al. (2018). POM & MOF-based electrocatalysts for energy-related reactions. *ChemCatChem* 10 (8): 1703–1730.

57 Bagotsky, V.S. (2006). *Fundamentals of Electrochemistry*, 2e. New Jersey: John Wiley & Sons, Inc.

58 Lakard, S., Pavel, I.A., and Lakard, B. (2021). Electrochemical biosensing of dopamine neurotransmitter: a review. *Biosens. Basel* 11 (6).

59 Fernandes, D.M., Barbosa, A.D.S., Pires, J. et al. (2013). Novel composite material polyoxovanadate@MIL-101(Cr): a highly efficient electrocatalyst for ascorbic acid oxidation. *ACS Appl. Mater. Interfaces* 5 (24): 13382–13390.

60 Liu, X., Cui, G., Dong, L. et al. (2022). Synchronous electrochemical detection of dopamine and uric acid by a PMo12@MIL-100(Fe)@PVP nanocomposite. *Anal. Biochem.* 648: 10.

61 Ho, W.H., Chen, T.Y., Otake, K. et al. (2020). Polyoxometalate adsorbed in a metal-organic framework for electrocatalytic dopamine oxidation. *Chem. Commun.* 56 (79): 11763–11766.

62 Zhao, T.T., Cui, L., Yu, K. et al. (2022). Porous $P_6Mo_{18}O_{73}$-type poly (oxometalate) metal-organic frameworks for improved pseudocapacitance and electrochemical sensing performance. *ACS Appl. Mater. Interfaces* 13.

63 Zhang, W., Jia, G., Li, Z. et al. (2017). Selective electrochemical detection of dopamine on polyoxometalate-based metal-organic framework and its composite with reduced graphene oxide. *Adv. Mater. Interfaces* 4 (10): 7.

64 Zhou, K.F., Han, H., Sha, J. et al. (2021). Synthesis of POMOFs with 8-fold helix and its composite with carboxyl functionalized SWCNTs for the voltammetric determination of dopamine. *Anal. Bioanal. Chem.* 413 (21): 5309–5320.

65 Hao, Y.J., Feng, S., Liu, Y. et al. (2017). Electrochemical sensor based on indium tin oxide glass modified with poly(ethyleneimine)/phosphomolybdic acid composite multilayers. *Electroanalysis* 29 (4): 1188–1196.

66 Zhu, W., Zhang, W., Li, S. et al. (2013). Fabrication and electrochemical sensing performance of a composite film containing a phosphovanadomolybdate and cobalt(II) tetrasulfonate phthalocyanine. *Sens. Actuators B Chem.* 181: 773–781.

67 Li, S., Ma, H., O'Halloran, K.P. et al. (2013). Enhancing characteristics of a composite film by combination of vanadium-substituted molybdophosphate and platinum nanoparticles for an electrochemical sensor. *Electrochim. Acta* 108: 717–726.

68 Zhou, C.L., Li, S., Zhu, W. et al. (2013). A sensor of a polyoxometalate and Au–Pd alloy for simultaneously detection of dopamine and ascorbic acid. *Electrochim. Acta* 113: 454–463.

69 Wang, Q.W., Khungwa, J., Li, L. et al. (2018). Fabrication of polyoxometalate/GO/PDDA hybrid nanocomposite modified electrode and electrocatalysis for nitrite ion, ascorbic acid and dopamine. *J. Electroanal. Chem.* 824: 91–98.

70 Jiang, J., Liu, L., Liu, G. et al. (2020). Organic-inorganic hybrid cerium-encapsulated selenotungstate including three building blocks and its electrochemical detection of dopamine and paracetamol. *Inorg. Chem.* 59 (20): 15355–15364.

71 Jiao, J., Zuo, J., Pang, H. et al. (2018). A dopamine electrochemical sensor based on Pd–Pt alloy nanoparticles decorated polyoxometalate and multiwalled carbon nanotubes. *J. Electroanal. Chem.* 827: 103–111.

72 Bai, Z.Y., Gao, N., Xu, H. et al. (2020). Construction of an ultra-sensitive electrochemical sensor based on polyoxometalates decorated with CNTs and AuCo nanoparticles for the voltammetric simultaneous determination of dopamine and uric acid. *Microchim. Acta* 187 (8): 12.

73 Thakur, N., Adhikary, S.D., Kumar, M. et al. (2018). Ultrasensitive and highly selective electrochemical detection of dopamine using poly(ionic liquids)-cobalt polyoxometalate/CNT composite. *Acs Omega* 3 (3): 2966–2973.

74 Shafiei, H. and Hassaninejad-Darzi, S.K. (2023). Simultaneous Electrocatalytic Measurement of Dopamine and Acetaminophen by Nanosensor Based on Ag@Polyoxometalate@ Reduced Graphene Oxide and Ionic Liquid. *Electrocatalysis* 18.

75 Ma, C.A., Xu, P., Chen, H. et al. (2022). An electrochemical sensor based on reduced graphene oxide/β-cyclodextrin/multiwall carbon nanotubes/polyoxometalate tetracomponent hybrid: simultaneous determination of ascorbic acid, dopamine and uric acid. *Microchem. J.* 180: 8.

76 Arrigoni, O. and De Tullio, M.C. (2002). Ascorbic acid: much more than just an antioxidant. *BBA-Gen. Subjects* 1569 (1-3): 1–9.

77 Li, F.R., Ji, T., and Chen, W.L. (2022). A tri-vanadium-capped Keggin phosphomolybdate: synthesis, characterization, photocatalytic and bifunctional electrocatalytic properties. *Tungsten* 4 (2): 99–108.

78 Fernandes, D.M., Granadeiro, C.M., de Sousa, P.M.P. et al. (2014). SiW11Fe@MIL-101(Cr) composite: a novel and versatile electrocatalyst. *ChemElectroChem* 1 (8): 1293–1300.

79 Liu, X.Z., Cui, L., Yu, K. et al. (2021). Cu/Ag complex modified keggin-type coordination polymers for improved electrochemical capacitance, dual-function electrocatalysis, and sensing performance. *Inorg. Chem.* 60 (18): 14072–14082.

80 Zhong, R., Cui, L., Yu, K. et al. (2021). Wells-Dawson arsenotungstate porous derivatives for electrochemical supercapacitor electrodes and electrocatalytically active materials. *Inorg. Chem.* 60 (13): 9869–9879.

81 Yang, A.S., Cui, L., Yu, K. et al. (2022). Supramolecular host-guest assembly based on phosphotungstate nanostructures for pseudocapacitive and electrochemical sensing applications. *Acs Appl. Nano Mater.* 5 (8): 10452–10461.

82 Zhang, L.J., Wang, X., Yang, P. et al. (2022). A lindquist-type polyoxometalate-based metal-organic framework as electrochemical sensor and efficient catalyst for selective oxidation of thioether. *Inorg. Chim. Acta* 542: 6.

83 Yu, Y., Pang, H., Ma, H. et al. (2013). A novel 12-tungstovanadate doubly-grafted by Zn-complexes: synthesis, crystal structure and properties. *J. Clust. Sci.* 24 (1): 17–29.

84 Ammam, M. and Easton, E.B. (2011). Selective determination of ascorbic acid with a novel hybrid material based 1-butyl-3-methylimidazolium tetrafluoroborate ionic liquid and the Dawson type ion P2Mo18O62 (6-) immobilized on glassy carbon. *Electrochim. Acta* 56 (7): 2847–2855.

85 Zhang, Y.C., Chen, Y.Z., Chang, Z.H. et al. (2022). Metal-directed two new Anderson-type polyoxometalate-based metal-organic complexes with different electrocatalytic sensing performance. *Polyhedron* 221: 7.

86 Zhou, W.L., Peng, J., Zhang, Z. et al. (2015). Bifunctional electrocatalysts based on the Keggin/Dawson-type polyoxotungstates and Cu-tris(imidazolyl) complexes. *Electrochim. Acta* 180: 887–893.

87 Zhou, W.L., Liu, P., Zheng, Y. et al. (2019). Four new dual-functional electrocatalysts formed from small molybdenum clusters and Cu-pyridyl complexes. *Dalton Trans.* 48 (43): 16350–16357.

88 Zheng, Y.P., Zhou, W., Liu, X. et al. (2021). Experimental and theoretical study of bifunctional electrocatalysts constructed from different polyoxometalates and Ag-bimpy segments. *Electrochim. Acta* 391: 10.

89 Zheng, Y.P., Tan, Y., Zhou, W. et al. (2021). Three polyoxovanadates-based organic-inorganic hybrids: structural variation, bifunctional electrocatalytic activities, and computational studies. *Inorg. Chem.* 60 (16): 12323–12330.

90 Li, F.R., Lv, J., Yu, K. et al. (2017). Two extended Wells-Dawson arsenomolybdate architectures directed by Na(I) and/or Cu(I) organic complex linkers. *CrstEngComm* 19 (17): 2320–2328.

91 Zhou, W.L., Zheng, Y.P., and Peng, J. (2018). Cu-containing Keggin-type polyoxometalates-based organic-inorganic hybrids with double electro-catalytic behaviors. *J. Solid State Chem.* 258: 786–791.

92 Zhou, W.L., Zheng, Y., Yuan, G. et al. (2019). Three polyoxometalates-based organic-inorganic hybrids decorated with Cu-terpyridine complexes exhibiting dual functional electro-catalytic behaviors. *Dalton Trans.* 48 (8): 2598–2605.

93 Zhang, X.W., Lai, G., Yu, A. et al. (2013). A glassy carbon electrode modified with a polyaniline doped with silicotungstic acid and carbon nanotubes for the sensitive amperometric determination of ascorbic acid. *Microchim. Acta* 180 (5-6): 437–443.

94 Zhang, W.Y., Du, D., Gunaratne, D. et al. (2014). Polyoxometalate-graphene nanocomposite modified electrode for electrocatalytic detection of ascorbic acid. *Electroanalysis* 26 (1): 178–183.

95 Muqaddas, S., Aslam, H., Hassan, S.U. et al. (2023). Electrochemical sensing of glucose and ascorbic acid via POM-based CNTs fiber electrode. *Mater. Sci. Eng. B: Adv. Funct. Solid-State Mater.* 293: 11.

96 Ma, H.Y., Shi, S., Zhang, Z. et al. (2010). An thin film of the di-decatungstosilicate with a tetra-ruthenium(IV)-oxo core and its electrochemical properties. *J. Electroanal. Chem.* 648 (2): 128–133.

97 Chen, C.Y., Song, Y.H., and Wang, L. (2009). Electrochemical behavior and its electrocatalytic properties of chemically modified electrode with Keggin-type SiNi(H_2O)W11O39 (6-). *Electrochim. Acta* 54 (5): 1607–1611.

98 Guo, D.X., Zhou, C., Tan, L. et al. (2020). Electrochemical ascorbic acid sensor of composite film based on Keggin-type Vanadium-substituted Polyoxometalates decorated with graphene and Ru $(bpy)_3^{2+}$. *Colloids Surf A: Physicochem. Eng. Asp.* 592: 7.

99 Zhang, L., Li, S., O'Halloran, K.P. et al. (2021). A highly sensitive non-enzymatic ascorbic acid electrochemical sensor based on polyoxometalate/tris(2,2′-bipyridine)ruthenium(II)/chitosan-palladium inorganic-organic self-assembled film. *Colloids Surf A: Physicochem. Eng. Asp.* 614: 8.

100 Glantzounis, G.K., Tsimoyiannis, E.C., Kappas, A.M. et al. (2005). Uric acid and oxidative stress. *Curr. Pharm. Des.* 11 (32): 4145–4151.

101 Bai, Z.Y., Zhou, C., Xu, H. et al. (2017). Polyoxometalates-doped Au nanoparticles and reduced graphene oxide: a new material for the detection of uric acid in urine. *Sens. Actuators B Chem.* 243: 361–371.

102 Bao, Y.Y., Li, Z., Wang, H. et al. (2020). Electrochemical reduction-assisted in situ fabrication of a graphene/Au nanoparticles@polyoxometalate nanohybrid film: high-performance electrochemical detection for uric acid. *Langmuir* 36 (26): 7365–7374.

103 Liu, C., Xu, M., Tan, Z. et al. (2020). Assembly of Wells-Dawson polyoxometalate based crystal compound for uric acid electrochemical detection. *Z. Anorg. Allg. Chem.* 646 (11-12): 489–494.

104 Xu, M.Q., Liu, C., Wang, Y. et al. (2021). A new Well-Dawson polyoxometalate based compound containing helix for electrochemical uric acid biosensor. *J. Clust. Sci.* 32 (6): 1543–1552.

105 Fernandes, D.M. et al. (2017). $PMo_{11}V$@N-CNT electrochemical properties and its application as electrochemical sensor for determination of acetaminophen. *J. Solid State Electrochem.* 21 (4): 1059–1068.

106 Han, H., Liu, C., Sha, J. et al. (2021). Ferrocene-reduced graphene oxide-polyoxometalates based ternary nanocomposites as electrochemical detection for acetaminophen. *Talanta* 235: 8.

Part III

Polyoxometalates for Reductive Electrocatalysis

5

POM-based Electrocatalysts for Inorganic Water Contaminants and Hydrogen Peroxide Reduction

Diana M. Fernandes

Universidade do Porto, LAQV-REQUIMTE, Departamento de Química e Bioquímica, Faculdade de Ciências, Rua do Campo Alegre, s/n, Porto 4169-007, Portugal

5.1 Introduction

Over the last two decades there has been an increased interest in the electrocatalytic activity of polyoxometalates (POMs). Heteropolyanions and in particular their metal-substituted derivatives are at the top of the list due to their fascinating properties. These include the high stability of most of their redox states, the ability to undergo reversible multi-valence reductions/oxidations, forming mixed-valence species, the possibility to tune their redox potentials by changing the addenda atoms or the heteroatoms without affecting their structure and the number of transition-metal cations that can be incorporated into their structure.

Their reduced forms, owing to their electron and proton transfer and/or storage abilities, may behave as donors or acceptors of several electrons while maintaining their structure. Therefore, most POMs may be considered as close mimics of mixed metal oxides and, consequently, be studied as their soluble analogues, at least as concerns their electrochemistry. The high redox activity of POMs makes them ideal for reductive and oxidative electrocatalysis which is extremely attractive for the development of multi-electron transfer reactions that are generally hampered by high overpotentials leading to low overall conversion efficiencies. In fact, the possibilities of electrocatalysis of reactions at electrodes are among the powerful incentives for the electrochemical study of POMs. Furthermore, the design of suitable POM-based electrocatalysts can improve not only the reactivity but also the product selectivity.

Several POMs and POM-based materials have been used as reductive electrocatalysts, and a huge number of papers can be found in the literature regarding this subject. Some of these examples were covered in two reviews in 1998 [1] and 2007 [2] and one book chapter in 2016 [3].

This chapter will focus on the POM-assisted electrochemical reduction of nitrite, bromate, iodate and hydrogen peroxide.

Applied Polyoxometalate-based Electrocatalysis, First Edition. Edited by Diana M. Fernandes.

5.2 Nitrite Reduction

Nitrites are primarily used as food preservative, e.g. for the rapid curing of meat and as fertilizing agent but can also be applied as anticorrosive agents in cooling fluids. Nitrite is also the metabolite of nitrite oxide in biological systems. Thus, it is extensively present in waters, soil, foods and physiological systems and is a key species in the global nitrogen cycle [4]. However, nitrite is highly toxic because it interacts with amines to form carcinogenic nitrosamines in stomach [5], also with the Fe^{III} of the haemoglobin to form methaemoglobin [6], which has no oxygen-carrying ability. The excessive nitrite in food products, drinking water, vegetables and beverages is a serious threat to human health as it can cause severe respiratory disease, skin and eye irritation and even fatality at exceptionally high levels of exposure. As a consequence, this has become a global issue, and over the years numerous analytical techniques have been developed for its detection. The most common are spectrophotometry, chromatography, fluorescence and flow injection analysis, but the use of these methods requires tedious and time-consuming sample pre-treatment and expensive reagents. Therefore, the development of simple and low-cost methods is of utmost importance. Electrochemical techniques, which are characterized by their easy operation, low cost and high sensitivity, have gained widespread interest.

The fact that POMs are able to undergo reversible multi-steps, multi-electrons transfer reactions make them ideal candidates for electroreduction of numerous NO_x like nitrite and nitric oxide. For nitrite, the difficulty arises from the question of which are the actual reactive species in acidic solution. Equations (5.1) and (5.2) suggest that HNO_2 and/or NO should be the reactive species at least up to pH 3.3 [1]. HNO_2 disproportionates in fairly acidic solutions even though the rate of this process is known to be low.

$$HNO_2 \rightleftarrows H^+ + NO^{2-} \tag{5.1}$$

$$3HNO_2 \rightleftarrows HNO_3 + 2NO + H_2O \tag{5.2}$$

The pioneering works on the electrocatalysis of nitrite (which is actually a mixture of HNO_2 and NO, for pH < 4) by POMs were demonstrated in the late 1980s by Toth and Anson [7] and Keita et al. [8], and since then, this reaction has become a classical test for the electrocatalytic properties of POMs.

Numerous POMs in solution have been explored for nitrite electroreduction, and these are summarized in Table 5.1. As the majority of the results obtained until 2005 have already been discussed in two review papers by Sadakane and Steckhan [1] and Keita and Nadjo [2], the electrocatalytic reduction of nitrite by POMs dissolved in solution will not be discussed here.

One of the major problems encountered in the direct reduction of nitrite is the fact that it requires very large overpotential at most conventional electrodes. Consequently, the immobilization of POMs at the electrode surfaces was proposed to overcome this problem. POMs have the ability to be attached onto electrode

Table 5.1 Electroreduction of nitrite by homogeneously dissolved POMs.

Electrocatalyst	Conditions	References
$[SiW_{12}O_{40}]^{4-}$	Acidic solutions with pH < 2.0	[9]
α-,β-$[SiW_{12}O_{40}]^{4-}$	Acidic solution	[10]
$[SiMo_{12}O_{40}]^{4-}$	0.2 M Na_2SO_4 + H_2SO_4, pH 2.0	[11]
$[SiMo_{12}O_{40}]^{4-}$	0.5 M H_2SO_4 solution	[12]
$[PW_{12}O_{40}]^{3-}$	1.0 M H_2SO_4 solution	[13]
[α-$SiVW_{11}O_{40}]^{5-}$	0.5 M CH_3COONa/CH_3COOH	[14]
[γ-(1,2)-$SiV_2W_{10}O_{40}]^{6-}$	0.5 M CH_3COONa/CH_3COOH	[14]
$[XW_{11}Fe^{III}(H_2O)O_{39}]^{n-}$, X = P, Si, As, Ge	Phosphate and acetate buffer solution with different pH values	[7]
$[SiW_{11}Fe^{III}(H_2O)O_{39}]^{5-}$	$H_2SO_4/CH_3COONa/CH_3COOH$ buffer, pH 2.0	[15]
$[P_2W_{18}O_{62}]^{6-}$	H_2SO_4 solutions with different pH values	[16]
$[P_2V_2W_{16}O_{62}]^{8-}$	0.5 M Na_2SO_4 + H_2SO_4, pH 1.0	[17]
$[P_2W_{17}Fe^{III}(H_2O)O_{61}]^{8-}$	CH_3COONa/CH_3COOH, pH 5.0	[18]
$[H_4AsW_{18}O_{62}]^{7-}$	CH_3COONa/CH_3COOH, pH 5.0	[19]
$[M_4(H_2O)_2(H_4AsW_{15}O_{56})_2]^{18-}$, M = Zn, Cu	CH_3COONa/CH_3COOH, pH 5.0	[19]
$[Ln(As_2W_{17}O_{61})_2]^{17-}$, Ln = La, Pr, Sm, Eu, Gd, Tb, Dy, Tm, Yb	CH_3COONa/CH_3COOH, pH 5.5	[20]
$[Cu_4(H_2O)_2(X_2W_{15}O_{56})_2]^{16-}$, X = P or As	CH_3COONa/CH_3COOH, pH 5.0	[21]

surfaces as a result of interactions between them. Rong and Anson reported that $[PMo_{12}O_{40}]^{3-}$, $[P_2Mo_{18}O_{62}]^{6-}$, $[P_2W_{18}O_{62}]^{6-}$ and $[PW_{11}Fe^{III}(H_2O)O_{39}]^{4-}$ adsorb spontaneously on glassy carbon and edge pyrolytic graphite electrodes and that all POM-modified electrodes were active towards nitrite reduction [22]. A glassy carbon electrode (GCE) modified with a Keggin-type $[SiW_{11}Fe^{III}(H_2O)O_{39}]^{5-}$ also showed to be an excellent catalyst [15].

An alcohol-solvated polyoxometalate, $[NH_4]_4[SiW_{12}O_{40}]·3C_2H_5OH$ with excellent electrocatalytic activity for nitrite electroreduction, was also reported in 2017 by Zonoz and co-workers [23]. Due to its low solubility in water and most organic solvents as a result of the high hydrogen bonding in this POM, the authors were able to immobilize it in a carbon paste electrode (CPE) and access its electrochemical and electrocatalytic behaviour. The POM revealed four reversible redox peaks assigned to W centres (peaks I, II and IV) and C_2H_5OH molecules (peak III), with the half-wave potentials at −0.23 (I, 1e$^-$), −0.47 (II, 1e$^-$), −0.67 (III, 2e$^-$) and −0.84 V (IV, 2e$^-$). The electrocatalytic activity for nitrite reduction was evaluated at different pH values (1.5, 2, 2.5, 3 and 3.5), and results showed that peak IV presented the higher activity and that the best results were achieved for pH = 1.5 with a catalytic efficiency (CAT) of 2056%, which is defined by the following equation: CAT = $100\% \times [i_p(\text{POM, substrate}) - i_p(\text{POM})]/i_p$ (POM), where i_p (POM) and i_p

(POM, substrate) are the current intensities of POMs in the absence and presence of the substrates, respectively. The authors attributed the excellent performance of this POM to the crystallization solvent. The most recent work, in 2018, was reported by Xu et al. [24] and presented a new sandwich-type polyoxometalate compound built on Strandberg-type anions and divalent transition-metal cation ($(NH_4)_{10}[Cu(H_2O)_4(P_2Mo_5O_{23})_2]\cdot 6H_2O$) with excellent electrocatalytic performance towards NO_2^- reduction in 1 mol dm^{-3} H_2SO_4 and 0.2 mol dm^{-3} Na_2SO_4 solution (pH = 0.8).

The latest work was published by Badets and co-workers [25] and is quite different from the previous ones. It reports the reduction of nitrite catalysed by a set of POMs presenting different central atom which results in both different total negative charge and redox properties ($[PW_{12}O_{40}]^{3-}$, $[SiW_{12}O_{40}]^{4-}$, $[BW_{12}O_{40}]^{5-}$ and $[H_2W_{12}O_{40}]^{6-}$). Besides, the potential of the reduction peaks, the number of peaks, the electrons involved and the pH dependence are also different in these selected POMs. The authors used for the first time a differential electrochemical mass spectrometry (DEMS) analysis to identify the volatile products resulting from the reduction of nitrite and nitrous acid (NO^{2-} or HNO_2). The advantage of this technique is that it is able to identify the in situ formation of gaseous products during one single potential scan. Additionally, it allows a direct relationship between the applied electrochemical potential and formed products. Experiments were conducted at both pH 2 and 5 in order to compare the differences on the products and kinetics as a function of proton concentration and the analyte (nitrite or nitrous acid). At both pH values, the main gaseous products found were only ^{15}NO and $^{15}N_2O$. At pH 2, all POMs were able to reduce nitrite or nitrous acid to NO during the first redox peak and to produce N_2O in the second one. This correlated splendidly with the number of electrons transferred to the POM in each peak (1 for first peak and 2 for second peak). In addition, a direct correlation between the redox potential of the POM and the production of NO was also observed for both pH values evaluated.

Numerous organic–inorganic POM-based composites have also been applied towards nitrite reduction. This approach is advantageous as it combines the properties of both parts (organic and inorganic) and promotes a favourable synergic effect [26]. Zhang et al. [10] showed the nitrite electrocatalysis by a $[Ru(bipy)_3]_2[\alpha\text{-}SiW_{12}O_{40}]$-modified electrode where the process occurred at the W-centres and was independent of proton concentration, rather than exhibiting first-order dependence as observed for the POM in solution. McCormac and co-workers [27] also show that a significant catalytic current was obtained at the W-centres (pH 4.5) when a multilayer $[Fe(bipy)_3]_3[P_2W_{18}O_{62}]$-modified electrode (bipy = 2,2′-bipyridine) was used for nitrite reduction. In 2008, Ma's group [28] reported an inorganic–organic complex with a vanadium-substituted polyoxometalate $[Cu(phen)_2]_2PVW_{11}O_{40}$ (phen = 1,10-phenanthroline) where the reduction of NO_2^- was successfully electroreduced by the V-centre. Other examples of organic–inorganic hybrid compounds are the $[Cu_2(Htib)_2(\alpha\text{-}SiW_{12}O_{40})]$, $[Cu_2(Htib)_2(H_2tib)_2(H_2O)_4(\alpha\text{-}HBW_{12}O_{40})_2]\cdot 8H_2O$ and $[Cu_2(H_2tib)_4(H_2O)_2(\alpha\text{-}HP_2W_{18}O_{62})_2]\cdot 14H_2O$ (tib = 1,3,5-tris(1-imidazoly) benzene), obtained through hydrothermal methods [29]. Cyclic voltammetry (CV) experiments revealed that the NO_2^- was effectively catalysed by the W-centres of POMs and that the Dawson-type exhibited better performance.

In 2016, a 1 : 2-type inorganic–organic hybrid heterometallic germanotungstate was also obtained, under hydrothermal conditions, by simultaneously introducing Cu^{2+} and Ce^{4+} cations to the trivacant Keggin-type $[A\text{-}\alpha\text{-}GeW_9O_{34}]^{10-}$ system with the support of $Na_2WO_4{\cdot}2H_2O$ and the ethylenediamine ligand [30]. The tungsten centres were also the responsible active sites for nitrite electroreduction.

In 2018, Wang et al. [31] prepared by hydrothermal method three novel polyoxo metalate-based metal–organic complexes (MOCs) constructed from rigid pyridyl-triazole-tetrazole and pyridyl-bis(triazole) ligands $[Co_3(Htyptz)_2(\gamma\text{-}Mo_8O_{26})(H_2O)_6]{\cdot}4H_2O$ (1), $H_2[Co_3(Htyptz)_2(TeMo_6O_{24})(H_2O)_8]{\cdot}6H_2O$ (2), $[Co_3(Hpyttz)_2(\gamma\text{-}Mo_8O_{26})(H_2O)_8]{\cdot}2H_2O$ (3) (H_2typtz = 3-(tetrazol-5-yl)-5-(pyrid-3-yl)-1,2,4-triazolyl, H_2-pyttz = 3-(pyrid-4-yl)-5-(1*H*-1,2,4-triazol-3-yl)-1,2,4-triazolyl). The authors evaluated the effects of different ligands and of the type of POM used and observed quite different structures (Figure 5.1a). Complex (1) presented a three-dimensional (3D) framework, while (2) and (3) a two-dimensional (2D) network.

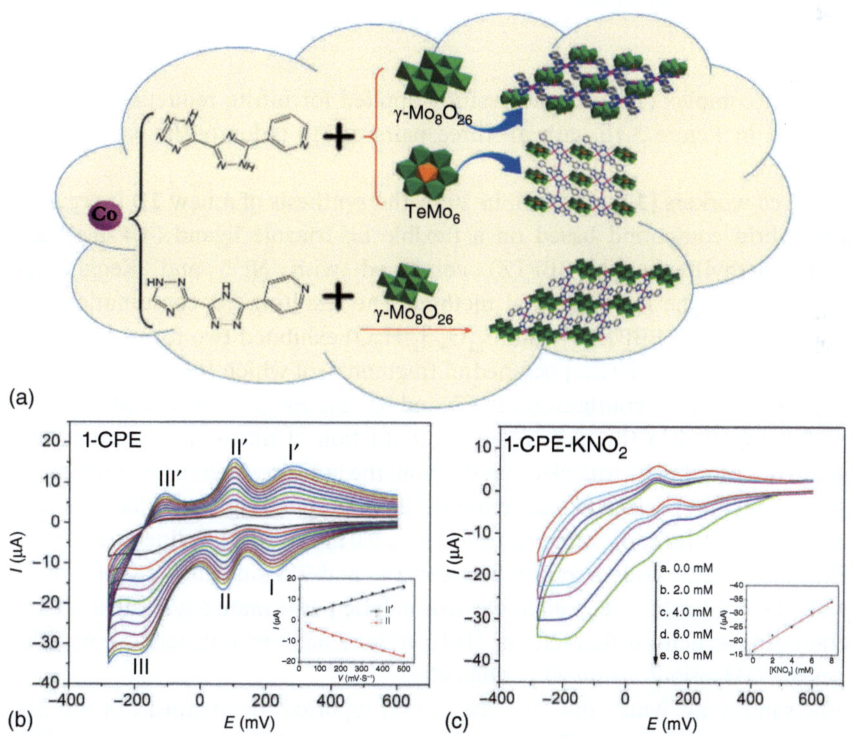

Figure 5.1 (a) Schematic effect of different ligands and POMs on the structures of the POM-based MOCs; (b) CVs of 1-CPE in 0.1 mol dm^{-3} H_2SO_4 + 0.5 mol dm^{-3} Na_2SO_4 electrolyte at different scan rates. Inset: dependence of i_{pc} and i_{pa} with scan rate; (c) CVs in the same electrolyte containing 0.0–8.0 mmol dm^{-3} of KNO_2. Inset: The linear dependence of the cathodic current of peak III on KNO_2 concentration. Source: Wang et al. [31]/with permission of Royal Society of Chemistry.

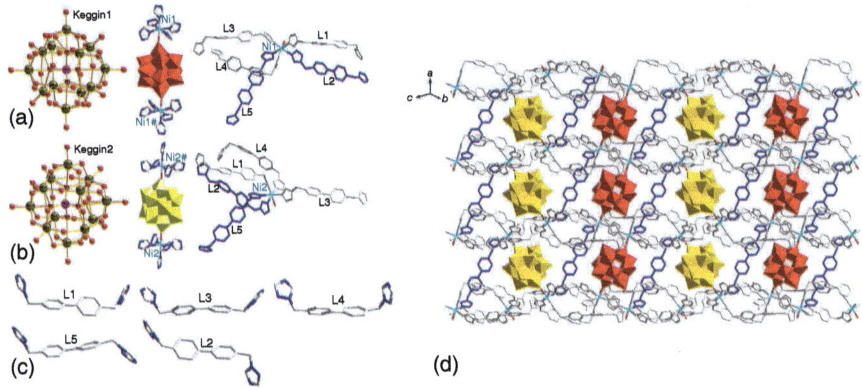

Figure 5.2 (a) Keggin 1 unit, the coordination environment of Keggin 1 and Ni_1^{2+} in the compound; (b) Keggin 2 unit, the coordination environment of Keggin 2 and Ni_2^{2+}; (c) 'U'-type (L_U) and 'S'-type (L_S) configurations of BBPTZ ligand in the compound. Red polyhedra: Keggin 1; Yellow polyhedra: Keggin 2; grey stick: 'U'-type BBPTZ ligand; blue stick: 'S'-type (L_S) BBPTZ ligand; (d) Polyhedral representation of the 2D layer. Source: Ding et al. [32]/with permission of Elsevier.

The former (complex (1)) was successfully applied for nitrite reduction as it can be observed in Figure 5.1b with all three pairs of Mo redox peaks participating (Figure 5.1c).

Liu and co-workers [32] reported, in 2021, the synthesis of a new 2D inorganic–organic hybrid compound based on a flexible bis-triazole ligand (4,4′-bis(1,2,4-triazol-1-ylmethyl)biphenyl, BBPTZ) combined with Ni^{2+} and Keggin-type $[SiMo_{12}O_{40}]^{4-}$ by the hydrothermal method. The resulting 2D coordination network structure ($[Ni_2(BBPTZ)_5]_2[SiMo_{12}O_{40}]_2 \cdot H_2O$) exhibited two forms of Keggin anions and two $[Ni_2(BBPTZ)_5]$ octahedral fragments, of which five BBPTZ ligands showed two types of configurations (S and U, Figure 5.2). The material was successfully applied in the electrocatalytic reduction of nitrite and bromate (see Section 5.3). Regarding nitrite electroreduction, the authors observed an increase in the three reduction peaks and a decrease in the corresponding oxidation ones with the increase of nitrite concentration from 1.0 to 5.0 mmol dm^{-3}. A linear range was obtained for a nitrite concentration between 166 and 913 μmol dm^{-3} with a sensitivity of 735.56 μA mM^{-1}. The good electrocatalytic performance was attributed to the interesting structural character of 2D layer, in which two independent negative charges of POM anions arranged periodically.

In the same year, Wang and co-workers [33] reported the synthesis of the first example of a coordination polymer derived from the Wells–Dawson POM anion $[P_2W_{18}O_{62}]^{6-}$ and di-nuclear nickel (II)-glycine cations. The organic–inorganic hybrid was prepared through hydrothermal method and resulted in a fascinating 3D supramolecular architecture where the POM fragments and the nickel counter-cations are connected through hydrogen bonds. The electrochemical behaviour and the potential electrocatalytic activity for both nitrite and iodate reduction were evaluated in 0.5 mol dm^{-3} Na_2SO_4/H_2SO_4 using a CPE. The material presented three reduction peaks at −0.444 (I), −0.641 (II) and −0.862 (III) V *vs.* Ag/AgCl and three anodic peaks at −0.387 (I′), −0.589 V (II′) and −0.803 (III′) V *vs.* Ag/AgCl. Upon the addition of the analyte (NO_2^- or IO_3^-) there is an enhancement of the cathodic peak current of peak (III) and a reduction of the corresponding anodic peak. The catalytic efficiencies (CAT) were determined in the concentration ranges from 1 to 9×10^{-3} mol dm^{-3} achieving values of CAT from 6.13% to 50.78% for nitrite and 47.22% to 117.36% to iodate.

In 2022, Li et al. [34] used a solvothermal approach to prepare three new Keggin-type polyoxometalate-based MOCs, $[Co(HL^1)_2(H_2O)_2][SiW_{12}O_{40}]$ (1), $[Co_2(L^2)_2(H_2O)_2(SiW_{12}O_{40})]$ (2) and $[Co_2(L^3)_2(H_2O)_2(SiW_{12}O_{40})]$ (3). These were prepared with cobalt ions, $[SiW_{12}O_{40}]^{4-}$ anion and different amide derivative ligands, *N*-(quinoline-5-yl)-1*H*-imidazole-4-carboxamide (L^1), 4-[(1*H*-imidazole-4-carboxamide)ethyl]phenyl-1*H*-imidazole-4-carboxamide (L^2) and bis[4-(1*H*-imidazole-4-carboxamide)phenyl]methanone (L^3) using mixed solvent conditions (5 : 3, 5 : 5 and 3 : 5 of H_2O:EtOH for L^1, L^2 and L^3, respectively). The MOCs (1) and (3) were evaluated towards nitrite electroreduction but also bromate and exhibited good performances with excellent selectivity.

Another type of POM-based composites has also emerged in the last decade (POMOF) as a promising class of materials for electrocatalysis, in which the properties of the active support (MOF = metal organic framework) can be enhanced by the incorporated guest (POM). Some have been employed in the electrocatalysis of nitrite with good performances, namely the $[Cu_2(H_2O)_2(bpp)_2Cl][PM_{12}O_{40}]\cdot xH_2O$ (M = W, Mo; bpp = 1,3-bis(4-pyridyl)propane) [35] and the $[Co_2(bpy)_3(ox)][As_2W_{18}O_{62}]\cdot 2(H_2bpy)\cdot 2H_2O$ (bpy = 4,4′-bipyridine; ox = oxalate) [36]. In 2014, Fernandes et al. [37] reported the application of a $SiW_{11}Fe$@MIL-101(Cr) basal plane pyrolytic graphite-modified electrode towards nitrite electroreduction with excellent performance. The electrocatalyst showed good stability, with the catalytic current increasing linearly with the NO_2^- concentration up to 0.6×10^{-3} mol dm^{-3} and with a detection limit (DL) of 3.2×10^{-5} mol dm^{-3} (at pH = 2.5 H_2SO_4/Na_2SO_4 buffer solution). The catalyst showed an electrocatalytic efficiency CAT of 164.4%.

Composite electrocatalysts resulting from the combination of POMs and carbon materials (carbon nanotubes and graphene) have also emerged as promising materials. The immobilization of POMs on carbon materials is expected to increase the number of active centres and therefore improve their electrochemical and electrocatalytic properties. Song and co-workers [38] immobilized a phosphotungstic acid on multi-walled carbon nanotubes (MWCNT) grafted with a poly(4-vinylpyridine) linker ($PW_{12}O_{40}$/PVP-g-MWCNT), and the new material exhibited a linear range for a nitrite concentration between 1.2×10^{-6} and 1.75×10^{-5} mol dm^{-3} with a detection limit of 2×10^{-7} mol dm^{-3}. A lacunary phosphomolybdate $[PMo_{11}O_{39}]^{7-}$ (PMo_{11}) was also used to prepare different glassy carbon (GC)-modified electrodes with carbon materials [39]. This POM was confined within a matrix of either oxidized single-walled carbon nanotubes (ox-SWCNTs) or reduced graphene oxide (rGO) to improve the electron transfers. Five different configurations were prepared with and without intermittent chitosan (CS) layers (PMo_{11}/CS/GCE, PMo_{11}/ox-SWCNT/GCE, PMo_{11}/rGO/GCE, PMo_{11}/CS/rGO/GC and PMo_{11}/CS/ox-SWCNT/GC). The best-performing material was the PMo_{11}/ox-SWCNT/GCE with a sensitivity of 44.41 ± 1.16 mA M^{-1} followed by PMo_{11}/CS/ox-SWCNT/GC with 37.96 ± 0.981 mA M^{-1}. PMo_{11}/ox-SWCNT/GCE also presented a linear range between 3.0×10^{-5} and 1.6×10^{-2} mol dm^{-3} and a detection limit of 3.0×10^{-5} mol dm^{-3}.

In 2017, Keggin-type $[PMo_{12}O_{40}]^{3-}$ (PMo_{12}) was used to prepare a new nanocomposite using a chemical reduction method [40]. Through the optimization of the amount of reducing agent (thiourea), the POM was reduced, and then reacted with graphene oxide (GO) to assemble the POM clusters onto the rGO by the strong interaction between rGO and PMo_{12}.The PMo_{12}-rGO nanocomposite was successfully applied for the electrochemical reduction of nitrite, and the authors attributed its good performance to different factors: (i) the high conductivity of rGO accelerates the electron transfer between target molecule and electrode, (ii) the well-dispersed PMo_{12} with small size is beneficial for electrocatalytic activities, (iii) more importantly, the crumpled, porous and paper-like morphology structure of the as-prepared nanocomposite improved the electroactive surface area, enhancing the catalytic current densities and improving the electrocatalytic activity.

Ternary composite materials based on carbon nanotubes, POMs and ionic liquids (IL) have also been reported as electrocatalysts for nitrite electroreduction [41, 42]. Wang et al. [42] prepared three new composites (SWCNTs-ILC$_n$-PMo_{12}, $n = 4, 8, 12$). In order to achieve it, the SWCNT-substituted ionic liquids were prepared by esterification reaction of the carboxylated SWCNT with hydroxyl ammonium ionic liquids $(CH_3N(CH_2CH_2OH)_2(C_nH_{2n+1})Br$ with $n = 4, 8, 12$). These were then reacted with the POM leading to the final materials. These materials were applied towards the reduction/detection of nitrite, iodate and bromate with excellent results. Regarding nitrite, linear ranges of 0.02–0.99, 0.02–1.58 and 0.02–4.33 mmol dm^{-3}, with a sensitivity of 0.43, 0.29 and 0.19 μA M m^{-1} for SWCNTs-ILC$_n$-PMo_{12} with $n = 4, 8, 12$, were found, respectively. The better performance of SWCNTs-ILC$_{12}$-PMo_{12} was attributed to the longer length of alkyl carbon chain which led to a better-ordered arrangement. Practically at the same time, Feizy and Haghighi reported a MWCNT coated with PMo_{12} and the ionic liquid (*n*-dodecyl pyridinium hexafluorophosphate, $[C_{12}Py][PF_6]$) [41]. The electrocatalytic properties of this new composite were evaluated towards different species including nitrite with good performance (sensitivity of 8.2 μA mM^{-1} cm^{-2}, linear range between 0.1 and 23 mmol dm^{-3} and a limit of detection of 57 μmol dm^{-3}). The reason for this electrocatalytic activity was the strong interaction between the hydrophobic alkyl chain of the ionic liquid $[C_{12}Py]$ and the MWCNTs which improved the charge transfer properties of the POM units.

Another very popular method for POMs immobilization onto electrode surfaces is the layer-by-layer (LbL) one. It allows the preparation of highly uniform, versatile and stable films [43]. Ma's group employed this method to immobilize two Keggin-type POMs $[\alpha\text{-SiVW}_{11}O_{40}]^{5-}$ and $[\gamma(1,2)\text{-SiV}_2W_{10}O_{40}]^{6-}$ on a 4-aminobenzoic acid (4-ABA)-modified GCE with polyethylenimine (PEI) as counterions, and the $(SiVW_{11}/PEI)_{14}$/4-ABA/GCE multilayer film showed good electrocatalytic response towards the reduction of NO_2^- (in acetate buffer, pH 3.81) [44]. Later, the same group obtained a CAT of 283% towards NO_2^- reduction using a $[PEI/(P_2MoW_{17}/PDDA–AuNPs)_6/P_2MoW_{17}]$ film (pH = 2.0, PDDA = Poly-(diallyldimethylammonium chloride), AuNPs = gold nanoparticles) [45]. Fernandes et al. have also prepared several POM-based LbL films with excellent electrocatalytic activity towards nitrite reduction.

In these works the films were prepared using different materials as cationic counterparts like polyelectrolytes, cationic dyes and metallopolymers. Multilayer films based on Keggin-type $[SiW_{11}Co^{II}(H_2O)O_{40}]^{6-}$ and PEI [46] showed a linear range for the catalytic current up to a nitrite concentration of 1.2×10^{-3} mol dm^{-3}, a detection limit of 7.2×10^{-5} mol dm^{-3} and a CAT = 114% (in pH = 4 acetate buffer). Iron-substituted POMs $[PW_{11}Fe^{III}(H_2O)O_{39}]^{4-}$ ($PW_{11}Fe$) and $[SiW_{11}Fe^{III}(H_2O)O_{39}]^{5-}$ ($SiW_{11}Fe$) were also applied in the preparation of films with PEI (Figure 5.3a) [47]. The GCE/$(PEI/SiW_{11}Fe)_7$-modified electrode presented excellent electrocatalytic activity in pH 4.0 buffer (Figure 5.3b).

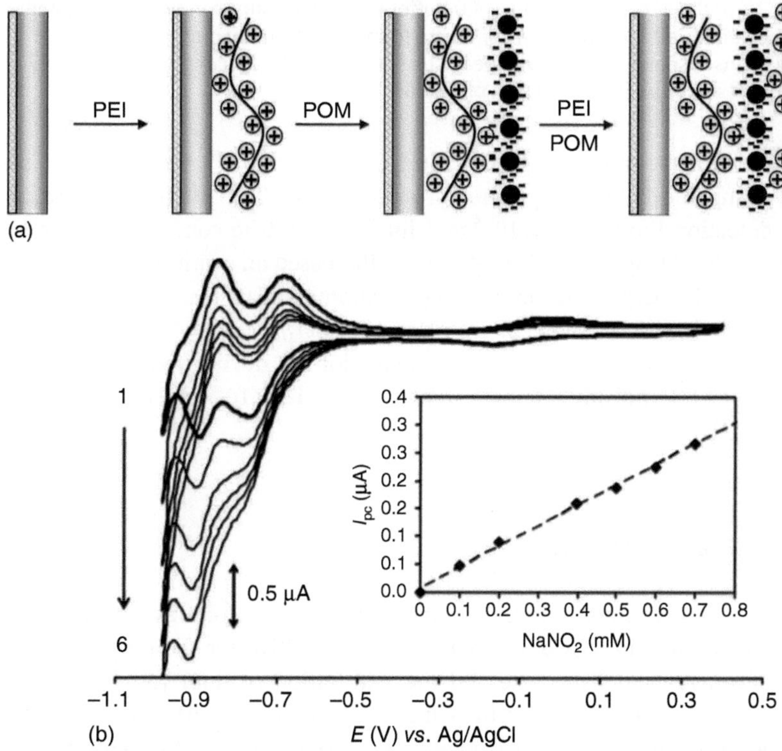

Figure 5.3 (a) Schematic diagram of the formation of the multilayer structure via alternate adsorption of cationic PEI and anionic POM on a substrate; (b) CVs of GCE/$(PEI/SiW_{11}Fe)_7$ in pH 4.0 buffer solution at a scan rate of 10 mV s^{-1} obtained in the absence and in the presence of added concentrations of nitrite: (1) 0; (2) 0.2; (3) 0.4; (4) 0.5; (5) 0.6 and (6) 0.7 mM. The inset shows the catalytic peak current at −0.77 V *vs.* the concentration of nitrite. Source: Fernandes et al. [47]/with permission of Elsevier.

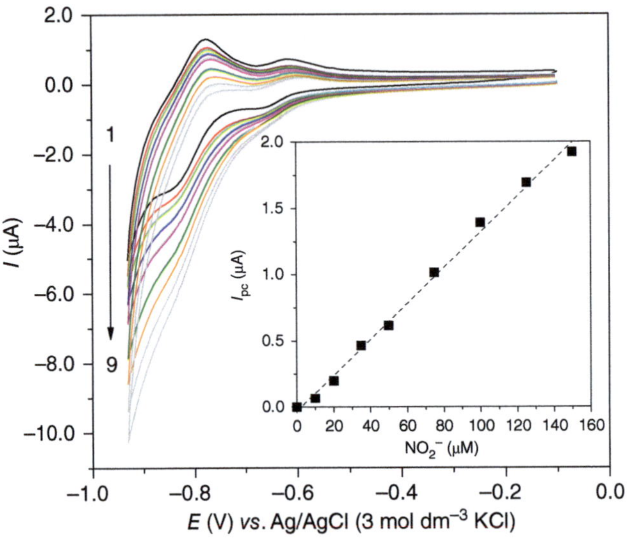

Figure 5.4 CVs of $[MG/Co_4(PW_9)_2]_n$ in pH 4.0 buffer solution obtained in the absence and in the presence of added concentrations of nitrite from 0 to 150 μmol dm^{-3}; scan rate 0.050 Vs^{-1}. The inset shows the catalytic peak current *vs.* analyte concentration at −0.849 V. Source: Novais et al. [48]/with permission of Elsevier.

In another work, the sandwich-type $[Co_4(H_2O)_2(PW_9O_{34})_2]^{10-}$ was immobilized on a GCE using the cationic dye methyl green (MG). The modified electrode obtained a detection limit of 7.5×10^{-6} mol dm^{-3} for a nitrite concentration up to 1.50×10^{-4} mol dm^{-3} (Figure 5.4) [48]. A hybrid film based on osmium metallopolymer $[Os(bpy)_2(PVP)_{10}Cl]Cl$ and a lanthano-phosphomolybdate $[Eu(PMo_{11}O_{39})_2]^{11-}$ was also prepared and it showed, in pH 2.5 H_2SO_4/Na_2SO_4 buffer solution, Mo-based electrocatalytic activity towards reduction of nitrite with a detection limit of 4.96×10^{-5} mol dm^{-3} (linear range up to 1.3×10^{-3} mol dm^{-3}) and a CAT = 234% [49].

5.3 Bromate Reduction

Bromate (BrO_3^-) is commonly found in drinking water samples as a by-product of ozone disinfection and is widely used as a food additive for the production of fermented beverages and fish paste and the maturation of flour [50]. However, bromate has been declared as carcinogenic and associated with the renal cell tumours, mesotheliomas of the peritoneum and follicular cell tumours by the presence of potassium bromate in drinking water. As a consequence, its determination has gained much interest, and several analytical methods have been proposed. Electrochemical reduction of bromate provides a suitable method for its determination and removal from aqueous media since it is highly selective, relatively clean, environment-friendly and cost efficient.

Even though several POM-based composites have been applied towards bromate reduction, to the best of our knowledge only one report can be found in literature regarding the application of pristine POM. Weng et al. [51] prepared four isostructural POMs consisting of Nb/W mixed-addenda Dawson-type subunits and lanthanide ions (Ln = La, Ce, Pr and Nd) leading to the general formula $K_4Na_8H_4([Nb_4O_6(H_2O)_4Na_4(H_2O)_8][LnP_2W_{12}Nb_6O_{61}(H_2O)_7]_4)\cdot xH_2O$. All POMs presented two pair of redox peaks attributed to the W-centres, and both these peaks were active for bromate electroreduction. The CAT values were determined and for La-based POM the values increased from 30% to 157% for I, and 54% to 258% for II when BrO_3^- concentration increases from 2 to 10 mmol dm^{-3}. For Pr, Nd and Ce-based POM the values went from 25% to 200% (I) and 50% to 245% (II), 25% to 240% (I) and 60% to 250% (II) and 60% to 150% (I) and 40% to 220% (II), respectively. Since in all cases the activity of redox peak II was higher, the authors suggested that the reduced Dawson-type unit reduced the bromate to bromide.

The most common approach to prepare successful bromate electrocatalysts with POMs is by far the synthesis of MOCs and inorganic–organic hybrids. Over the years dozens of reports have been published in literature regarding MOCs. Wang group prepared many MOCs using different POM units, namely $[\gamma\text{-}Mo_8O_{26}]^{4-}$, $[CrMo_6(OH)_5O_{19}]^{4-}$ ($CrMo_6$), $[CoMo_6(OH)_6O_{18}]^{3-}$ ($CoMo_6$), $[TeMo_6O_{24}]^{6-}$ ($TeMo_6$) and $[AlMo_6(OH)_5O_{19}]^{4-}$ ($AlMo_6$) combined with different pyridyl ligands [31, 34, 52–56]. Four polyoxometalate-based complexes were also reported by the same group using different POMs ($PMo_{12}O_{40}$, $SiW_{12}O_{40}$, $P_2W_{18}O_{62}$ and Mo_6O_{20}) copper nitrate and 2,20-dimethyl-4,40-bithiazole (dm4bt) [57]. As referred in the previous section, Liu and co-workers [32] reported a 2D coordination network structure $[Ni_2(BBPTZ)_5]_2[SiMo_{12}O_{40}]_2\cdot H_2O$ which was also evaluated towards bromate electroreduction in 0.5 mol dm^{-3} Na_2SO_4/0.1 mol dm^{-3} H_2SO_4. The results showed that the currents of the reduction peaks II and III increased significantly with the increase of bromate concentration from 0 to 5.0 mmol dm^{-3}, while the corresponding oxidation peak currents were reduced. In the concentration range from 550 to 3750 µmol dm^{-3}, a linear correlation was obtained leading to a sensitivity of 387.56 µA mM^{-1}. The same group reported a very similar material but using octamolybdate $[\beta\text{-}Mo_8O_{26}]^{4-}$ ($[Ni_2(BBPTZ)_4][\beta\text{-}Mo_8O_{26}]$). As for the previous example, the reduction peaks II and III were active for bromate reduction over a linear range between 1.35 and 4.75 mmol dm^{-3} with a sensitivity of 62.39 µA cm^{-2} mM^{-1}.

Several other POM-based inorganic–organic hybrid materials have also been reported as efficient bromate electrocatalysts. Wang et al. [58] used an inorganic–organic hybrid material $(C_{21}H_{38}N)_3PMo_{12}O_{40}$ as a bulk modifier to fabricate a renewable three-dimensional chemically modified CPE, and the proposed modified electrode showed excellent electrocatalytic activity towards the reduction of bromate. The same POM was used by Dong et al. [59] that prepared a $[C_8mim]_3PMo_{12}O_{40}$-modified ionic liquid CPE ($[C_8mim]_3PMo_{12}O_{40}$-ILCPE, C_8mim = 1-methyl-3-octylimidazolium). In both cases the molybdenum atoms were responsible for the bromate reduction. Vanadium-centred tungstovanadate [60] $[Zn(phen)_2(H_2O)]_2[VW_{12}O_{40}]$ (phen = 1,10-phenanthroline) and Dawson-type molybdosulphate [26] $(MB)_5[S_2Mo_1VMo_{17}VO_{62}]\cdot CH_3CN$ (MB = $C_{16}H_{18}N_3S$) were also used to fabricate modified CPEs with good electrocatalytic activity towards reduction of bromate. A gold electrode modified with $[SiNi(H_2O)W_{11}O_{39}]^{6-}$/cysteamine [61] was also used in the bromate reduction in acidic solution, and a linear range of the BrO_3^- detection was from 1.4×10^{-5} to 1.36×10^{-2} mol dm^{-3} and the detection limit was estimated to be 14.88×10^{-6} mol dm^{-3}, at a signal-to-noise ratio of 3.

As referred in Section 5.2, another type of POM-based composites that has emerged in the last decade with very promising electrocatalytic properties are the POMOF [62–64]. In 2016, Dong et al. [62] prepared a two-dimensional (2D) network of $[TBA]_3[H_2PMo_8{}^{V}Mo_4{}^{VI}O_{40}Zn_4][C_7H_6(COO)_2]$ with an angular bridging ligand of 5-mip (5-methylisophthalic acid). The electrocatalyst showed remarkable electrocatalytic activity toward bromate reduction but also nitrite reduction and ascorbic acid oxidation. In the following year, the same authors reported a new POMOF, $[TBA]_9([H_3PMo_8{}^{V}Mo_4{}^{VI}O_{40}]_2[H_2PMo_4{}^{V}Mo_8{}^{VI}O_{40}])[Zn_{12}(L^1)_4]$ using the methodology of extending the reduced transition-metal-grafted ε-Keggin polyoxoanions with terphenyl-based tricarboxylates of H_3L^1 (H_3L^1 = [1,1′;3′,1″-terphenyl]-3,5′,3″-tricarboxylic acid) [63]. The POMOF exhibited, in 0.1 mol dm^{-3} H_2SO_4, a sensitivity towards bromate detection of 422.42 μA cm^{-2} mM^{-1} over a linear range between 0.10 and 1.00 mmol dm^{-3}. Two electroactive POMOFs, NENU-3 and NENU-5, were also synthesized by a facile one-step solvothermal method containing $Cu(NO_3)_2$, 1,3,5-benzentricarboxylate (H_3BTC, also known as HKUST-1) and Keggin-type PMo_{12} and PW_{12} [65]. The POMs were wrapped into MOFs structure to increase its stability. Both POMOFs immobilized onto a carbon cloth presented good electrocatalytic activity towards bromate. NENU-3/CC showed a low detection limit of 0.55 μmol dm^{-3}, a linear range between 5 and 560 μmol dm^{-3} and a high sensitivity of 45.11 μA cm^{-2} mM^{-1}, while NENU-5/CC showed a detection limit of 1.18 μmol dm^{-3}, good linearity range from 15 to 380 μmol dm^{-3} and a high sensitivity of 18.83 μA cm^{-2} mM^{-1}.

Lu et al. [66] prepared a special kind of POMOF, a POM-based MONT where the last one stands for metal–organic nanotubes. MONTs have the advantages of carbon nanotubes and MOFs, like an open nanoporous structure, large specific surface area and crystallinity. The developed material ($[[(Cu_3(\mu 3\text{-}O))_2(trz)_6 Cu_2(H_2O)_{13}][H_{1.73}P_2As_{1.73}W_{16.27}O_{62}]\cdot 8.25H_2O$) (Figure 5.5a) is the first example of a POM-based MONT templated by twin-Dawson $P_2(As/W)_{18}$ clusters (Figure 5.5b). The material was then used to modify a GCE and was successfully applied towards bromate reduction (Figure 5.5c) achieving a CAT of 848.4% for a bromate concentration of 0.5 mmol dm^{-3}.

Immobilization of POMs in films is another very promising approach to overcome their problem of solubility. Hassan et al. [67] also demonstrated that a PMo_{12}/PEDOT/AuNP/GCE (PEDOT = poly(3,4-ethylenedioxythiophene)) had good electrocatalytic properties towards BrO_3^- reduction in acidic solution (0.1 mol dm^{-3} H_2SO_4), with LR = 2.50×10^{-4} to 3.0×10^{-3} mol dm^{-3} and sensitivity = 68 μA cm^{-2} mM^{-1}. PEDOT has also been used to prepare a film by chronocoulometry technique with Wells–Dawson-type POM $[S_2W_{18}O_{62}]^{4-}$ by McCormac and co-workers [68]. The resulting film modified electrode was

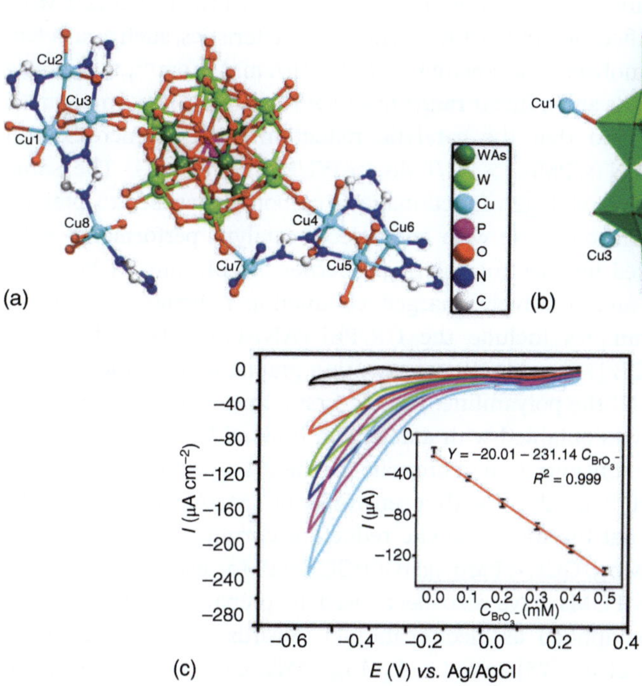

Figure 5.5 (a) The basic crystallographic unit in POM-based MONT; (b) the polyhedral/ball representation of the coordination mode of $P_2(As/W)_{18}$ cluster; (c) reduction of BrO_3^- in $1\,mol\,dm^{-3}$ H_2SO_4 solution. The inset shows a linear dependence of the cathodic catalytic current of peak IV with bromate. Source: Lu et al. [66]/with permission of Royal Society of Chemistry.

applied towards bromate reduction and monitored by voltammetry and amperometry. In the voltammetric mode, the linear range was observed from 2000 to 10,000 $\mu mol\,dm^{-3}$ with a sensitivity of about $229\,\mu A\,cm^{-2}\,mM^{-1}$ and limit of detection of $380\,\mu mol\,dm^{-3}$. In the chronoamperometric mode, the linear range was from 100 to 2000 $\mu mol\,dm^{-3}$ with a sensitivity of $97.59\,\mu A\,cm^{-2}\,mM^{-1}$ and a limit of detection (LOD) of $4\,\mu mol\,dm^{-3}$. The same group immobilized the Ni-Crown $[Ni_4(P_8W_{48}O_{184})(WO_2)]^{28-}$ into the polypyrrole matrix using the electropolymerization method [69]. In this case, only the first tungsten reduction peak (four electrons) was involved in the electrocatalytic reduction of bromate achieving a sensitivity of $69.40\,\mu A\,cm^{-2}\,mM^{-1}$ with a limit of detection of $0.2\,\mu mol\,dm^{-3}$.

Immobilization of POM using the LbL method has also been used, and a good example is the multilayer films based on polyethylenimine and Co-substituted silicotungstate, $GCE/(PEI/SiW_{11}Co)_n$ [46]. Results indicated that the second cathodic wave was responsible for the catalytic activity. The linear range for the catalytic current for bromate was up to $0.6 \times 10^{-3}\,mol\,dm^{-3}$ with a detection limit of $3.5 \times 10^{-5}\,mol\,dm^{-3}$.

Bromate electrocatalysts based on POMs and carbon materials have also been developed, and several examples concerning the use of phosphomolybdic acid can be found in the literature. In 2006, Li et al. [70] prepared a PG/MWCNTs/PMo_{12} pyrolytic-graphite-modified electrode with excellent characteristics, such as a detection limit of 5.0×10^{-7} mol dm^{-3}, a sensitivity of 760.9 μA mM^{-1} cm^{-2}, a response time less than two seconds and a linear range from 5×10^{-6} to 1.5×10^{-2} mol dm^{-3}. The authors also observed that the catalytic reduction currents increased in the sequence PG/MWCNTs/PMo_{12} > PG/PMo_{12} > PG/MWCNTs > PG. The same POM was immobilized on viologen-functionalized carbon nanotubes to prepare a MWNT-V@PMo_{12} nanocomposite with good electrocatalytic performance [71]. Skunik et al. [72] showed the electrocatalytic properties of LbL films of MWCNT modified with PMo_{12} and positively charged conducting polymer (PEDOT or polypyrrole). Some examples include the GR/PEI-PANI-PMo_{12}/GCE [73] and PMo_{12}@rGO-PDDA/GCE [74], with GR representing graphene, rGO the reduced graphene oxide and PANI the polyaniline. For the first a linear range of 7.5×10^{-6} to 5.0×10^{-4} mol dm^{-3} was obtained with a detection limit of 3.0×10^{-6} mol dm^{-3}. The linear range for the concentration of bromate for the PMo_{12}@rGO-PDDA/GCE was from 0.02 to 10×10^{-3} mol dm^{-3} with a sensitivity of 454 μA cm^{-2} mM^{-1}. The authors also reported that the total catalytic reduction currents increased in the following order: PMo_{12}/GCE < PMo_{12}@rGO/GCE < PMo_{12}@rGO-PDDA/GCE at 0.05 V. Dawson-type POMs have also been used to prepare composites with carbon materials to be applied as electrocatalysts towards bromate electrocatalytic reduction. Zhou et al. [75] tested a P_2Mo_{18}/OMC/GCE (OMC – ordered mesoporous carbon) and obtained a detection limit of 9.22×10^{-7} mol dm^{-3} (LR = 2.77×10^{-6} to 4.00×10^{-3} mol dm^{-3}), while Sosnowska et al. [76] prepared a CNTs/PDDA/$P_2W_{17}VO_{62}$-modified electrode with high electrocatalytic activity.

5.4 Iodate Reduction

Iodine is a vital micronutrient, which is an important part of the thyroid hormones that play a key role in the development of cell growth and brain function [77]. Deficiency of iodine can result in a serious delay in neurologic development. Otherwise, an excess of iodine or iodide can produce goitre and hypothyroidism as well as hyperthyroidism [78].

In natural waters, iodide concentration is normally low (<10 μg dm^{-3}), but due to seawater intrusion or even special geological formations it can reach values higher than 50 μg dm^{-3}. Potassium iodate has been extensively used for iodination of commercial table salts as a source of iodine and is an environmental pollutant and suspected carcinogen formed, e.g. in drinking water upon ozone treatment [77]. Still, it is quickly reduced to iodide in the body unlike bromate.

Apparently, only one paper has been reported regarding the electroreduction of iodate by POMs in solution [79]. Kemmegne-Mbouguen and co-workers studied the electrochemical and electrocatalytic properties of a series of iso-structural POMs $[SiW_{10}O_{36}(M_2O_2E_2)]^{6-}$ (M = Mo or W and E = S or O) [79]. The three POMs ($[SiW_{10}–Mo_2O_4]$, $[SiW_{10}–Mo_2O_2S_2]$ and $[SiW_{10}–W_2O_2S_2]$) were evaluated towards the reduction of iodate into iodide by CV in 0.1 mol dm^{-3} $Li_2SO_4 + H_2SO_4$

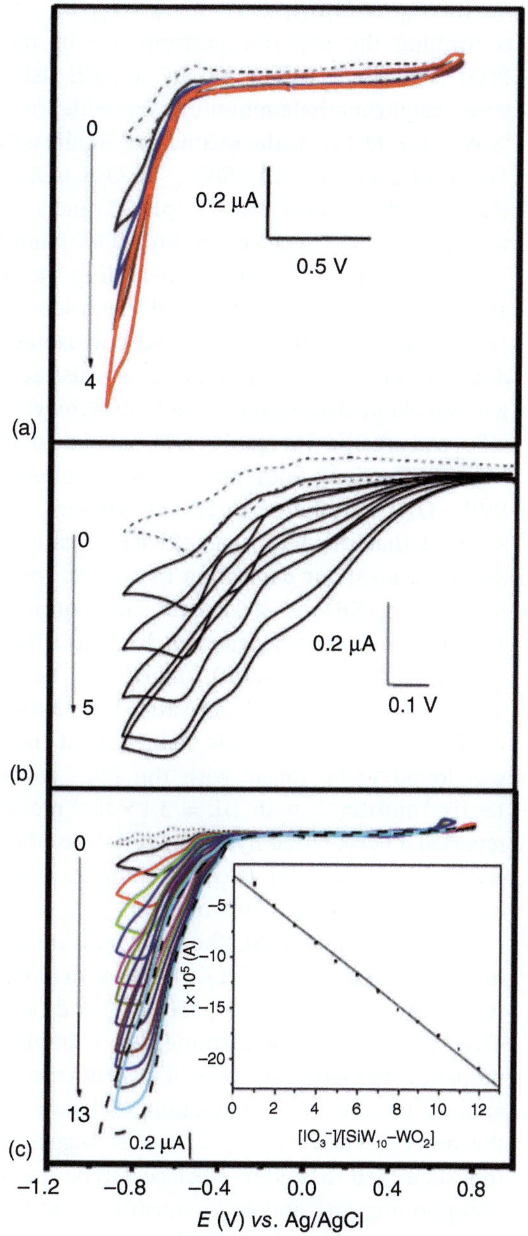

Figure 5.6 Cyclic voltammograms for (a) $[SiW_{10}-Mo_2O_4]$, (b) $[SiW_{10}-Mo_2O_2S_2]$ and (c) $[SiW_{10}-W_2O_2S_2]$ obtained in the absence and in the presence of added concentrations of iodate in pH 3 buffer (0.1 mol dm^{-3} Li_2SO_4/H_2SO_4). Inset shows the influence of γ on the electrocatalytic current. Arrows on the left part of the figure indicate the range of the γ values. Source: Kemmegne-Mbouguen et al. [79]/Académie des sciences/CC BY 4.0.

(pH 3). For all POMs, there was an increase in peak current intensities with increasing IO_3^- concentration as it can be observed in Figure 5.6. The reduction current for $[SiW_{10}–W_2O_2S_2]$ (at −0.750 V) increased linearly with increasing IO_3^- concentration up to $\gamma = 12$ (with $\gamma = [IO_3^-]/[POM]$), while for $[SiW_{10}–Mo_2O_4]$ (at −0.840 V) and $[SiW_{10}–Mo_2O_2S_2]$ (at −0.770 V) the reduction current increased up to $\gamma = 4$. The authors also estimated the CAT values for $\gamma = 4$ and obtained 1187%

for $[SiW_{10}–W_2O_2S_2]$, 219% for $[SiW_{10}–Mo_2O_4]$ and 332% for $[SiW_{10}–Mo_2O_2S_2]$ confirming the superior performance of the first POM. The best-performing POM was also used to modify a GCE using the LbL method. For this purpose, cetyltrimethylammonium bromide (CTAB) was used as the first layer $[SiW_{10}–Mo_2O_2S_2]$, as the second and finally a film of cationic biopolymer chitosan. The resulting GCE/CTA/$[SiW_{10}–W_2O_2S_2]$/Chitosan was evaluated in 0.1 mol dm^{-3} H_2SO_4 + 0.1 mol dm^{-3} Na_2SO_4 (pH 2.5) and a linear range from 37 to 709 μmol dm^{-3} was obtained with a detection limit of 6.2 μmol dm^{-3}.

The majority of published regarding electroreduction of iodate with POMs are based on CPEs or modified electrodes with composite materials or films. For example, the alcohol-solvated polyoxometalate, $[NH_4]_4[SiW_{12}O_{40}]·3C_2H_5OH$ described previously in Section 5.2, was also applied towards iodate electroreduction with excellent electrocatalytic activity achieving a CAT of 134% at pH = 3.5 [23].

Several organic–inorganic POM-based composites have also been applied towards iodate reduction. Ma's group [28] reported composites based on Keggin-type $PVW_{11}O_{40}$ coordinated to two $[Cu(phen)_2]^{2+}$ units. Cyclic voltammetry studies revealed that the IO_3^- was efficiently catalysed by the W-centres. Sun's group [80] used a vapour deposition method to immobilize $CoW_{11}Co$ and poly(4-vinylpyridine) in a TiO_2 sol–gel matrix. The resulting modified electrode ($CoW_{11}Co$/PVP/TiO_2/GCE) showed a detection limit of 8.0×10^{-7} mol dm^{-3} with a linear range from 2.0×10^{-6} to 2.8×10^{-4} mol dm^{-3}. Keggin-type α-PW_{12} [81] supported on the surface of silica gel derivatized by 3-aminopropyl(triethoxy)silane was also used in the electrocatalytic reduction of iodate, and the catalytic peak current was found to be linear with the IO_3^- concentration in the range 5×10^{-6} to 1×10^{-3} mol dm^{-3}, with DL = 3.1×10^{-6} mol dm^{-3}. Kakhki and Shams have also reported a POM-based hybrid material based on N,N_0-bis(salicylidene)-1,2-phenylenediaminocobalt(III) (CoSal) complex and SiW_{12} ($(CoSal)_4SiW_{12}$), where a detection limit of 4.8×10^{-8} mol dm^{-3} was obtained [82].

More recently, in 2020, Zhou and co-workers [83] reported three hybrid composites, $(H_2en)[(K(H_2O)_{0.5})_2(K_2(H_2O)_3)(Ni(H_2O)(en)_2)_2(Ni_4(H_2O)_2(PW_9O_{34})_2)]$ (1), $[Cu_6(Himi)_6\text{-}(As^{III}W_9O_{33})_2]·5H_2O$ (2) and $(H_2btp)_4[Fe^{III}{}_2Fe^{II}{}_2(H_2O)_2(AsW_9O_{34})_2]·4H_2O$ (3) (en = ethanediamine; imi = imidazole; btp = 1,3-bis(1,2,4-triazol-1-yl) propane), prepared via in situ hydrothermal reaction. These were successfully applied for the electroreduction of iodate with catalytic efficiencies following the order: 3-GCE > 1-GCE > 2-GCE, suggesting that tetra-Fe-substituted arsenotungstate had a better electrocatalytic performance than tetra-Ni-substituted phosphotungstate and 6-Cu-substituted arsenotungstate. The better performance of (3) was attributed to the higher redox peak position of iron centre and the higher synergistic effect between the W centre in the POM and the mixed valence tetranuclear iron cluster.

As referred previously, the construction of POM-based films is another common approach. Fernandes et al. have reported several POM-based modified electrodes using the LbL methodology. In 2011, they reported a $(PEI/SiW_{11}Co)_n$/GCE [46] where the linear range for the catalytic current for iodate was up to 0.5×10^{-3} mol dm^{-3} and a detection limit 5.8×10^{-5} mol dm^{-3}. Later, a (Os-poly/Eu

$(PMo_{11})_2)_n$ multilayer film composed of osmium metallopolymer $[Os(bpy)_2(PVP)_{10}Cl]Cl$ (Os-poly) and europium phosphomolybdate $[Eu^{III}(PMo_{11}O_{39})_2]^{11-}$ was reported [49]. Electrochemical studies showed that the second Mo-centred reduction process was significantly enhanced by the addition of iodate with a CAT of 89%. The linear range for the catalytic current for iodate was up to 4×10^{-4} mol dm^{-3}, with a DL of 2.95×10^{-5} mol dm^{-3}. The authors also used two organic dyes, pararosaniline (PR) and MG, to prepare LbL films with a vanadium-substituted phosphomolybdate $[PMo_{11}VO_{40}]^{5-}$ [84] and a sandwich-type polyoxometalate $K_{10}[Co_4(H_2O)_2(PW_9O_{34})_2]$ [48], respectively. In the films with $PMo_{11}V$, the Mo centres were responsible for electrocatalysis of iodate (LR up to 0.8×10^{-4} mol dm^{-3}, DL = 4.3×10^{-6} mol dm^{-3}), while for the films with $Co(PW_9)_2$ the iodate reduction was W-centred (LR up to 2.25×10^{-4} mol dm^{-3}, DL = 5.9×10^{-6} mol dm^{-3}).

Ma's group also employed the LbL method: (i) to immobilize two Keggin-type POMs $[\alpha\text{-}SiVW_{11}O_{40}]^{5-}$ and $[\gamma(1,2)\text{-}SiV_2W_{10}O_{40}]^{6-}$ on a 4-aminobenzoic-acid-modified GCE with poly(ethylenimine) as counterions where the $SiVW_{11}$ multilayer films showed much higher electrocatalytic response than SiV_2W_{10} multilayer films [44]; and (ii) to prepare a multilayer film with Dawson-type P_2MoW_{17} and PDDA which was further decorated by gold nanoparticles, denoted as $(PEI/(P_2MoW_{17}/PDDA\text{–}AuNPs)_6/P_2MoW_{17})$ [45]. The CAT obtained was CAT = 414%. McCormac group have also reported POM-based films either using LbL [85] method or by entrapment in a polymeric film by electropolymerization [86]. In the first example, a $(Cu_{20}P_8W_{48}/RuDen)$ film was fabricated using a ruthenium (II) metallodendrimer (RuDen) and the wheel-shaped Cu_{20}-tungstophosphate $[Cu_{20}Cl(OH)_{24}(H_2O)_{12}(P_8W_{48}O_{184})]^{25-}$. The catalytic current increased linearly with iodate concentration up to 1 mmol dm^{-3}. In the other work, a Dawson-type sulphate $[S_2W_{18}O_{62}]^{4-}$ was used to prepare POM-based films through the entrapment in polypyrrole (PPy) films on GCE surfaces through pyrrole electropolymerization [86]. A linear range from 100 μmol dm^{-3} to 0.7 mmol dm^{-3} was obtained with a detection limit of 0.5 μmol dm^{-3} and a sensitivity of 0.769 μA cm^{-2} μM^{-1}.

A robust and stable film comprising *n*-octylpyridinum hexafluorophosphate ($[C_8Py][PF_6]$) and phosphomolybdic acid was prepared by Haghighi et al. [78] on GCEs modified with multi-walled carbon nanotubes (GCE/MWCNTs). Electrocatalytic tests showed a good linear range (2×10^{-5} to 2×10^{-3} mol dm^{-3}), where the sensitivity and detection limit were 190 μA mM^{-1} cm^{-2} and 15×10^{-6} mol dm^{-3}, respectively. The electrocatalytic activity towards iodate was also addressed by Guo et al. [87] who reported a modified electrode where CNTs were dispersed in polycationic chitosan films which were then used to electrostatically immobilize Dawson anions P_2W_{18}. Electrochemical studies demonstrated that the P_2W_{18}/CNTs–chitosan electrode exhibits fast response and good electrocatalytic activity for the reduction of iodate anions (LR = 1.96–35.3×10^{-6} mol dm^{-3}; DL = 2.17×10^{-7} mol dm^{-3}), and it was reported that the sensor works under acidic conditions also, broadening its applicability under industrial conditions.

More recently, in 2021, Dianat and co-workers [88] reported a $(bdpy)SiW_{11}Ni/GO$ nanohybrid compound deposited on a GCE by one-step drop-casting, followed by the electrochemical reduction of graphene oxide through chronoamperometry at a constant potential of −1.5 V (*vs.* Ag/AgCl), for 60 seconds, in a phosphoric acid solution (0.1 mol dm^{-3}) containing bdpy (1.0 mmol dm^{-3}). According to the authors, the (bdpy)$SiW_{11}Ni$ interacted with phosphorus-doped electrochemically reduced graphene oxide (P-ERGO) via $Pd^{\delta+}$ and π-electrons of (1,10-(1,4-butanediyl) dipyridinium) ionic liquid (bdpy) without affecting the electronic structure of POM. The presence of bdpy allowed a higher loading of POM and improved the nanocomposite film stability due to the strong electrostatic attraction between $[SiW_{11}Ni]^{6-}$ and positively charged bdpy. The modified electrode was successfully applied towards iodate electroreduction with a high sensitivity of 28.1 mA mM^{-1}, good selectivity and a low limit of detection (4.7×10^{-4} μmol dm^{-3}) within a linear range 10–1600 μmol dm^{-3}. Later, the same group used the same methodology to prepare a nanocomposite based on PW_{12}, a cobalt-based MOF and graphene oxide (PW_{12}/MOF/GO) [89]. The proposed sensor showed a linear range of 10–2000 μmol dm^{-3} with a limit of detection of 3.86 μmol dm^{-3}, a sensitivity of 20.0 μA mM^{-1} and excellent selectivity for IO_3^- detection in the presence of other potential interfering species.

5.5 Hydrogen Peroxide Reduction Reaction

Hydrogen peroxide (H_2O_2) is a strong oxidant, widely applied in food and pharmaceutical proceedings as sterilizing agent and released from several industrial processes, whereby its food content and environmental presence need to be strictly controlled [90, 91]. Therefore, the development of H_2O_2 electrochemical sensors is of great interest. Commonly, these sensors are based on the hydrogen peroxide reduction reaction. However, despite the positive standard potential of H_2O_2/H_2O couple ($E° = 1.76$ V *vs.* NHE), the H_2O_2 reduction is not conducted at significant rates at most electrode surfaces. This can be overcome using electrocatalysts. Among the several materials that can be used to catalyse this reaction, POMs appeared as good candidates as these are readily reduced at quite positive potentials. Furthermore, the ability of acid/base and redox properties of POMs to work cooperatively in the catalytic processes can be quite helpful, since the H_2O_2 reduction is a two-electron/two-proton process ($H_2O_2 + 2e^- + 2H^+ \rightarrow 2H_2O$). During the electrocatalytic reduction of H_2O_2 by POMs, the formation of a peroxo-complex between POM and H_2O_2 as intermediate is accepted, which should lead to the weakening of the O—O bond and its cleavage [92].

Martel and Kuhn [92] studied the electrocatalytic properties of $[P_2Mo_{18}O_{62}]^{6-}$ both in solution and when adsorbed as a monolayer onto a GCE. The authors observed that, in 0.5 mol dm^{-3} H_2SO_4, the H_2O_2 reduction occurs at $E = 0.35$ V (*vs.* Ag/AgCl) in both cases, but the CAT was higher for the adsorbed POM (6.1) in comparison with the POM in solution (2.2). A GCE modified with a Keggin-type palladium-substituted polyoxotungstate ($[SiW_{11}Pd^{II}O_{39}]^{6-}$) also showed to be an excellent electrocatalyst [93], with the Pd centre playing a key part in the process. The electrocatalyst showed good durability and sensitivity, with the catalytic current increasing linearly with the H_2O_2 concentration up to 0.15 mol dm^{-3} and with a detection limit of 4.0×10^{-6} mol dm^{-3} (at pH = 4).

As for the other reactions, in H_2O_2 reduction, the use of POM-based MOCs and organic–inorganic composites is highly promising and has been used to overcome the issue of poor solubility of POMs in aqueous solutions and some organic solvents. Therefore, taking advantage of the coordination ability of POMs, the combination of transition-metal complexes and POMs has been explored to provide new organic–inorganic hybrids. Examples of these hybrids were prepared by Ma's group, with the Keggin-type anion $[PVW_{11}O_{40}]^{4-}$ coordinated to two $[Cu(phen)_2]^{2+}$ units [28], with a dimeric Keggin $[(PW_{11}CuO_{39})_2]^{10-}$ cluster linked to $[Cu(en)_2]^{2+}$ (en = ethylenediamine) complex to form a 3D framework [94] and between the $[PW_{12}O_{40}]^{3-}$ and $[Co(H_2O)_2(bimb)_2]_2$ (bimb = 1,3-bis(1-imidazoly)benzene) [95], employing hydrothermal conditions. In all cases, the CV experiments showed that POM W-centres were responsible for electrocatalytic activity towards H_2O_2 reduction. In the same context, a composite between $PMo_{12}O_{40}$ and tris-(2,2′-bipyridine) ruthenium(II) was prepared ($[Ru(bpy)_3]_3[PMo_{12}O_{40}]_2 \cdot 4H_2O$) and used to modify a CPE [96]. Its good electrocatalytic performance, in the range between 100 and 700 mV *vs.* SCE, was attributed to PMo_{12} counterpart. The same POM was used to prepare a new supramolecular compound by a hydrothermal method using Ag^+ cations and dpdo molecules (dpdo = 4,4′-dipyridine-*N*,*N*′-dioxide) [97]. The electrocatalyst presented good electrocatalytic activity with a CAT of 116.7%. Similar inorganic–organic hybrids were prepared with 5-phenyl-2-(4-pyridinyl) pyridine (pbpy) [98] and between bipyridine and 12-tungstovanadate ($[VW_{12}O_{40}]$.4(2,2′-Hbipy)) [99]. In 2012, a composite based on tris-(2,2′-bipyridine) ruthenium(II) and the Dawson-dissymmetrical $[H_4PW_{18}O_{62}]^{7-}$ was prepared [100]. The resulting H_2O_2 sensor showed a reductive detection at $E = -0.4$ V (*vs.* Ag/AgCl) in a linear range from 0.5 to 0.090 mol dm^{-3} of H_2O_2 and high selectivity towards interferences such as uric acid and ascorbic acid. The Wang group have also prepared a number of POM-based MOCs using different POM units, namely $[CrMo_6(OH)_5O_{19}]^{4-}$ ($CrMo_6$), $[CoMo_6(OH)_6O_{18}]^{3-}$ ($CoMo_6$) and $[TeMo_6O_{24}]^{6-}$ ($TeMo_6$) combined with different pyridyl ligands [31, 52, 53].

The latest work was published in 2022 and reports the synthesis, under hydrothermal conditions, of a Keggin-type arsenotungstate supramolecular assembly modified with MOCs, $[Zn(dpb)_2]_2[AsW^{VI}{}_{11}W^{V}O_{40}] \cdot 4H_2O$ where dpb = 2,6-di(2-pyridine)-4,40-bipyridine [101]. The electrocatalyst showed, in 0.1 mol dm^{-3} H_2SO_4/Na_2SO_4, four pairs of reversible peaks labelled as I–I′, II–II′, III–III′ and IV–IV′ (Figure 5.7a) which were attributed to the W centres in the $[AsW_{12}]$ anion. All these peaks presented electrocatalytic activity towards H_2O_2 electroreduction (Figure 5.7b). The CAT for different H_2O_2 concentrations was calculated (Figure 5.7c) with the highest value (71.55%) achieved for 0.8 mmol dm^{-3}. The sensing performance was also determined by amperometry with the proposed electrocatalyst presenting a low detection limit (0.43 μmol dm^{-3}) and wide linear range from 1.28 μmol dm^{-3} to 2.23 mmol dm^{-3}.

Electrocatalysts based on POM immobilization on carbon-based materials have showed to allow maximize the amperometric response and maintaining the POM

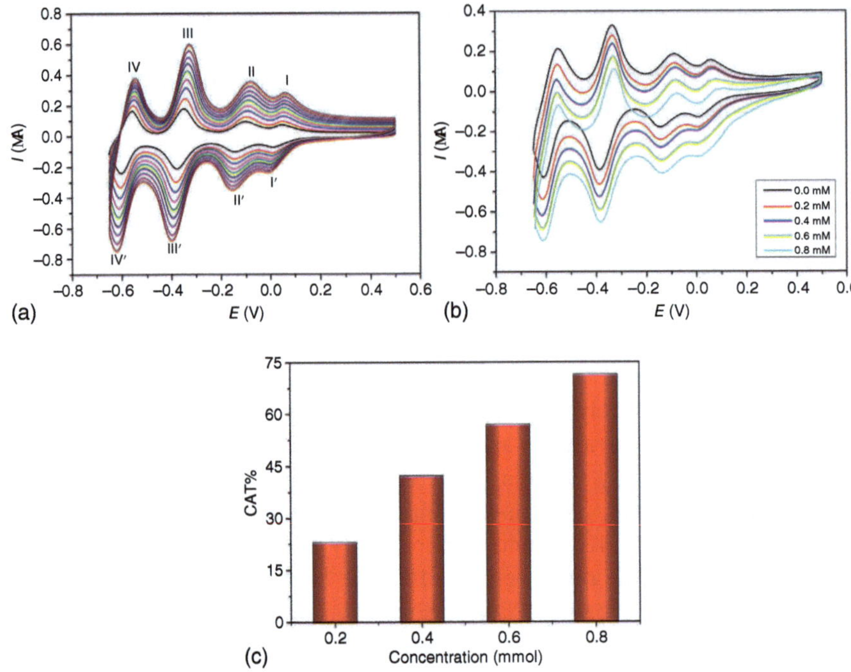

Figure 5.7 (a) The CV curves of $[Zn(dpb)_2]_2[AsW^{VI}_{11}W^{V}O_{40}]$/GCE at different scan rates; (b) The CV in 0.1 mol dm^{-3} H_2SO_4/Na_2SO_4 solution containing H_2O_2 at different concentrations. (c) Chart of the CAT *vs.* concentration of H_2O_2. I_p values of cathodic peak VI′. Source: Liu et al. [101]/with permission of Elsevier.

dispersion to reach low analyte detection limits [90]. Considering this, Zhou et al. [75] reported the immobilization of an OMC functionalized with P_2Mo_{18} onto a GCE. Other three-component H_2O_2 sensors were developed by the GCE modifications with SWCNTs, $SiMo_{12}$, and the copper complex $[Cu(bpy)_2]Br$ [102] and with Au NPs, PW_{12} and graphene [103]. Compared with non-POM-functionalized electrodes, the POM-modified hybrids gave excellent electrocatalytic activity towards H_2O_2 electroreduction, with good sensitivity in a wide linear range of H_2O_2 concentrations and with long-term stabilities. This is attributed to high dispersion and adsorption levels of POMs in these carbon materials through stable interactions. An electrocatalyst prepared by the formation of a film comprising an ionic liquid and PMo_{12} onto a GCE modified with MWCNT was also used towards H_2O_2 electroreduction showing an elevated catalytic reaction rate (3.7×10^8 cm^3 mol^{-1} s^{-1}) [78].

In 2018, Zhang et al. [104] reported the one-pot syntheses of large quantity of highly stable MoS/rGO and MoS/Au nanoparticle nanohybrids in aqueous solution using the polyoxomolybdate, $[Mo^{V}_4O_8(OH)_2(H_2O)_2(C_4O_4)_2]^{2-}$ (MoS), as reducing and stabilizing agent. The MoS/rGO-modified electrode was evaluated towards H_2O_2 reduction by CV and amperometry in 0.5 mol dm^{-3} $Na_2SO_4 + H_2SO_4$ (pH = 2). Results showed that the reduction of hydrogen peroxide is efficiently triggered by the Mo-waves of MoS/rGO with the gradual increase of the reduction currents upon increasing H_2O_2 concentration. Amperometric sensor showed a response range from 5×10^{-5} to 2×10^{-2} mol dm^{-3} with a limit of detection of 2 μmol dm^{-3} and a sensitivity of 33.2 μA cm^{-2} mM^{-1}. The authors showed the ability of a LbL film using the same POM, graphene oxide and PDDA ($[PDDA/MoS/PDDA/GO]_{17}$) to reduce H_2O_2. The LbL is commonly used as it allows obtaining highly uniform and stable films with a high degree of structural and functional control [105]. McCormac's group reported the assemblage of the crown-type POM – $P_8W_{48}Fe_{16}$ with cationic ruthenium metallodendrimer [105] and methylene-blue [106] onto a GCE, as effective electrocatalysts for H_2O_2 electroreduction. Another POM ($[Fe_4(C_2O_4)_4(H_2O)_2(SbW_9O_{33})_2]^{14-}$) was also combined with a ruthenium metallodendrimer [107] with good results. This suggested that the POM clusters are the ones responsible for the H_2O_2 electroreduction. Ammam et al. [108] also fabricated multilayer films using Wells–Dawson derived sandwich-type $[Fe^{III}{}_4(H_2O)_2(As_2W_{15}O_{56})_2]^{12-}$ and mono-substituted-type α_2-$[P_2W_{17}O_{61}(OH)_2Fe^{III}]^{7-}$ and poly(allylamine hydrochloride). Although the POMs can suffer degradation during their incorporation in multilayer films as result of the electrostatic repulsions between the positively charged metal-cation-substituted (Fe^{III}) of the POM and the positively charged poly(allylamine hydrochloride) (PAH), leading to the formation of lacunary-type POMs, the number of POM units that remain unchanged showed some electrocatalytic response for H_2O_2 reduction. The CAT was determined at −0.4 V, and values of 500% and 2400% were obtained for concentration of H_2O_2 of 60 and 100 μmol dm^{-3}, respectively.

5.6 Conclusions

Polyoxometalate-based composite materials have become, in last years, a significant class of promising candidates for functional materials in numerous applications as a result of their outstanding electrochemical performances. In this chapter, the electrocatalytic reduction of nitrites, bromates, iodates and hydrogen peroxide is reviewed. A great variety of modified electrodes with POMs and POM-based (nano) composites have been explored in the selected reactions with improved selectivity and sensitivity. In particular, the preparation of MOCs and the immobilization of POMs in high surface materials such as carbon-based nanomaterials (e.g. graphene and CNT) have proved to be the best approaches to enhance the electrocatalytic performance. In general, for all systems studied, the W-, Mo- and V-centres of the POM moieties are responsible for the electrocatalytic activities in sensor applications.

Exploring new materials with better performance, stability and proton and electron conductivity is always imperative in practical applications and will contribute to take the application of POMs in electrocatalysis into a next level. Still, there are several aspects to take into account. First, it is important to design and produce new POM-based materials with unique structures by introducing different metal components such as high nuclear transition metals. This will contribute to increased electrocatalytic performances due to possible synergetic effect between these transition-metal centres within the POM structure and the excellent electron and proton transfer capacity of the POM skeletons. POM-based organic–inorganic hybrids are also a relevant class of materials, and there is still plenty of room for improvement. Different organic components may provide different coordination modes to construct more remarkable structures, which will potentially exhibit unexpected electrocatalytic activities. Developing new synthetic strategies for POM-based composites is also an important key point as it will enable the appearance of novel materials with enhanced electrocatalytic properties.

Acknowledgment

This work received support and help from FCT/MCTES (LA/P/0008/2020 DOI 10.54499/LA/P/0008/2020, UIDP/50006/2020 DOI 10.54499/UIDP/50006/2020 and UIDB/50006/2020 DOI 10.54499/UIDB/50006/2020), through national funds. D.M. Fernandes also thanks FCT for funding through the Individual Call to Scientific Employment Stimulus 2021.00771.CEECIND/CP1662/CT0007.

List of Abbreviations

$[C_8Py][PF_6]$	*n*-octylpyridinum hexafluorophosphate
$[Ru(bpy)_3]$	tris-(2,2′-bipyridine) ruthenium(II)
2D	two dimensional
3D	three dimensional
4-ABA	4-aminobenzoic acid
5-mip	5-methylisophthalic acid
Ag/AgCl	Ag/AgCl reference electrode
AuNPs	gold nanoparticles
BBPTZ	4,4′-bis(1,2,4-triazol-1-ylmethyl)biphenyl
bdpy	1,10-(1,4-butanediyl)dipyridinium ionic liquid
bimb	1,3-bis(1-imidazoly)benzene
bipy	2,2′-bipyridine
bpp	1,3-bis(4-pyridyl)propane
bpy	4,4′-bipyridine
btp	1,3-bis(1,2,4-triazol-1-yl) propane
C_8mim	1-methyl-3-octylimidazolium
CAT	catalytic efficiency
CNT	carbon nanotubes
CoSal	N,N_0-bis(salicylidene)-1,2-phenylenediaminocobalt(III)
CPE	carbon paste electrode
CS	chitosan
CTAB	cetyltrimethylammonium bromide
CV	cyclic voltammetry
DEMS	differential electrochemical mass spectrometry

DL	detection limit
dm4bt	2,20-dimethyl-4,40-bithiazole
dpb	2,6-di(2-pyridine)-4,40-bipyridine
dpdo	4,4′-dipyridine-*N*,*N*′-dioxide
en	ethanediamine
GCE	glassy carbon electrode
GO	graphene oxide
GR	graphene
H_2-pyttz	3-(pyrid-4-yl)-5-(1*H*-1,2,4-triazol-3-yl)-1,2,4-triazolyl
H_2typtz	3-(tetrazol-5-yl)-5-(pyrid-3-yl)-1,2,4-triazolyl
H_3BTC	1,3,5-benzentricarboxylate (also known as HKUST-1)
H_3L^1	[1,1′;3′,1″-terphenyl]-3,5′,3″-tricarboxylic acid
IL	ionic liquid
imi	imidazole
L^1	*N*-(quinoline-5-yl)-1*H*-imidazole-4-carboxamide
L^2	4-[(1*H*-imidazole-4-carboxamide)ethyl]phenyl-1*H*-imidazole-4-carboxamide
L^3	bis[4-(1himidazole-4-carboxamide)phenyl]methanone
LbL	layer-by-layer
LOD	limit of detection
LR	linear range
MG	methyl green
MOCs	metal–organic complexes
MOF	metal organic framework
MONTs	metal–organic nanotubes
MWCNT	multi-walled carbon nanotubes
NPs	nanoparticles
OMC	ordered mesoporous carbon
ox	oxalate
ox-SWCNTs	oxidized single-walled carbon nanotubes
PAH	poly(allylamine hydrochloride)
PANI	polyaniline
pbpy	5-phenyl-2-(4-pyridinyl)pyridine
PDDA	poly(diallyldimethylammonium chloride)
PEDOT	poly(3,4-ethylenedioxythiophene)
PEI	polyethylenimine
P-ERGO	phosphorus-doped electrochemically reduced graphene oxide
PG	pyrolytic graphite
phen	1,10-phenanthroline
POMOF	POM–organic frameworks
POMs	polyoxometalates
PPy	polypyrrole
PR	pararosaniline
PVP	poly(4-vinylpyridine)

rGO	reduced graphene oxide
RuDen	ruthenium(II) metallodendrimer
SWCNT	single-walled carbon nanotubes
TBA	tetrabutylammonium
tib	1,3,5-tris(1-imidazoly)benzene

References

1 Sadakane, M. and Steckhan, E. (1998). Electrochemical properties of polyoxometalates as electrocatalysts. *Chem. Rev.* 98 (1): 219–237.

2 Keita, B. and Nadjo, L. (2007). Polyoxometalate-based homogeneous catalysis of electrode reactions: recent achievements. *J. Mol. Catal. A-Chem.* 262 (1–2): 190–215.

3 Freire, C., Fernandes, D.M., Nunes, M., and Araújo, M. (2016). Polyoxometalate-based modified electrodes for electrocatalysis: from molecule sensing to renewable energy-related applications. In: *Advanced Electrode Materials* (ed. A. Tiwari, F. Kuralay, and L. Uzun), 149–211. USA: Wiley Scrivener Publishing.

4 Cammack, R., Joannou, C.L., Cui, X.Y. et al. (1999). Nitrite and nitrosyl compounds in food preservation. *BBA-Bioenergetics* 1411 (2–3): 475–488.

5 Honikel, K.O. (2008). The use and control of nitrate and nitrite for the processing of meat products. *Meat Sci.* 78 (1–2): 68–76.

6 Yue, R., Lu, Q., and Zhou, Y.K. (2011). A novel nitrite biosensor based on single-layer graphene nanoplatelet-protein composite film. *Biosens. Bioelectron.* 26 (11): 4436–4441.

7 Toth, J.E. and Anson, F.C. (1989). Electrocatalytic reduction of nitrite and nitric-oxide to ammonia with iron-substituted polyoxotungstates. *J. Am. Chem. Soc.* 111 (7): 2444–2451.

8 Keita, B., Nadjo, L., Contant, R. et al. (1989). Process for the selective catalytic reduction of nitrogen compounds and catalysts used. (CNRS). French Patent FR2642987A1, filed 10 February 1989 and issued 17 August 1990.

9 Dong, S.J., Xi, X.D., and Tian, M. (1995). Study of the electrocatalytic reduction of nitrite with silicotungstic heteropolyanion. *J. Electroanal. Chem.* 385 (2): 227–233.

10 Zhang, J., Goh, J.-K., Tan, W.-T., and Bond, A.M. (2006). Mechanistic analysis of the electrocatalytic properties of dissolved alpha and beta isomers of $SiW_{12}O_{40}^{4-}$ and solid $[Ru(bipy)^3]_2[\alpha\text{-}SiW_{12}O_{40}]$ on the reduction of nitrite in acidic aqueous media. *Inorg. Chem.* 45 (9): 3732–3740.

11 Keita, B., Belhouari, A., Nadjo, L., and Contant, R. (1995). Electrocatalysis by polyoxometalate polymer systems – reduction of nitrite and nitric-oxide. *J. Electroanal. Chem.* 381 (1–2): 243–250.

12 Debiemme-Chouvy, C., Cachet, H., Folcher, G., and Deslouis, C. (2007). Electrocatalytic reduction of HNO_2 by a silicomolybdate polyanion: a differential electrochemical mass spectrometry study. *Electroanalysis* 19 (2–3): 259–262.

13 Ojani, R., Rahmanifar, M.S., and Naderi, P. (2008). Electrocatalytic reduction of nitrite by phosphotungstic heteropolyanion. Application for its simple and selective determination. *Electroanalysis* 20 (10): 1092–1098.

14 Li, C.X., Zhang, Y., O'Halloran, K.P. et al. (2009). Electrochemical behavior of vanadium-substituted Keggin-type polyoxometalates in aqueous solution. *J. Appl. Electrochem.* 39 (3): 421–427.

15 Fernandes, D.M., Simões, S.M.N., Carapuça, H.M., and Cavaleiro, A.M.V. (2008). Functionalisation of glassy carbon electrodes with deposited tetrabutylammonium microcrystalline salts of lacunary and metal-substituted alpha-Keggin-polyoxosilicotungstates. *Electrochim. Acta* 53 (22): 6580–6588.

16 Xi, X.D. and Dong, S.J. (1996). Electrocatalytic reduction of nitrite using Dawson-type tungstodiphosphate anions in aqueous solutions, adsorbed on a glassy carbon electrode and doped in polypyrrole film. *J. Mol. Catal. A-Chem.* 114 (1–3): 257–265.

17 Keita, B., Mbomekalle, I.-M., Nadjo, L. et al. (2005). Vanadium-substituted Dawson-type polyoxometalates as versatile electrocatalysts. *C. R. Chim.* 8 (6–7): 1057–1066.

18 Dong, S.J. and Liu, M.J. (1994). Electrochemical and electrocatalytic properties of iron(III)-substituted dawson-type tungstophosphate anion. *J. Electroanal. Chem.* 372 (1–2): 95–100.

19 Keita, B., Mbomekalle, I.-M., Nadjo, L., and Contant, R. (2001). $[H_4AsW_{18}O_{62}]^{7-}$, A novel Dawson heteropolyanion and two of its sandwich-type derivatives $[Zn_4(H_2O)_2(H_4AsW_{15}O_{56})_2]^{18-}$, $[Cu_4(H_2O)_2(H_4AsW_{15}O_{56})_2]^{18-}$: cyclic voltammetry and electrocatalytic properties towards nitrite and nitrate. *Electrochem. Commun.* 3 (6): 267–273.

20 Xi, X.D., Wang, G., Liu, B., and Dong, S. (1995). Electrochemical-behavior of bis(2/17-arsenotungstate) lanthanates and their electrocatalytic reduction for nitrite. *Electrochim. Acta* 40 (8): 1025–1029.

21 Keita, B., Mbomekalle, I.M., and Nadjo, L. (2003). Redox behaviours and electrocatalytic properties of copper within Dawson structure-derived sandwich heteropolyanions $[CU_4(H_2O)_2(X_2W_{15}O_{56})_2]^{16-}$ (X = P or As). *Electrochem. Commun.* 5 (9): 830–837.

22 Rong, C.Y. and Anson, F.C. (1996). Spontaneous adsorption of heteropolytungstates and heteropolymolybdates on the surfaces of solid electrodes and the electrocatalytic activity of the adsorbed anions. *Inorg. Chim. Acta* 242 (1–2): 11–16.

23 Jamshidi, A., Zonoz, F.M., Wei, Y., and Maleki, B. (2018). An organic-inorganic nano-hybrid material containing a mixed-addenda Keggin-type polyoxometalate, piperazine: synthesis, characterization, its electrochemical investigation. *Inorg. Chim. Acta* 477: 233–241.

24 Xu, M., Li, F., Wang, T., and Xu, L. (2018). Single ion sandwich-type polyoxomolybdates based on Strandberg anions and their electrocatalytic properties. *Inorg. Chem. Commun.* 94: 123–126.

25 Zhou, Y.L., Bihl, F., Bonnefont, A. et al. (2022). Selectivity and efficiency of nitrite electroreduction catalyzed by a series of Keggin polyoxometalates. *J. Catal.* 405: 212–223.

26 Cao, G.L., Xiong, J., Xue, Q. et al. (2013). Organic-inorganic heteropoly blue based on Dawson-type molybdosulfate and organic dye and its characterization and application in electrocatalysis. *Electrochim. Acta* 106: 465–471.

27 Fay, N., Dempsey, E., and McCormac, T. (2005). Assembly, electrochemical characterisation and electrocatalytic ability of multilayer films based on $[Fe(bpy)_3]^{2+}$, and the Dawson heteropolyanion, $[P_2W_{18}O_{62}]^{6-}$. *J. Electroanal. Chem.* 574 (2): 359–366.

28 Li, C.X., Cao, R., O'Halloran, K.P. et al. (2008). Preparation, characterization and bifunctional electrocatalysis of an inorganic-organic complex with a vanadium-substituted polyroxometalate. *Electrochim. Acta* 54 (2): 484–489.

29 Zhou, W.L., Peng, J., Zhang, Z. et al. (2015). Bifunctional electrocatalysts based on the Keggin/Dawson-type polyoxotungstates and Cu-tris(imidazolyl) complexes. *Electrochim. Acta* 180: 887–893.

30 Han, Q., Wen, Y., Liu, J. et al. (2016). Synthesis, structure and electrochemical properties of an inorganic-organic hybrid Cu^{II}–Ce^{III} heterometallic germanotungstate. *Inorg. Chem. Commun.* 71: 54–60.

31 Wang, X.L., Bai, X., Lin, H. et al. (2018). Novel polyoxometalate-based cobalt complexes based on rigid pyridyl-triazole-tetrazole and pyridyl-bis(triazole) ligands. *CrystEngComm* 20 (41): 6438–6448.

32 Ding, Y.X., Peng, M.-T., Zou, K.-F. et al. (2021). A new inorganic-organic hybrid compound based on Keggin and 4,4′-bis(1,2,4-triazol-1-ylmethyl)biphenyl: crystal structure and electrocatalytic performance. *J. Mol. Struct.* 1239: 6.

33 Li, L.F., Zhao, H.Y., Liu, J.M. et al. (2021). A functional electrocatalyst of coordination polymer derived from di-nuclear nickel(II)-glycine cations and Wells-Dawson polyoxoanions. *J. Clust. Sci.* 32 (6): 1499–1506.

34 Li, L., Zhang, Y., Wang, X. et al. (2022). Various amide derivatives induced Keggin-type $SiW_{12}O_{40}{}^{4-}$-based cobalt complexes: assembly, structure, electrochemical sensing and dye adsorption properties. *CrystEngComm* 24 (6): 1195–1202.

35 Wang, X.L., Bi, Y., Chen, B. et al. (2008). Self-assembly of organic-inorganic hybrid materials constructed from eight-connected coordination polymer hosts with nanotube channels and polyoxometalate guests as templates. *Inorg. Chem.* 47 (7): 2442–2448.

36 Yang, Y., Liu, S., Li, C. et al. (2012). A metal-organic framework based on Wells–Dawson polyoxometalates $[As_2W_{18}O_{62}]^{6-}$ template. *Inorg. Chem. Commun.* 17: 54–57.

37 Fernandes, D.M., Granadeiro, C.M., Paes de Sousa, P.M. et al. (2014). $SiW_{11}Fe$@MIL-101(Cr) composite: a novel and versatile electrocatalyst. *ChemElectroChem* 1 (8): 1293–1300.

38 Yuan, J.H., Jin, X., Li, N. et al. (2011). Large scale load of phosphotungstic acid on multiwalled carbon nanotubes with a grafted poly(4-vinylpyridine) linker. *Electrochim. Acta* 56 (27): 10069–10076.

39 Boussema, F., Haddad, R., Ghandour, Y. et al. (2016). Polyoxometalate $[PMo_{11}O_{39}]^{7-}$/carbon nanocomposites for sensitive amperometric detection of nitrite. *Electrochim. Acta* 222: 402–408.

40 Wei, X.Y., Jiawei, W., Liping, H. et al. (2017). Facile fabricating the polyoxometalates functionalized graphene nanocomposite applied in electrocatalytic reduction. *Inorg. Chem. Commun.* 81: 10–14.

41 Feizy, S. and Haghighi, B. (2019). A survey on the effect of ionic liquid on electrochemical behavior and electrocatalytic activity of a phosphomolybdic acid-ionic liquid-MWCNT-modified glassy carbon electrode. *J. Solid State Electrochem.* 23 (5): 1339–1350.

42 Wang, T., Zhou, Q., Ren, X. et al. (2019). Incorporation of Keggin-type phosphomolybdic acid, ionic liquid and carbon nanotube leading to formation of multifunctional ternary composite materials: fabrication, characterization and electrochemical reduction/detection of iodate, borate and nitrite. *J. Clust. Sci.* 30 (4): 973–984.

43 Liu, S.Q. and Tang, Z.Y. (2010). Polyoxometalate-based functional nanostructured films: current progress and future prospects. *Nano Today* 5 (4): 267–281.

44 Li, C.X., Wang, X., Ma, H. et al. (2008). Fabrication and electrochemical behavior of vanadium-substituted keggin-type polyoxometalates multilayer films on 4-aminobenzoic acid modified glassy carbon electrodes. *Electroanalysis* 20 (10): 1110–1115.

45 Ma, H.Y., Gu, Y., Zhang, Z. et al. (2011). Enhanced electrocatalytic activity of a polyoxometalates-based film decorated by gold nanoparticles. *Electrochim. Acta* 56 (21): 7428–7432.

46 Fernandes, D.M., Brett, C.M.A., and Cavaleiro, A.M.V. (2011). Layer-by-layer self-assembly and electrocatalytic properties of poly(ethylenimine)-silicotungstate multilayer composite films. *J. Solid State Electrochem.* 15 (4): 811–819.

47 Fernandes, D.M., Carapuça, H.M., Brett, C.M.A., and Cavaleiro, A.M.V. (2010). Electrochemical behaviour of self-assembly multilayer films based on iron-substituted alpha-Keggin polyoxotungstates. *Thin Solid Films* 518 (21): 5881–5888.

48 Novais, H.C., Fernandes, D.M., and Freire, C. (2015). Hybrid methyl green/cobalt-polyoxotungstate nanostructured films: self-assembly, electrochemical and electrocatalytic properties. *Appl. Surf. Sci.* 347: 40–47.

49 Fernandes, D.M., Vos, J.G., and Freire, C. (2014). Europium phosphomolybdate and osmium metallopolymer multi-functional LbL films: redox and electrocatalytic properties. *J. Colloid Interface Sci.* 420: 127–135.

50 Khan, N. and Sultana, S. (2004). Abrogation of potassium bromate-induced renal oxidative stress and subsequent cell proliferation response by soy isoflavones in Wistar rats. *Toxicology* 201 (1–3): 173–184.

51 Weng, Z.W., Ren, Y., Li, X. et al. (2021). Aggregation of Nb/W mixed addenda Dawson-type polyoxometalates units and lanthanide ions: synthesis, characterization and electrocatalysis. *Polyhedron* 195: 7.

52 Wang, X.L., Sun, J., Lin, H. et al. (2016). A series of Anderson-type polyoxometalate-based metal-organic complexes: their pH-dependent electrochemical behaviour, and as electrocatalysts and photocatalysts. *Dalton Trans.* 45 (31): 12465–12478.

53 Wang, X.L., Sun, J., Lin, H. et al. (2017). A series of novel Anderson-type polyoxometalate-based Mn^{II} complexes constructed from pyridyl-derivatives: assembly, structures, electrochemical and photocatalytic poperties. *CrystEngComm* 19 (23): 3167–3177.

54 Wang, X.L., Bai, X., Lin, H. et al. (2018). A series of new polyoxometalate-based metal-organic complexes with different rigid pyridyl-bis(triazole) ligands: assembly, structures and electrochemical properties. *RSC Adv.* 8 (40): 22676–22686.

55 Zhang, J.Y., Chang, Z.-H., Wang, X.-L. et al. (2021). Different Anderson-type polyoxometalate-based metal-organic complexes exhibiting –OH group-directed structures and electrochemical sensing performance. *New J. Chem.* 45 (6): 3328–3334.

56 Zhang, Y.C., Chen, Y.-Z., Chang, Z.-H. et al. (2022). Metal-directed two new Anderson-type polyoxometalate-based metal-organic complexes with different electrocatalytic sensing performance. *Polyhedron* 221: 7.

57 Wang, X., Lin, L., Zhang, R., and Wang, X. (2017). Polyoxometalate-based supramolecular hybrids composed of Cu/2,2'-dimethyl-4,4'-bithiazole fragments extended via non-bonding S···O interactions. *Transit. Met. Chem.* 42 (3): 203–210.

58 Wang, X.L., Wang, E.-B., Lan, Y., and Hu, C.-W. (2002). Renewable PMo_{12}-based inorganic-organic hybrid material bulk-modified carbon paste electrode: preparation, electrochemistry and electrocatalysis. *Electroanalysis* 14 (15–16): 1116–1121.

59 Dong, T., Du, J., Cao, M., and Hu, C. (2010). The electrochemical properties of 12-molybdophosphoric acid modified ionic liquid carbon paste electrode. *J. Clust. Sci.* 21 (2): 155–162.

60 Yu, Y., Pang, H., Ma, H. et al. (2013). A novel 12-tungstovanadate doubly-grafted by Zn-complexes: synthesis, crystal structure and properties. *J. Clust. Sci.* 24 (1): 17–29.

61 Chen, C.Y., Song, Y.H., and Wang, L. (2009). Electrochemical behavior and its electrocatalytic properties of chemically modified electrode with Keggin-type $[SiNi(H_2O)W_{11}O_{39}]^{6-}$. *Electrochim. Acta* 54 (5): 1607–1611.

62 Dong, B.X., Wu, Y.-C., Tian, H. et al. (2016). Synthesis, crystal structure and electrochemical properties of a new 2D network containing linear $\{\varepsilon\text{-}H_2PMo_8{}^{V}Mo_4{}^{VI}O_{40}Zn_4\}_\infty$ inorganic chain. *J. Clust. Sci.* 27 (1): 361–371.

63 Dong, B.X., Bu, F.-Y., Wu, Y.-C. et al. (2017). Construction of (3,4)-connected polyoxometalate-based metal-organic frameworks (POMOFs) from triangular carboxylate and tetrahedral $Zn_{4\text{-}\varepsilon}$-Keggin. *Cryst. Growth Des.* 17 (10): 5309–5317.

64 Li, S.B., Lu, B., Xin, J. et al. (2019). A novel polyoxometalate-encapsulating 3D polycatenated metal-organic framework. *J. Solid State Chem.* 278: 6.

65 Zhang, Y.N., Zhang, Y., Li, L. et al. (2020). One-step *in situ* growth of high-density POMOFs films on carbon cloth for the electrochemical detection of bromate. *J. Electroanal. Chem.* 861: 10.

66 Lu, B.R., Li, S., Zhang, X. et al. (2019). A novel polyoxometalate-based metal-organic nanotube framework templated by twin-Dawson clusters: synthesis, structure and bifunctional electrocatalytic properties. *New J. Chem.* 43 (39): 15804–15810.

67 Hassan, S.S., Liu, Y., Sirajuddin et al. (2013). Phosphomolybdate-doped-poly(3,4-ethylenedioxythiophene) coated gold nanoparticles: synthesis, characterization and electrocatalytic reduction of bromate. *Anal. Chim. Acta* 803: 41–46.

68 Sakthinathan, I., Yamasaki, N., Barreca, D. et al. (2023). Wells-Dawson type polyoxometalate, $[S_2W_{18}O_{62}]^{4-}$-doped poly (3,4-ethylenedioxythiophene) films: voltammetric behaviour and applications to selective bromate detection. *Electrochim. Acta* 462: 13.

69 Ali, B., Laffir, F., Kailas, L. et al. (2019). Electrochemical characterisation of Ni^{II}-crown-type polyoxometalate-doped polypyrrole films for the catalytic reduction of bromate in water. *Eur. J. Inorg. Chem.* 2019 (3–4): 394–401.

70 Li, Z.F., Chen, J., Pan, D. et al. (2006). A sensitive amperometric bromate sensor based on multi-walled carbon nanotubes/phosphomolybdic acid composite film. *Electrochim. Acta* 51 (20): 4255–4261.

71 Liu, J., Wang, J., Chen, M., and Qian, D.-J. (2017). Fabrication, electrochemical and catalytic properties of the nanocomposites composed of phosphomolybdic acid and viologen-functionalized multi-walled carbon nanotubes. *J. Nanopart. Res.* 19 (8): 11.

72 Skunik, M. and Kulesza, P.J. (2009). Phosphomolybdate-modified multi-walled carbon nanotubes as effective mediating systems for electrocatalytic reduction of bromate. *Anal. Chim. Acta* 631 (2): 153–160.

73 Papagianni, G.G., Stergiou, D.V., Armatas, G.S. et al. (2012). Synthesis, characterization and performance of polyaniline-polyoxometalates (XM_{12}, X = P, Si and M = Mo, W) composites as electrocatalysts of bromates. *Sens. Actuators, B* 173: 346–353.

74 Ding, L., Liu, Y., Guo, S.-X. et al. (2014). Phosphomolybdate@poly(diallyl-dimethylammonium chloride)-reduced graphene oxide modified electrode for highly efficient electrocatalytic reduction of bromate. *J. Electroanal. Chem.* 727: 69–77.

75 Zhou, M., Guo, L.-P., Lin, F.-Y., and Liu, H.-X. (2007). Electrochemistry and electrocatalysis of polyoxometalate-ordered mesoporous carbon modified. *Anal. Chim. Acta* 587 (1): 124–131.

76 Sosnowska, M., Goral-Kurbiel, M., Skunik-Nuckowska, M. et al. (2013). Hybrid materials utilizing polyelectrolyte-derivatized carbon nanotubes and vanadium-mixed addenda heteropolytungstate for efficient electrochemical charging and electrocatalysis. *J. Solid State Electrochem.* 17 (6): 1631–1640.

77 Burgi, H., Schaffner, T., and Seiler, J.P. (2001). The toxicology of iodate: a review of the literature. *Thyroid* 11 (5): 449–456.

78 Haghighi, B., Hamidi, H., and Gorton, L. (2010). Formation of a robust and stable film comprising ionic liquid and polyoxometalate on glassy carbon electrode modified with multiwalled carbon nanotubes: toward sensitive and fast detection of hydrogen peroxide and iodate. *Electrochim. Acta* 55 (16): 4750–4757.

79 Kemmegne-Mbouguen, J.C., Floquet, S., and Cadot, E. (2021). Electrochemical properties of the $SiW_{10}O_{36}(M_2O_2E_2)^{6-}$ polyoxometalates series (M = Mo(V) or W(V); E = S or O) in aqueous medium: application to the electroanalysis of iodates. *C. R. Chim.* 24 (2): 91–101.

80 Li, Y.C., Bu, W., Wu, L., and Sun, C. (2005). New amperometric sensor for the determination of bromate, iodate and hydrogen peroxide based on titania sol-gel matrix for immobilization of cobalt substituted Keggin-type cobalt-tungstate anion by vapor deposition method. *Sens. Actuators, B* 107 (2): 921–928.

81 Hamidi, H., Shams, E., Yadollahi, B., and Esfahani, F.K. (2008). Fabrication of bulk-modified carbon paste electrode containing alpha-$PW_{12}O_40^{3-}$ polyanion supported on modified silica gel: preparation, electrochemistry and electrocatalysis. *Talanta* 74 (4): 909–914.

82 Kakhki, S. and Shams, E. (2013). A new bifunctional electrochemical sensor for oxidation of cysteine and reduction of iodate. *J. Electroanal. Chem.* 704: 249–254.

83 Ma, X.Y., Yu, K., Yuan, J. et al. (2020). Multinuclear transition metal sandwich-type polytungstate derivatives for enhanced electrochemical energy storage and bifunctional electrocatalysis performances. *Inorg. Chem.* 59 (7): 5149–5160.

84 Fernandes, D.M., Teixeira, A., and Freire, C. (2015). Multielectrocatalysis by layer-by-layer films based on pararosaniline and vanadium-substituted phosphomolybdate. *Langmuir* 31 (5): 1855–1865.

85 Ali, B., Imar, S., Laffir, F., and McCormac, T. (2019). Electrochemical, surface and electrocatalytic properties of electrode multilayer assemblies composed of a ruthenium metallodendrimer and a wheel-shaped Cu-20 tungstophosphate. *J. Electroanal. Chem.* 836: 77–84.

86 Yaqub, A., Vagin, M., Walsh, J.J. et al. (2022). Organic-inorganic hybrid films of the sulfate Dawson polyoxometalate, $[S_2W_{18}O_{62}]^{4-}$, and polypyrrole for iodate electrocatalysis. *ACS Omega* 7 (48): 43381–43389.

87 Guo, W.H., Xu, L., Li, F. et al. (2010). Chitosan-assisted fabrication and electrocatalytic activity of the composite film electrode of heteropolytungstate/carbon nanotubes. *Electrochim. Acta* 55 (5): 1523–1527.

88 Sharifi, M., Dianat, S., and Hosseinian, A. (2021). Electrochemical investigation and amperometry determination iodate based on ionic liquid/polyoxotungstate/P-doped electrochemically reduced graphene oxide multi-component nanocomposite modified glassy carbon electrode. *RSC Adv.* 11 (15): 8993–9007.

89 Ravanbakhsh, H. and Dianat, S. (2023). Voltammetric and amperometric determination of iodate using a modified glassy carbon electrode based on PW_{12}/MOF/P@ERGO nanohybrid. *Sens. Bio-Sens. Res.* 40: 8.

90 Ji, Y.C., Huang, L., Hu, J. et al. (2015). Polyoxometalate-functionalized nanocarbon materials for energy conversion, energy storage and sensor systems. *Energy Environ. Sci.* 8 (3): 776–789.

91 Wang, B., Meng, R.-Q., Bi, L.-H., and Wu, L.-X. (2011). A novel detection of hydrogen peroxide based on a luminescent polyoxometalate. *Dalton Trans.* 40 (19): 5298–5301.

92 Martel, D. and Kuhn, A. (2000). Electrocatalytic reduction of H_2O_2 at $P_2Mo_{18}O_{62}{}^{6-}$ modified glassy carbon. *Electrochim. Acta* 45 (11): 1829–1836.

93 Xie, Y., Sun, W., Liu, H. et al. (1998). Palladium-substituted Keggin type polyoxotungstate chemically modified electrode and its catalytic activity toward hydrogen peroxide reduction. *Anal. Lett.* 31 (12): 2009–2024.

94 Li, S.B., Zhu, W., Ma, H. et al. (2013). Structure and bifunctional electrocatalytic activity of a novel 3D framework based on dimeric monocopper-substituted polyoxoanions as ten-connected linkages. *RSC Adv.* 3 (25): 9770–9777.

95 Wu, D., Wang, K., Ma, H. et al. (2014). Two new hybrid compounds assembled from Keggin phosphotungstates, cobalt/zinc cations and V-shaped imidazole ligands. *Solid State Sci.* 35: 39–44.

96 Wang, X.L., Han, Z.-B., Wang, E.-B. et al. (2003). A bifunctional electrocatalyst containing tris(2,2'-bipyridine) ruthenium(II) and 12-molybdophosphate bulk-modified carbon paste electrode. *Electroanalysis* 15 (18): 1460–1464.

97 Zhang, Z.F., Ma, H., Pang, H. et al. (2016). Assembly of a new Keggin-templated supramolecular compound by silver salt and 4,4′-dipyridine-*N*,*N*′-dioxide ligand. *Inorg. Chem. Commun.* 72: 91–95.

98 Han, Z.G., Zhao, Y., Peng, J. et al. (2005). Inorganic-organic hybrid polyoxometalate: preparation, characterization and electrochemistry properties. *J. Solid State Chem.* 178 (5): 1386–1394.

99 Yang, M.X., Lin, S., Chen, X.-H. et al. (2010). Two organic-inorganic hybrid compounds constructed from 12-tungstovanadate and bipyridine. *J. Coord. Chem.* 63 (3): 406–417.

100 Ammam, M. and Easton, E.B. (2012). Novel organic-inorganic hybrid material based on tris(2,2'-bipyridyl)dichlororuthenium(II) hexahydrate and Dawson-type tungstophosphate K-7 $H_4PW_{18}O_{62}$ center dot $18H_2O$ as a bifuctional hydrogen peroxide electrocatalyst for biosensors. *Sens. Actuators, B* 161 (1): 520–527.

101 Liu, X.Z., Cui, L.-P., Yu, K. et al. (2022). A 3D supramolecular assembly based on a $\{AsW_{12}\}$ cluster and in-situ ligand modified metal-organic complexes for photocatalytic properties and electrocatalytic sensing for detection of hydrogen peroxide. *J. Solid State Chem.* 306: 7.

102 Salimi, A., Korani, A., Hallaj, R. et al. (2009). Immobilization of $[Cu(bpy)_2]Br_2$ complex onto a glassy carbon electrode modified with alpha-$SiMo_{12}O_{40}{}^{4-}$ and single walled carbon nanotubes: application to nanomolar detection of hydrogen peroxide and bromate. *Anal. Chim. Acta* 635 (1): 63–70.

103 Liu, R.J., Li, S., Yu, X. et al. (2012). Facile synthesis of Au-nanoparticle/polyoxometalate/graphene tricomponent nanohybrids: an enzyme-free electrochemical biosensor for hydrogen peroxide. *Small* 8 (9): 1398–1406.

104 Zhang, S.S., Liu, R., Li, S. et al. (2018). Simple and efficient polyoxomolybdate-mediated synthesis of novel graphene and metal nanohybrids for versatile applications. *J. Colloid Interface Sci.* 514: 507–516.

105 Naseer, R., Mal, S.S., Ibrahim, M. et al. (2014). Redox, surface and electro-catalytic properties of layer-by-layer films based upon Fe(III)-substituted crown polyoxometalate $[P_8W_{48}O_{184}Fe_{16}(OH)_{28}(H_2O)_4]^{20-}$. *Electrochim. Acta* 134: 450–458.

106 Anwar, N., Vagin, M., Naseer, R. et al. (2012). Redox switching of polyoxometalate-methylene blue-based layer-by-layer films. *Langmuir* 28 (12): 5480–5488.

107 Anwar, N., Mccormac, T., Compain, J.D. et al. (2012). Surface immobilisation of the sandwich type $Na_{14}Fe_4(Ox)_4(H_2O)_2(SbW_9O_{33})_2{\cdot}60H_2O$ polyoxometalate. *Electrochim. Acta* 59: 1–7.

108 Ammam, M., Keita, B., Nadjo, L. et al. (2010). Attempts to immobilize catalytically active substituted-heteropolytungstates in multilayer film of charged polyelectrolyte poly(allylamine hydrochloride). *J. Electroanal. Chem.* 645 (2): 65–73.

6

POM-based Electrocatalysts for Carbon Dioxide Reduction

Bruno Fabre[1], *Clément Falaise*[2], *and Emmanuel Cadot*[2]

[1] *University of Rennes, CNRS, ISCR (Institut des Sciences Chimiques de Rennes)-UMR 6226, Campus de Beaulieu, Avenue du Général Leclerc, Rennes F-35000, France*
[2] *UVSQ, Université Paris-Saclay, Institut Lavoisier de Versailles (UMR-CNRS 8180), 45 Avenue des Etats-Unis, Versailles 78000, France*

6.1 Introduction

Anthropogenic emissions of carbon dioxide (CO_2), mostly due to the utilization of fossil sources for electricity production, have been driving long-term global warming which impact our environment and ecosystem negatively. To tackle climate change, the European Union and 194 states have ratified the Paris Agreement aiming to keep the rise in mean global temperature well below 2 °C. This context urges the development of more sustainable energy production and processes to convert CO_2 into useful chemicals or fuels. Among the different approaches converting CO_2 into molecules with high-energy densities, such as alcohols or CH_4, the electrochemical CO_2 reduction reaction (CDRR) represents certainly the most promising route.

Over decades of research, a broad spectrum of electrocatalytic homogeneous and heterogeneous systems have been investigated. The CDRR using metallic electrodes leads to CO as the primary reduction product when the electrode is based on Zn, Au or Ag, while electrodes of Sn, Pb, Hg, In and Bi mainly produce formic acid or its base conjugate [1]. The Cu-based electrodes are unique because Cu is the only metallic element capable of reducing CO_2 beyond two electrons, offering a way to produce for instance C_2 compounds (ethylene, ethanol and acetate) in a considerable quantity [2]. Thanks to their tuneable nature and electronic features, molecular engineering allowed access to a large variety of catalysts with excellent activity and selectivity. The most active molecular metal complexes can be roughly divided into three families depending on the nature of the organic ligands: macrocyclic ligands (e.g. porphyrin, cyclam, phthalocyanine), bipyridine ligands or phosphine ligands [3]. Although polyoxometalates (POMs) have been often described as inorganic-porphyrin analogues [4, 5], their utilization as CDRR catalysts remains relatively unexplored, and so far, their efficiency is still low.

This chapter intends to provide a complete description of advances in using POMs as electro-active molecules for electrochemical reduction of CO_2. After introducing

Applied Polyoxometalate-based Electrocatalysis, First Edition. Edited by Diana M. Fernandes.

the fundamentals of electrochemical CO_2 reduction, we will give an overview of the most striking properties of POMs that could make them relevant for the electrocatalytic process, and we will cover the different studies showing POMs are capable of interacting covalently with CO_2. Then we will provide a state of the art about the utilization of POMs in homogeneous as well as in heterogeneous electrocatalytic CDRR systems. Finally, the last section will give some concluding remarks and propose future investigation routes to overcome the current issues related to the use of POMs in CDRR. Due to the limited scope of this book, we will not discuss the use of POMs in photocatalytic reduction of CO_2; however, the readers are invited to refer to a recent review on this topic [6].

6.2 Thermodynamics of CO_2 Reduction

The one-electron reduction of CO_2 to produce the radical $CO_2^{\cdot-}$ requires nearly 2 eV of free energy; therefore, proton-assisted approach is required to lower the thermodynamic barrier significantly. Depending on the number of protons and electrons transferred, CO_2 can be reduced to different C1, C2 and C3 products, including CO, oxalic acid, formic acid, formaldehyde, methanol, methane, ethanol or *n*-propanol. A list of the selected proton-assisted reactions is reported in Table 6.1 [7]. Importantly, the hydrogen evolution reaction (HER) is a two-electron process that often competes with the CDRR processes. Considering exclusively the redox potential E^0, the CDRR process towards hydrocarbon or alcohol products should be thermodynamically more favourable than for the production of carbon monoxide, acetic acid or formaldehyde. However, CDRR has also an important kinetic dependence related

Table 6.1 Electrochemical CDRR and HER reactions with equilibrium potentials at pH 7.

Reactions	E^0/(V *vs.* NHE)	Products
$CO_2 + 4H^+ + 4e^- \rightarrow C + 2H_2O$	−0.20	Graphite
$CO_2 + 8H^+ + 8e^- \rightarrow CH_4 + 2H_2O$	−0.24	Methane
$2CO_2 + 14H^+ + 14e^- \rightarrow C_2H_6 + 4H_2O$	−0.27	Ethane
$3CO_2 + 18H^+ + 18e^- \rightarrow C_3H_7OH + 5H_2O$	−0.31	Propanol
$2CO_2 + 12H^+ + 12e^- \rightarrow C_2H_5OH + 3H_2O$	−0.32	Ethanol
$2CO_2 + 12H^+ + 12e^- \rightarrow C_2H_4 + 4H_2O$	−0.34	Ethylene
$CO_2 + 6H^+ + 6e^- \rightarrow CH_3OH + H_2O$	−0.38	Methanol
$2H^+ + 2e^- \rightarrow H_2$	−0.42	Hydrogen
$CO_2 + 4H^+ + 4e^- \rightarrow HCHO + H_2O$	−0.48	Formaldehyde
$CO_2 + 2H^+ + 2e^- \rightarrow CO + H_2O$	−0.53	Carbon monoxide
$CO_2 + 2H^+ + 2e^- \rightarrow HCOOH$	−0.61	Formic acid
$2CO_2 + 2H^+ + 2e^- \rightarrow (COOH)_2$	−0.91	Oxalic acid

Source: Bard et al. [7]/with permission of Routledge.

to the concentration of available protons in solution; thus, the hydrogenation of adsorbed C1 intermediates on the catalytic site is kinetically easier than the formation of C—C bonds, limiting generally the rate and selectivity of C2 and higher hydrocarbon production.

6.3 Appealing Properties of POMs for CO_2 Reduction

In this part, we recall the main properties of POMs that are relevant to face the challenges related to the CO_2 reduction to added-value products. For more details about the electrochemical properties of POMs, the reader is invited to refer to older and recent reviews on this topic [8–10].

6.3.1 A Reservoir of 'Hopping' Electrons

POMs in which all metalate centres are in octahedral sites with a single unshared oxygen atom exhibit reversible or pseudo-reversible redox activity that arises from the empty and non-bonding d orbitals of the metal centres. Such characteristics make them appealing to develop stable POM-based electrocatalytic systems. Among the library of electro-active POMs, the archetypal Keggin- and Dawson-type POMs and their derivatives, including metallo-substituted and hybrid forms, represent the most studied POMs in the CDRR context. When homometallic POMs are reduced by one electron, this additional electron is delocalized, partially or totally, on the POM structure. This property is crucial for the development of POM-based electrocatalysts because the rapid mobility of the electrons can make their transfer towards the catalytic sites efficient, allowing multi-electron transfer reactions. The additional electron is delocalized owing to fast thermal hopping at room temperature over the entire POMs when all metal sites are equivalent (case of Lindqvist- or Keggin-type POMs). Otherwise, the electron is delocalized only on a fragment of the POM as observed for Dawson-type POMs [11, 12]. Weakly reduced POMs therefore belong to the class II mixed-valence compounds according to the Robin–Day classification [13]. As with general trends, the delocalization is easier (i) on W-based POMs than in Mo-based POMs and (ii) through corner-sharing octahedra due to more favourable $d\pi$-$p\pi$ orbital overlap. Although the POMs reduced by one electron are paramagnetic entities, the two electron-reduced POMs are often diamagnetic compounds due to the existence of antiferromagnetic spin coupling between the pair of delocalized electrons [14]. The electron delocalization is not limited to a small amount of electrons; indeed, recent observations suggest a huge number of electrons can be delocalized on POM structures. For instance, it appears that the Preyssler-type POM formulated $[P_5W_{30}O_{110}]^{15-}$ can hold up to 23 delocalized electrons [15]. However, massive electron delocalization cannot be always achieved. Indeed, recent observations indicate that two-electron reduced derivatives of the trivacant-based Keggin-type POMs release H_2 upon protonation in organic media [16]. The electron hopping is not restrained to intra-molecular delocalization on metal–oxo frameworks but can also occur through organic fragments in the case of the hybrid POMs and simply through space when inter-molecular electron exchange occurs either in solution [17–19] or in the solid state [20]. In solution, the electron self-exchange is significantly affected by the coulombic forces between the acceptor and donor entities. In solids, the inter-electron transfer between POMs leads to a conductivity increase of the insulating porous catalytic matrices.

6.3.2 Proton-coupled Electron Transfer in POMs

The proton-coupled electron transfer in POMs is a very common phenomenon observed both in organic solvents and in aqueous solutions [9]. A very classical example of evidence of the ability of POM to undergo proton-coupled electron transfers is the electrochemical behaviour of Keggin-type dodecatungstosilicate α-$[SiW_{12}O_{40}]^{4-}$ [9]. In this chapter, we present a more recent example corresponding to the sandwich-like POM $[(VO)_3(AsW_9O_{33})_2]^{9-}$ (denoted hereafter $\{V^V{}_3\}^{9-}$; see structure on Figure 6.1a) of which the electrochemical behaviour varies strongly also in function of pH as evidenced by cyclic voltammetry investigations [21]. In highly acidic media, this POM is able to exchange three electrons corresponding to the simultaneous reduction of all vanadium centres. As the pH increases, the three-electron wave is gradually split into three successive monoelectronic events (Figure 6.1b). At pH = 4.5, three quasi-reversible and successive mono-electronic redox waves are observed clearly. The potential-pH diagram delimits six domains numbered from I to VI (Figure 6.1c). The domains labelled II ($\{V^V{}_3\}^{9-}$ + 1 e$^-$ $\rightarrow \{V^{IV}{}_1V^V{}_2\}^{10-}$), IV ($\{V^V{}_2V^{IV}{}_1\}^{10-}$ + 1 e$^-$ $\rightarrow \{V^V{}_1V^{IV}{}_2\}^{11-}$), VI ($\{V^V{}_1V^{IV}{}_2\}^{11-}$ + 1 e$^-$ $\rightarrow \{V^{IV}{}_3\}^{12-}$) correspond to three pH-independent monoelectronic redox couples. Conversely, the domains labelled I ($\{V^V{}_3\}^{9-}$ + 3 e$^-$ + 3 H$^+$ $\rightarrow \{H_3V^{IV}{}_3\}^{9-}$), III ($\{V^{IV}{}_1V^V{}_2\}^{10-}$ + 2 e$^-$ + 2 H$^+$ $\rightarrow \{H_2V^{IV}{}_3\}^{10-}$) and V ($\{V^V{}_1V^{IV}{}_2\}^{11-}$ + 1 e$^-$ + H$^+$ $\rightarrow \{HV^{IV}{}_3\}^{11-}$) correspond to three pH-dependent redox processes which involve trielectronic, bielectronic and monoelectronic processes, respectively. The origin of this ability to undergo a 3e$^-$/3H$^+$ proton-coupled electron transfer is certainly related to the fact that vanadium centres in sandwich-type POM

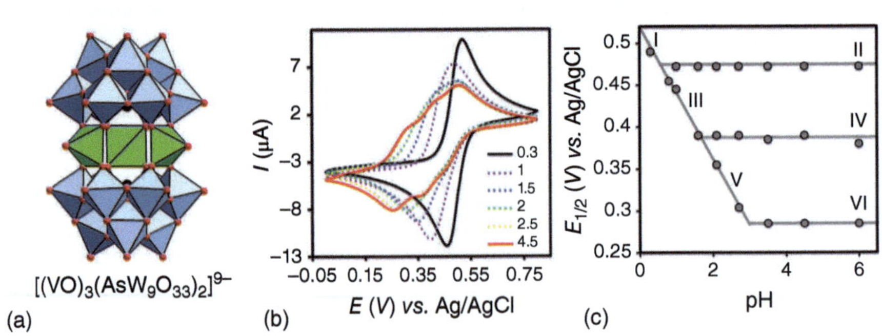

Figure 6.1 (a) Polyhedral representation of the vanadium-containing Hervé-type POMs built from B-type trivacant Keggin derivatives $[AsW_9O_{33}]^{9-}$. (b) Evolution of the cyclic voltammograms of $[(VO)_3(AsW_9O_{33})_2]^{9-}$ at different pH (from 0.3 to 4.5). (c) Variation of apparent redox potentials of the different redox events as a function of acidity. Source: Bamba et al. [21]/with permission of American Chemical Society.

$[(VO)_3(AsW_9O_{33})_2]^{9-}$ are physically far from each other (>5 Å). Although not yet mastered, playing with the charge density of the POMs and the proton concentration must offer a way to control to a large extent the multi-electron transfer coupled with the protonation process required to push CDRR processes towards hydrocarbon or alcohol products.

6.3.3 Tuning of the Reducibility of the POMs

The reduction potential of Keggin-type POMs is usually controlled by the overall negative ionic charge of the polyanion; therefore, a simple way to tune the reducibility of Keggin-type polyoxotungstates is to change the nature of the central heteroatom X. Thus the reduction potential of $[XW_{12}O_{40}]^{n-}$ POM in aqueous solution decreases linearly when n increases. This variation is about −180 mV per unit charge for the Keggin anion. The magnitude of the effect of the overall charge of POMs on the first electron-reduction process appears to be correlated with the POM size, and as a consequence, the tuning of the first redox potential is much less affected for large POMs like the Preyssler-type POMs (only 45 mV per charge unit). As discussed in the previous section, the half-wave potentials of POMs are very sensitive to pH, and therefore this represents another way to tune their reducibility. Playing with the metalate composition is certainly the best way to dramatically influence the electrochemical behaviour of POMs. Replacing one W centre with a Mo centre in Keggin POM $[SiW_{11}MO_{40}]^{4-}$ with M = Mo(VI) or W(VI) leads to a large positive shift of the first reduction potential (from −180 mV *vs.* Ag/AgCl for $[SiW_{12}O_{40}]^{4-}$ to +335 mV *vs.* Ag/AgCl for $[SiW_{11}MoO_{40}]^{4-}$). This effect is even more pronounced by replacing one W^{6+} centre by one V^{5+} centre (+440 mV *vs.* Ag/AgCl for $[SiW_{11}VO_{40}]^{5-}$), despite an increase in the global anionic charge density. The redox properties of POMs are also strongly influenced by the nature of the solvent. Usually, the first redox wave shifts linearly to more negative potentials with decreasing solvent acceptor number; thus, transferring POMs from aqueous solution to dimethylformamide can lead to a variation up to 1 V. Last but not least, the voltammetric behaviour of POMs can be also influenced by the nature of the electrolyte cation. This represents another appealing way to deeply tune the redox properties of POMs (half-wave potential and number of electrons per wave) by adding alkali cations (Li^+, Na^+, etc.) to the organic solutions. For instance, Ueda, Bond and co-workers [22] reported that the first reduction of the couple $V^{5+/4+}$ of a vanadium centre embedded in a Keggin-type POM can be drastically shifted by the addition of Li^+ cations in acetonitrile (Figure 6.2), becoming easier to reduce. They demonstrated that the magnitude of this positive shift appears to be dependent on the overall charge of the POMs. In short, the Keggin-type POMs with an overall anionic charge of −3 are not affected by the presence of Li^+, while a shift of +230 mV can be detected for those with an overall anionic charge of −4.

6.3.4 Massive Electron Storage in POMs

Many of the CO_2 reduction processes involve a higher number of electrons, indicating such processes require the fast delivery of several electrons. In context,

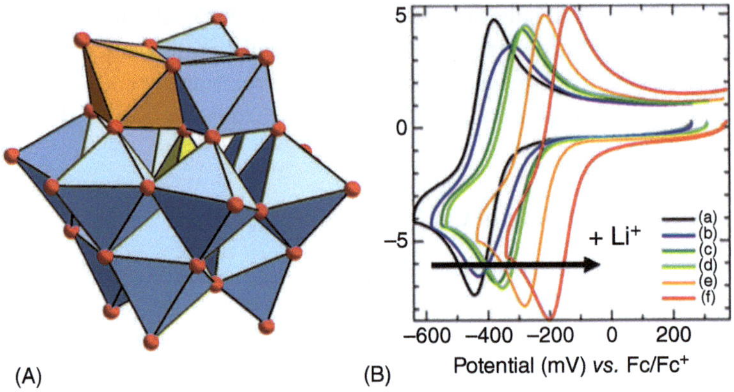

Figure 6.2 (A) View of the vanadium mono-substituted Keggin-type POMs. (B) Cyclic voltammograms obtained for the $V^{V/IV}$ process with $[PVMo_{11}O_{40}]^{4-}$ in $CH_3CN + 0.1$ M $TBAPF_6$ in the presence of (a) 0, (b) 0.2, (c) 0.6, (d) 1.0, (e) 10 and (f) 100 mM $LiClO_4$. Source: Konishi et al. [22]/with permission of American Chemical Society.

POMs can maybe play a role since they are able to be reduced by a large number of electrons. For instance, it was proposed based on polarographic studies that the metatungstate anion $[H_2W_{12}O_{40}]^{6-}$ can be reduced by 32 electrons [23]. The chemistry of super-reduced POMs is not new, but so far, it remains largely unknown despite their recent renewal of interest due to their potential applications in electrochemical-energy storage systems such as redox flow batteries [24]. As an illustrative example of the lack of knowledge about the super-reduced POMs, no structural characterization (either by X-ray diffraction or solution-state NMR) of POMs electrochemically reduced by more than six electrons has been carried out. All published POMs reduced by six electrons exhibit a Keggin structure in which the added electrons appear located within three metal–metal bonds forming an electron-rich triad [25–28]. Consequently, and in contrast to the weakly reduced POMs, these POMs with metal-metal bounded triad belong to the class I mixed-valence compounds according to the Robin–Day classification [25]. Very recent works have stated that the super-reduction properties are not restrained to Keggin-type POMs. Indeed, it was demonstrated that Dawson- or Preyssler-type POMs can accept up to 18 and 23 electrons, respectively [15, 29].

6.3.5 A Versatile Platform

The POM chemistry is extremely rich due to the capabilities of POMs to include in their inorganic metal–oxo framework almost all the elements of the periodic table belonging to the s-, d-, f- or p-blocks, offering the possibility to tune in large extent the properties of POMs [30]. This is not limited to the alteration of the redox activity since combining covalently POMs with active catalytic centres or complexes allows also the coordination of catalytic substrates. In the last two decades, many new POM architectures, often exhibiting highly complex structural organization, have been discovered through 'one-pot' solvothermal synthesis. While this approach has led to thousands of POMs, we decided to not discuss it herein since it remains quite unpredictable due to the high dynamical character of the metal–oxo bonds. We mostly dedicated this part to the rational design of POMs built from the sequential steps consisting in the reaction of preformed archetypal POM structures that are easily prepared on a large scale [31].

A very popular and extremely efficient strategy to integrate rationally catalytically active metal sites in POMs, allowing potential synergy between the redox activity of the POM framework and the catalytic properties of the transition metals, is to employ preformed lacunary POMs which contain nucleophilic oxo groups capable to bind metallic cations. To design monosubstituted POMs containing di- or trivalent transition metals, POMs chemists typically take benefit of the chelating properties of the monovacant polyoxotungstates (Keggin- or Dawson-type POMs) to insert cationic transition-metal species ($M^{2+/3+}$) in the POM frameworks. This process generally leads to the regeneration of the plenary POMs formulated $[XW_{11}O_{39}M(H_2O)]^{n-}$ or $[P_2W_{17}O_{61}M(H_2O)]^{n-}$, in which the coordination sphere of $M^{2+/3+}$ contains an aqua group. This aqua group can be removed in an organic solvent, offering a coordinatively unsaturated site that can be useful for coordinating catalytic substrates [32]. The formation of POMs containing di- or trivalent transition metals can be extended to other polyvacant polyoxometalates to produce saturated species as well as sandwich-like compounds. Furthermore, the macrocyclic anion $[H_7P_8W_{48}O_{184}]^{33-}$ (denoted $\{P_8W_{48}\}$) represents a very appealing and hydrolytically robust supervacant POM that can lead to the formation of large polynuclear clusters (e.g. $\{Cu_{20}\}$, $\{Al_{16}\}$, $\{Fe_{16}\}$ and $\{Ag_{30}\}$) embedded within the redox-active macrocyclic anion [33–36]. Although some of these entities could exhibit excellent catalytic activity, for instance, $\{Ag_{30}\}@\{P_8W_{48}\}$ for the selective reduction of nitrobenzene to aniline [36], their employment in the field of CO_2 reduction has been so far unexplored. Another avenue to develop efficient POMs for electrocatalytic reduction consists of associating mono- or polyvacant POMs with preformed cluster units, such as metal-sulfide clusters. In that context, the best achievement has been realized using the cationic $\{Mo_3S_4\}^{4+}$ unit that is a cluster capable of reacting with POMs, leading to various assemblies [37]. Although this class of POMs represents one of the best POM-based electrocatalysts for the HER reaction [38], their utilization in the field of CO_2 reduction has not yet been explored.

An elegant route to design POM-based electrocatalysts is to use hybrid POMs. There are many synthetic pathways to develop such POMs. If the reader is interested to know more about this research topic, we invite him/her to read the reviews that have been recently published [39, 40]. In very short, this approach provides an attractive alternative route for the grafting of covalently POMs with catalytically active complexes. Although the art of the synthesis of hybrid POMs is much more synthetically demanding than the preparation of systems based on purely inorganic components, these species not only allow the combination of the POM's properties with those of the appended catalytic subunit and in some cases, novel properties can emerge.

6.4 Coordination of CO_2 by POM Compounds

The combination of CO_2 with POMs in the form of coordination complexes could constitute a valuable and promising approach for CO_2 fixation with challenging outcomes in the synthesis of fine chemicals and clean fuels, but also in controlling the selectivity of the electrocatalytic CDRR. However, it is somewhat surprising to notice that the number of POM complexes formed with CO_2 or its carbonate CO_3^{2-} derivative is very limited. For CO_3^{2-}, it was reported that a large variety of POMs ranging from basic structures to giant Keplerate anionic nanocapsules could covalently link to or encapsulate carbonate ligands using only aqueous, room temperature and open-air chemistry [41, 42]. Regarding CO_2, the first example of coordination complexes between CO_2 and POMs was reported in 1998 by Kozik and co-workers [43]. The reactivity of various transition-metal-substituted lacunary α-Keggin ($[\alpha\text{-}XW_{11}O_{39}M(\_)]^{n-}$), α-Wells–Dawson ($[P_2W_{17}O_{61}M(\_)]^{m-}$) POMs and sandwich-like structures ($[P_2W_{18}O_{68}Co_4(\_)_2]^{10-}$) with CO_2 was examined in non-polar solvents (e.g. toluene). The presence of coordinatively unsaturated sites on the surface of the POMs denoted '(_)' in the formula was caused by transferring the hydrated forms in nonpolar solvents using phase-transfer agents (e.g. tetraheptylammonium) and then heating wet toluene under vacuum (Figure 6.3a). The resulting polyanions usually showed considerable metalloporphyrin-like reactivity in nonaqueous solvents including interactions with small molecules (O_2, SO_2, CO_2) [32, 43, 44].

Spectroscopic measurements evidenced that certain transition-metal-substituted heteropolyanions, such as Co(II), Ni(II) and Mn(II), reacted with CO_2, forming reversible complexes (Figure 6.3b), while Mn(III) and Cu(II) derivatives were non-reactive. It was demonstrated that the stability of the complexes formed with CO_2 was highly dependent on (i) the facility with which the terminal water ligand on the substituted metal is lost, (ii) the redox potential of the substituted metal and (iii) the geometry around the metal. However, any of these complexes was not obtained in the form of crystalline samples.

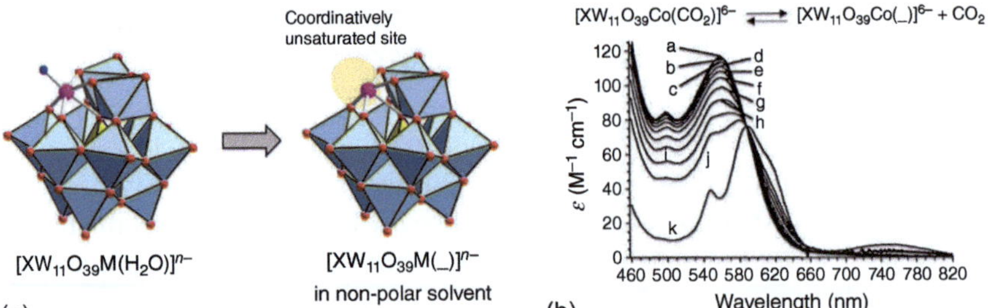

Figure 6.3 (a) Representation of transition-metal-substituted lacunary α-Keggin that exhibits an unsaturated site (_) in a non-polar solvent. (b) Evolution of the UV–vis spectra of $\alpha\text{-}[SiW_{11}O_{39}Co(CO_2)]^{6-}$ in function of the temperature. Letters correspond to temperature ($a = 0\,°C$ up to $k = 90\,°C$). Source: Szczepankiewicz et al. [43]/with permission of American Chemical Society.

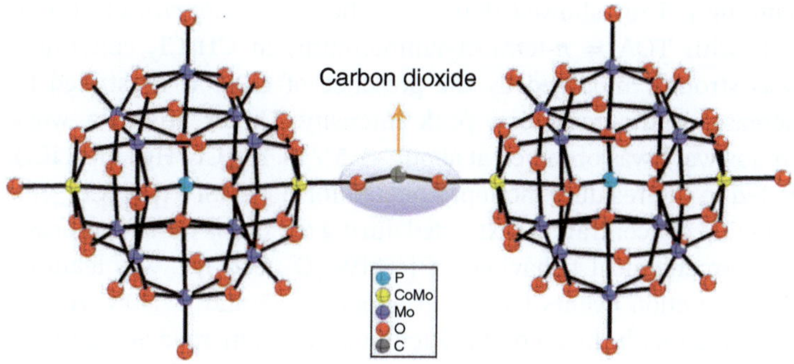

Figure 6.4 Ball-stick representation of polymeric chain structure of $[PMo_{11}CoO_{38}(CO_2)]_n$. Source: Gao et al. [45]/with permission of American Chemical Society.

Much later, significant advances in the complexation of CO_2 by POMs have been made. Indeed, stable and crystallized complexes have been readily prepared in water from the reaction of CO_2 with either $[H_2PMo_{11}CoO_{40}]^{5-}$ or $[H_2SiMo_{11}CoO_{40}]^{6-}$ [45]. X-ray diffraction data revealed that these complexes were composed of $[PMo_{11}CoO_{38}(CO_2)]_n$ and $[SiMo_{11}CoO_{38}(CO_2)]_n$ linear polymeric chains with one CO_2 molecule bridging two POMs in an axial direction (Figure 6.4). This work constitutes an eco-friendly chemical approach to CO_2 coordination chemistry.

Using a multi-component solvothermal reaction, Cheng et al. prepared a three-dimensional CO_2-coordinated POM polymer composed of both CO_2 ligands and Zn-ε-Keggin with the general formula $(TBA)_2PMo_{12}O_{40}Zn_4(CO_2)$ (with TBA = tetra-*n*-butylammonium) [46]. Such an unparalleled architecture featuring 1D open channels is particularly appealing for catalysis, as demonstrated in this work by the high catalytic activity of this system for the cycloaddition of bound CO_2 with epoxides into cyclic carbonates.

The CO_2 coordination by POMs has also been investigated by R. Neumann' group. The ligation of CO_2 to Zn(II)-substituted Wells–Dawson and Ru(III)-substituted Keggin type POMs yielded $[\alpha_2\text{-}P_2W_{17}O_{61}Zn(CO_2)]^{8-}$ [47] and $[SiW_{11}O_{39}Ru(CO_2)]^{6-}$ [48], respectively. Interestingly, the second one catalysed the photochemical reduction of CO_2 to CO in the presence of a sacrificial tertiary amine. The formation of an adduct between CO_2 and the nitrogen atom of the amine was found to be responsible for the observed photocatalytic activity.

6.5 Electrocatalytic Reduction of CO_2 with Dissolved POMs

6.5.1 3D Transition-metal-substituted POMs as Electrocatalysts in Organic Solvents

Inspired by the pioneering work of Kozik who showed that CO_2 interacts with the one coordinatively unsaturated site of transition-metal-substituted lacunary α-Keggin, the group of Anna Proust investigated the CDRR by $[\alpha\text{-}SiW_{11}O_{39}Co(\_)]^{6-}$

in dichloromethane [49]. They showed that the cyclic voltammogram of $(TOA)_6[\alpha\text{-}SiW_{11}O_{39}Co(\_)]$ (with TOA = *n*-tetraoctylammonium) in CH_2Cl_2 containing 0.1 M $TBABF_4$ was strongly modified by the presence of CO_2, as illustrated by the dramatic increase of the reduction peak intensity. Under argon, a weak quasi-reversible redox wave was observed at about −1.5 V *vs.* Hg_2Cl_2/Hg (1 M LiCl), which was attributed to the residual monoprotonated form $[\alpha\text{-}HSiW_{11}O_{39}Co(\_)]^{5-}$ (about 10% of total POM concentration) extracted during the phase transfer process. Under CO_2, the electrochemical behaviour of $[\alpha\text{-}SiW_{11}O_{39}Co(\_)]^{6-}$ was featured by two irreversible reduction events located at about −1.55 and −1.90 V. A new anodic process associated with the second reduction process emerged at −1.1 V. In the presence of proton and absence of CO_2, the reduction waves of POM appeared at −1.55 and −1.25 V and were only slightly influenced by the presence of CO_2 (positive shift of the reduction peak and decrease in intensity of the associated anodic peak). Such findings were consistent with the absence of an electrocatalytic process involving CO_2 but rather with the occurrence of an interaction between the reduced protonated POM and CO_2. After bulk electrolysis performed at 1.5 V, the resulting products were analysed by gas chromatography (GC) (quantification of H_2, CO, CH_4), ionic chromatography (quantification of formate and oxalate) and Nash colorimetric test (quantification of formaldehyde). Regardless of the experimental conditions, CO was always detected after a lag time that is related to the time needed to reduce the POM. This system appeared quite robust, remaining active after more than two days; however, the system was slow yielding a very small turnover number (TON) of 3.7 after 66 hours. Two reasons could be at the origin of this weak activity: (i) a slow electron transfer between reduced POM and CO_2 or (ii) the need for a reaction between two CO_2 molecules ($2CO_2 + 2e^- \rightarrow CO_3^{2-} + CO$) when there is no proton in the electrolyte. Concomitant to the production of CO, the authors noted that formaldehyde was also formed during the electrolysis performed in the presence of protons and water. The production of formaldehyde was even observed in the absence of CO_2, meaning that a part of the reduced POMs could indeed promote dehalogenation reactions of CH_2Cl_2. Nevertheless, the amount of formaldehyde was significantly more important when the electrolysis was performed under CO_2. The pioneering study of Proust demonstrated for the first time that POM promoted the electrocatalytic reduction of CO_2 to CO and could even be used to perform the four-electron/four-proton reduction of formaldehyde.

The group of Ronny Neumann studied the electrocatalytic performance of trimetallo-substituted Keggin-type POMs towards the CO_2 reduction in acetonitrile [50]. The trimetallo-substituted Keggin-type POMs $\beta\text{-}[SiW_9O_{37}M'_2M''(H_2O)_3]^{q-}$ (M′ or M″ = Fe^{3+}, Ni^{2+} and Cu^{2+}; denoted $\{SiW_9M'_2M''\}$ hereafter) were prepared from the reaction of metal acetate or preformed heterotrinuclear acetate with a tri-lacunary POM $\beta\text{-}[SiW_9O_{34}H]^{9-}$. Although no structural characterizations (X-ray diffraction and NMR data) of the trimetallo-substituted Keggin-type POMs were carried out, the IR spectra and mass spectra suggested the desired compounds were successfully obtained. Then, the trimetallo derivatives in the form of tetrahexylammonium salts were prepared by metathesis to obtain the compounds soluble in

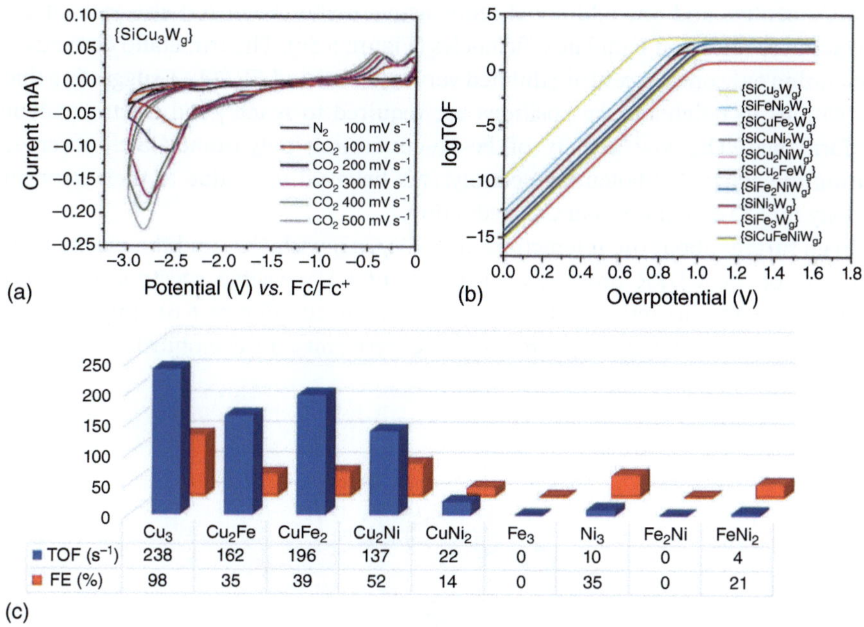

Figure 6.5 (a) Cyclic voltammograms of $\{SiW_9Cu_3\}$ at different scan rates and under 1 bar of CO_2 or N_2 (2 mM [POM] in acetonitrile containing 0.1 M $TBAPF_6$). (b) Catalytic Tafel plots for CO_2 reduction by various POM-based electrocatalysts. (c) Results of the controlled potential electrolysis in an undivided cell for the reduction of CO_2. Source: Azaiza-Dabbah et al. [50]/with permission of John Wiley & Sons.

acetonitrile. Although the exact nature of the coordination sphere of the 3d transition metals within $\{SiW_9M'M'_2\}$ dissolved in acetonitrile can be questionable, the CDRR activity of these POMs for CO_2 reduction in acetonitrile was systematically evaluated.

The CV of $\{SiW_9Cu_3\}$ performed under either N_2 or CO_2 evidenced the presence of an electrocatalytic peak at approximately −2.5 V *vs.* Fc/Fc^+ that has been associated with the CO_2 reduction. Furthermore, the authors suggested that the small cathodic peak at around −2.0 V *vs.* Fc/Fc^+ corresponded to the CO_2 coordination (Figure 6.5a). To compare the electrocatalytic activity of $\{SiW_9Cu_3\}$ with other trimetallo-substituted Keggin-type POMs, the authors determined the maximum turnover frequencies (TOF^{max}) by calculating TOF^{max} and the overpotential (η) from a catalytic Tafel plot. This comparison showed that $\{SiW_9Cu_3\}$ was the best catalyst of the series and exhibited a relatively low overpotential $\eta = 0.77$ V at its TOF^{max} (Figure 6.5b). Then, bulk controlled-potential electrolysis (−2.5 V *vs.* Fc/Fc^+ for 15 hours under 1 bar of CO_2) using $\{SiW_9Cu_3\}$ as electrocatalyst was carried out to identify and quantify the product formation. Gas chromatography analysis revealed that CO was the only gaseous product formed (no H_2), while analysis of the electrolyte showed no discernible formation of any products. The Faradaic efficiency (FE) for CO was very high (98%), and using $^{13}CO_2$ as the reactant showed that all CO was produced from the reactant. Other trimetallo derivatives containing two Cu centres and one additional redox-active metal (Fe or Ni) also yielded CO but at lower rates and Faradaic efficiencies (Figure 6.5c). The trimetallo derivatives containing only one Cu centre exhibited very low TOF and FE (14%), suggesting that Cu cations in neighbouring positions are required to reach good electrocatalytic performance. The low activity of Fe–Ni-containing polyoxometalates observed during the controlled-potential electrolysis is believed to be due to their stronger activity for CO oxidation than CO_2 reduction.

To go further about the interactions of $\{SiW_9Cu_3\}$ with CO_2 and the mechanistic pathway of the CDRR process, an experimental UV–visible spectroscopy study combined with theoretical calculations was realized (Figure 6.6). Important to note, the electrocatalytic experiments were performed in acetonitrile containing

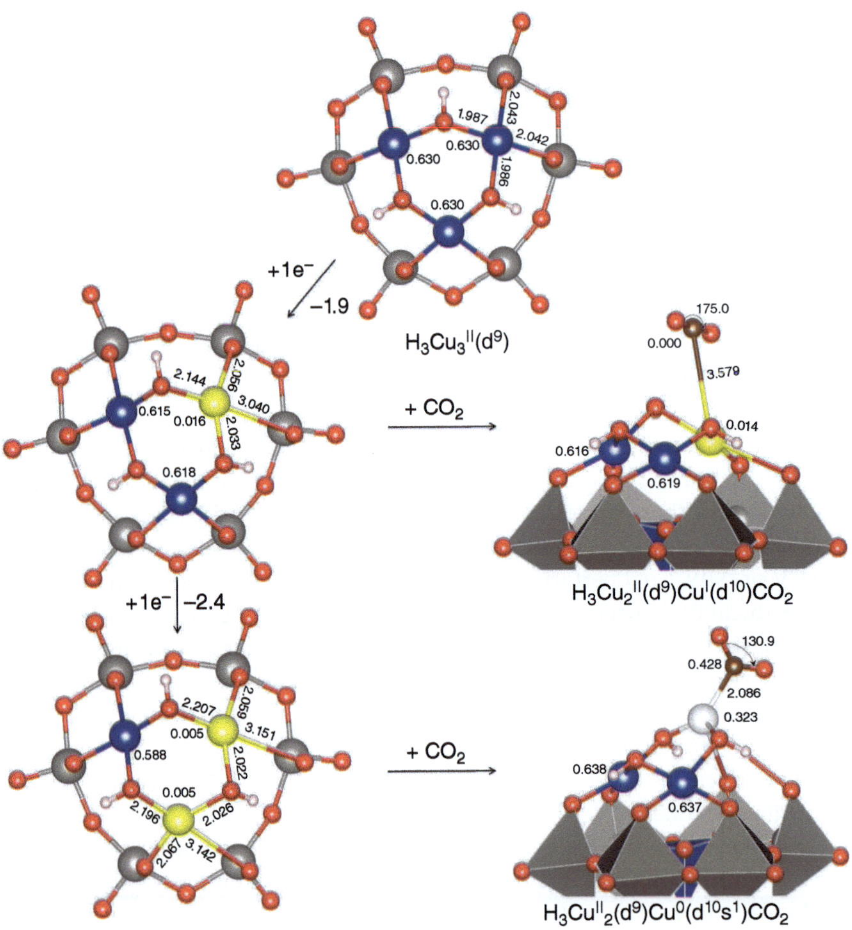

Figure 6.6 Theoretical analysis of the binding of CO_2 to $\{SiW_9Cu_3\}$ in function of the reduction state of the POMs. The angles (in °) and main distances (in Å) are shown in black, and the spin densities are in bold and italic. Colour code: blue (Cu^{II}), yellow (Cu^{I}), white (Cu^{0}), red (O), pink (H) and grey (W). Source: Azaiza-Dabbah et al. [50]/with permission of John Wiley & Sons.

1% water, allowing the tri-protonation of the POM. The one-electron reduced POM $H_3\{SiW_9Cu^{II}{}_2Cu^{I}\}$ showed no binding interaction with CO_2. Indeed, the optimized distance between Cu^I and CO_2 is 3.6 Å, and the O=C=O molecule is almost linear with an angle of 175°, suggesting that only one Cu^I ion is not able to interact with a CO_2 molecule. In contrast, the two-electron reduced POM $H_3\{SiW_9Cu^{I}{}_2Cu^{II}\}$ interacted strongly with CO_2, as supported by UV–visible spectroscopy, and formed a polyanion with a $H_3\{SiW_9Cu^0Cu^{II}{}_2\}$ electronic configuration wherein CO_2 was bound to the Cu^0 atom, according to theoretical calculations. This step is endergonic (+7.8 kcal mol^{-1}) indicating that it may be involved in the reaction rate-determining step. It is also important to note that only one CO_2 molecule was capable of binding $H_3\{SiW_9Cu^0Cu^{II}{}_2\}$; therefore, no C2 products were formed during the electrocatalytic reduction contrary to what was observed with Cu_x-based extended heterogeneous electrocatalysts.

6.5.2 Platinoid-containing Hybrid POMs as Electrocatalysts in Organic Solvents

Many platinoid complexes are quite efficient for CO_2 reduction. Therefore, their grafting on POMs can be an appealing way to boost their performance or to modify their selectivity. Proust and co-workers have explored this concept first [51]. They selected the organometallic derivative $[Cp^*Rh^{III}(bpy)Cl]^+$, a well-known CO_2 reduction catalyst that produces $CHOO^-$ over H_2 and CO, and used the mono-lacunary POM $[PW_{11}O_{39}]^{7-}$ as a redox-active ligand to replace the bipyridine. The compound $(TBA)_3[\alpha\text{-}H_2PW_{11}O_{39}\{Rh^{III}Cp^*(OH_2)\}]$ was prepared in dichloromethane from $[Cp^*Rh(CH_3CN)_3](BF_4)_2$ and $(TBA)_5[\alpha\text{-}H_2PW_{11}O_{39}]$ in the presence of triethylamine. Under argon, the hybrid POMs exhibited a complex cyclic voltammogram showing several redox processes which were sometimes irreversible and difficult to attribute. Upon the addition of water, there was an overall shift of the redox events towards less negative potentials and a concomitant increase in their intensity, especially for the most negative waves. Controlled-potential electrolysis performed at −1.8 V in acetonitrile-containing water led to the production of H_2 and partial reduction of the POMs. Replacing argon with CO_2 led to an increase in the electrochemical response below −1.5 V indicating the probable interactions between the reduced hybrid POMs and CO_2. Again, the addition of a small amount of water provoked a shift towards less reducing potentials and an increase of the current. Bulk electrolysis performed at controlled potential (−1.6 V) in the presence of 5% water under a CO_2 atmosphere, led, after five hours electrolysis, to the production of H_2 as the majority product (TON of 12 with 68% FE) and $HCOO^-$ as the very minority reduction product (TON of 0.8 with 4.5% FE). Importantly, in similar conditions, the electrolysis of the molecular complex $[Cp^*Rh^{III}(bpy)Cl]^+$ led to the generation of $HCOO^-$ (TON of 18 with 60% FE) as the majority product, and only H_2 as the minority product (TON of 9 with 30% FE). Thus, the use of POM-containing Rh^{III} is quite disappointing for the CDRR since its use as a bidentate ligand instead of the bipyridine reverses the selectivity of the Rh^{III}-based molecular catalyst in favour of the HER process.

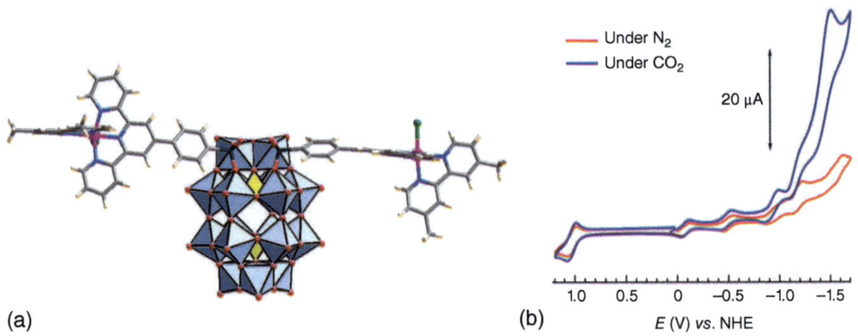

Figure 6.7 (a) X-ray diffraction structure of the hybrid polyanion $[(Ru(dmbpy)(ppt)Cl)_2P_2W_{17}O_{57}]^{4-}$ (dmbpy: 4,4′-dimethyl-2,2′-dipyridyl and ppt: 4′-(4-phosphonophenyl)-2,2′ : 6′,2″-terpyridine). (b) Comparison of the cyclic voltammograms of $[(Ru(dmbpy)(ppt)Cl)_2P_2W_{17}O_{57}]^{4-}$ in *N*-methyl-2-pyrrolidone under either N_2 or CO_2 (1 mM [POM] in *N*-methyl-2-pyrrolidone containing 0.1 M $TBAPF_6$). Source: Kuramochi et al. [53]/with permission of Royal Society of Chemistry.

Molecular polypyridyl complexes of Ru are also known to be capable to reduce CO_2 to CO [52]. Newton and co-workers designed a hybrid POM cluster containing Ru-polypyridyl moieties $(TBA)_3H[(Ru(dmbpy)(ppt)Cl)_2P_2W_{17}O_{57}]$ (dmbpy: 4,4′-dimethyl-2,2′-dipyridyl and ppt: 4′-(4-phosphonophenyl)-2,2′ : 6′,2″-terpyridine) in which two ruthenium antenna complexes are associated with an electroactive lacunary Dawson-type polyoxotungstate [53]. Under N_2, the electrochemical behaviour of this hybrid POM in *N*-methyl-2-pyrrolidone is particularly rich (Figure 6.7) involving the Ru centres at high potential (Ru^{2+}/Ru^{3+} process at +1.1 V *vs.* NHE), the W-oxo fragment of the POMs (four monoelectronic redox events located between −0.1 and −1.5 V *vs.* NHE) and the organic ligands of Ru centres (observed between −1.2 and −1.6 V *vs.* NHE). Under CO_2, the CV recorded between −1.0 and +1.2 *vs.* NHE was similar to that collected under N_2; however, as the potential was decreased below −1.00 V *vs.* NHE, the current rapidly increased. Spectroelectrochemical measurements evidenced that the organic anion radicals reacted rapidly in the presence of CO_2 and that the chloride ligand was lost over time. Under a saturated CO_2 atmosphere, bulk controlled-potential electrolysis was performed at −1.56 V *vs.* NHE in acetonitrile containing 0.1 M $TBAPF_6$, and the gas-phase products of the electrocatalytic CO_2 reduction were detected by GC and capillary electrophoresis. After one hour electrolysis, a mixture of H_2 and CO was detected with TON values of 38 and 1.4, respectively. In the presence of 10 equivalents of H_2O, similar results were obtained, suggesting that the proton source of the produced H_2 was likely water impurities. Interestingly, the authors discovered that the addition of acetic acid drastically changed the selectivity. Bulk electrolysis experiments performed in the presence of 10 equivalents of acetic acid showed that HER as well as the CO production were almost stopped, while formic acid became the majority product of the CO_2 reduction (TON of 22 with 66% FE after one hour of electrolysis at −1.56 V *vs.* NHE). It is worth noting that this example represents a perfect interplay between the POM unit, acting as an electron/proton reservoir, and a catalytic unit. Indeed, H_2 remained the main reduction product when the Ru-polypyridyl moieties were not grafted to the POM unit.

6.5.3 POMs as Electron Relays in Aqueous Solution

The metal copper-based electrocatalysts represent an important class of cathode materials for CO_2 reduction because of their relatively high Faradaic efficiency and their ability to produce C2 derivatives. In contrast, most of the other metal catalysts favour the production of C1 compounds. Therefore, obtaining C2 by electrocatalysis using metal-based electrodes represents an important challenge in which POMs can be employed due to their ability to undergo multiple electron transfers and store protons. In 2020, the group of Chunxiang Li reported that the POM $[SiW_{11}MnO_{39}(H_2O)]^{5-}$ dissolved in an aqueous electrolyte (pH = 4) strongly influenced the electrochemical reduction of CO_2 on metal indium-based electrode yielding the formation of formate [54]. Metal indium catalysts usually favour the formation of C1 products with reasonably high efficiency but are reluctant to produce C2 compounds. They showed that the use of POM $[SiW_{11}MnO_{39}(H_2O)]^{5-}$ as a redox mediator allowed to achieve superior selectivity towards HCOOH and CH_3COOH formation within the optimized conditions, namely 2 mM of POM electrolyzed at −1.0 V *vs.* Ag/AgCl at pH 3.9. Faradaic efficiency for the formation of acetic acid reached 72%. Furthermore, the authors noticed that the concentration of $[SiW_{11}MnO_{39}(H_2O)]^{5-}$ had little effect on the production of formic acid, but had an obvious effect on the formation of acetic acid. They observed that the surface of the indium electrode evolved during the electrolysis process. Indeed, the energy dispersive X-ray spectroscopy (EDS) analysis evidenced the deposition of tungsten oxide on the indium surface, while Mn was not detected.

Chunxiang Li's group continued their research on CDRR using an indium electrode and POM as the redox mediator and published a series of articles. Using the vanadium-containing POM $[SiW_9V_3O_{40}]^{7-}$ in similar conditions (2 mM of POM electrolyzed at −1.15 V *vs.* Ag/AgCl at pH 4), they showed that this electrocatalytic system was very efficient for acetate production (FE > 95%) and robust [55]. EDS analysis of the electrode after one hour electrolysis revealed the presence of V and W at the surface of the indium electrode. They also studied the mixed-metal Keggin anion $[PMo_{12-x}V_xO_{40}]^{n+3-}$ by varying the content of vanadium centres in the molybdenum-oxo frameworks [56]. The systematic analysis of the products formed during the bulk electrolysis carried out at different potentials (from 0.3 to 1 V *vs.* Ag/AgCl) revealed that the products were H_2, CO and ethanol. When a low potential (+0.3 V) was applied, the main product formed during the bulk electrolysis was ethanol with excellent efficiency. For example, a FE value of 93.4% was determined for the vanadium tri-substituted phosphomolybdate. The amount of ethanol produced by CO_2 reduction decreased gradually as a function

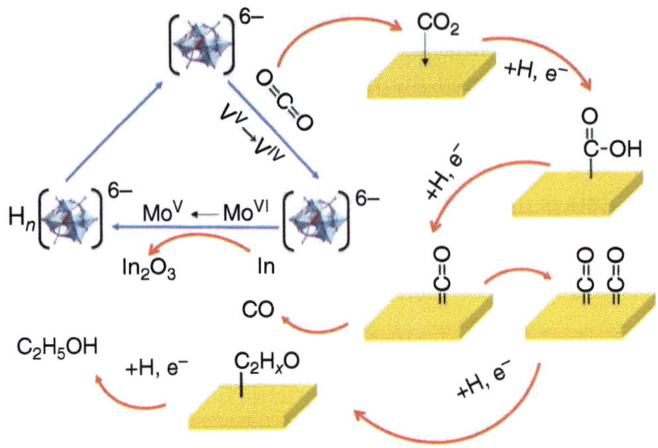

Figure 6.8 Proposed mechanism for CDRR from POM and indium synergistic catalysis. Source: Sun et al. [56]/with permission of Elsevier.

of the electrolysis potential when moving to more negative values. The authors proposed a mechanism to explain the formation of ethanol following several steps (Figure 6.8): (i) the vanadium centres were first reduced through the proton-coupled electron transfer, (ii) the generated reduced POM reacted with CO_2 to produce HCO_3^-, (iii) the latter step enhanced the concentration of CO_2 by promoting equilibrium reaction between CO_2 and HCO_3^-, (iv) this favoured the adsorption of CO_2 on the indium surface, (v) the adsorbed CO_2 was reduced at In^0 active site, (vi) *CO at the adjacent site generated ethanol through C–C coupling and multistep proton-coupled electron transfer (PCET) processes and then (vii) POM continued the PCET process during which the indium electrode was oxidized to In_2O_3 [56].

Chunxiang Li and co-workers also investigated a similar concept but in organic media using the tetrabutylammonium salt of POMs [57, 58]. They showed that the tandem indium-based electrode and $[SW_{11}VO_{40}]^{3-}$ worked as synergistic catalysts firstly for the electrocatalytic reduction of CO_2 to acetic acid and ethanol with Faradaic efficiencies of 67.4% (at −0.5 V *vs.* Ag/Ag⁺) and 19.4% (at −1.1 V *vs.* Ag/Ag⁺), respectively [57]. Again, the presence of the POM was found to be beneficial to prevent the formation of C1 products by regulating single-proton transfer. Very recently, the same team reported the electrocatalytic reduction of CO_2 in similar conditions but using the POM $[SMo_{11}VO_{40}]^{3-}$, and interestingly it was observed that the product selectivity significantly differed [58]. Indeed, the presence of $[SMo_{11}VO_{40}]^{3-}$ shifted the reduction pathway of CO_2 to produce ethanol with good Faradaic efficiency (69.5%) at −1.1 V *vs.* Ag/Ag^+, which was not observed for previously reported In-based materials or exceeded largely that obtained when the POM $[SW_{11}VO_{40}]^{3-}$ was employed [57].

Although not clearly understood, Chunxiang Li's work demonstrated unambiguously that the use of POMs as soluble electron/proton relays in electrolyte allowed to modify the CDRR pathways of indium electrodes. This discovery is encouraging and should motivate further utilization of POMs in this field.

6.6 Electrocatalytic Reduction of CO_2 at POMs-modified (Semi)conducting Electrode Surfaces

6.6.1 Immobilization of POMs on Electrodes

At the outset, it is worth recalling some general principles about electrocatalyst-modified (photo)electrodes. The immobilization of an electrocatalyst, such as POM, onto a (photo)electrode surface is essentially motivated to take advantage of its intrinsic redox and electrocatalytic properties observed in solution [59]. The transfer from a homogeneous to a supported electrocatalytic reaction using an immobilized catalyst is also beneficial for synthetic and sensing applications. Indeed, the electrode modification allows the use of an ultra-small amount of catalyst, the recyclability of the catalyst and the easy separation of the electrogenerated products (which makes purification easier). The first two aspects are particularly relevant when costly or high-value catalysts that have been sometimes synthesized in several steps are involved. Once immobilized, it is essential that the electrocatalyst durably remains on the electrode surface and that its electrocatalytic characteristics are not electrochemically degraded upon repeated oxidation and reduction, but also not chemically altered upon the action of some oxidizing species (e.g. O_2) or unfavourable experimental conditions (e.g. pH or temperature). In addition, the overall efficiency of the electrocatalytic process at modified electrodes depends on three main factors: (i) the rate of the reaction occurring at the immobilized electrocatalyst, (ii) the permeation of the substrate and the electrogenerated products through the film and (iii) the transport of electrons across the film from the bulk electrolyte solution to the electrode surface [60].

Overall, the immobilization methods of POMs reported over the last 15 years can be broadly categorized into three types [61]: (i) covalent or electrostatic attachment to organic/organometallic monolayers or 3D metal–organic frameworks (MOFs), (ii) chemisorption on planar and micro-/nanostructured metallic or carbon structures and (iii) electrostatic entrapment within charged polymer films (redox and electronically conducting polymers). Depending on the method, the POM assembly is either directly immobilized on the (semi)conducting electrode surface or dispersed in an appropriate solvent before being deposited by, e.g. drop-casting or spin-coating. In some cases, carbon black is added to the dispersion to improve the electron conductivity of the assembly.

Concerning the electrode surface to be modified, a conducting and catalytically inert material, such as glassy carbon or graphite, is commonly used so that the contribution of the underlying electrode in the overall electrocatalytic process can be neglected. The situation is completely different when the electrode surface is a photoactive semiconductor, such as doped silicon. Indeed, unlike metallic and carbonaceous surfaces, the conductive properties of semiconductors can be modified upon light illumination [62–64]. The irradiation of a semiconductor with light of appropriate wavelength (higher than the semiconductor bandgap) and intensity leads to the generation of a high concentration of minority charge carriers (i.e. electrons or holes for *p*- or *n*-type semiconductor, respectively) which can activate the electrochemical process. As a result of the photogenerated minority charge carrier-induced activation of the redox process, the electrocatalytic reaction at illuminated catalyst-modified photoanode or photocathode is often observed at less positive or negative potentials, respectively, compared to the case of similarly modified conducting electrodes [63–65]. Consequently, the use of photoactive semiconducting electrodes could be a real advantage in the field of electrochemical catalysis in terms of energy gain.

The aforementioned immobilization procedures have been successfully used to produce a wide variety of robust POMs-modified electrodes which were catalytically effective for a broad diversity of reactions of interest, such as cathodic HER, reduction of oxyanionic species (nitrite, iodate, bromate, etc.), anodic oxygen evolution reaction (OER) and oxidation of ascorbic acid [10]. Nevertheless, in comparison, the involvement of POMs-modified electrodes in CDRR has been underexplored. Also surprisingly, there is no reported example of a POM-modified semiconducting electrode which is photoelectrocatalytically active for CDRR to the best of our knowledge.

6.6.2 POMs-modified Electrodes Electrocatalytically Active for CO_2 Reduction

Owing to the anionic nature of POMs, the immobilization procedure by electrostatic incorporation constitutes the most straightforward approach for producing functional POM-modified electrodes. So, layer-by-layer assemblies sequentially associating multimetallic porphyrins [Mn(III), Ni(II) and Zn(II) μ-(meso-5,10,15,20 tetra(pyridyl)porphyrin)tetrakis {bis(bipyridine)chloride ruthenium(II)}] and Keggin-type polyoxotungstate $[SiW_{12}O_{40}]^{4-}$ have been durably immobilized on optically transparent electrode surfaces and further tested for electrochemical and photoelectrochemical CDRR [66]. The three modified electrodes were catalytically active for CO_2 reduction in aqueous solution below ca. −0.7 V *vs.* Ag/AgCl, but the highest cathodic current intensities were measured for the assembly incorporating the Mn(III) cation. Controlled-potential electrolysis tests showed that these systems were not selective for CDRR as CO, formic acid and methanol were detected as the reaction products, the relative concentrations of which were dependent on the nature of the metal coordinated in the porphyrin. Interestingly, the light illumination of the electrodes led to an improvement in the electronic transfer thanks to the photoactive Ru(II) centres which promoted the formation of the four-electron product, namely formaldehyde.

In another example of electrostatic assemblies, POMs and metalloporphyrins have been combined to elaborate MOFs [67]. Such compounds represent a class of coordination polymers that are built from the assembly of transition metals, POMs and organic ligands [68]. Due to the insulating character of these extended 3D networks, the integration of electron reservoirs such as POMs capable of quickly exchanging a large number of electrons is a relevant approach for efficiently and selectively promoting the multielectron CDRR electrocatalysis. In the report from Wang et al. [67], a series of stable materials with the general formula $[PMo^{V}_{8}Mo^{VI}_{4}O_{35}(OH)_5Zn_4]_2[M\text{-}TCPP][2H_2O][1.5TBAOH]$ (with M = Fe, Co,

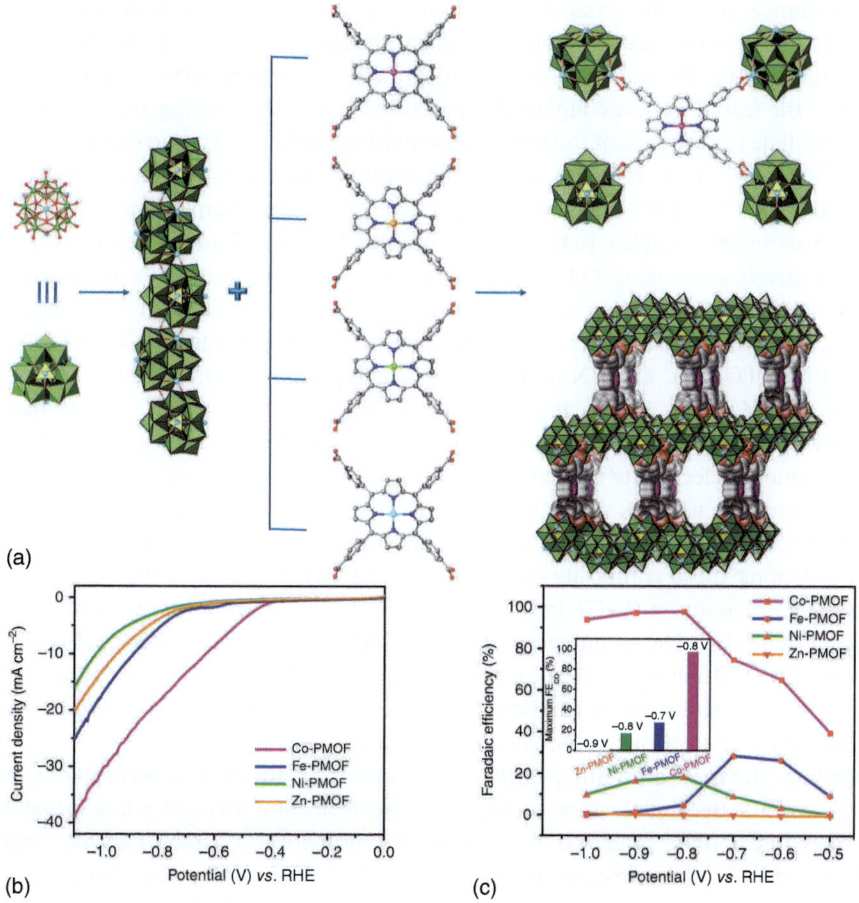

Figure 6.9 (a) Construction of stable polyoxometalate–metalloporphyrin organic frameworks (M-PMOFs with M = Co, Fe, Ni or Zn) by associating tetrakis[4-carboxyphenyl]-porphyrin-M (M-TCPP) linkers and reductive Zn-ε-Keggin clusters ($\{\varepsilon\text{-}PMo_8{}^{V}Mo_4{}^{VI}O_{40}Zn_4\}$ including eight Mo^V atoms). Electrocatalytic performance of M-PMOFs in CO_2-saturated 0.5 M $KHCO_3$ (pH 7.2) electrolyte: (b) linear sweep voltammetric (LSV) curves at 5 mV s^{-1}. (c) FEs for CO. Source: Wang et al. [67]/with permission of Springer Nature.

Ni and Zn; TCPP = tetrakis[4-carboxyphenyl]-porphyrin and TBAOH = tetrabutylammonium hydroxide) has been prepared through an in situ hydrothermal method (Figure 6.9). After their immobilization on carbon cloth electrodes, the so-modified electrodes showed as expected an electrocatalytic activity for CDRR in an aqueous electrolyte (Figure 6.9b). However, clear differences were observed depending on the nature of the porphyrin metal cation. The best-performing electrocatalyst, in terms of onset potential, current densities and electron transfer kinetics, was the material incorporating a Co-porphyrin. The latter was able to selectively convert CO_2 to CO with a remarkable Faradaic efficiency ranging from 94% to 99%, a high TOF of 1656 h^{-1} (at −0.8 V *vs.* reversible hydrogen electrode RHE) and excellent stability in operation for more than 36 hours (Figure 6.9c). The superior electrocatalytic performance of this material was supported by density functional theory (DFT) calculations that provided evidence for the harder desorption of CO from the M–CO intermediate (in the case of the system integrating the Fe-porphyrin) and the slowness of the initial proton-coupled electron transfer to CO_2 forming the M-COOH intermediate (in the case of the systems integrating the Ni- or Zn-porphyrin).

In 2020, Du et al. reported another innovative approach to the electrostatic assembly of a POM with a catalytically active molecular complex, namely manganese carbonyl complex [69]. Such a complex is a well-known electrocatalyst for selectively converting CO_2 to CO in organic solution [70–72]. In an aqueous medium, this catalyst is much less selective owing to the occurrence of the unavoidable competitive HER. In contrast, the composite compounds $[Mn^I(bipy)(CO)_3(CH_3CN)]_n(POM)\cdot n'\ CH_3CN$ (with bipy = 2,2′-bipyridine, POM = $[SiW_{12}O_{40}]^{4-}$, $n = 4$, $n' = 5$; $[PW_{12}O_{40}]^{3-}$, $n = 3$, $n' = 2$; $[PMo_{12}O_{40}]^{3-}$, $n = 3$, $n' = 2$) formed between a Keggin-type POM and the Mn complex enabled to electrocatalyse CO_2 to CO in aqueous electrolyte with high selectivity. The transposition of electrocatalysis from organic to aqueous solution was ascribed to the modification of the electron transfer process between POM and the Mn complex depressing the competitive HER. Among these composite catalysts, the one incorporating $[SiW_{12}O_{40}]^{4-}$ was the most efficient for CDRR in an aqueous electrolyte with particularly high FE for CO more than 95% at an overpotential of 0.61 V, which was nearly double that of the parent Mn complex (Figure 6.10). Moreover, a high stability of 12 hours in operation was observed. Theoretical calculations demonstrated that $[SiW_{12}O_{40}]^{4-}$ delivered faster electrons to the Mn complex for further catalytic CDRR.

Besides POMs-incorporating electrostatic assemblies, the deposition of POMs on catalytically active metal nanostructures has also been demonstrated to be a valuable strategy to elaborate protected and stabilized nanomaterials exhibiting synergistic effects for various electrocatalytic reactions [73]. Following this approach, Zhang and co-workers have electrodeposited Ag-POM nanocomposites on graphite electrodes from the electrochemical reduction of Ag^+ in the presence of $[PMo_{12}O_{40}]^{3-}$ [74]. The modified electrodes showed electrocatalytic performance for CO_2-to-CO reduction superior to those of metal and POM moieties tested alone, as assessed by a lower overpotential, higher current densities and better stability during long-term electrolysis tests. In another study from the same group, Ag nanoclusters (NCs) were first capped with bovine serum albumin (BSA) and then coated with a $[SiW_{12}O_{40}]^{4-}$ deposit [75]. BSA is a well-known stabilizer of metal nanostructures, but it also behaves as an insulating network in Ag NCs-BSA that impedes an efficient electronic communication between the metal catalyst and the underlying electrode. Therefore, the addition of POM to this type of structure to reestablish effective electronic communication makes sense because the POM can act as an electron mediator able to fasten electron transfer by hopping. Although the Ag NCs-BSA/$[SiW_{12}O_{40}]^{4-}$ assembly showed an electrocatalytic activity for CO_2-to-CO reduction higher than that of a bulk Ag electrode, the FE for CO was somewhat modest (about 75%) and H_2 was detected as a by-product with a FE of about 15%. Furthermore, compared with Ag-$[PMo_{12}O_{40}]^{3-}$ nanocomposites from the same group [74], the beneficial effects of BSA on the electrocatalytic reaction have not been really demonstrated.

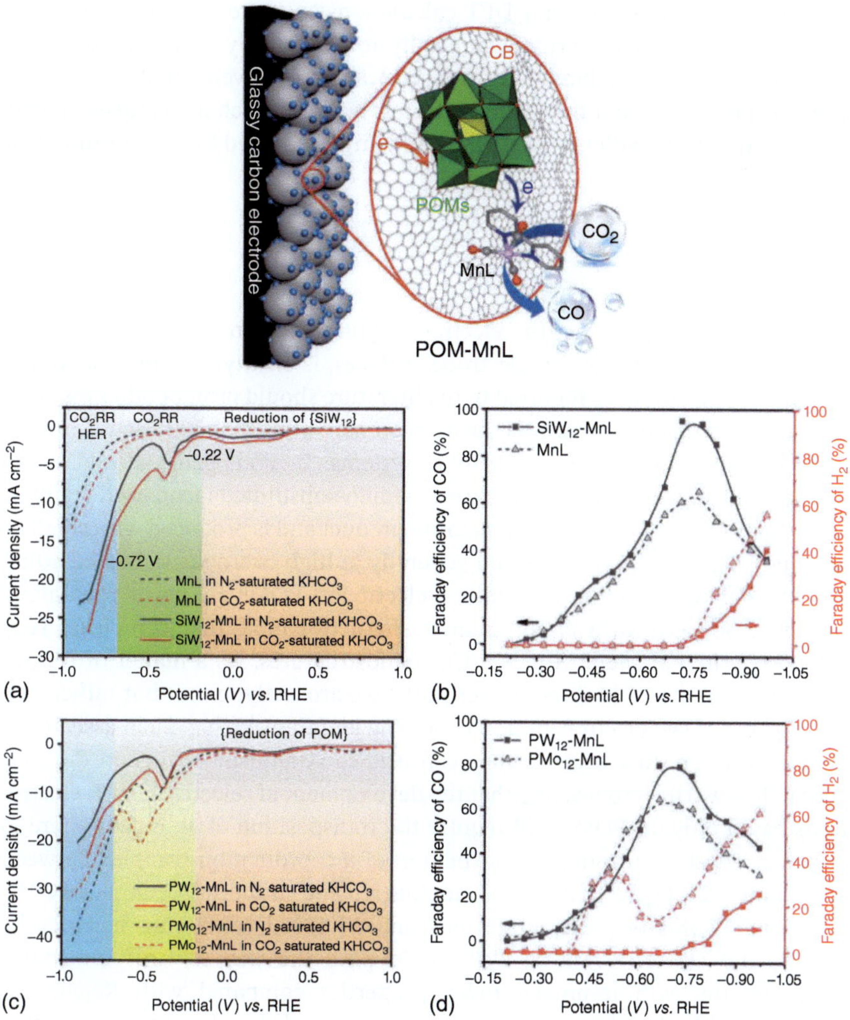

Figure 6.10 CO_2-to-CO reduction electrocatalysed at $[Mn^I(bipy)(CO)_3(CH_3CN)]_n(POM)\bullet\ n'$ CH_3CN/carbon black (POM–MnL/CB). (a) LSV curves of $SiW_{12}O_{40}$–MnL ($n = 4$, $n' = 5$)/CB and MnL/CB in 0.5 M N_2-(black curves) or CO_2-(red curves) saturated $KHCO_3$ electrolyte and (b) corresponding FE values for CO at different potentials (the line is drawn only to guide the eye). (c) LSV curves of $PW_{12}O_{40}$–MnL ($n = 3$, $n' = 2$)/CB and $PMo_{12}O_{40}$–MnL ($n = 3$, $n' = 2$)/CB in 0.5 M N_2-(black curves) or CO_2-(red curves) saturated $KHCO_3$ electrolyte and (d) corresponding FE values for CO at different potentials (the line is drawn only to guide the eye). Source: Du et al. [69]/with permission of Royal Society of Chemistry.

Another interesting example of metal–POM nanocomposites electrocatalytically active for CDRR has been recently reported by Zang et al. [76]. In their work, copper nanocubes have been first modified with $Cu_2Mo_8O_{26}$, and this nanomaterial was then electrodeposited on TiO_2 nanotubes-coated titanium foil. The so-functionalized electrode achieved the reduction of CO_2 to acetate with a state-of-the-art FE of 48.7% and current density exceeding $100\,mA\,cm^{-2}$ at

−1.13 V *vs.* RHE. Complementary DFT calculations suggested the formation of a Cu–Mo–O interface which thermodynamically and kinetically favoured the acetate production. It is worth recalling that the electrocatalytic conversion of CO_2 to high added-value products, other than CO and formic acid, is still challenging even now. Consequently, such a result is encouraging and unprecedented for a POM-modified electrode.

6.7 Conclusions

In this chapter, it appears that the electrocatalytic reduction of CO_2 from POMs, either in solution or attached to electrode surfaces, is clearly underexplored, and the few promising examples reported in the literature should prompt other research groups to pursue this route towards the development of novel highly active, robust and selective POMs-based electrocatalytic systems. In homogeneous phase, the few reported examples focus essentially on metallo-substituted lacunary Keggin- or Dawson-type POMs. CO is often the majority product and is produced with small or quite modest TON or TOF values and generally at high overpotentials. Regarding the electrocatalytic reduction at modified electrodes, POMs are never used alone but always associated with either catalytically active molecular complexes (e.g. metalloporphyrins) or metal (e.g. Ag, Cu) nanostructures. As a matter of fact, in these assemblies, the POM does not act as the electrocatalytic unit but rather as a moiety capable of fast electron/proton collecting and transferring. Such assemblies are, however, interesting because they can lead to synergistic effects on the CO_2 reduction. It is worth emphasizing that the development of selective CDRR electrocatalysts based only on POMs will require the transposition of an outer-sphere to inner-sphere catalysis through the formation of a coordination complex between the POM unit and CO_2. In that context, engineered and structurally complex POMs, such as giant POMs (e.g. Keplerate ions and 3d metal containing macrocyclic cryptates), and cluster-POMs as well, could pave the way towards innovative catalytic systems with improved redox properties compared with Keggin- or Dawson-derived basic structures. Finally, the use of photoactive semiconducting electrodes as POMs-immobilization substrates has never been explored for the photoelectrocatalytic reduction of CO_2 and therefore merits exploration. Besides the energy gain for electrocatalysis provided by the semiconductor, the thoughtful mix between the photocatalytic activity of some POMs with the photoactivity of semiconductors could also yield particularly innovating modified photoelectrodes.

References

1 Bagger, A., Ju, W., Varela, A.S. et al. (2017). *ChemPhysChem* 18: 3266–3273.
2 Nitopi, S., Bertheussen, E., Scott, S.B. et al. (2019). *Chem. Rev.* 119: 7610–7672.
3 Elouarzaki, K., Kannan, V., Jose, V. et al. (2019). *Adv. Energy Mater.* 9: 1900090.
4 Mansuy, D., Bartoli, J.F., Battioni, P. et al. (1991). *J. Am. Chem. Soc.* 113: 7222–7226.

5 Khenkin, A.M. and Hill, C.L. (1993). *J. Am. Chem. Soc.* 115: 8178–8186.

6 Gu, J., Chen, W., Shan, G.-G. et al. (2021). *Mater. Today Energy* 21: 100760.

7 Bard, A.J., Parsons, R., and Jordan, J. (1985). *Standard Potentials in Aqueous Solution*. New York: M. Dekker.

8 Ueda, T. (2018). *ChemElectroChem* 5: 823–838.

9 Sadakane, M. and Steckhan, E. (1998). *Chem. Rev.* 98: 219–238.

10 Fabre, B., Falaise, C., and Cadot, E. (2022). *ACS Catal.* 12: 12055–12091.

11 Sanchez, C., Livage, J., Launay, J.P. et al. (1982). *J. Am. Chem. Soc.* 104: 3194–3202.

12 Sanchez, C., Livage, J., Launay, J.P., and Fournier, M. (1983). *J. Am. Chem. Soc.* 105: 6817–6823.

13 Robin, M.B. and Day, P. (1968). *Advances in Inorganic Chemistry and Radiochemistry* (ed. H.J. Emeléus and A.G. Sharpe), 247–422. Academic Press.

14 Suaud, N., Gaita-Ariño, A., Clemente-Juan, J.M. et al. (2002). *J. Am. Chem. Soc.* 124: 15134–15140.

15 Zhao, T., Bell, N.L., Chisholm, G. et al. (2023). *Energy Environ. Sci.* 16: 2603–2610.

16 Bidi, L.K., Desjonquères, A., Izzet, G., and Guillemot, G. (2023). *Inorg. Chem.* 62: 1935–1941.

17 Kozik, M. and Baker, L.C.W. (1990). *J. Am. Chem. Soc.* 112: 7604–7611.

18 Geletii, Y.V., Hill, C.L., Bailey, A.J. et al. (2005). *Inorg. Chem.* 44: 8955–8966.

19 Fatima, A., Smortsova, Y., Falaise, C. et al. (2023). *Chem. Commun.* 59: 10988–10991.

20 Turo, M.J., Chen, L., Moore, C.E., and Schimpf, A.M. (2019). *J. Am. Chem. Soc.* 141: 4553–4557.

21 Bamba, I.F., Falaise, C., Marrot, J. et al. (2022). *Inorg. Chem.* 61: 8309–8319.

22 Konishi, T., Kodani, K., Hasegawa, T. et al. (2020). *Inorg. Chem.* 59: 10522–10531.

23 Launay, J.P. (1976). *J. Inorg. Nucl. Chem.* 38: 807–816.

24 Chen, J.-J., Symes, M.D., and Cronin, L. (2018). *Nat. Chem.* 10: 1042–1047.

25 Piepgrass, K. and Pope, M.T. (1987). *J. Am. Chem. Soc.* 109: 1586–1587.

26 Falaise, C., Mpacko Priso, G., Leclerc, N. et al. (2023). *Inorg. Chem.* 62: 2494–2502.

27 Yamase, T. and Ishikawa, E. (1996). *J. Chem. Soc. Dalton Trans.* 1619–1627.

28 Jeannin, Y., Launay, J.P., and Sedjadi, M.A.S. (1980). *Inorg. Chem.* 19: 2933–2935.

29 Chen, J.-J., Vilà-Nadal, L., Solé-Daura, A. et al. (2022). *J. Am. Chem. Soc.* 144: 8951–8960.

30 Long, D.-L., Tsunashima, R., and Cronin, L. (2010). *Angew. Chem. Int. Ed.* 49: 1736–1758.

31 Hervé, G., Tézé, A., and Contant, R. (2003). *Polyoxometalate Molecular Science* (ed. J.J. Borrás-Almenar, E. Coronado, A. Müller, and M. Pope), 33–54. Dordrecht: Springer Netherlands.

32 Katsoulis, D.E. and Pope, M.T. (1984). *J. Am. Chem. Soc.* 106: 2737–2738.

33 Mal, S.S. and Kortz, U. (2005). *Angew. Chem. Int. Ed.* 44: 3777–3780.

34 Yang, P., Alsufyani, M., Emwas, A.-H. et al. (2018). *Angew. Chem. Int. Ed.* 57: 13046–13051.

35 Mal, S.S., Dickman, M.H., Kortz, U. et al. (2008). *Chem. Eur. J.* 14: 1186–1195.

36 Yonesato, K., Yanai, D., Yamazoe, S. et al. (2023). *Nat. Chem.* 15: 940–947.

37 Cadot, E., Sokolov, M.N., Fedin, V.P. et al. (2012). *Chem. Soc. Rev.* 41: 7335–7353.

38 Tourneur, J., Fabre, B., Loget, G. et al. (2019). *J. Am. Chem. Soc.* 141: 11954–11962.

39 Anyushin, A.V., Kondinski, A., and Parac-Vogt, T.N. (2020). *Chem. Soc. Rev.* 49: 382–432.

40 Cameron, J.M., Guillemot, G., Galambos, T. et al. (2022). *Chem. Soc. Rev.* 51: 293–328.

41 Garai, S., Haupt, E.T.K., Bögge, H. et al. (2012). *Angew. Chem. Int. Ed.* 51: 10528–10531.

42 Bandeira, N.A.G., Garai, S., Müller, A., and Bo, C. (2015). *Chem. Commun.* 51: 15596–15599.

43 Szczepankiewicz, S.H., Ippolito, C.M., Santora, B.P. et al. (1998). *Inorg. Chem.* 37: 4344–4352.

44 Katsoulis, D.E., Tausch, V.S., and Pope, M.T. (1987). *Inorg. Chem.* 26: 215–216.

45 Gao, G., Li, F., Xu, L. et al. (2008). *J. Am. Chem. Soc.* 130: 10838–10839.

46 Cheng, W., Xue, Y., Luo, X.-M., and Xu, Y. (2018). *Chem. Commun.* 54: 12808–12811.

47 Chen, B. and Neumann, R. (2018). *Eur. J. Inorg. Chem.* 2018: 791–794.

48 Khenkin, A.M., Efremenko, I., Weiner, L. et al. (2010). *Chem. Eur. J.* 16: 1356–1364.

49 Girardi, M., Blanchard, S., Griveau, S. et al. (2015). *Eur. J. Inorg. Chem.* 2015: 3642–3648.

50 Azaiza-Dabbah, D., Vogt, C., Wang, F. et al. (2022). *Angew. Chem. Int. Ed.* 61: e202112915.

51 Girardi, M., Platzer, D., Griveau, S. et al. (2019). *Eur. J. Inorg. Chem.* 2019: 387–393.

52 Johnson, B.A., Maji, S., Agarwala, H. et al. (2016). *Angew. Chem. Int. Ed.* 55: 1825–1829.

53 Kuramochi, S., Cameron, J.M., Fukui, T. et al. (2023). *Chem. Commun.* 59: 10801–10804.

54 Li, C., Zha, B., and Li, J. (2020). *J. CO_2 Util.* 38: 299–305.

55 Zha, B., Li, C., and Li, J. (2020). *J. Catal.* 382: 69–76.

56 Sun, W., Yao, D., Tai, Y. et al. (2023). *J. Colloid Interface Sci.* 650: 121–131.

57 Sun, W., Li, C., and Zhao, H. (2022). *J. Power Sources* 517: 230665.

58 Sun, W., Tai, Y., Tian, W. et al. (2023). *Electrochim. Acta* 445: 142067.

59 Murray, R.W. (ed.) (1992). *Techniques of Chemistry: Molecular Design of Electrode Surfaces*. New York, NY: Wiley-Interscience.

60 Savéant, J.M. (2006). *Elements of Molecular and Biomolecular Electrochemistry: An Electrochemical Approach to Electron Transfer Chemistry*. Hoboken, NJ: Wiley-Interscience.

61 Cherevan, A.S., Nandan, S.P., Roger, I. et al. (2020). *Adv. Sci.* 7: 1903511.

62 Zhang, X.G. (2001). *Electrochemistry of Silicon and Its Oxide*. New York: Kluwer Academic/Plenum Publishers.

63 Tan, M.X., Laibinis, P.E., Nguyen, S.T. et al. (1994). *Progress in Inorganic Chemistry*, 21–144. Wiley.

64 Bard, A.J., Stratmann, M., and Licht, S. (ed.) (2002). *Semiconductor Electrodes and Photoelectrochemistry ,Encyclopedia of Electrochemistry*, vol. 6. Weinheim: Wiley-VCH.

65 Fabre, B. (2016). *Chem. Rev.* 116: 4808–4849.

66 García, M., Aguirre, M.J., Canzi, G. et al. (2014). *Electrochim. Acta* 115: 146–154.

67 Wang, Y.-R., Huang, Q., He, C.-T. et al. (2018). *Nat. Commun.* 9: 4466.

68 Du, D.-Y., Qin, J.-S., Li, S.-L. et al. (2014). *Chem. Soc. Rev.* 43: 4615–4632.

69 Du, J., Lang, Z.-L., Ma, Y.-Y. et al. (2020). *Chem. Sci.* 11: 3007–3015.

70 Neri, G., Walsh, J.J., Teobaldi, G. et al. (2018). *Nat. Catal.* 1: 952–959.

71 Bourrez, M., Molton, F., Chardon-Noblat, S., and Deronzier, A. (2011). *Angew. Chem. Int. Ed.* 50: 9903–9906.

72 Torralba-Peñalver, E., Luo, Y., Compain, J.-D. et al. (2015). *ACS Catal.* 5: 6138–6147.

73 Wang, Y. and Weinstock, I.A. (2012). *Chem. Soc. Rev.* 41: 7479–7496.

74 Guo, S.-X., Li, F., Chen, L. et al. (2018). *ACS Appl. Mater. Interfaces* 10: 12690–12697.

75 Guo, S.-X., MacFarlane, D.R., and Zhang, J. (2016). *ChemSusChem* 9: 80–87.

76 Zang, D., Li, Q., Dai, G. et al. (2021). *Appl. Catal. B Environ.* 281: 119426.

Part IV

Polyoxometales for Fuel Cells and Electrolysers

7

POM-based Electrocatalysts for Oxygen Evolution Reaction

Víctor K. Abdelkader-Fernández[1], Pablo Garrido-Barros[1], Marta Nunes[2], and Marcos Gil-Sepulcre[3]

[1]University of Granada (UGR), Department of Inorganic Chemistry, Science Faculty, Av. Fuentenueva s/n, Granada 18071, Spain
[2]Associated Laboratory for Green Chemistry of the Network of Chemistry and Technology (LAQV-REQUIMTE), Department of Chemistry and Biochemistry, Science Faculty of the University of Porto, Rua do Campo Alegre s/n, Porto 4169-007, Portugal
[3]Max Planck Institute for Chemical Energy Conversion, Department of Inorganic Spectroscopy, Stiftstraße 34-36, Mülheim an der Ruhr 45470, Germany

7.1 Introduction: The OER Process

The oxygen evolution reaction (OER) is an energy-related process that involves the generation of molecular oxygen, O_2, via the electrochemical oxidation of water. Together with the hydrogen evolution reaction (HER, which consists in the production of molecular hydrogen, H_2, from water), it forms the highly strategic process known as water splitting or water electrolysis. As can be observed in Figure 7.1, OER takes place at the anode of electrolyser systems, while HER occurs at the cathode.

Ideal water splitting demands a potential of 1.23 V *vs.* RHE (reversible hydrogen electrode) under standard thermodynamic conditions. However, in practice, the activation barrier associated with the chemical reaction and ohmic loss significantly increase the required potential for water splitting. It has been observed that OER at the anode is the rate-determining step of the overall process and requires higher overpotential when compared to HER since this is a two-electron process, whereas OER involves four electrons. Consequently, from the catalyst perspective, the performance and efficiency of the water electrolysis cells are intricately related to the OER electrocatalyst performance at the anode [2].

Pablo Garrido-Barros, Marta Nunes and Marcos Gil-Sepulcre have contributed equally to this work.

Applied Polyoxometalate-based Electrocatalysis, First Edition. Edited by Diana M. Fernandes.

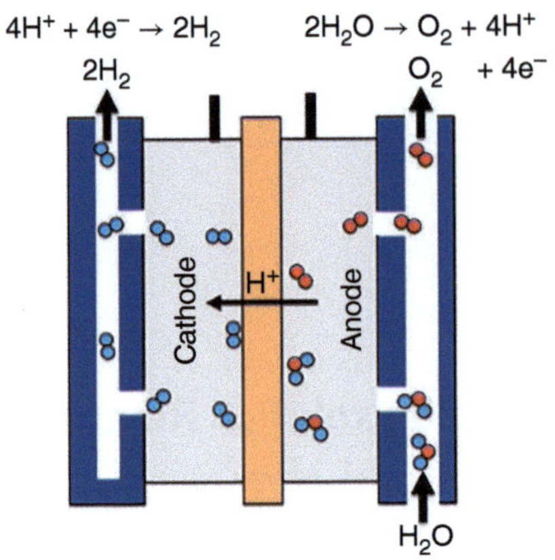

Figure 7.1 Scheme for a basic electrolyser device. Source: O.O.E.E.R. Energy [1]/U.S. Department of Energy./Public domain.

Focusing on the OER process, in neutral and acid electrolytes, two molecules of H_2O are oxidized into four protons and an O_2 molecule, while four hydroxyl groups are oxidized giving rise to H_2O and O_2 in alkaline electrolytes [3].

$$2H_2O\ (l) \rightarrow 4H^+ + O_2\ (g) + 4e^-\ \text{(acidic medium)}$$

$$4OH^- \rightarrow 2H_2O\ (l) + O_2\ (g) + 4e^-\ \text{(alkaline medium)}$$

Regarding the mechanistic aspects, it is generally accepted that OER mechanisms include a complex multistep proton-coupled electron transfer (PCET) and the formation of oxygen—oxygen (O—O) bonds, entailing several intermediates [4]. Therefore, the overall reaction rate depends on the kinetic constraints of the elementary reaction steps, which depends on the relative stability of each intermediate and the activation barriers between them [5] and mostly results in sluggish kinetics and elevated overpotentials [3]. More mechanistic details will be exposed in Section 7.2.1.

To fairly evaluate the performance of OER electrocatalysts, some kinetic parameters are commonly used. Firstly, the overpotential (η_{10}) required by the catalyst to achieve a current density (j) of $10\,\text{mA cm}^{-2}$. For OER (and HER), $10\,\text{mA cm}^{-2}$ is the desired value since in the solar water splitting process this is the current density expected in a solar to hydrogen device for a 12.3% efficiency, which would be required for cost-competitive photoelectrochemical water splitting [6]. Tafel slope (TS) is another widely used kinetic parameter. As shown in Figure 7.2, TS is obtained by the fit of the linear part of the Tafel plot (η *vs.* logarithm of j) by applying the $\eta = a + b \log j$ equation (Tafel eq.) and is related to the mechanism of the electrocatalysed reaction. A high current density – and low overpotential and TS – are desired indicators of a highly performing OER catalyst [7]. Apart from j, η_{10} and TS, an additional figure of merit is the turnover frequency (TOF), which is the quantity of desired product that a catalyst can produce per active site and per unit of time [8]. It is worth mentioning that TOF values are imprecise in some cases because the catalysts are typically heterogeneous and it is challenging to estimate the number of active sites. Nevertheless, these results can be useful and pertinent if analogous

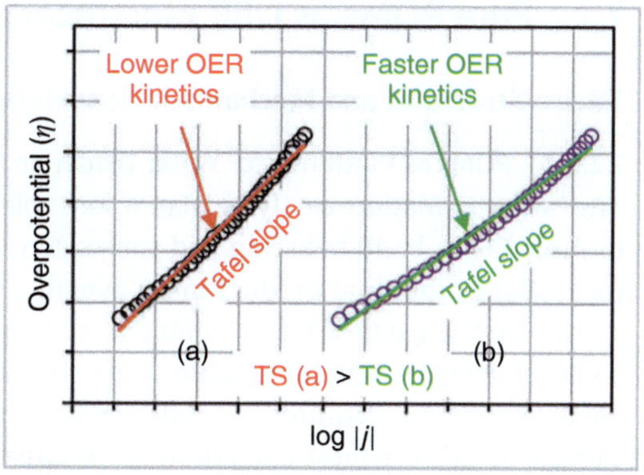

Figure 7.2 Generic Tafel plot examples for two OER electrocatalysts – (a) and (b) – showing different Tafel slopes (TS) at the same overpotential range and, consequently, different kinetics. Source: Own elaboration.

materials are compared. Finally, the long-term stability of OER electrocatalysts is also vital with a view to their applicability [8]. It can be evaluated by the measurement of the current density at an applied potential for an extended period of time (from tens to hundreds of hours), or by potential cycling (from hundreds to thousands of cycles).

Presently, the materials that show the highest electrocatalytic activity towards OER – being commonly accepted as the state-of-the-art benchmarks – are ruthenium(IV) and iridium(IV) oxides, RuO_2 and IrO_2 [9], with the latter being the one that presents the best relationship between electrocatalytic performance and stability. Since ruthenium and iridium are platinum group metals (PGM) and usually show prices c. 15 and 150 € g^{-1}, respectively [10], the development of efficient and cost-effective alternative OER electrocatalysts is essential. In this context, this chapter will provide an overview about the application of polyoxometalates (POMs) and POM-containing nanocomposites in the developing of electrocatalysts with notable OER performances at affordable costs. The growing interest in POM-based OER catalysts is manifested by the significant increment of scientific articles focused on this topic – particularly, from 2017 ahead (Figure 7.3).

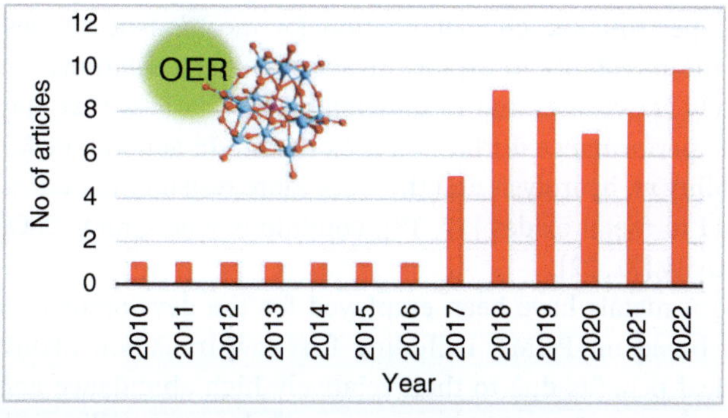

Figure 7.3 Publication trend of articles reporting the application of POM-based electrocatalysts to OER. Own elaboration from results retrieved by Web of Science^{TM}. Source: Image of POM structure elaborated by Rtomsa – Own work, CC BY-SA 3.0, https://commons.wikimedia.org/w/index.php?curid=12443580.

7.2 Pure POMs as OER Electrocatalysts

7.2.1 Structural and Mechanistic Considerations

7.2.1.1 POMs as Platforms for Water Oxidation Electrocatalysis

The water oxidation catalysts play a key role in activating the water molecules towards O—H bond breaking and subsequent O—O bond formation to release O_2 (or H_2O_2) as the final product in an overall $4\,H^+/4\,e^-$ (or $2\,H^+/2\,e^-$) process [11]. To this end, a catalyst must be able to accumulate oxidative equivalents without compromising its structural nature due to potential oxidative degradation under the anodic conditions required for water oxidation. An important related aspect is the capacity to achieve these oxidations in a relatively narrow redox potential range that allows minimizing the energy required. In addition, due to the elevated number of protons being released with the subsequent decrease in the local pH, and the benefits of performing proton reduction (the other half-reaction of water splitting) in acidic conditions, the catalyst candidate must display resistance to acid-promoted degradation (e.g. hydrolysis) [12–14]. Transition metals have been extensively studied for water oxidation catalysis, either as part of molecular platforms or solid-state materials since they can coordinate and activate water molecules and promote their oxidation [15, 16]. High valent, and thus electron-deficient metal centres are required to polarize the O—H bonds and generate electrophilic O atoms that are capable of O—O bond formation.

In this regard, POMs demonstrated to be competent catalysts for water oxidation [17, 18]. Their structural nature, containing high valent metal centres connected by oxo-hydroxo bridges, provides active sites on the surface for water coordination and activation [19]. In addition, their multimetallic composition allows for the accumulation of oxidative equivalents within the clusters via chemical or electrochemical oxidation without suffering important structural changes and potential oxidative degradation [20]. Compared to their solid analogues – metal oxides – POMs present distinct benefits. Their molecular nature allows for exquisite control of their structural and electronic properties and favours the solubility of these materials in water, simplifying the spectroscopic characterization for mechanistic studies. These aspects determine the possibility to explore structure-activity relationships to elucidate the catalytically active sites and to understand the features that play an important role in POMs performance during water oxidation. In general, POMs also exhibit enhanced stability *vs.* hydrolysis and thus, are more resistant in neutral to acid media as compared to metal oxides [17, 18], conditions where only CoO_x has shown remarkable activity [21, 22].

To date, a wide variety of metals have been employed for the development of water oxidation catalysts based on POMs, including first-row transition metals that typically enjoy practical benefits due to their relatively high abundance and environmentally benign character. These POMs are generally heterometallic, with a catalytically active metal-core (containing metals such as Ru or Fe) embedded into a metal oxide framework (typically tungstate or molybdate), which act as an oxidatively robust inorganic ligand. In addition, most POMs include a *p* or *d* block heteroatom such as Si or Ge that provides higher stability in a wider pH range.

7.2.1.2 Water Oxidation Mechanism of POMs

The general mechanism for water oxidation on POMs involves three main steps: oxidative activation of water, O—O bond formation and O_2 release. While the latter is typically more favourable from a thermodynamic and kinetic perspective, the two first steps determine the required operation conditions and the catalytic performance [23].

Water molecules are coordinated to the active metal centres in a POM structure in the terminal positions and, depending on the nature and oxidation state of the metal, the pK_a for the M^n—OH_2/M^n—OH deprotonation will vary giving rise to certain speciation [18, 24]. To generate an oxygen atom electrophilic enough to engage in O—O bond formation, POMs have to undergo oxidation steps leading to high valent metal centres that allow heterolytic O—H bond breaking and polarization of the resulting M=O bond. An important mechanistic aspect of this oxidative activation is the involvement of PCET (Figure 7.4a) [17]. PCETs are essential steps in a wide variety of biological processes and synthetic transformations where the transfer of a proton is intimately paired with the transfer of an electron [26]. An important

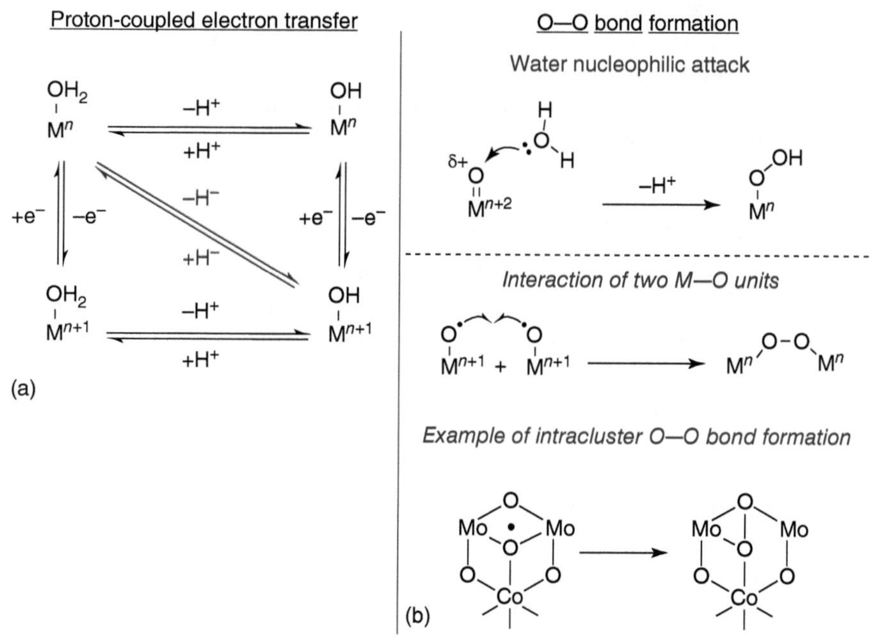

Figure 7.4 (a) Schematic representation of the pathways for oxidative activation of water by a transition metal with electron transfer (vertical arrows), proton transfer (horizontal arrows) and proton-coupled electron transfer (diagonal arrow, purple). (b) Mechanistic alternative for the O—O bond formation step during water oxidation catalysis. Source: Own elaboration, except for the bottom example Taira et al. [25]/with permission of American Chemical Society.

consequence of PCET is the redox levelling upon oxidation: removal of negative charges (electrons) from a redox system is coupled with the removal of positive charges (protons), maintaining the charge neutrality of the process. Oxidative activation of a number of POMs has been shown to follow PCET mechanisms involving transformation between $M^n—OH_2/M^{(n+1)}—OH/M^{(n+2)}=O$ units with no building up in the overall charge, allowing to access a wide set of oxidation states within a relatively narrow potential range; this translates into operational advantages regarding the applied redox potential required to promote water oxidation [27].

Once reactive M=O or M—OH units are generated with electrophilic oxygen atoms, two alternatives can be envisioned for the O—O bond formation (Figure 7.4b). The first one involves a nucleophilic attack from an exogenous water molecule to the M=O/M—OH unit, constituting the so-named water nucleophilic attack (WNA) pathway. This step, typically followed by a proton transfer, results in the formation of a hydroperoxo intermediate M—OOH where the metal is reduced by two electrons. Alternatively, the interaction of two M=O units (I2M) can also lead to the formation of an O—O bond within the resulting peroxo-bridged M—O—O—M species, where each metal centre is reduced by one electron. This mechanism is typically associated with a metal-oxyl character, M—O. While WNA has been found to be the most common pathway for the O—O bond formation, a recent work has evidenced the viability of an intracluster 'I2M type' mechanism in a Co-POM water oxidation catalyst [25].

As aforementioned, once the O—O bond is formed, further oxidation steps that are usually downhill under the catalytic conditions lead to the increase in the O—O bond order, resulting in the formation and release of O_2 with concomitant regeneration of the POM in its initial oxidation state.

7.2.2 Homogeneous Electrocatalysis

Due to the molecular character of POMs, they have been employed as OER homogeneous electrocatalysts. In these cases, a catalytically inert electrode (typically based on glassy carbon) is used to drive the oxidative processes required for the water activation and O_2 release steps. Homogeneous electrocatalysis also provides the additional benefit of a more accessible mechanistic evaluation and subsequent optimization.

An early example of POM as OER catalysts is based on the heterometallic $Na_{14}[Ru^{III}{}_2Zn_2(H_2O)_2(ZnW_9O_{34})_2]$, where the Ru atoms are proposed to be the catalytically active centres [28]. Electrochemical characterization of this POM revealed three oxidation events associated with the subsequent oxidation of Ru^{II} to Ru^{V} within a potential range of −0.15 to 0.76 V *vs.* NHE (Normal Hydrogen Electrode). Interestingly, a comparison with a previously reported mono-ruthenium POM, $[PW_{11}O_{39}Ru^{III}(H_2O)]^{4-}$ (with no electrocatalytic activity) suggests the importance of the two Ru atoms positioned in close proximity.

The interest in Ru-POMs continued. A mechanistically well-characterized example is $[Ru_4(H_2O)_4(\mu\text{-}O)_4(\mu\text{-}OH)_2(\gamma\text{-}SiW_{10}O_{36})_2]^{10-}$, which was initially studied towards chemical water oxidation in 2008 by Bonccio's group [29]. Further study on this POM demonstrated its electrocatalytic OER activity at 1.15 V *vs.* Ag/AgCl (Silver-Silver Chloride Electrode) at a glassy carbon electrode [29], revealing four redox processes that were ascribed to subsequent oxidation of the initial Ru_4^{IV} core to a catalytically active Ru_4^{V} intermediate. These four redox processes take place in a relatively narrow potential range (from 0.45 to 1.15 V *vs.* Ag/AgCl) and were accompanied by proton release to furnish $[Ru_4(OH)_4(\mu\text{-}O)_4(\mu\text{-}OH)_2(\gamma\text{-}SiW_{10}O_{36})_2]^{10-}$ species with four highly electrophilic $Ru^{V}(OH)$ centres, shown (by density functional theory, DFT) to be active towards WNA. Therefore, this system highlighted the benefits of the multimetallic nature of POMs and the involvement of PCET processes towards the accumulation of the oxidative equivalents required in OER within an accessible range of redox potentials.

Despite previous findings using a tetranuclear Ru motif, in 2011 Fukuzumi and coworkers demonstrated the viability of single-site Ru-POMs, $[Ru^{III}(H_2O)SiW_{11}O_{39}]^{5-}$ and $[Ru^{III}(H_2O)GeW_{11}O_{39}]^{5-}$, for OER using sacrificial chemical oxidants [30]. Although electrocatalytic water oxidation was not demonstrated with this platform, cyclic voltammetry (CV) analysis evidenced the accessibility to four oxidation states based on a single Ru redox active metal, where a $Ru^{III}(H_2O)$ is oxidized to the $Ru^{V}(O)$ intermediate, and the presence of an electrocatalytic wave suggestive of water oxidation. Once again, redox levelling via PCET along the oxidative processes was essential to access catalytically active intermediates within a potential range of around 1 V. In addition, this platform demonstrated the capability of a single active metal centre to bear all the oxidative equivalents required to perform water oxidation.

Among all the first-row transition metal (TM) POMs studied in OER catalysis, those based on Co presented the most remarkable activity. In 2010, Hill et al. reported the catalytic activity of $[Co_4(H_2O)_2(PW_9O_{34})_2]^{10-}$ for OER catalysis using sacrificial chemical oxidants ($[Ru(bpy)_3]^{3+}$) [31]. This POM contained a tetranuclear cluster with CoO_6 octahedra embedded by two Keggin-type $[PW_9O_{34}]^{9-}$ fragments, known as the Weakley sandwich. Remarkably, the other seven related Co-based POMs (Co-POM) also analysed in this work were inactive, an aspect that was potentially related to their stability under different pH values. While the initial report also provided electrochemical characterization of the oxidative processes associated with this platform that suggested potential electrocatalytic activity, in 2011, it was identified the presence of Cobalt oxide (CoO_x) as the dominant electrocatalyst under the tested conditions, raising the question about the real active species (Figure 7.5) [32]. This Co-POM showed an electrocatalytic wave at potentials above 1.1 V *vs.* Ag/AgCl that subsequently increased over time after being dissolved; in addition, upon electrolysis under such oxidative conditions, a constant increase in current was obtained and a CoO_x film was deposited on the surface of the electrode. The formation of this film was attributed to the hydrolytic decomposition of the Co-POM and the release of Co^{2+} to the solution, which was converted to CoO_x under anodic potentials. This CoO_x showed in turn a high catalytic performance as measured from the current density at potentials above 1 V *vs.* Ag/AgCl. More recent work revisited the activity of this POM under chemical

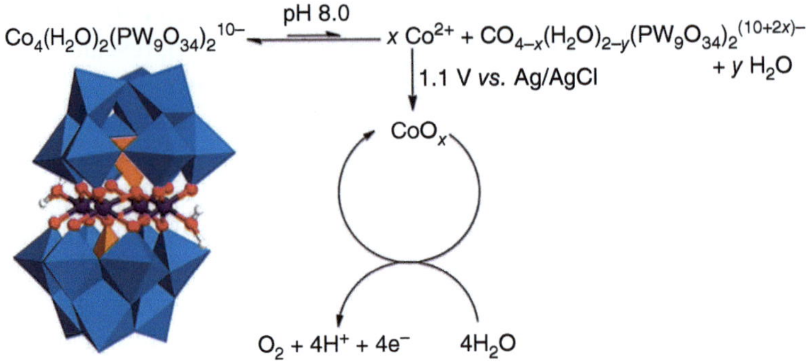

Figure 7.5 Structure of $Co_4(PW_9)_2$ and mechanistic scenario for OER catalysis. Source: Stracke and Finke [32]/with permission of American Chemical Society.

catalysis, providing a complete picture of their different mechanistic pathways, including both molecular POM and CoO_x as the active catalysts [33].

Co-POMs continued to be explored in the context of OER catalysis. In 2012, Galán-Mascarós group reported the electrocatalytic activity of $[Co_9(H_2O)_6(OH)_3(HPO_4)^{2-}(PW_9O_{34})_3]^{16-}$, a POM with higher nuclearity than those previously explored and related structural features with few distinctions: it contains three triads of $Co^{II}O_6$ octahedra that are interconnected through three OH^- and two $HPO_4{}^{2-}$ bridges instead of only oxo bridges and are supported by three Keggin-type $[PW_9O_{34}]^{9-}$ units [34]. Under electrocatalytic conditions, at an applied potential of 1.41 V *vs.* NHE in an aqueous solution at pH 7, this POM generated CoO_x on the surface of the working FTO (fluorine-doped tin oxide) electrode which was responsible of the observed electrocatalytic response. However, the addition of 2,2′-bipyridine – a scavenger for the Co^{2+} ions dissolved from the POM – prevented the formation of the CoO_x, supporting the intrinsic catalytic activity of the POM precursor.

Previous results evidencing the interplay of free Co^{2+} in solution as source of CoO_x formation prompted a careful (re)evaluation of Co-POMs as OER electrocatalysts. A series of Co-POM precatalysts has been examined under different pH (5.8–9) values at different aging times to assess the potential release of Co^{2+} and subsequent formation of CoO_x under anodic conditions [35]. This study includes a diverse range of POM structures and compositions: $[Co_4(H_2O)_2(PW_9O_{34})_2]^{10-}$, $[Co_9(H_2O)_6(OH)_3(HPO_4)_2(PW_9O_{34})_3]^{16-}$, $[\beta\beta\text{-}Co_4(H_2O)_2(P_2W_{15}O_{56})_2]^{16-}$, $[Co(H_2O)PW_{11}O_{39}]^{5-}$, $[\alpha_1\text{-}Co(H_2O)P_2W_{17}O_{61}]^{8-}$ and $[\alpha_2\text{-}Co(H_2O)P_2W_{17}O_{61}]^{8-}$. Out of all the samples and conditions, only $[Co(H_2O)PW_{11}O_{39}]^{5-}$ and $[\alpha_1\text{-}Co(H_2O)P_2W_{17}O_{61}]^{8-}$ were shown to generate sufficiently small amounts of free Co^{2+} (around 1.5% and 0.25% of the total cobalt in the POM) in sodium phosphate buffer solutions at pH = 5.8 so that the electrocatalytic activity observed can be attributed to the molecular Co-POM instead of in situ generated CoO_x. Unfortunately for the field, it was shown that CoO_x still outperforms these two molecular Co-POMs by ≈75× current levels.

Other first-row TMs – Fe, Cu and Mn – have also been studied as OER homogeneous POM electrocatalysts. Among these, Fe is a highly desired element due to its high abundance and environmentally friendly character. Thus, $[Fe_{11}(H_2O)_{14}(OH)_2(W_3O_{10})_2(\alpha\text{-}SbW_9O_{33})_6]^{27-}$ was explored due to its demonstrated activity under photochemical/chemical oxidative conditions at basic pH values [36]. Cyclic voltammetry showed a lack of electrocatalytic activity that highly contrasted with its performance under chemical and photochemical conditions. This was associated with the absence of a key Fe^{III}/Fe^{IV} oxidation wave under electrochemical conditions.

Mn-POMs have also been tested towards OER. $[Mn_3(H_2O)_3(SbW_9O_{33})_2]^{12-}$, $[Mn_3(H_2O)_5(PW_9O_{34})_2]^{9-}$ and $[Mn_3(H_2O)_3(AsW_9O_{33})_2]^{12-}$ showed electrocatalytic activity in a pH 9 borate buffer with an onset potential of around 1.1 V *vs.* Ag/AgCl [37]. Electroactivity at 1.2 *vs.* Ag/AgCl revealed the formation of O_2 and the absence of any electroactive layer on the surface of the FTO electrode employed, suggesting the homogeneity of this system. This set of POMs also provided an interesting point of comparison regarding the influence of the heteroatom included in the Keggin ligands: Sb, P or As. The POM containing Sb displayed the best catalytic performance evidenced by the lowest TS value, which was associated with the smaller charge transfer resistance of this POM, – a property measured from Electrochemical Impedance Spectroscopy. This observation suggested the influence of rate contributing electron transfer steps, but further mechanistic analysis was warranted.

On the other hand, late TMs were also explored as part of POM platforms for OER catalysis. The sandwich-type $[(\alpha\text{-}SbW_9O_{33})_2Cu_3(H_2O)_3]^{12-}$ POM, with two Keggin fragments connected by three Cu atoms, was the first Cu-POM reported as OER electrocatalyst; it showed the best performance among other five Cu-POMs explored in the same study containing a different number of Cu atoms [38]. This system, which operated under neutral pH with an onset potential of 1.2 V *vs.* Ag/AgCl, showed evidence to perform OER as a homogeneous catalyst with an observed kinetic constant of $0.7\,s^{-1}$ and an apparent first order in catalyst concentration. As the pH is increased from 5 to 10, this Cu-POM displayed onset potentials cathodically shifted with enhanced j. While $CuCl_2$ as a source of Cu^{2+} ions also displayed catalytic activity, potentially due to the formation of CuO_x on the surface of the electrode, the studied Cu-POM showed higher and more stable current density over time, although with relatively similar onset potentials.

7.2.3 Heterogeneous Electrocatalysis

The use of POMs as heterogeneous electrocatalysts by their deposition on the surface of electrodes is a much less explored field than homogeneous electrocatalysis. One general approached pursued to study heterogeneous electrocatalysis with molecular platforms, like POMs, is based on the preparation of Nafion-carbon black inks containing an insoluble molecular candidate and subsequent deposition on the surface of working electrodes (e.g. glassy carbon) by drop-casting methods. This is the case of a work involving Co-POMs ($[Co_4(H_2O)_2(PW_9O_{34})_2]^{10-}$)

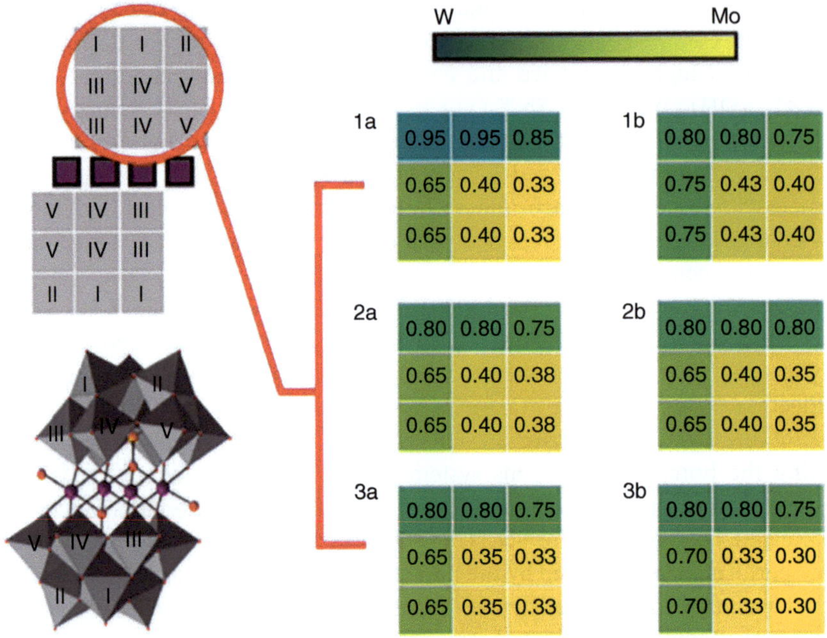

Figure 7.6 Structure of the Co-POM $[Co_4(H_2O)_2(PW_9O_{34})_2]^{10-}$ and crystallographic data showing the Mo doping pattern in the tungstate for the different compounds prepared using different Mo:W source ratios. Source: Martin-Sabi et al. [39]/with permission of Springer Nature.

doped with different proportions of molybdenum (Figure 7.6) [39]. Tests performed in phosphate buffer (pH = 7), in the potential range of 0.8–1.4 V *vs.* Ag/AgCl, showed enhanced electrocatalytic activity for all the Mo-doped Co-POMs in comparison with the parent undoped Co-POM. As the proportion of Mo increased, the onset potential decreased (up to 188 mV of difference at the same *j*), and the current density increased. Analysis of the stability for the POMs with the highest content in Mo revealed constant current densities for over 12 hours with no sign of decomposition and CoO_x formation.

A similar approach based on insoluble salts of POMs (Ba^{2+}) has been followed but using instead commercial carbon paste oil (CPO) as conductive ink to deposit the active material on the electrodes surface [40]. Using this method, the Co-POM $[[Co_4(OH)_3(PO_4)]_4(SiW_9O_{34})_4]^{32-}$ and the structurally analogous Fe-POMs $[[Fe_2Co_2(OH)_3PO_4]_4(SiW_9O_{34})_4]^{24-}$ and $[[FeCo_3(OH)_3PO_4]_4(SiW_9O_{34})_4]^{28-}$, obtained by substitution of Co by Fe, have been used as OER heterogeneous electrocatalysts. Comparative electrochemical analysis of these compounds revealed the benefits derived from Co by Fe substitution, as the overpotential was lowered up to 66 mV relative to the Co-POM (0.385 V at 10 mA cm^{-2}). The best performance was achieved with the $FeCo_3$ core-containing POM, upon substitution of only one Co atom by a Fe atom. Further comparison of the FeCo-POMs with benchmark IrO_2 revealed the similar activity achieved despite the potential difference in the concentration of active sites in both materials, and thus the promise of this platform as OER electrocatalyst.

Related work has shown the effects of Co-by-Fe substitution in the traditional Weakley sandwich Co-POM using $[(Fe^{III}OH_2)_2Fe^{III}{}_2\,(B\text{-}\alpha\text{-}PW_9O_{34})_2]^{6-}$ as heterogeneous catalyst via its insoluble barium salt [41]. This POM showed electrocatalytic water oxidation activity in neutral pH with an overpotential ≈50 mV higher than the Co-analogue. Experimental and computational analysis of the mechanism using this platform revealed an initial state of the POM where the terminal water molecules at the Fe atoms were deprotonated, forming Fe^{III}—OH at neutral pH. Further computational modelling evidenced a potential limiting step consisting in a PCET from Fe^{III}—OH to Fe^{IV}=O and a subsequent rate-limiting chemical step involving the O—O bond formation via a WNA mechanism.

7.3 POM-containing (Nano)composites as OER Electrocatalysts

7.3.1 POM/Carbon (Nano)composites

In 2010, Toma et al. [42] reported one of the first examples regarding the application of POMs combined with carbon materials as efficient OER electrocatalysts. In their work, they assembled a POM cluster with a conducting bed of multiwalled carbon nanotubes (MWCNTs). The POM used was a tetraruthenate cluster belonging to the class of POM anions $[Ru_4(H_2O)_4(\mu\text{-}O)_4(\mu\text{-}OH)_2\,(\gamma\text{-}SiW_{10}O_{36})_2]^{10-}$, which presented oxygen-evolving activity (Figure 7.7). The MWCNTs acted as conductive nanowire scaffolds, and their presence allowed to improve the electrical contact between the redox-active centre and the electrode surface, once (i) provided a heterogeneous support to POM, (ii) controlled the material morphology, (iii) increased the surface area and (iv) tapered the sequential electron transfer to the electrode, and thus favoured energy dispersion and relieved catalytic fatigue. The assembly of organic and inorganic domains was achieved through electrostatic scavenging at pH 5 by exposure of POM to water-soluble MWCNTs decorated with polyamidoamine ammonium dendrimers. The hybrid electrocatalytic activity was accessed using indium tin oxide (ITO)-modified electrodes (in phosphate buffer, pH 7) and showed an onset catalytic current at applied voltages >1.10 V caused by water oxidation. Furthermore, it exhibited good robustness and a remarkable TOF of 36 h^{-1} at $\eta = 0.35$ V, reaching 306 h^{-1} at $\eta = 0.60$ V.

After this, several types of carbon materials have been used as POMs support. For example, amorphous carbon paste electrodes were modified with Co-POMs to improve the POM stability and electrocatalytic activity in solid state [43–45]. In the work of Blasco-Ahicart et al. [44], the authors reported the interesting OER electrocatalytic activity of a barium salt of a cobalt-phosphotungstate polyanion supported on a carbon-paste electrode in acidic medium, which outperformed the state-of-the-art IrO_2 catalyst (even at pH < 1), with an η of 189 mV at 1 mA cm^{-2} and a Tafel slope of 97 mV dec^{-1}, indicating a competition between a chemical and

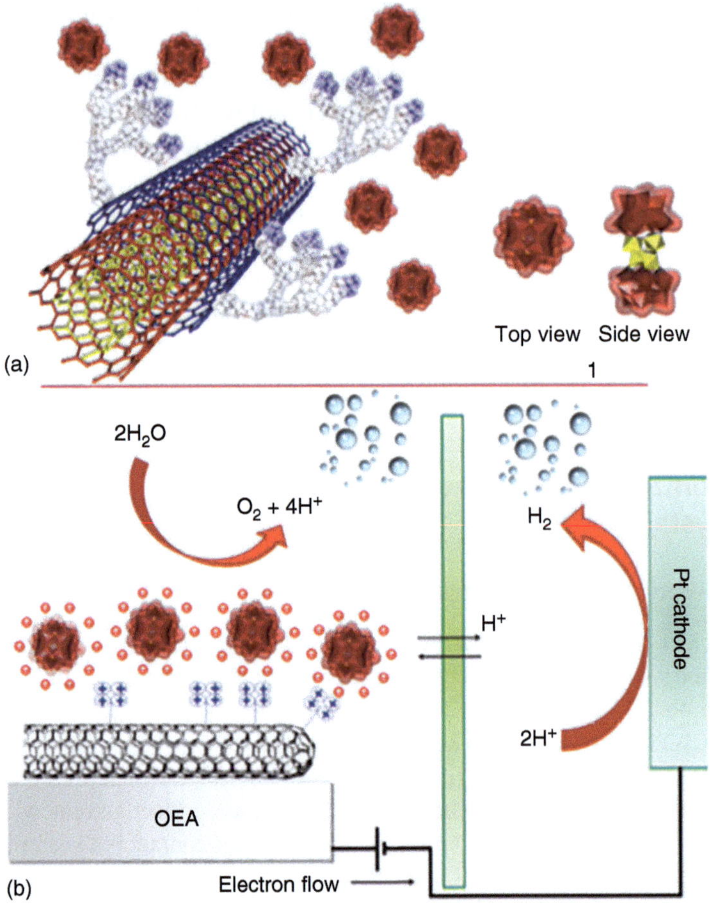

Figure 7.7 (a) Electrostatic capture of polyanionic ruthenium-containing clusters (negatively charged, red surface) by polycationic dendrons on the MWCNT surface (positively charged, blue surface) and polyhedric structure showing the side and front view of the POM (red)-embedded tetraruthenate core (yellow). (b) General scheme for a water-splitting electrocatalytic cell with integrated nanostructured oxygen-evolving anodes. Source: Toma et al. [42]/with permission of Springer Nature.

an electron-transfer limiting step. Its electrocatalytic activity was six times better than that of a similar composite but with caesium counter ion – turnover number (TON) of 47 *vs.* 8. In terms of stability, the composite showed a slow decrease of j during the first 4 hours to ~0.36 mA cm^{-2}, which was maintained for over 24 hours without significant changes. These results showed that the OER electrocatalytic activity was influenced by the barium cations but also that arises from a better matching between the carbon support and the Co-POM catalysts when compared with IrO_2, for instance.

Graphene has also been selected as a conductive carrier to prepare POM-based heterogeneous electrocatalysts due to its excellent electrical conductivity, large surface area and high chemical stability [46, 47] Li et al. [46] reported the synthesis of two Keggin-type POM-based Ag-containing biimidazole (H_2biim) complexes ([Ag(H_2biim)$_2$]$_2$[Ag_4-(H_2biim)$_4$][($PW_{12}O_{40}$)$_2$] (1) and [Ag(H_3biim)$_2$] [$PW_{12}O_{40}$]·$4H_2O$ (2)) under hydrothermal conditions, and their use as precursors (combined with positively charged graphene modified with poly(diallyldimethyl-ammonium chloride) and *N,N*-dimethylformamide [DMF] as reflux solvent and reducing agent) to prepare new materials denoted as PW_{12}/Ag/graphene-a and PW_{12}/Ag/graphene-b – including (1) and (2), respectively. The preparation of these composites allowed taking advantage of the good redox properties of POMs and Ag nanoparticles (NPs), but to prevent their tendency to aggregate and poor conductivity. During the reflux process, PW_{12} anions were uniformly dispersed on the graphene surface by electrostatic interaction, while Ag NPs grew up by continuous release and reduction of Ag^+ from the H_2biim-Ag complex. The PW_{12}/Ag/graphene-a and PW_{12}/Ag/graphene-b nanocomposites were assessed as OER catalysts in phosphate-buffered saline (PBS) solution, pH 7, and exhibited low η_{10} (540 and 610 mV, respectively) and TS of 190 and 192 mV dec^{-1}, respectively. Their electrocatalytic activity was better than those of the two-phase composites (Ag/graphene and PW_{12}/graphene), owing to a synergistic role of the three components.

Heteroatom-doped MWCNTs have also been studied [48, 49]. Lee et al. [49] reported a linker-free spontaneous binding of tetracobalt-based POMs (Co_4POMs, [$Co_4(H_2O)_2(PW_9O_{34})_2$]10$^-$) on N-doped nanotubes (NCNTs). Their protonated N-dopant sites offered chemically favourable direct binding sites for electrostatic immobilization of the Co_4POMs, while simultaneously increased the electrical conductivity and fluent electron transfer in the resultant Co_4POM/NCNT hybrids. This linker-free binding of Co_4POMs on NCNTs was beneficial for OER when compared with similar hybrids employing cationic polymer linkers, because the last ones can be degraded under oxidizing conditions. Thus, the Co_4POM/NCNT hybrid was tested as OER catalyst at neutral pH, exhibiting a low η_{10} of 370 mV, a Tafel slope of 203 mV dec^{-1} and a TOF of 0.211 s^{-1} at 2.01 V *vs.* RHE. The measured catalytic activity was comparable to the results from Ru-based POM/CNT hybrid catalysts.

Another carbon allotrope, the graphdyine (GDY), was combined with POMs due to its interesting structure, which comprises sp^2- and sp-hybridized C atoms with high degrees of π-conjugation that feature uniformly distributed pores, unevenly distributed surface charges, natural band gap and impressive electrical conductivity. In 2022, Sun et al. [50] synthesized a composite composed of POMs and GDY and applied it to OER. $Co_4(PW_9)_2$@GDY was prepared through a two-step process: first, microcrystals of $K_{10}[Co_4(PW_9O_{34})_2]\cdot 24H_2O$ ($Co_4(PW_9)_2$) were synthesized on commercial cobalt foam electrodes under mild hydrothermal conditions, and then GDY nanosheets were grown on the microcrystals, forming the core–shell structure. $Co_4(PW_9)_2$@GDY exhibited excellent OER activity in alkaline media, with a low E_{10} of 1.52 V (*vs.* RHE), a small TS of 80 mV dec^{-1}, a large electrochemically active surface area (ECSA) of 182.5 cm^2, a high TOF value of 0.86 s^{-1} and long-term stability for 30 hours. The interactions between POMs and GDY in the $Co_4(PW_9)_2$@GDY system were significant to the electrochemical performance, in which GDY supplied more pores for electron storage and transfer and increased the conductivity of POMs.

Recently, Abdelkader-Fernández et al. prepared a ternary nanocomposite consisting of the sandwich-type POM $[Co_4(H_2O)_2(PW_9O_{34})_2]^{10-}$ ($Co_4(PW_9)_2$), metal nanoparticles and a carbon matrix ($Co_4(PW_9)_2$@NP@C) [51]. Using an innovative approach for the 'bulk' deposition of the POM salt nanocrystals, the authors decorated four different MOF-74-derived nanocarbons. The nanocarbons embedding different metal NPs were prepared through the carbonization of MOF-74 (formed by divalent metal ions linked by 2,5-dihydroxyterephtalate) precursors with different metal compositions (Co, Ni and bimetallic Co/Ni composition) and doping heteroatoms (dual N,S-doping). The comparative studies between the four nanocomposites obtained ($Co_4(PW_9)_2$@Co@C, $Co_4(PW_9)_2$@Co/Ni@C, $Co_4(PW_9)_2$@N,S-Co@C and $Co_4(PW_9)_2$@N,S-Co/Ni@C) revealed that the nanocarbon doping and metal composition had drastic effects on the achieved POM loadings, ECSAs and OER behaviours. In particular, the $Co_4(PW_9)_2$@Co/Ni@C nanocomposite was intrinsically different from the other materials, since the highly nanostructured Co/Ni@C support induced structural changes into the deposited POM nanocrystals, which resulted in an active surface area increment but not traduced in a proportional nominal current density enhancement. Globally, the N,S-doping of the nanocarbon supports led to the immobilization of higher $Co_4(PW_9)_2$ amounts (diminishing their ECSAs), but resulted in a substantial increase of their intrinsic OER activities. In absolute terms, $Co_4(PW_9)_2$@Co/Ni@C and $Co_4(PW_9)_2$@N,S-Co@C nanocomposites presented the best OER metrics: moderate η_{10} (c. 400 mV), high nominal j at 1.75 V (c. 70 mV cm^{-2}) and fast kinetics at intermediate overpotentials (<67 mV dec^{-1}). Except for $Co_4(PW_9)_2$@Co@C, all the nanocomposites presented good levels of stability, completing the 12 hours of chronoamperometric tests with current levels higher than the initial ones.

Beyond the water splitting, the OER (concomitantly with oxygen reduction, ORR) is also a significant process in other important technologies, such as fuel cells and metal–air batteries. Thus, considering a practical application, Lee et al. [52] reported recently the application of an oxo-bridged tetraruthenium POM (RuPOM) with excellent OER (and ORR) electrocatalytic activity with minimal use of noble metal elements. The RuPOM was deposited on carbon felt electrodes with conducting carbon Ketjen black (KB) by slurry coating method (using polyvinylidene fluoride as a polymeric binder) to develop efficient seawater batteries (SWBs, a specific type of Na-air batteries). The deposited RuPOM improved the energy efficiency of SWBs by lowering the OER (and ORR) overpotentials. The RuPOM/KB composite exhibited important performance comparable to commercial catalysts, with a whole charging-discharging η (the sum of OER and ORR overpotentials) of 0.76 V at 0.1 mA cm^{-2} and maximum power density of 12.3 mW cm^{-2}. The SWB remained stable at least for 50 cycles. Although a POM containing an expensive Ru element was used in this work, its specific activity per precious metal was significantly higher than Pt/C and RuO_2 (11 and 25 times higher, respectively, Figure 7.8) due to its molecular nature and minimal use.

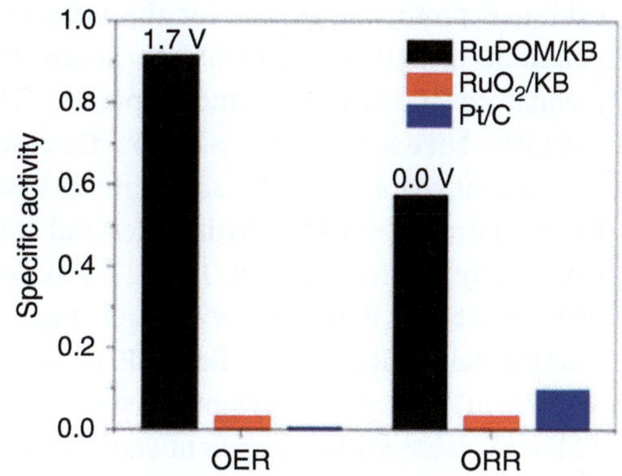

Figure 7.8 Comparison between specific catalytic activities of RuPOM/KB, RuO_2/KB and Pt/C in OER (and ORR) (the specific activity was determined by calculating the rate of electron transfer in OER and ORR per unit mass [g] of each precious metal element [i.e. Ru or Pt]). Source: Lee et al. [52]/with permission of American Chemical Society.

7.3.2 POMs Combined with Metals/Metal Oxides/Metal Hydroxides/Metal Complexes

Considering the technological application of POMs as OER electrocatalysts, one important step forward is their stable deposition on commercial metal electrodes, avoiding the detachment of the POM catalysts from the electrode under operating conditions and the loss of catalytic activity. In this respect, the deposition of POMs on nickel foam electrodes has been explored [53–55]. Luo et al. [53] described a one-step hydrothermal route to deposit a Dexter–Silverton POM ($[Co_{6.8}Ni_{1.2}W_{12}O_{42}(OH)_4(H_2O)_8]$, NiCo-POM) on a commercial Ni foam (NF) electrode. The OER activity of the NiCo-POM/Ni electrode was assessed in alkaline electrolyte, exhibiting low η_{10} of 360 mV, a Tafel slope of 126 mV dec^{-1} and high faradaic efficiency (FE), 96 ± 5%. Post-catalytic analyses after 10 hours of working time showed no mechanical or chemical degradation and no physical detachment of the microcrystals from the electrode. With the same purpose of increasing the POM stability, Mahanta et al. [56] prepared spherical hollow POM nanoparticles ([HPMo]NPs) from the Keggin POM $[H_3PMo_{12}O_{40}]$ in the presence of the surfactant sodium dodecyl sulphate (SDS) and polyvinylpyrrolidone (PVP) in aqueous medium. They were used to modify a gold working electrode and applied for OER in a phosphate buffer (pH 7). The [HPMo] NPs modified Au electrode showed excellent activity, with a low η_{10} of ~350 mV and a j of ~1.7 mA cm^{-2}, good stability under exhaustive electrolysis conditions and also showed long term stability (30 days). The electrocatalytic performance was attributed to the large surface of the hollow [HPMo] NPs, which adsorbed the water molecule through hydrogen bonding interaction; this facilitated the PECT process and the cleavage of O—H bonds in water molecule.

In 2021, Ojani's group reported a tri-metallic OER electrocatalyst, prepared through the incorporation of Ni, Co and Fe cations on Triton X-100/phosphotungstic acid organic-inorganic composite [57]. The organic-inorganic catalyst support was constructed via the simple coupling of the nonionic surfactant Triton X-100 (TX) as the organic unit and the phosphotungstic acid (PW) as the inorganic unit on the surface of the CNT paste electrode (TX/PW-CNTPE). Then, Ni, Co and Fe cations were adsorbed on the as-prepared catalyst support. The obtained NiCoFe/TX/PW-CNTPE exhibited promising results for the OER in alkaline medium, requiring an overpotential of 210 and 330 mV to attain 10 and 100 mA cm^{-2}, respectively. At j equal to 100 mA cm^{-2}, the apparent OER performance followed the order: NiCoFe > NiFe > Ni > CoFe > NiCo > Co > Fe.

In recent years, naturally abundant TM-based compounds (specifically based on Co, Ni, Fe and Zn oxides, hydroxides and complexes) and their mixtures have been extensively studied for OER. In 2022, Gautam et al. [58] reported a heterostructure of POM ($(NH_4)_6P_2Mo_{18}O_{62}$)/$ZnFe_2O_4$ nanoplates as OER and HER (bifunctional) electrocatalyst. Regarding the OER tests, POM/$ZnFe_2O_4$ furnished low η of 220 and 290 mV to achieve current densities of 20 and 50 mA cm^{-2}, respectively, a TS of 124.6 mV dec^{-1} and an excellent stability over 20 hours of operation (at 1.5 V) with no significant current loss. This electrocatalytic performance was attributed to (i) a unique heterostructure of porous POM/$ZnFe_2O_4$ nanoplates with a heterogeneous outer surface that enhanced the wettability, ECSA and O_2 release process, (ii) heterojunctions of POM/$ZnFe_2O_4$ nanohybrids that increased the number of catalytically active sites, leading to faster electron transfer and (iii) the presence of different Zn and Fe valence states in nanohybrids, that accelerated the more suitable A—O binding energy, facilitating the generation of OER intermediates. In addition, taking advantage of the OER/HER bifunctional electrocatalytic activity, a two-electrode alkaline electrolyser composed of POM/$ZnFe_2O_4$ as cathode and anode was assembled and required a low operating voltage of 1.53 V to deliver a current of 10 mA cm^{-2} (*vs.* 1.55 V using the state-of-the-art RuO and Pt/C as anode/cathode materials).

In other work, Shang et al. [59] prepared a TM substituted-POM-doped β-FeOOH nanocomposite (Co_4POM@FeOOH-P). For the nanocomposite synthesis, firstly a Ni foam-supported Co_4POM-doped β-FeOOH was prepared by a hydrothermal method and then the obtained material was phosphorized to obtain the final nanocomposite. Compared with the non-substituted POM-doped β-FeOOH (e.g. PW_{12}@FeOOH-P, PMo_{12}@FeOOH-P and SiW_{12}@FeOOH-P), the Co_4POM@FeOOH-P nanocomposite showed superior OER performance, with η_{10} equal to 213 mV and a TS of 47 mV dec^{-1}. This was attributed to the multi-electronic structure of Co_4POM and the structural advantage of pores in β-FeOOH, resulting in an overall better performance. A TOF value of 1.064 s^{-1} (at η = 400 mV) was determined for Co_4POM@FeOOH-P.

Streb and Rau's group reported a POM-derived OER electrocatalyst with self-activation properties [60]. The POM-derived composite was prepared by immobilizing a cobalt-functionalized polyoxotungstate $[Co_4(H_2O)_2(PW_9O_{34})_2]^{10-}$ (Co_4POM) on commercial TiO_2 nanoparticles. The cationic polymer polyethyleneimine was used as an electrostatic anchoring agent, forming a cationic surface on the NPs that facilitated the electrostatic adsorption of the anionic Co_4POM. The OER performance was studied under alkaline conditions, by drop-casting the prepared nanocomposite on carbon paper (CP). During prolonged activity periods (>10 hours at 1.79 V *vs.* RHE), it was observed a continuous enhancement of the current densities with a decrease in the η_{10} (from 560 to 490 mV), resulting in a behaviour that was described as 'self-activation'. Furthermore, a TS of 74 mV dec^{-1}, a FE of 91%, an ECSA of $0.65\,cm^2$ and a charge transfer resistance of $23.75\,\Omega$ were determined for the nanocomposite.

Das's group reported the synthesis of a POM-supported nickel(II) coordination complex ($[Ni^{II}(2,2'\text{-bpy})_3]_3[[Ni^{II}(2,2'\text{-bpy})_2(H_2O)][HCo^{II}W^{VI}{}_{12}O_{40}]]_2\cdot 3H_2O$, where 2,2′-bpy = 2,2′-bipyridine) and its application as heterogeneous OER catalyst [61]. This compound was solvothermally synthesized. When evaluated towards OER in a neutral medium, the compound showed a moderate η_{10} of 475.6 mV, a very high TOF of $18.49\,s^{-1}$ and a FE of 88.7%. The compound showed to be very stable and robust under the tested electrochemical conditions, with negligible current changes. Ni^{II} from the $[Ni^{II}(2,2'\text{-bpy})_2(H_2O)]$ unit, attached to the POM cluster, acted as a catalytically active centre.

Although several POM-based nanocomposites have been reported as showing high OER activities due to synergistic effects among the components, the right role of each component remains unclear due to its complex structure. Recently, Shimoyama et al. used porous ionic crystals (PICs) based on Co-POMs, to establish a composition-structure-function relationship and to understand the origin of the synergistic catalysis in this type of composites [62]. It was the first time that a PIC was utilized as OER electrocatalyst, taking advantage of its high designability and well-defined structure. As PIC building blocks, the authors used a Keggin-type Co-POM $[\alpha\text{-CoW}_{12}O_{40}]^{6-}$ (CoW_{12}) and a Cr complex $[Cr_3O(OOCCH_2CN)_6(H_2O)_3]$ (Cr-CN) with cyano groups as macrocation. The CoW_{12} was successfully anchored in the crystalline PIC matrix via Coulomb interactions and hydrogen bonding induced by the N atom of the cyano groups of the Cr complex, resulting in the CoW_{12}-based PIC $K_2[C_{r3}O(OOCCH_2CN)_6(H_2O)_3]_4[\alpha\text{-CoW}_{12}O_{40}]\cdot 32H_2O$, (here denoted as **I**, Figure 7.9). Regarding the OER activity, while each building block was inactive, the PIC **I** (blended with carbon paste) exhibited efficient and sustained OER activity under acidic to neutral (pH 5.6) conditions. The onset potential of **I** at a current density of $1\,mA\,cm^{-2}$ was 1.89 V (*vs.* RHE), which was ~240 mV lower

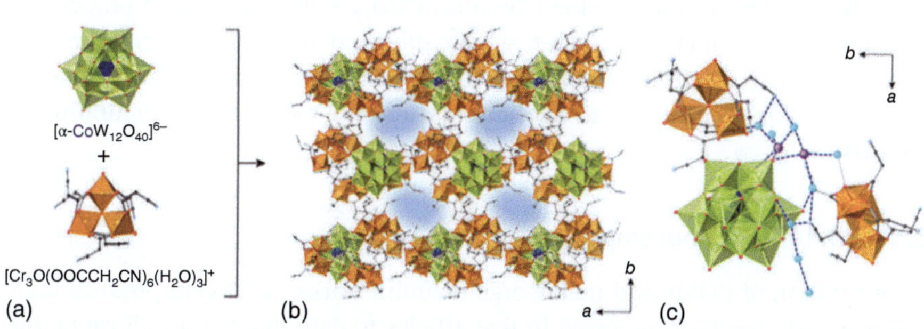

Figure 7.9 (a) Molecular structures of the building blocks of **I**. (b) Crystal structure of **I** along the *c*-axis. The 1D channels are shown in blue. (c) Local structure of **I**. Light green, orange and dark blue polyhedrons show the $[WO_6]$, $[CrO_6]$ and $[CoO_4]$ units, respectively. Light blue and purple spheres show the oxygen atoms of the water crystallization and the potassium ions, respectively. Blue dashed lines show the possible hydrogen bonds. Hydrogen atoms, crystallized water molecules and potassium ions were omitted for clarity. Source: Shimoyama et al. [62]/with permission of American Chemical Society.

than those of pristine CoW_{12} and Cr-CN (2.13 V *vs.* RHE). Furthermore, an η_{10} of 670 mV and a TS of 124 mV dec^{-1} were measured for **I**, suggesting rate-determining single-electron and proton transfer reactions. The analysis after chronoamperometry for five hour (or after 10 cycles in acidic conditions, pH 1.0) showed negligible leaching of Co and W from **I**, and thus good robustness. The electrochemical, X-ray photoelectron (XPS) and infrared (IR) spectroscopy results demonstrated that the charge transfer occurred from Cr-CN to CoW_{12}, suggesting that the increased electron density of CoW_{12} upon the formation of **I** was the key to the observed catalysis.

Owing to the unique layered structure and tuneable electronic structure, NiFe-based layered double hydroxides (NiFe LDHs) and their nanocomposites with POMs have been also pursued as OER electrocatalysts [63, 64]. However, the rational establishment of performance-influencing factors remains scarce, and their synergistic effects are not clear enough. Cai et al. [63] reported an all-in-one POMs etching approach to ingeniously reconstruct the archetypal NiFe LDH electrocatalyst. For that, the NiFe LDH was electrodeposited on Ni foam, followed by the etching treatment with different Keggin-type POMs (PW_{12}, SiW_{12} and PMo_{12}). The materials characterization showed cross-linked NiFe LDH nanosheets with <5 nm of layer thickness, uniformly and vertically grown on Ni foam. The OER performance of NiFe LDH-PMo_{12} showed the best enhancement (at pH 2.5), with η of 206 and 249 mV at j of 10 and 100 mA cm^{-2}, respectively (*vs.* 286 and 353 mV for IrO_2 reference). The NiFe LDH-PMo_{12} also yielded the lowest TS of 47.5 mV dec^{-1} (*vs.* 73.6 mV dec^{-1} for IrO_2). The stability test indicated robust electrochemical stability of the NiFe LDH-PMo_{12}, with an OER activity increase from 1.653 to 1.647 V (*vs.* RHE) after 24 hours, corresponding to a total cell voltage drop < 0.3%. The OER activity and stability of the NiFe LDH-PMo_{12} were mainly attributed to the following aspects: (i) the adjustment of the surface morphology, reduction of the nanosheets thickness to ~2.6 nm and exposure of more active sites; and (ii) the lattice of as-etched NiFe LDH that becomes more disordered and more α-$Ni(OH)_2$ phases were generated (active sites); besides, PMo_{12} etching produced proper and controllable O and metal vacancies, which subtly modulated local coordination environments and electronic structures of Fe and Ni cations; the surface-anchored PMo_{12} polyanionic clusters was also able to tune the electronic configuration of NiFe LDH.

7.3.3 POM/MOF Nanocomposites

The integration of POMs and metal–organic frameworks (MOFs) to produce OER electrocatalysts has been reported in five articles to date [65–69]. In all cases, the selected POMs presented the usual Keggin heteropolyanion structure, $[XM_{12}O_{40}]^{n-}$, while MOFs were zeolitic imidazolate frameworks (ZIFs). ZIFs are a specific MOF subgroup that presents crystalline nanoarchitectures formed by transition-metal ions (e.g. Fe^{2+}, Co^{2+}, Cu^{2+} and Zn^{2+}) connected by imidazolate linkers, mimicking the cage-channel structures of zeolites [70]. In three of the five articles [65, 68, 69], the Keggin–POM/ZIF integration consisted in a smart self-assembly approach

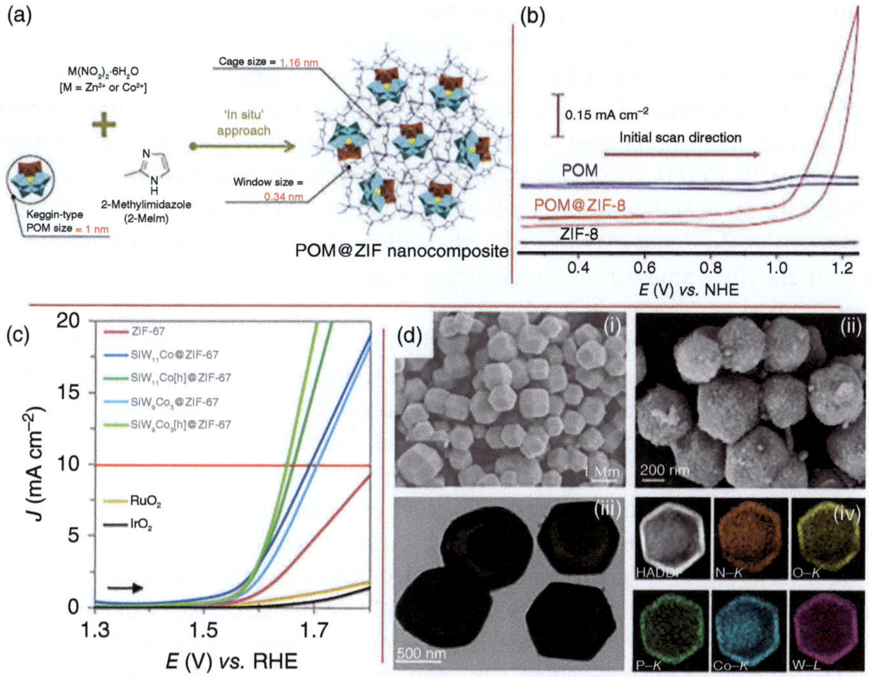

Figure 7.10 (a) In situ methodology for the preparation of POM@ZIF nanocomposites. Source: Abdelkader Fernández et al. [68]/with permission of American Chemical Society. (b) Cyclic voltammograms of ZIF-8, POM@ZIF-8 and POM ($K_6[CoW_{12}O_{40}]$), electrolyte: 0.1 M Na_2SO_4 solution (neutral pH). Source: Mukhopadhyay et al. [65]/with permission of John Wiley & Sons. (c) OER polarization curves obtained by linear sweep voltammetry (N_2-saturated KOH 0.1 M solution, scan rate = 5 mV s^{-1}, rotation speed = 1600 rpm) for pristine ZIF-67, POM@ZIF-67 nanocomposites and benchmark materials (current density normalized to geometric electrode area). Source: Abdelkader-Fernandez et al. [69]/with permission of Royal Society of Chemistry. and (d) Scanning electron microscopy (SEM) images of ZIF-67 (i) and ZIF-67@POM (ii), TEM images (iii) and elemental mappings (iv) of yolk/shell ZIF-67@POM nanoparticles. Source: Li et al. [66]/American Chemical Society.

through which the POM clusters/units (guest) were individually confined into the ZIF cages (host). This strategy is based on the in situ growth of ZIF surrounding the previously solved POM units by taking advantage of the size complementary between the Keggin POM units (c. 1 nm) and the sodalite (SOD)-type ZIF cages – with a diameter of 1.16 nm connected by smaller windows of 0.34 nm [70] – (Figure 7.10a).

The first work testing a guest(POM)@host(ZIF) nanocomposite in OER electrocatalysis was reported by S. K. Das and co-workers in 2018 [65]. The authors described the encapsulation of Keggin CoW_{12} units, $[CoW_{12}O_{40}]^{6-}$, inside the cages of ZIF-8, $[Zn(C_4H_6N_2)_2]_n$ and their OER activity in neutral pH. They demonstrated that CoW_{12} turned into an efficient and robust electrocatalyst (Figure 7.10b) after its confinement in the well-defined void space of ZIF-8. This interesting behaviour was attributed to the stabilization effect that the ZIF-cage confinement has on POM units – preventing them from leaching – and a 'POM activation' resulting from a ZIF-8 → CoW_{12} charge transfer. It was observed that CoW_{12}@ZIF-8 retained its initial activity after 1000 CV cycles and exhibited a TOF of $10.8\,s^{-1}$. Then, two articles published by C. Freire and co-workers [68, 69] continued to explore this approach. The first work [68] reported the characterization of two guest@host nanocomposites – SiW_9Co_3@ZIF-8 and SiW_9Co_3@ZIF-67, (POM formula: $[SiW_9Co_3(H_2O)_3O_{37}]^{10-}$)– as OER catalysts in alkaline medium. By encapsulating the same POM units in two isostructural ZIF hosts involving different metal nodes (ZIF-8, Zn^{2+} and ZIF-67, Co^{2+}), the authors demonstrated that the existence of Co in the ZIF framework is mandatory to improve the OER performance of these nanocomposites. Therefore, a SiW_9Co_3 ↔ ZIF-67 bidirectional synergy – on one hand, POM-induced structural defects in the ZIF-67 framework, and, on the other hand, a ZIF-67 → POM charge transfer – was detected, being responsible for the significant enhancement in the OER activity shown by SiW_9Co_3@ZIF-67, in which the confined SiW_9Co_3 units exhibited a 22-fold individual activity compared with the pristine (non-confined) POM. The next article [69] reported the preparation of two POM@ZIF-67 electrocatalysts with larger POM loadings: $SiW_{11}Co$[h]@ZIF-67 and SiW_9Co_3[h]@ZIF-67 – with ≈9 and 13 POM units per 100 ZIF-67 cages, respectively. As a consequence of these elevated POM loadings, the nanocomposites showed highly modified frameworks, ZIF-67 → POM charge transfers and improved OER activity: 150–160 mV decreased overpotentials (Figure 7.10c) and faster kinetics (≈35% decreased Tafel slopes) in relation to pristine ZIF-67, and intrinsic OER activities higher than those developed by the benchmarks (RuO_2 and IrO_2 powders).

Apart from the guest(POM)@host(ZIF) strategy, two studies published by H. Pang and co-workers [66, 67] in 2019 explored the coating of ZIF particles with the same POM compound, $H_3PW_{12}O_{40}$. The first article [66] reported the synthesis of yolk/shell ZIF-67@POM nanostructures and their testing as OER catalysts in alkaline medium. As it is displayed in Figure 7.10d, the covering of the ZIF-67 surfaces by a POM shell was visualized through electronic microscopy. The best-performing sample exhibited a low overpotential, η_{10}, of 287 mV, a Tafel slope of $58\,mV\,dec^{-1}$ and no obvious activity loss after 12 hours-chronoamperometry. Finally, the other work [67] reported the preparation of ZIF-8@ZIF-67@POM hybrids by growth ZIF-67 on the surface of previously synthesized ZIF-8 particles and, subsequent deposition of $H_3PW_{12}O_{40}$. Regarding OER, these catalysts showed an overpotential of 490 mV and a Tafel slope of $88\,mV\,dec^{-1}$ in 1 M KOH electrolyte. In both cases, the authors attributed the OER performance of these ZIF@POM catalysts to (i) the existence of numerous active sites exposed on the nanocomposite surfaces and (ii) the elevated surface areas that results in exceptional diffusion efficiency.

7.3.4 Other Nanomaterials

This miscellanea collects, ordered by chronology, four articles [71–74] that presented diverse OER electrocatalysts that do not fit properly in Sections 7.3.1–7.3.3.

In 2014, T. McCormac and co-workers [71] deposited Ru_4POM-doped polypyrrole films onto glassy carbon electrodes via layer-by-layer (LBL) technique. Ru_4POM

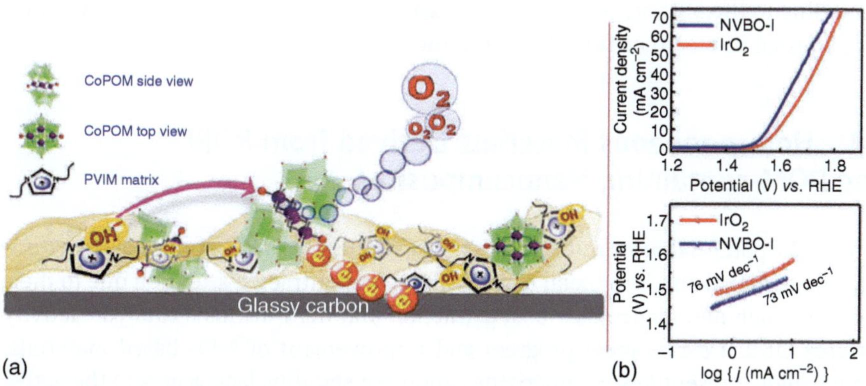

Figure 7.11 (a) Schematic representation of OER on PVIM-CoPOM nanocomposite. Source: Adhikary et al. [72]/with permission of American Chemical Society. (b) OER polarization curves (LSV) (top) and the corresponding Tafel plots (bottom) of compounds NVBO-I and IrO_2 in 1 M KOH electrolyte solution. Source: Rom et al. [74]/with permission of American Chemical Society.

denoted a polyoxometalate formed by two twisted dimeric structures with the formula Na_{10} $[Ru_4O_4(OH)_2(H_2O)_4](\gamma\text{-}SiW_{10}O_{36})_2]$. The experiments showed that the Ru_4POM-based LBL assemblies presented much better OER activity in neutral medium than those developed by Ru_4POM paste electrodes and LBL assemblies composed of the Dawson-type $K_8[P_2W_{18}O_{62}]$ moiety. Then, the incorporation of a Weakly sandwich-type POM (CoPOM: $[Co_4(H_2O)_2(PW_9O_{34})_2]^{10-}$) into another polymeric matrix, – in this case, an ionic polymer, poly(vinyl butyl imidazolium) (PVIM) – was reported by D. Mandal and co-workers [72] This PVIM-CoPOM OER catalyst (Figure 7.11a) exhibited, in 1 M NaOH electrolyte, a low overpotential of 0.20 V and a high current density of 250 mA cm^{-2} (at 1.75 V *vs.* RHE) with TOF of 52.8 s^{-1}. This remarkable electrocatalytic behaviour was raised from the stabilization effect that the ionic polymer has on CoPOM and its role as a conductive linker between the polyoxometalate and the electrode surface.

In another article, H. Pang and co-workers [73] described the hydrothermal preparation of efficient OER catalysts based on a hollow PBA@POM hybrid, in which the cubic Prussian Blue analogue (PBA) was etched by the POM – in this case, a strong Brønsted acid, $H_3PMo_{12}O_{40}$. They tested this material in alkaline medium, requiring an overpotential of 440 mV to achieve a current density of 10 mA cm^{-2}, with a fast kinetics: TS equal to 23.45 mV dec^{-1}. This noteworthy OER performance was attributed to the hollow shape of PBA@POM that exposed many active sites and simultaneously promoted electron transport. Finally, a study published in 2022 [74] reported the development of a 3D polyoxovanadoborate framework compound labelled NVBO-I. This completely inorganic framework was formed by $[V_{12}B_{18}]$ clusters linked by $[NaO_6]$ units (formula: $[Na_{10}(H_2O)_{18}][(VO)_{12}(\mu 3\text{-}OH)_6\text{-}(B_3O_7)_6]\cdot 5H_2O$) and exhibited a bifunctional character towards both OER and HER. As can be observed in Figure 7.11b, this electrocatalyst showed better OER activity than IrO_2, with lower overpotential, η_{10} = 324 mV, and Tafel slope, 73 mV dec^{-1}.

According to the authors, the presence of a mixed-valence state of V ions (V^{4+} and V^{5+}) boosted its electrochemical performance.

7.4 Heterogeneous Materials Derived from POM and POM-containing Nanocomposites

Among the potential substitutes of the expensive and scarce Ru- and Ir-based materials, cobalt, nickel and iron metal oxides (MOx) have attracted attention due to their relatively high abundance, low-cost extraction and high intrinsic catalytic activity towards OER. Despite great progress and improvement of MOx-based materials, these usually present low conductivities and poor stability, lack access to the active sites and their catalytic performance falls below the expected values for their applications on practical devices (>500 mA cm^{-2} at low η, <300 mV) [75]. Therefore, further synthetic advances are needed to structurally optimize them towards efficient OER. On this basis, three main strategies have been adopted to improve their performance:

- Preparation of mixed-valence metal oxide materials, which incorporates binary or ternary mixtures of metallic ions or oxides providing an excellent platform to tune the intrinsic activity of the systems by modifying the microstructure, conductivity and/or absorption/desorption of the intermediate catalytic products on the surface [76].
- Design of highly porous nanoscale structures that maximize the exposure of the active sites and the mass transport within the material [77, 78].
- Combination with highly conductive materials – 3D carbon-based electrodes, carbon nanotubes (CNTs), graphene – which can enhance the electron transport efficiency as well as disperse the active components in highly porous supports.

Within this context, the combination of POMs as well-defined metal precursors with other organic or inorganic materials has recently emerged as an interesting strategy to improve the catalytic performance of MOx-based materials obtained using traditional synthetic methodologies and to overcome their main difficulties.

7.4.1 Encapsulation of POMs into MOFs Structures as Precursors for WO Electrocatalysts

Of special interest is the incorporation of POMs into MOFs [79, 80]. As described in Section 7.3.3, MOF structures are capable of hosting POMs in their large hollow porous cages, providing a well-characterized guest@host POM@MOF nanocomposite. The so-called POM@MOF material contains a well-dispersed concentration of POMs that will behave as a source of metallic dopants or active sites (W, V and Mo, among others), a good density of metal ions that form the backbone of the MOF and a carbon-based backbone as a template, which can be transformed into porous carbon supports with high specific surface areas via carbonization process (Figure 7.12) [83, 84]. POM@MOF hybrids benefit from the properties of both POMs and MOFs, preventing aggregation phenomena, promoting synergy effects provided

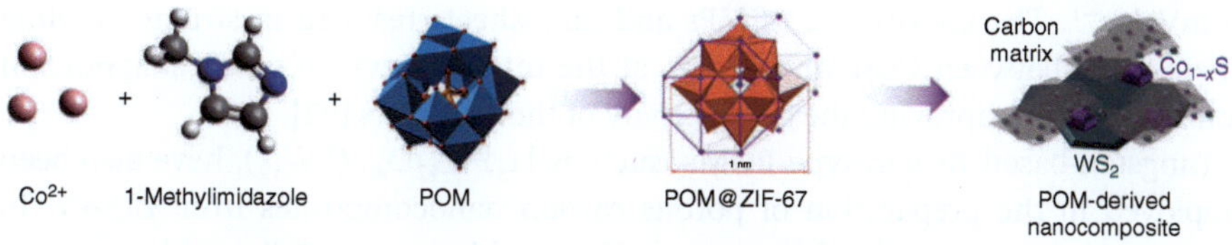

Figure 7.12 Schematic drawing of the preparation of POM-derived nanocomposites by using the POM@MOF strategy. Source: Adapted from Ref. [81, 82]; published by Elsevier Ltd., and RSC Publishing, respectively.

by multi-metallic active sites, and enhancing charge transfer processes, properties that are difficult to achieve through traditional synthetic routes [81, 85, 86].

The most widely used approach is the combination of POMs with ZIFs. The encapsulation of these metal clusters prevents the undesired leaching of POMs during catalytic or post-synthetic treatments, ensuring the integrity of the structure and preventing the aggregation of metal centres during the calcination process [82]. Among ZIFs, ZIF-67 – involving Co^{2+} nodes and 2-methylimidazole ligands – is extensively used. This framework has been successfully transformed into highly porous Co_3O_4 nanostructures with various morphologies and properties, such as nanopolyhedra, nanocubes or nanosheets [87–89]. The resulting composite's catalytic performance is highly dependent on its microstructure and components. Additionally, the Co^{2+} of the ZIF-67 can be partially substituted by other transition metals to prepare heterometallic constructs. On the other hand, Keggin- and Anderson-type POM clusters (usually containing W, Mo or V in high oxidation states) have been extensively employed due to their size, which is suitable for trapping them inside the ZIF-67 cages.

A recent example is the preparation of $Co_3O_4/CoMoO_4$ nanocages through the combination of a Keggin-type polyoxometalate with ZIF-67 via an in situ self-assembly route. The POM, $H_3PMo_{12}O_{40}$ (PMo_{12}) served as both a Mo^{6+} source and template to obtain the hollow hybrids after a carbonization step, which retained the structure of the ZIF-67 parent cage. The key to the superior reactivity of these systems compared to hybrids prepared in the absence of POM was the high surface area and favourable electronic conductivity provided by the hybrid heterometallic $Co_3O_4/CoMoO_4$-50 material. This catalyst reached a current of 10 mA cm^{-2} at a low overpotential of 318 mV displaying a TS of 63 mV dec^{-1}, which was significantly lower than other Co–Mo oxides [90]. The surface area and conductivity of Co–Mo POM-derived materials can be improved by using highly porous CNTs as graphitic support under the reaction conditions. This promotes the homogeneous formation of highly porous Mo-doped Co_3O_4 on the surface of the CNTs during the carbonization process, enriching the material in Co^{2+} active sites [91]. The POM-assisted formation of nanocomposites has been employed in the preparation of many types of OER catalysts. By using the same PMo_{12} precursor and ZIF-67, but following a selenization process, a heterostructured CoSe/$MoSe_2$ hybrid system was obtained. These systems contain well-dispersed CoSe nanoparticles on $MoSe_2$ nanosheets, which enhanced the intrinsic activity of OER, significantly improving the catalytic performance of the previously reported systems, achieving η_{10} = 262 mV and TS of

55 mV dec^{-1}. The coexistence of NPs and nanosheets resulted in strong coupling interactions between CoSe and $MoSe_2$ at the interface, which made a significant contribution to improving the conductivity of the materials [92].

Tungsten-based Keggin-type POMs, such as $H_3PW_{12}O_{40}$ (PW_{12}), have also been employed in the preparation of porous carbon nanocomposites from ZIF-67. In this case, the synthesis also follows the self-assembly strategy, followed by one-step sulphurization (heating in H_2S atmosphere) to produce a homogeneous dispersion of tungsten sulphide and cobalt sulphide particles on the ZIF-derived porous carbon matrix. The bimetallic Co–W material behaves not only as a powerful catalyst with TS of 53 mV dec^{-1} and η_{10} = 365 mV but also as a bifunctional material performing HER. The electron transfer between Co–W atoms via the S atoms played an essential role under turnover conditions, where a too high or low content limited the catalytic performance of the material [81]. The same group recently reported the preparation of highly active Co–Mo–S/N,S-doped porous carbon starting from ZIF-67 nanocubes and using PMo_{12} as an acid-etching agent. The POM-derived nanocomposite presented an evenly distributed ultra-small defect-rich Co–Mo–S phase, which enhanced the catalytic performance of the previously reported Co–W–S material in terms of η_{10} = 350 mV [93]. Therefore, this demonstrated the importance of including POMs as precursors to obtain well-distributed NPs on porous carbon supports to maximize the exposure of active sites and facilitate mass transport [94, 95].

Despite ZIF-67 being the most common template to produce POM-derived hybrids, other MOF architectures have been used to prepare highly performing OER catalysts. This is the case for the in situ formation of PBAs inside a NiMoBTC structure formed from Ni, trimesic acid (BTC) and $(NH_4)_6Mo_7O_{24}$ as POM, using the PBA@POMOF strategy. The resulting materials displayed a microsphere structure formed by a trimetallic Fe/Ni/Mo mixture, and after the annealing step, yielded NiMoFe@NC-type materials exhibiting low η_{10} of 254 mV with no signs of degradation for 11 hours [96].

7.4.2 Other POM-based Materials

Although most of the efforts have been focused on combining POMs with MOF structures to obtain highly porous materials, their applicability in catalysis is broader. As well-defined metal clusters, they are promising candidates to act as sacrificial metal sources to functionalize carbon and/or metal-based structures, perform as etching agents due to their strong acidity and/or serve as templates for the growth of highly nanostructured MOx.

POM structures are also suitable precursors to obtain metal-rich carbon-based supports for their ultimate functionalization with a well-dispersed metallic colloidal suspension [96]. This approach has been employed for the preparation of vanadium carbide (VC) using substituted Lindqvist-type POMs ($Na_2[V_6O_{13}((OCH_2)_3CCH_3)_2]$) – as V precursors – which was decorated with a colloidal suspension of Ir NPs to form Ir/VC/C nanocomposites. This material performed as an HER/OER bifunctional catalyst, with a η_{10} of 295 mV for the latter. As well-defined clusters,

POMs can also behave as a template for the in situ growth of metal oxide structures. An example is the encapsulation of $[Co(CN)_6]^{3-}$ in the PW_{12} POM nanochannels. This material can be pyrolysed in the presence of an inert atmosphere, whereas PW_{12} remains unaltered; the cobalt precursor decomposes into C, N and Co forming a decorated hollow dodecahedral structure. The POM-based yolk–shell nanoframes achieved a great activity towards OER, thanks to the low η_{10} = 241 mV and stable currents for 24 hours [97]. Recent studies have also demonstrated the ability of Keggin-type POMs to behave as etching agents towards the preparation of Fe–P nanotubes, controlling the porosity of the formed materials [98]. The pore size and increased surface area are strongly dependent on the acidity of the POM, where the stronger the acidity, the higher the porosity and the catalytic performance. The functionalization of a Ni foam with POM-Fe-P porous nanotubes achieved a η_{10} of 230 mV, a low TS of 55 mV dec^{-1} and a stable current for more than 300 hours.

Finally, Anderson-type POMs ($(NH_4)_3[FeMo_6O_{24}H_6]$) have been used as a metal precursor to accurately dope NF electrodes, forming multi-metallic Fe–S–$NiMoO_4$/MoO_3@NF electrodes through a one-step solvothermal reaction. A combination of analytical and computational techniques concluded that the Fe-doping provided by the POM induced a transformation of the MoO_3 crystal planes from (110) to (100), which was key for improved conductivity and stability. The system delivered a remarkable j of 500 mA cm^{-2} at an overpotential of 271 mV and was capable of reaching 10 mA cm^{-2} at 212 mV, making it suitable for large-scale industrial applications [99].

7.5 Concluding Remarks

This chapter highlights the potential role that polyoxometalate platforms play in the field of water oxidation electrocatalysis. The appealing electrochemical properties exhibited by POMs make them highly suitable for (i) getting insight into the mechanistic details of OER catalysis, (ii) developing promising POM-based OER electrocatalysts and (iii) serving as precursors/templates to obtain nanostructured metal oxide-based catalysts.

Studies focused on homogeneous OER catalysis using several POM architectures revealed the good performance of both multi- and mononuclear Ru-based POM platforms. While in the case of the Co-POMs, the formation of cobalt oxides under the operando conditions seems to be the main responsible for the developed OER electroactivity. Other POM species involving different transition metals, such as Fe, Mn and Cu, have demonstrated their significant performances in homogeneous water oxidation. In addition, the partial substitution of Co by Fe or Mo ions appears as a promising approach to enhance the POM performances in heterogeneous OER processes.

Regarding the effective integration of POMs into nanocomposites, the development of POM/carbon (nanotubes, graphene, etc.) and POM/Ni foams – among other metal-based nanostructures – is the most explored strategy. Additionally, POMs have been successfully assembled into MOFs – especially within ZIFs – and other polymeric compounds. The underlying idea of these combinations consists in maximizing the POM electrochemical properties to produce efficient, active, and durable OER electrocatalysts, i.e. increasing POM dispersion to facilitate the access of the reactant water molecules – or hydroxyl ions in alkaline medium – to the active sites, enhancing electrical conductivity and stability, etc. In this diverse field, numerous synergistic effects raised from POM–matrix interactions have been reported. However, the lack of systematicity in the studies and the elevated complexity of the POM-based nanocomposites avoid the extraction of general rules for the described catalytic behaviours.

Finally, this chapter also examined an alternative approach to producing OER-active nanocomposites in which POMs do not preserve their pristine structural integrity. This interesting strategy allows the synthesis of POM-derived metal oxides integrated into carbon matrixes or other metallic supports and takes advantage of the template effect of MOFs to prepare well-dispersed POM-based nanocomposites – and/or etching effect – shown by POMs. These effects are originated from the molecular nature of POMs and their acid character, respectively.

Regardless of the employed strategy, some studies claim the development of POM electrocatalysts with similar – or even, better – OER behaviours than the expensive RuO_2 and IrO_2 benchmarks, indicating the huge potential of POMs in this strategic field. Nevertheless, a more detailed and systematic approach is required to definitively boost OER performances under real working conditions.

Acknowledgements

VKAF and co-authors express their gratitude to the editor of this book, Dr. Diana Fernandes. VKAF acknowledges Junta de Andalucía (Spain) for its postdoctoral grant within the Programa de Jóvenes Doctores (PAIDI 2020). PGB acknowledges Ministerio de Ciencia e Innovación for a Ramon y Cajal Fellowship. MSN acknowledges FCT/MCTES for the work contract through the project PTDC/CTM-TEX/4126/2021. MGS acknowledges the support of the HORIZON-MSCA-2021-PF project TRUSol No. 101063820.

List of Abbreviations

BTC	trimesic acid
C	carbon (material)
CNTs	carbon nanotubes
CNTPE	carbon nanotube paste electrode
CP	carbon paper
CPO	carbon paste oil
CV	cyclic voltammetry
DFT	density functional theory
DMF	*N,N*-dimethylformamide

ECSA	electrochemically active surface area
FE	faradaic efficiency
FTO	fluorine-doped tin oxide
GDY	graphdyine
HER	hydrogen evolution reaction
H_2biim	biimidazole
ITO	indium tin oxide
IR	infrared (spectroscopy)
KB	conducting carbon Ketjen black
LBL	layer-by-layer (technique)
LDH	layered double hydroxide
LSV	linear sweep voltammetry
MOF	metal–organic framework
MWCNT	multiwalled carbon nanotube
NCNT	N-doped carbon nanotube
NF	Ni foam
NHE	normal hydrogen electrode
NP	nanoparticle
OER	oxygen evolution reaction
ORR	oxygen reduction reaction
PBA	Prussian blue analogue
PBS	phosphate-buffered saline (solution)
PCET	proton-coupled electron transfer
PGM	platinum group metal
PIC	porous ionic crystal
POM	polyoxometalate
POMOF	polyoxometalate-organic framework
PVIM	poly(vinyl butyl imidazolium)
PVP	polyvinylpyrrolidone
RHE	reversible hydrogen electrode
SEM	scanning electron microscopy
SDS	sodium dodecyl sulphate
SOD	SODalite
SWB	seawater battery
TM	transition metal
TOF	turnover frequency
TON	turnover number
TS	Tafel slope
TX	Triton X-100 (surfactant)
VC	vanadium carbide
WNA	water nucleophilic attack
WO	water oxidation
XPS	X-ray photoelectron spectroscopy
ZIF	zeolitic imidazolate framework

References

1 O.o.E.E.R. Energy (2023). Hydrogen production: electrolysis. https://www.energy.gov/eere/fuelcells/hydrogen-production-electrolysis (accessed 24 February 2023).

2 Saha, S., Kishor, K., Sivakumar, S., and Pala, R.G.S. (2016). Models and mechanisms of oxygen evolution reaction on electrocatalytic surface. *J. Indian Inst. Sci.* 96 (4): 325–349.

3 Tahir, M., Pan, L., Idrees, F. et al. (2017). Electrocatalytic oxygen evolution reaction for energy conversion and storage: a comprehensive review. *Nano Energy* 37: 136–157.

4 Suen, N.T., Hung, S.F., Quan, Q. et al. (2017). Electrocatalysis for the oxygen evolution reaction: recent development and future perspectives. *Chem. Soc. Rev.* 46 (2): 337–365.

5 Freire, C., Fernandes, D.M., Nunes, M., and Araujo, M. (2017). *Polyoxometalate-based Modified Electrodes for Electrocatalysis: From Molecule Sensing to Renewable Energy-related Applications, Advanced Electrode Materials*, 149–212. Scrivener Publishing LLC.

6 Benck, J.D., Hellstern, T.R., Kibsgaard, J. et al. (2014). Catalyzing the hydrogen evolution reaction (HER) with molybdenum sulfide nanomaterials. *ACS Catal.* 4 (11): 3957–3971.

7 Jiao, Y., Zheng, Y., Jaroniec, M.T., and Qiao, S.Z. (2015). Design of electrocatalysts for oxygen- and hydrogen-involving energy conversion reactions. *Chem. Soc. Rev.* 44 (8): 2060–2086.

8 Zou, X.X. and Zhang, Y. (2015). Noble metal-free hydrogen evolution catalysts for water splitting. *Chem. Soc. Rev.* 44 (15): 5148–5180.

9 Lee, Y., Suntivich, J., May, K.J. et al. (2012). Synthesis and activities of rutile IrO_2 and RuO_2 nanoparticles for oxygen evolution in acid and alkaline solutions. *J. Phys. Chem. Lett.* 3 (3): 399–404.

10 MetalsDaily (2023). Precious metals news, information and live prices. https://www.metalsdaily.com/live-prices/pgms/ (accessed 24 February 2023).

11 Sala, X., Maji, S., Bofill, R. et al. (2014). Molecular water oxidation mechanisms followed by transition metals: state of the art. *Acc. Chem. Res.* 47 (2): 504–516.

12 Garrido-Barros, P., Gimbert-Surinach, C., Matheu, R. et al. (2017). How to make an efficient and robust molecular catalyst for water oxidation. *Chem. Soc. Rev.* 46 (20): 6088–6098.

13 Siwal, S.S., Yang, W.Q., and Zhang, Q.B. (2020). Recent progress of precious-metal-free electrocatalysts for efficient water oxidation in acidic media. *J. Energy Chem.* 51: 113–133.

14 Li, P.P., Zhao, R.B., Chen, H.Y. et al. (2019). Recent advances in the development of water oxidation electrocatalysts at mild pH. *Small* 15 (13): 1805103.

15 Blakemore, J.D., Crabtree, R.H., and Brudvig, G.W. (2015). Molecular catalysts for water oxidation. *Chem. Rev.* 115 (23): 12974–13005.

16 Hunter, B.M., Gray, H.B., and Muller, A.M. (2016). Earth-abundant heterogeneous water oxidation catalysts. *Chem. Rev.* 116 (22): 14120–14136.

17 Li, N., Liu, J., Dong, B.X., and Lan, Y.Q. (2020). Polyoxometalate-based compounds for photo- and electrocatalytic applications. *Angew. Chem. Int. Ed.* 59 (47): 20779–20793.

18 Lv, H., Geletii, Y.V., Zhao, C. et al. (2012). Polyoxometalate water oxidation catalysts and the production of green fuel. *Chem. Soc. Rev.* 41 (22): 7572–7589.

19 Long, D.L., Burkholder, E., and Cronin, L. (2007). Polyoxometalate clusters, nanostructures and materials: from self assembly to designer materials and devices. *Chem. Soc. Rev.* 36 (1): 105–121.

20 Goberna-Ferron, S., Soriano-Lopez, J., Galan-Mascaros, J.R., and Nyman, M. (2015). Solution speciation and stability of cobalt-polyoxometalate water oxidation catalysts by X-ray scattering. *Eur. J. Inorg. Chem.* 2015 (17): 2833–2840.

21 Kanan, M.W. and Nocera, D.G. (2008). In situ formation of an oxygen-evolving catalyst in neutral water containing phosphate and Co^{2+}. *Science* 321 (5892): 1072–1075.

22 Esswein, A.J., Surendranath, Y., Reece, S.Y., and Nocera, D.G. (2011). Highly active cobalt phosphate and borate based oxygen evolving catalysts operating in neutral and natural waters. *Energy Environ. Sci.* 4 (2): 499–504.

23 Soriano-Lopez, J., Musaev, D.G., Hill, C.L. et al. (2017). Tetracobalt-polyoxometalate catalysts for water oxidation: key mechanistic details. *J. Catal.* 350: 56–63.

24 Fertig, A.A. and Matson, E.M. (2023). Connecting thermodynamics and kinetics of proton coupled electron transfer at polyoxovanadate surfaces using the Marcus cross relation. *Inorg. Chem.* 62 (5): 1958–1967.

25 Taira, N., Yamauchi, K., and Sakai, K. (2023). Intracluster O-O coupling pathway evidenced for an Anderson-type single-cobalt polymolybdate water oxidation catalyst. *ACS Catal.* 13 (5): 3211–3223.

26 Agarwal, R.G., Coste, S.C., Groff, B.D. et al. (2022). Free energies of proton-coupled electron transfer reagents and their applications. *Chem. Rev.* 122 (1): 1–49.

27 Sumliner, J.M., Lv, H.J., Fielden, J. et al. (2014). Polyoxometalate multi-electron-transfer catalytic systems for water splitting. *Eur. J. Inorg. Chem.* 2014 (4): 635–644.

28 Howells, A.R., Sankarraj, A., and Shannon, C. (2004). A diruthenium-substituted polyoxometalate as an electrocatalyst for oxygen generation. *JACS* 126 (39): 12258–12259.

29 Sartorel, A., Miro, P., Salvadori, E. et al. (2009). Water oxidation at a tetraruthenate core stabilized by polyoxometalate ligands: experimental and computational evidence to trace the competent intermediates. *JACS* 131 (44): 16051–16053.

30 Murakami, M., Hong, D.C., Suenobu, T. et al. (2011). Catalytic mechanism of water oxidation with single-site ruthenium-heteropolytungstate complexes. *JACS* 133 (30): 11605–11613.

31 Yin, Q.S., Tan, J.M., Besson, C. et al. (2010). A fast soluble carbon-free molecular water oxidation catalyst based on abundant metals. *Science* 328 (5976): 342–345.

32 Stracke, J.J. and Finke, R.G. (2011). Electrocatalytic water oxidation beginning with the cobalt polyoxometalate $[Co_4(H_2O)_2(PW_9O_{34})_2]^{10-}$: identification of heterogeneous CoO_x as the dominant catalyst. *JACS* 133 (38): 14872–14875.

33 Vickers, J.W., Lv, H.J., Sumliner, J.M. et al. (2013). Differentiating homogeneous and heterogeneous water oxidation catalysis: confirmation that $[Co_4(H_2O)_2(alpha\text{-}PW_9O_{34})_2]^{10-}$ is a molecular water oxidation catalyst. *JACS* 135 (38): 14110–14118.

34 Goberna-Ferron, S., Vigara, L., Soriano-Lopez, J., and Galan-Mascaros, J.R. (2012). Identification of a nonnuclear {Co-9(II)} polyoxometalate cluster as a homogeneous catalyst for water oxidation. *Inorg. Chem.* 51 (21): 11707–11715.

35 Folkman, S.J., Soriano-Lopez, J., Galan-Mascaros, J.R., and Finke, R.G. (2018). Electrochemically driven water-oxidation catalysis beginning with six exemplary cobalt polyoxometalates: is it molecular, homogeneous catalysis or electrode-bound, heterogeneous CoOx, catalysis? *JACS* 140 (38): 12040–12055.

36 Du, X.Q., Zhoufeng, X.F., Gong, Y.Q., and Ding, Y. (2017). Polyoxometalate-based catalysts for photocatalytic, chemical catalytic and electrocatalytic water oxidation. *Int. J. Hydrogen Energy* 42 (38): 24169–24175.

37 Yu, L., Ding, Y., and Deng, M. (2017). Polyoxometalate-based manganese clusters as catalysts for efficient photocatalytic and electrochemical water oxidation. *Appl. Catal., B* 209: 45–52.

38 Yu, L., Lin, J.Q., Zheng, M. et al. (2018). Homogeneous electrocatalytic water oxidation at neutral pH by a robust trinuclear copper(II)-substituted polyoxometalate. *Chem. Commun.* 54 (4): 354–357.

39 Martin-Sabi, M., Soriano-Lopez, J., Winter, R.S. et al. (2018). Redox tuning the Weakley-type polyoxometalate archetype for the oxygen evolution reaction. *Nat. Catal.* 1 (3): 208–213.

40 Han, X.B., Wang, D.X., Gracia-Espino, E. et al. (2020). Fe-substituted cobalt-phosphate polyoxometalates as enhanced oxygen evolution catalysts in acidic media. *Chin. J. Catal.* 41 (5): 853–857.

41 Azmani, K., Besora, M., Soriano-Lopez, J. et al. (2021). Understanding polyoxometalates as water oxidation catalysts through iron vs. cobalt reactivity. *Chem. Sci.* 12 (25): 8755–8766.

42 Toma, F.M., Sartorel, A., Iurlo, M. et al. (2010). Efficient water oxidation at carbon nanotube–polyoxometalate electrocatalytic interfaces. *Nat. Chem.* 2 (10): 826–831.

43 Soriano-López, J., Goberna-Ferrón, S., Vigara, L. et al. (2013). Cobalt polyoxometalates as heterogeneous water oxidation catalysts. *Inorg. Chem.* 52 (9): 4753–4755.

44 Blasco-Ahicart, M., Soriano-Lopez, J., Carbo, J.J. et al. (2018). Polyoxometalate electrocatalysts based on earth-abundant metals for efficient water oxidation in acidic media. *Nat. Chem.* 10 (1): 24–30.

45 Arens, J.T., Blasco-Ahicart, M., Azmani, K. et al. (2020). Water oxidation electrocatalysis in acidic media with Co-containing polyoxometalates. *J. Catal.* 389: 345–351.

46 Li, Y., Chang, X.-R., Sang, X.-J. et al. (2019). Keggin-type polyoxometalate modified Ag/graphene composite materials for electrocatalytic water oxidation. *Eur. J. Inorg. Chem.* 2019 (31): 3597–3604.

47 Xing, X., Wang, M., Liu, R. et al. (2016). Highly efficient electrochemically driven water oxidation by graphene-supported mixed-valent Mn_{16}-containing polyoxometalate. *Green Energy Environ.* 1 (2): 138–143.

48 Marques, I.S., Jarrais, B., Mbomekallé, I.-M. et al. (2022). Synergetic effects of mixed-metal polyoxometalates@carbon-based composites as electrocatalysts for the oxygen reduction and the oxygen evolution reactions. *Catalysts* 12 (4): 440.

49 Lee, G.Y., Kim, I., Lim, J. et al. (2017). Spontaneous linker-free binding of polyoxometalates on nitrogen-doped carbon nanotubes for efficient water oxidation. *J. Mater. Chem. A* 5 (5): 1941–1947.

50 Sun, H., Jing, C., Shang, W. et al. (2022). Polyoxometalate-based composite cluster with core–shell structure: $Co_4(PW_9)_2$@graphdiyne as stable electrocatalyst for oxygen evolution and its mechanism research. *New J. Chem.* 46 (24): 11553–11561.

51 Abdelkader-Fernández, V.K., Fernandes, D.M., Cunha-Silva, L. et al. (2021). Decorating MOF-74-derived nanocarbons with a sandwich-type polyoxometalate to enhance their OER activity: exploring the underestimated bulk-deposition approach. *Electrochim. Acta* 389: 138719.

52 Lee, C., Jeon, D., Park, J. et al. (2020). Tetraruthenium polyoxometalate as an atom-efficient bifunctional oxygen evolution reaction/oxygen reduction reaction catalyst and its application in seawater batteries. *ACS Appl. Mater. Interfaces* 12 (29): 32689–32697.

53 Luo, W., Hu, J., Diao, H. et al. (2017). Robust polyoxometalate/nickel foam composite electrodes for sustained electrochemical oxygen evolution at high pH. *Angew. Chem. Int. Ed.* 56 (18): 4941–4944.

54 Ding, Y., Li, H., and Hou, Y. (2018). Robust polyoxometalate-loaded nickel foam for electrocatalytic oxygen evolution reaction. *Mater. Lett.* 221: 264–266.

55 Ahmed, T., Asghar, M.A., Ali, A. et al. (2022). High-nuclearity cobalt(II)-containing polyoxometalate anchored on nickel foam as electrocatalyst for electrochemical water oxidation studies. *J. Alloys Compd.* 909: 164709.

56 Mahanta, A., Barman, K., and Jasimuddin, S. (2019). Electrocatalytic oxidation of water at a polyoxometalate nanoparticle modified gold electrode. *RSC Adv.* 9 (66): 38713–38717.

57 Imani, A.H., Ojani, R., and Raoof, J.-B. (2021). Facile synthesis of polyoxometalate-based composite with doped ternary NiCoFe cations as electrocatalyst for oxygen evolution reaction. *Int. J. Hydrogen Energy* 46 (1): 449–457.

58 Gautam, J., Kannan, K., Meshesha, M.M. et al. (2022). Heterostructure of polyoxometalate/zinc-iron-oxide nanoplates as an outstanding bifunctional electrocatalyst for the hydrogen and oxygen evolution reaction. *J. Colloid Interface Sci.* 618: 419–430.

59 Shang, W., Wang, Y., Jiang, Y. et al. (2022). Nanocomposite: Co_4-substituted polyoxometalate@β-FeOOH as high-performance electrocatalysts for oxygen evolution reaction in alkaline conditions. *Appl. Catal., A* 644: 118810.

60 Gong, R., Gao, D., Liu, R. et al. (2021). Self-activation of a polyoxometalate-derived composite electrocatalyst for the oxygen evolution reaction. *ACS Appl. Energy Mater.* 4 (11): 12671–12676.

61 Singh, C., Mukhopadhyay, S., and Das, S.K. (2018). Polyoxometalate-supported bis(2,2′-bipyridine)mono(aqua)nickel(II) coordination complex: an efficient electrocatalyst for water oxidation. *Inorg. Chem.* 57 (11): 6479–6490.

62 Shimoyama, Y., Ogiwara, N., Weng, Z., and Uchida, S. (2022). Oxygen evolution reaction driven by charge transfer from a Cr complex to Co-containing polyoxometalate in a porous ionic crystal. *JACS* 144 (7): 2980–2986.

63 Cai, Z., Wang, P., Zhang, J. et al. (2022). Reinforced layered double hydroxide oxygen-evolution electrocatalysts: a polyoxometallic acid wet-etching approach and synergistic mechanism. *Adv. Mater.* 34 (26): 2110696.

64 Xue, X., Yu, F., Li, J.-G. et al. (2020). Polyoxometalate intercalated NiFe layered double hydroxides for advanced water oxidation. *Int. J. Hydrogen Energy* 45 (3): 1802–1809.

65 Mukhopadhyay, S., Debgupta, J., Singh, C. et al. (2018). A Keggin polyoxometalate shows water oxidation activity at neutral pH: POM@ZIF-8, an efficient and robust electrocatalyst. *Angew. Chem. Int. Ed.* 57 (7): 1918–1923.

66 Li, Q.Y., Zhang, L., Xu, Y.X. et al. (2019). Smart yolk/shell ZIF-67@POM hybrids as efficient electrocatalysts for the oxygen evolution reaction. *ACS Sustainable Chem. Eng.* 7 (5): 5027–5033.

67 Wang, Y., Wang, Y.Y., Zhang, L. et al. (2019). Core-shell-type ZIF-8@ZIF-67@POM hybrids as efficient electrocatalysts for the oxygen evolution reaction. *Inorg. Chem. Front.* 6 (9): 2514–2520.

68 Abdelkader Fernández, V.K., Mesquita Fernandes, D., Silva Balula, S. et al. (2020). Oxygen evolution reaction electrocatalytic improvement in POM@ZIF nanocomposites: a bidirectional synergistic effect. *ACS Appl. Energy Mater.* 3 (3): 2925–2934.

69 Abdelkader-Fernandez, V.K., Fernandes, D.M., Balula, S.S. et al. (2020). Advanced framework-modified POM@ZIF-67 nanocomposites as enhanced oxygen evolution reaction electrocatalysts. *J. Mater. Chem. A* 8 (27): 13509–13521.

70 Zhong, G.H., Liu, D.X., and Zhang, J.Y. (2018). The application of ZIF-67 and its derivatives: adsorption, separation, electrochemistry and catalysts. *J. Mater. Chem. A* 6 (5): 1887–1899.

71 Anwar, N., Sartorel, A., Yaqub, M. et al. (2014). Surface immobilization of a tetra-ruthenium substituted polyoxometalate water oxidation catalyst through the employment of conducting polypyrrole and the layer-by-layer (LBL) technique. *ACS Appl. Mater. Interfaces* 6 (11): 8022–8031.

72 Adhikary, S.D., Tiwari, A., Nagaiah, T.C., and Mandal, D. (2018). Stabilization of cobalt-polyoxometalate over poly(ionic liquid) composites for efficient electrocatalytic water oxidation. *ACS Appl. Mater. Interfaces* 10 (45): 38872–38879.

73 Wang, Y.Y., Wang, Y., Zhang, L. et al. (2019). PBA@POM hybrids as efficient electrocatalysts for the oxygen evolution reaction. *Chem. Asian J.* 14 (16): 2790–2795.

74 Rom, T., Biswas, R., Haldar, K.K. et al. (2022). Vanadate encapsulated polyoxoborate framework with $V_{12}B_{18}$ clusters: an efficient bifunctional electrocatalyst for oxygen and hydrogen evolution reactions. *Cryst. Growth Des.* 22 (8): 4666–4672.

75 Liu, X., Guo, R.T., Ni, K. et al. (2020). Reconstruction-determined alkaline water electrolysis at industrial temperatures. *Adv. Mater.* 32 (40): 2001136.

76 Song, F., Bai, L.C., Moysiadou, A. et al. (2018). Transition metal oxides as electrocatalysts for the oxygen evolution reaction in alkaline solutions: an application-inspired renaissance. *JACS* 140 (25): 7748–7759.

77 Tao, L., Duan, X.D., Wang, C. et al. (2015). Plasma-engineered MoS_2 thin-film as an efficient electrocatalyst for hydrogen evolution reaction. *Chem. Commun.* 51 (35): 7470–7473.

78 De-Mello, G.B., Smith, L., Rowley-Neale, S.J. et al. (2017). Surfactant-exfoliated 2D molybdenum disulphide (2D-MoS_2): the role of surfactant upon the hydrogen evolution reaction. *RSC Adv.* 7 (58): 36208–36213.

79 Freire, C., Fernandes, D.M., Nunes, M., and Abdelkader, V.K. (2018). POM & MOF-based electrocatalysts for energy-related reactions. *ChemCatChem* 10 (8): 1703–1730.

80 Guo, P., Wang, Z.J., Ge, S.H. et al. (2020). In situ coupling reconstruction of cobalt-iron oxide on a cobalt phosphate nanoarray with interfacial electronic features for highly enhanced water oxidation catalysis. *ACS Sustainable Chem. Eng.* 8 (12): 4773–4780.

81 Huang, Z., Yang, Z.X., Hussain, M.Z. et al. (2020). Polyoxometallates@zeolitic-imidazolate-framework derived bimetallic tungsten-cobalt sulfide/porous carbon nanocomposites as efficient bifunctional electrocatalysts for hydrogen and oxygen evolution. *Electrochim. Acta* 330: 135335.

82 Stuckart, M. and Monakhov, K.Y. (2018). Polyoxometalate encapsulation into metal-organic frameworks: the way towards functional nanomaterials for water splitting. *J. Mater. Chem. A* 6 (37): 17849–17853.

83 Wu, H.B., Xia, B.Y., Yu, L. et al. (2015). Porous molybdenum carbide nano-octahedrons synthesized via confined carburization in metal-organic frameworks for efficient hydrogen production. *Nat. Commun.* 6: 6512.

84 Tang, Y.J., Gao, M.R., Liu, C.H. et al. (2015). Porous molybdenum-based hybrid catalysts for highly efficient hydrogen evolution. *Angew. Chem. -Int. Ed.* 54 (44): 12928–12932.

85 Xiao, X., Zhang, G.X., Xu, Y.X. et al. (2019). A new strategy for the controllable growth of MOF@PBA architectures. *J. Mater. Chem. A* 7 (35): 20436–20437.

86 Liang, Q.N., Chen, J.M., Wang, F.L., and Li, Y.W. (2020). Transition metal-based metal-organic frameworks for oxygen evolution reaction. *Coord. Chem. Rev.* 424: 213488.

87 He, W.H., Ifraemov, R., Raslin, A., and Hod, I. (2018). Room-temperature electrochemical conversion of metal-organic frameworks into porous amorphous metal sulfides with tailored composition and hydrogen evolution activity. *Adv. Funct. Mater.* 28 (18): 1707244.

88 Zheng, Y., Gao, R., Zheng, L.R. et al. (2019). Ultrathin Co_3O_4 nanosheets with edge-enriched {111} planes as efficient catalysts for lithium-oxygen batteries. *ACS Catal.* 9 (5): 3773–3782.

89 Xu, H., Shang, H.Y., Wang, C. et al. (2020). Three-dimensional open CoMoOx/CoMoSx/CoSx nanobox electrocatalysts for efficient oxygen evolution reaction. *Appl. Catal., B* 265: 118605.

90 Zhang, L.J., Mi, T.Q., Ziaee, M.A. et al. (2018). Hollow POM@MOF hybrid-derived porous Co_3O_4/$CoMoO_4$ nanocages for enhanced electrocatalytic water oxidation. *J. Mater. Chem. A* 6 (4): 1639–1647.

91 Lu, K.L., Gu, T.T., Zhang, L.J. et al. (2021). POM-assisted coating of MOF-derived Mo-doped Co_3O_4 nanoparticles on carbon nanotubes for upgrading oxygen evolution reaction. *Chem. Eng. J.* 408: 127352.

92 Yuan, M.L., Dipazir, S., Wang, M. et al. (2019). Polyoxometalate-assisted formation of CoSe/$MoSe_2$ heterostructures with enhanced oxygen evolution activity. *J. Mater. Chem. A* 7 (7): 3317–3326.

93 Huang, Z., Yang, Z.X., Jia, Q.L. et al. (2022). Bimetallic Co-Mo sulfide/carbon composites derived from polyoxometalate encapsulated polydopamine-decorated ZIF nanocubes for efficient hydrogen and oxygen evolution. *Nanoscale* 14 (12): 4726–4739.

94 Wang, H.Q., Xu, Z.F., Zhang, Z.F. et al. (2020). Addressable surface engineering for N-doped WS_2 nanosheet arrays with abundant active sites and the optimal local electronic structure for enhanced hydrogen evolution reaction. *Nanoscale* 12 (44): 22541–22550.

95 Wang, L.X., Wang, L., Meng, X.J., and Xiao, F.S. (2019). New strategies for the preparation of sinter-resistant metal-nanoparticle-based catalysts. *Adv. Mater.* 31 (50): 1901905.

96 Yu, Z.Y., Lin, T., Zhu, C.F. et al. (2021). Design of trimetallic NiMoFe hollow microspheres with polyoxometalate-based metal-organic frameworks for enhanced oxygen evolution reaction. *ChemElectroChem* 8 (7): 1316–1321.

97 Kang, Q.L., Lai, D.W., Su, M.F. et al. (2022). Tailored dodecahedral polyoxometalates nanoframes with in situ encapsulated Co, N, C for oxygen evolution reaction. *Chem. Eng. J.* 430 (4): 133116.

98 Lin, H., Wang, J.W., Guo, X.W. et al. (2018). Phosphorized polyoxometalate-etched iron-hydroxide porous nanotubes for efficient electrocatalytic oxygen evolution. *J. Mater. Chem. A* 6 (47): 24479–24485.

99 Zhang, Y., Guo, H.R., Ren, J.K. et al. (2021). MoO_3 crystal facets modulation by doping heteroatom Fe from polyoxometalate for quasi-industrial oxygen evolution reaction. *Appl. Catal., B* 298: 120582.

8

POM-based Electrocatalysts for Hydrogen Evolution Reaction

Inês S. Marques, Renata Matos, and Diana M. Fernandes

Universidade do Porto, LAQV-REQUIMTE, Departamento de Química e Bioquímica, Faculdade de Ciências, Rua do Campo Alegre, s/n, Porto, 4169-007, Portugal

8.1 Introduction: HER Process

The consumption of fossil fuels has been increasing exponentially in the last decades, resulting from the population growth and the development of new technologies. Despite the global efforts made towards the transition to greener energy, it is still in development stage, in part due to the lack of suitable energy carriers or the difficulty in obtaining these carriers in a sustainable and cost-effective way.

Molecular hydrogen is considered a clean fuel and a real competitor to fossil fuels, due to its high energy density [1, 2]. Also, H_2 has a fundamental role in the actual quest for energy transition and developing a method to produce it in a sustainable way is critical.

The hydrogen evolution reaction (HER) corresponds to the electrochemical process of generation of molecular hydrogen from the reduction of water. It is the cathodic half-reaction of the water-splitting process, which also involves the oxidation of water to molecular O_2 (oxygen evolution reaction, OER). Water electrolysers are able to produce H_2 with high purity, under mild operating conditions, constituting a sustainable alternative to the generation of H_2 from petrochemical resources [1, 3, 4].

In acidic medium, the HER process involves the transfer of two electrons to reduce H^+ ions directly to H_2 molecules (Eq. 8.1). This reaction can occur by two pathways, namely Volmer–Heyrovsky or Volmer–Tafel (Figure 8.1) [5]. In both cases, the first step is the formation of an adsorbed hydrogen atom (H^*) at the electrode surface (Eq. 8.2), followed by the generation of H_2 by the reaction between H^* and H^+ (Heyrovsky step, Eq. 8.3) or the bond of two H^* (Tafel step, Eq. 8.4) [6]. Tafel slope (TS) values around 40 mV dec^{-1} suggest that the Heyrovsky step is the rate-limiting step, while the Tafel step is the limiting one if the values are close to 30 mV dec^{-1}, giving therefore an indication regarding the most probable pathway for H_2 production [7].

These authors have contributed equally to this work.

Applied Polyoxometalate-based Electrocatalysis, First Edition. Edited by Diana M. Fernandes.

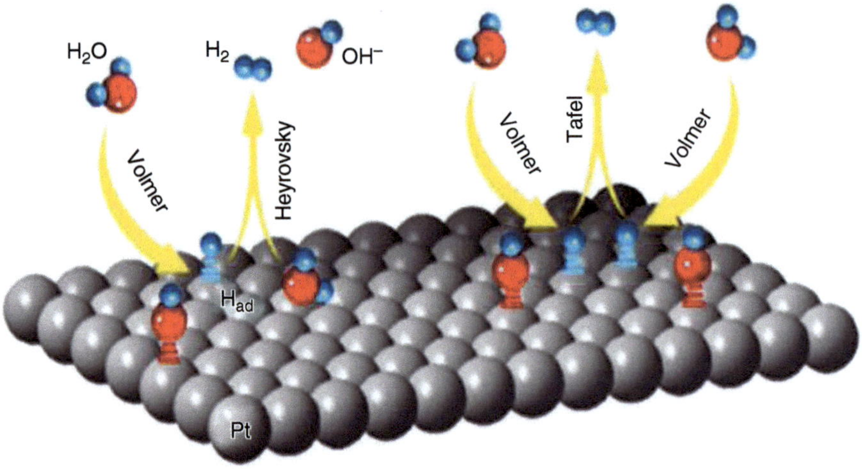

Figure 8.1 Schematic representation of Volmer–Heyrovsky and Volmer–Tafel pathways. Source: Ruqia and Choi [5]/with permission of John Wiley and Sons.

$$2H^+ + 2e^- \rightarrow H_2 \tag{8.1}$$

$$H^+ + e^- + * \rightarrow H^* \tag{8.2}$$

$$H^* + H^+ + e^- \rightarrow H_2 + {}^* \tag{8.3}$$

$$2H^* \rightarrow H_2 + 2^* \tag{8.4}$$

Currently, the application of water electrolysers for H_2 production is severely limited by the large overpotentials and sluggish kinetics associated with HER [2, 6]. Therefore, it is fundamental to use high-performance electrocatalysts in order to overcome these limitations. Platinum (Pt) and other noble-metal-based materials are the state-of-the-art electrocatalysts for HER, showing excellent catalytic activity and good long-term stability [2]. However, their natural scarcity (and consequently high cost) and poor durability hinder their large-scale utilization. On the other hand, cost-effective electrocatalysts such as carbon-based materials [8], transition-metal (TM) compounds (including nanoparticles (NPs), oxides, sulphides, carbides, nitrides, phosphides and selenides) [9], polyoxometalates (POMs) [10] and metal organic frameworks (MOFs) [11] have been successfully applied in this reaction, displaying very promising results. A series of detailed reviews regarding novel electrocatalysts for HER can be consulted elsewhere [12–15].

To evaluate the HER performance, some electrocatalytic parameters are considered, including overpotential (η), TS, current density (j), exchange current density (j_0), Faradaic efficiency and turnover frequency (TOF). Electrochemical active surface area (ECSA) can also be determined, displaying an important role in electrocatalysis. Finally, it is critical to evaluate the long-term stability of the electrocatalyst, by chronoamperometry over an extended period of time (over 10 hours) and/or by performing several (in the order of thousands) consecutive cyclic voltammetry (CVs) or linear sweep voltammetry (LSVs). Nevertheless, most studies only report the overpotential at a specific current density (usually 10 mA cm^{-2}, denoted as η_{10}) and the TS; for this reason, only these parameters will appear on tables during this chapter.

8.2 Pure POMs as HER Electrocatalysts

POMs, which have TM centres of uniform size and well-defined structures, offer the opportunity to extract highly intriguing materials for various applications with protected shells extracted from the functional carbon framework of POMs [16]. POMs are the most promising class of hydrogen evolution electrocatalysts because they share many characteristics with metal oxide semiconductors due to their construction from TM oxygen anion clusters with a d^0 electronic structure [17].

The electrocatalytic activity for HER of a POM (α-$[SiW_{12}O_{40}]^{4-}$) was first reported in 1985 by Keita and Nadjo [18, 19]. However, it was later discovered that the high catalytic performance reported at the time was also a result of the co-deposition of amounts of Pt on the working electrode, which resulted from the anodic dissolution of the Pt counter electrode [20]. Rong and Pope, in 1992, also noted that a Ru-substituted POM ($[PW_{11}Ru^{III}(H_2O)O_{39}]^{4-}$) promoted hydrogen evolution in an acidic aqueous solution [21]. In 2010, Biboum et al. [22] reported POMs-stabilized palladium (Pd^0@POMs) using the redox reaction and self-assembly behaviour of $K_9(H_4V^{IV}PW_{11}O_{62})$ (abbreviated as HPV^{IV}) and K_2PdC_{14}. The electrocatalysts exhibit effective HER performance in an acidic with TS of 52–83 mV dec^{-1}. Also, in 2011, the same group [23] prepared a double-cuboid-shaped copper(II)-containing polyoxo-22-palladate(II), $[Cu^{II}{}_2Pd^{II}{}_{22}P^{V}{}_{12}O_{60}(OH)_8]^{20-}$ using a simple, one-pot procedure in aqueous solution and mild reaction conditions. The $Na_{20}[Cu^{II}{}_2Pd^{II}{}_{22}P^{V}{}_{12}O_{60}(OH)_8]\cdot 58H_2O$ (labelled as Na-$Pd_{22}Cu_2P_{12}$) presented a TS of 65 mV dec^{-1} in neutral electrolyte.

More recently, in 2022, Wang et al. [24] developed a new crown ether-polyoxometalate assembly $[K(H_{24}C_{12}O_6)]_3[K(H_{24}C_{12}O_6)(CH_3CN)][SiMo_{12}O_{40}]$ based on the Keggin-type polyanion $[SiMo_{12}O_{40}]^{4-}$ by a one-pot strategy in an acetonitrile solution. The Keggin polyanion is connected to two [K([18]-crown-6]$^+$ supramolecular cations to form a spindle-shaped discrete structure. This POM was used to modify a carbon paper (CP) and was applied as HER electrocatalyst in acid medium (0.5 M H_2SO_4) with η_{10} = 120 mV (Figure 8.2a) and TS of 112 mV dec^{-1} (Figure 8.2b). After 2000 cycles, the electrocatalyst also demonstrated long-term stability (Figure 8.2c).

8.3 Composite/Hybrid Materials

8.3.1 Carbon/POM

POM immobilization onto carbon-based materials has recently attracted a lot of interest in several fields of research including in clean energy technologies

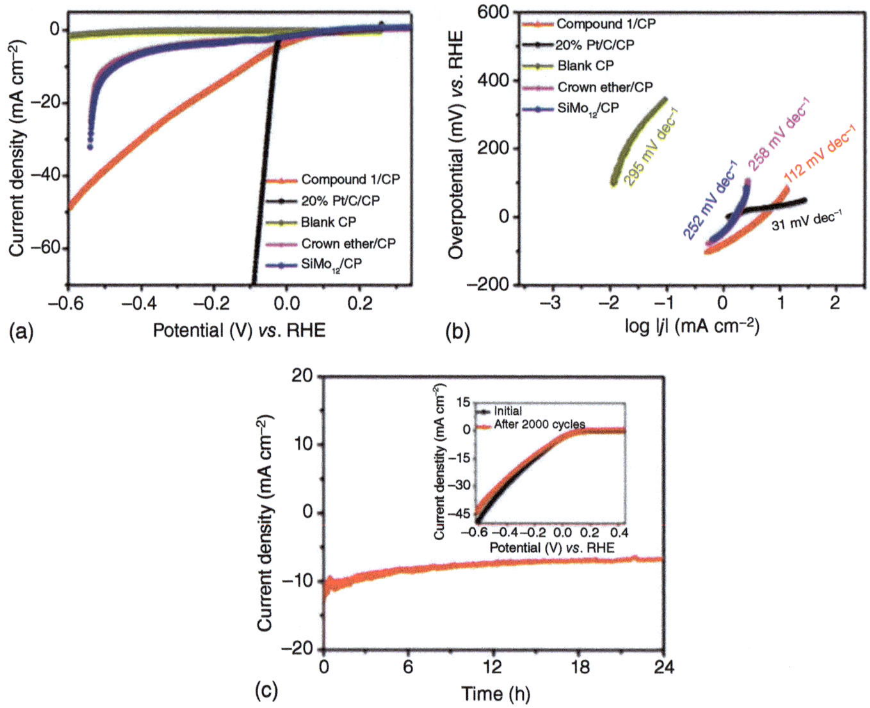

Figure 8.2 (a) LSV curves of complex 1/CP and other control samples in 0.5 M H_2SO_4; (b) The corresponding Tafel curves; (c) I–t curve at a fixed overpotential of 130 mV for 24 accelerate degradation test for 2000 CV cycles (0 to −0.2 V *vs.* RHE). Source: Wang et al. [24]/with permission of Elsevier.

involving HER electrocatalysts. Due to the increase in active centres and consequent improvement in electrocatalytic activity, these hybrids have advantageous electrocatalytic properties, such as high surface area which expose more catalytic active sites, high electrical conductivity and a reduced electrical resistance at the support–catalyst interface [25, 26]. Table 8.1 summarizes the HER activities of the present POM/carbon composites.

Keita et al. [34], in 2007, developed HER electrocatalysts based on $[H_7P_8W_{48}O_{184}]^{33-}$ (or P_8W_{48}), $[(Co_3(B\text{-}\beta\text{-}SiW_9O_{33}(OH))(B\text{-}\beta\text{-}SiW_8O_{29}OH)_2)_2]^{22-}$ (or Co_{15}) and $[Co_6(H_2O)_{30}(Co_9Cl_2(OH)_3(H_2O)_9(\beta\text{-}SiW_8O_{31})_3)]^{5-}$ (or Co_6). The selected hybrid was absorbed on the electrode (glassy carbon electrode, GCE) surface or was directly entrapped in polyvinylpyridine film on the electrode. By using CV, the authors found that the hydrogen-evolution-onset potential gradually shifted to more positive potentials as a function of time and that current density significantly increased. It was proposed that the proton and electron reservoir-like behaviours of those molecular oxides were responsible for this electrode activation for HER.

A nanocomposite multilayer based on P_5W_{30} ($[NaP_5W_{30}O_{110}]^{14-}$) and reduced graphene oxide (rGO) was fabricated through a two-step method that included

Table 8.1 HER performance of POM/carbon composites.

Catalyst	Electrolyte	Tafel slope (mV dec^{-1})	η_{10} (mV)	References
P_8W_{48}/rGO	0.5 M H_2SO_4	38	28	[27]
PEI/P_5W_{30}-rGO	0.5 M H_2SO_4	—	—	[28]
$[PW_{11}Pt\text{-}NiO_{39}]^{5-}$@PDDA-CNT	0.5 M H_2SO_4	164.5	260–270	[29]
$[PW_{11}Pt\text{-}NiO_{39}]^{5-}$@PDDA-rGo	0.5 M H_2SO_4	152.8	260–270	[29]
$[PW_{11}CuO_{39}]^{5-}$@Ru-rGO	0.5 M H_2SO_4	36.56	41	[30]
$[PW_{11}CoO_{39}]^{5-}$@Ru-rGO	0.5 M H_2SO_4	42.60	32	[30]
$[PW_{11}NiO_{39}]^{5-}$@Ru-rGO	0.5 M H_2SO_4	40.42	30	[30]
P_2W_{18}@rGF_ox	0.5 M H_2SO_4	37	35	[31]
P_5W_{30}@rGF_ox	0.5 M H_2SO_4	33	33	[31]
P_8W_{48}@rGF_ox	0.5 M H_2SO_4	41	44	[31]
PtW_6O_{24}/C	0.5 M H_2SO_4	29.8	22	[32]
$Pt_2(W_5O_{18})_2$/C	0.5 M H_2SO_4	29.5	26	[32]
$\{W_{17}Ni\}$CNT	0.5 M H_2SO_4	—	—	[33]
$\{W_{30}Ni_4\}$CNT	0.5 M H_2SO_4	—	—	[33]
$\{W_{30}Ni_2\}$CNT	0.5 M H_2SO_4	—	—	[33]
$[PW_{11}CuO_{39}]^{5-}$@Ru-rGO	2 M KOH	107.95	165	[30]
$[PW_{11}CoO_{39}]^{5-}$@Ru-rGO	2 M KOH	59.76	155	[30]
$[PW_{11}NiO_{39}]^{5-}$@Ru-rGO	2 M KOH	68.96	158	[30]

layer-by-layer (LbL) self-assembly using protonated poly(ethylamine) and postphotoreduction by Jiang et al. in 2014 [28]. The resultant nanocomposite acted as electrocatalysts for HER (and oxygen reduction reaction (ORR)) in an acidic solution (0.5 M H_2SO_4) and showed a rapid increase in the cathodic peak current at $E = -0.6$ V (*vs.* Ag/AgCl). According to this author, HER electrocatalysis is in part due to the rGO in multilayer.

In 2015, it was also reported the one-step electrochemical reduction synthesis of cyclic 48-tungsto-8-phosphate (P_8W_{48}) on rGO sheets. These composites were created by reacting P_8W_{48}'s 16-electron reduced form with GO in a pH 1.0 sulphate solution [27]. The electrons held in P_8W_{48} (16 e^-) are easily transferred to GO during this reduction process, resulting in the reduction of GO into rGO and the recovery of P_8W_{48}. The developed nanocomposite was applied as HER electrocatalyst in acid medium (0.5 M H_2SO_4) with η_{10} = 28 mV, η_{50} = 70 mV and a TS of 38 mV dec^{-1}. Also, after 1000 continuous cycles, only a slight decrease in the current density was observed. The electrocatalyst performance was attributed to a combination of factors and what the authors refer to as 'microenvironment effect', which significantly strengthens the advantageous synergistic effect of P_8W_{48} and rGO. This impact is significant because it promotes the proton and electron buildup of the P_8W_{48} framework, which increases the HER.

In 2016, Ensafi et al. [29] prepared a new composite based on Pt-modified $[PW_{11}NiO_{39}]^{5-}$ on rGO or multiwall carbon nanotubes (CNTs) by electrodeposition. First, polydiallydimethylammonium (PDDA) modified rGO (PDDA-rGO) and PDDA-CNT were prepared, the surfaces of which were then decorated with POM $[PW_{11}NiO_{39}]^{5-}$. Finally, $[PW_{11}NiO_{39}]^{5-}$@PDDA-rGO and $[PW_{11}NiO_{39}]^{5-}$@PDDA-CNT were decorated with platinum NPs to fabricate $[PW_{11}Pt\text{-}NiO_{39}]^{5-}$@PDDA-rGO and $[PW_{11}Pt\text{-}NiO_{39}]^{5-}$@PDDA-CNT, respectively. The prepared composites showed remarkable HER activity in acidic electrolyte (0.5 M H_2SO_4) with overpotentials varying between 0.55–0.70 V for $[PW_{11}NiO_{39}]^{5-}$@PDDA-rGO and $[PW_{11}NiO_{39}]^{5-}$@PDDA-CNT and 0.26–0.27 V for $[PW_{11}Pt\text{-}NiO_{39}]^{5-}$@PDDA-rGO and $[PW_{11}Pt\text{-}NiO_{39}]^{5-}$@PDDA-CNT. The TSs were 5.54 and 3.98 V dec^{-1} for $[PW_{11}NiO_{39}]^{5-}$@PDDA-CNT and $[PW_{11}NiO_{39}]^{5-}$@PDDA-rGO, respectively, and, 16.45 and 15.28 V dec^{-1} for $[PW_{11}Pt\text{-}NiO_{39}]^{5-}$@PDDA-CNT and $[PW_{11}Pt\text{-}NiO_{39}]^{5-}$@PDDA-rGO, respectively. The better HER performance for $[PW_{11}Pt\text{-}NiO_{39}]^{5-}$@PDDA-CNT and $[PW_{11}Pt\text{-}NiO_{39}]^{5-}$@PDDA-rGO was attributed to the nanostructure morphology and the sub-monolayer nature of Pt coated on the surface of the other two materials.

Later, the same group developed three new nanocomposites by depositing $[PW_{11}MO_{39}]^{5-}$ on the surface of Ru-reduced graphene oxide (Ru-rGO) but this time with different earth-abundant TMs (M: Co, Ni and Cu) [30]. The three as-prepared $[PW_{11}MO_{39}]^{5-}$@Ru-rGO were applied as HER electrocatalysts in both acidic (0.5 M H_2SO_4) and alkaline (2 M KOH) solutions. In acidic electrolyte, the η_{10} was 41, 32 and 30 mV and value of TS was 36.56, 42.60 and 40.42 mV dec^{-1} for $[PW_{11}CuO_{39}]^{5-}$@Ru-rGO, $[PW_{11}CoO_{39}]^{5-}$@Ru-rGO and $[PW_{11}NiO_{39}]^{5-}$@Ru-rGO, respectively. The same assessments were made for alkaline media and $[PW_{11}CuO_{39}]^{5-}$@Ru-rGO, $[PW_{11}CoO_{39}]^{5-}$@Ru-rGO and $[PW_{11}NiO_{39}]^{5-}$@Ru-rGO exhibited TSs of 107.95, 59.76 and 68.96 mV dec^{-1} and overpotentials of 165, 155 and 158 mV at a current density of 10 mA cm^{-2}, respectively. Different catalytic behaviours in acidic and alkaline solutions were shown by the POMs ability towards HER. In contrast to the catalytic behaviour of POM, which predominated in an acidic solution, the TM's catalytic behaviour predominated in an alkaline solution.

In 2018, Fernandes et al. [31] reported a new study in which three new nanocomposites based on reduced oxidized graphene flakes (rGF_ox) and three different POMs (P_2W_{18}, P_5W_{30} and P_8W_{48}) were produced by one-step electroreduction synthesis. The main point was to study the shape/size/composition/charge and activity relationships of the three different tungstophosphates on rGF_ox. The HER studies were conducted in an acidic solution (0.5 M H_2SO_4) and the nanocomposites showed similar overpotentials for a current density of 10 mA cm^{-2} (Figure 8.3a) and TSs (Figure 8.3b). The electrocatalyst with P_8W_{18} reached an overpotential of 44 mV and P_2W_{48} and P_5W_{30} an overpotential of 35 and 33 mV, respectively. The TSs obtained were 41, 37 and 33 mV dec^{-1} for P_8W_{48}, P_2W_{48} and P_5W_{30}, respectively. The nanocomposite P_5W_{30}@rGF_ox showed the best HER performance followed by P_2W_{18}@rGF_ox and P_8W_{48}@rGF_ox. P_5W_{30}@rGF_ox also presented a long-term electrochemical stability and reproducibility for up to 1000 potential sweeps. These

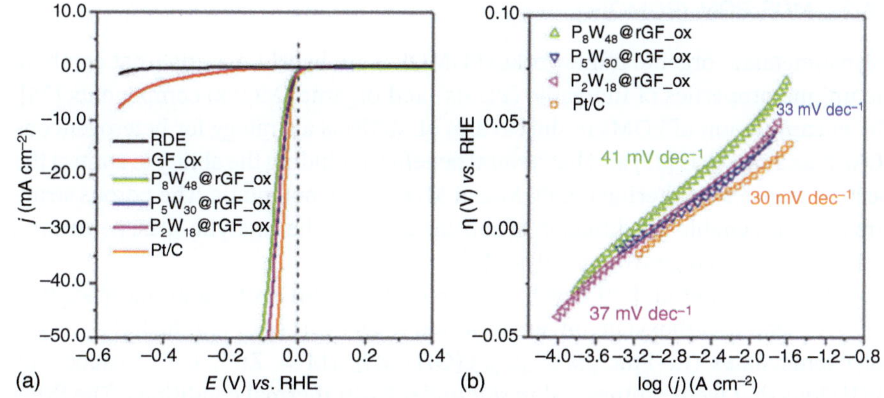

Figure 8.3 (a) Linear polarization curves for bare RDE and the different electrocatalysts and (b) the corresponding Tafel plots in 0.5 M H_2SO_4 at a $v = 0.005\ V\,s^{-1}$. Source: Fernandes et al. [31]/with permission of John Wiley and Sons.

promising results were attributed to the good stability of P_5W_{30} in highly acidic media (pH = 0.3) together with a synergistic effect between POM and rGF_ox, which act as a suitable and conductive support for the uniform distribution and immobilization of individual polyanions.

Later, in 2020, Yu et al. [32] synthesized two well-defined POMs containing platinum. The state-of-the-art electrocatalyst for HER is platinum, and some studies reveal that when Pt is oxidized, the HER activity increased compared with Pt^0 metal. The synthesis of the new hybrids was made by dispersing the POMs (PtW_6O_{24} or $Pt_2(W_5O_{18})_2$) in water with Ketjen black carbon and melamine-formaldehyde. The electrocatalytic performance was evaluated in 0.5 M H_2SO_4 and the electrochemical experiments showed that PtW_6O_{24}/C and $Pt_2(W_5O_{18})_2/C$ reached the overpotential of 22 and 26 mV at a current density of 10 mA cm^{-2} and TSs of 29.8 and 29.5 mV dec^{-1}, respectively. According to the authors, Pt—O bound should be the HER active site since Pt primarily functions as an electron-obtaining centre and O as a proton-adsorption centre. The extended Pt—O site speeds up the coupling of electron and proton when extra electrons and protons are added during an electrochemical reaction, which causes a quick release of H_2 from the Pt—O bond. Pt—O can therefore be used as a new active site for HER. This research clarifies the crucial function that O atoms play in the oxidized platinum-based electrocatalytic HER, which may soon lead to new insights that will help in the development of more effective Pt-based electrocatalysts.

Jawale et al. [33], in 2022, developed a nanohybrid (POM@CNT) *via* LbL approach. This method fully preserves the integrity of the carbonaceous support and allows for dense, homogeneous and stable anchoring of the active components on the surface of the CNTs. The resulting nanohybrid was applied as HER electrocatalysts in an acidic medium (0.5 M H_2SO_4) and the catalytic efficiency followed the order: $\{W_{17}Ni\}CNT < \{W_{30}Ni_4\}CNT < \{W_{30}Ni_2\}CNT$ with j_{max} of 5.09, 2.64 and 2.21 mA cm^{-2}, respectively.

8.3.2 MOF/POM (POMOFs)

Polyoxometalate-organic frameworks (POMOFs) are hybrid material that combine the unique properties of inorganic (POMs) and organic (MOFs) components [35]. The encapsulation of POMs in the cavities of MOFs is a strategy for heterogeneous POM-based catalysis [35] and has several benefits, including the ability to control the electron transfer by altering the POM and MOF redox potentials. The porous structure may also enable the diffusion of substances from the catalytic sites, permitting their reuse as heterogeneous catalysts [36].

In 2011, Nohra et al. [37] developed three new POMOFs by grafting triangular 1,3,5-benzene tricarboxylate linkers (referred to as trim) onto tetrahedral ε-Keggin polyoxometalates ($\{(\varepsilon\text{-Pmo}^{V}{}_{8}\text{Mo}^{VI}{}_{4}\text{O}_{40-x}(\text{OH})_{x}\text{M}_{4}\}$, (M = Zn^{II}, La^{III})) capped by Zn(II) ions that were synthesized *in situ* under hydrothermal conditions. The three differ in the following ways (Figure 8.4): $(\text{TBA})_3[\text{PmoV}_8\text{Mo}^{VI}{}_4\text{O}_{36}(\text{OH})_4\text{Zn}_4][\text{C}_6\text{H}_3(\text{COO})_3]_{4/3}\cdot 6\text{H}_2\text{O}$ **(ε(trim)$_{4/3}$)** which is a 3D open framework made of trim linkers, with channels occupied by tetrabutylammonium (TBA). In $(\text{TBA})_3[\text{PmoV}_8\text{Mo}^{VI}{}_4\text{O}_{37}(\text{OH})_3\text{Zn}_4][\text{C}_6\text{H}_3(\text{COO})_3]$ **(ε$_2$(trim)$_2$)**, a dimerized version of this POM serves as

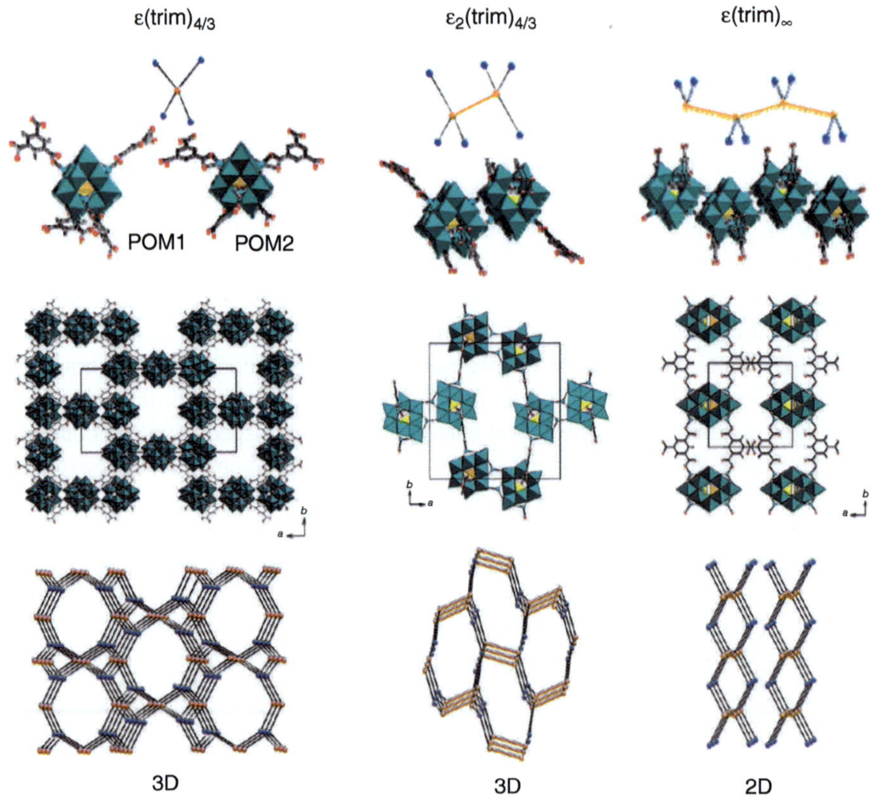

Figure 8.4 POM building blocks of ε(trim)$_{4/3}$, ε(trim)$_2$ and ε(trim)$_\infty$, a view of their unit cell and their schematic representation. Source: Nohra et al. [37]/with permission of American Chemical Society.

the building block rather than the molecule Keggin. Through trim linkers, they are connected to create a 3D structure with TBA cations filling the channels. In $(TBA)_3[PMo^{V}_8Mo^{VI}_4O_{37}(OH)_3Zn_4][C_6H_3(COO)_3]_3{\cdot}8H_2O$ (**[ε(trim)]**$_\infty$**)**, organic linkers join zigzag chains to create 2D grids. The TOF value for HER in 1 M $H_2SO_4$was $6.7\,s^{-1}$ at $\eta = 0.20\,V$. The POMOF structure and confinement effects, according to the authors, were responsible for this reliable performance.

Qin et al. [38] have also used the ε-Keggin polyoxometalate and TBA to develop two new POMOFs – NEMU-501 $[TBA]_3[\varepsilon\text{-}PMo_8{}^{V}Mo_4{}^{VI}O_{36}(OH)_4Zn_4][BPT]$ and NEMU-500 $[TBA]_3[\varepsilon\text{-}PMo_8{}^{V}Mo_4{}^{VI}O_{36}(OH)_4Zn_4][BTB]_{4/3}{\cdot}x$, where BPT = [1,1′-biphenyl]-3,4′,5-tricarboxylat and BTB = benzene tribenzoate. The NEMU-500 needed 237 mV to reach a current density of $10\,mA\,cm^{-2}$ while NEMU-501 needed 392 mV, respectively, in 0.5 M H_2SO_4 electrolyte. The TSs were 96 and 137 $mV\,dec^{-1}$ for NEMU-500 and NEMU-501, respectively. The two POMOFs maintained their electrocatalytic activity after 2000 cycles. Due to its good stability, porosity and exposed active sites, NENU-500 was considered the electrocatalyst with the best HER performance.

Later, in 2021, Shah et al. [39] performed a comparative study between two different MOFs (**(a)** MIL-100(Fe) and **(b)** UIO-67(Zr)) and three different nickel POMs (**(1)** $K_6[NiSiW_{11}O_{39}]{\cdot}15H_2O$, **(2)** $K_5[NiPW_{11}O_{39}]{\cdot}15H_2O$ and **(3)** $K_6[NiGeW_{11}O_{39}]{\cdot}15H_2O$). The authors chose these Ni-POMs because they are comparable in size to the cavities of the MOFs, and they are also stable under acidic conditions. MIL-100 (Fe) and UIO-67, on the other hand, were chosen because of their remarkable stability and enormous surface area in terms of cavities to support considerable POM concentration. The combination resulted in six different POMOFs (1a, 1b, 2a, 2b, 3a and 3b), and their activity was evaluated in an acidic medium (pH = 4), in an acetate buffer solution (50 mM). Between 1a and 1b, for the POMOF 1b the HER started at lower onset potential (0.72 V). The POMOF 2a presented an overpotential of 322 mV, while 2b presented an overpotential of 420 mV and TS of 65 and 110 $mV\,dec^{-1}$, respectively. The POMOF 3a and 3b presented overpotentials of 325 and 350 mV and TSs of 68 and 92 $mV\,dec^{-1}$, respectively. POMs act as a sink for the excited electrons of MOFs, which are effectively dissipated into them, facilitating the excitation of MOFs and decreasing the electron–hole recombination process within the MOF unit. While the binding energies of W are reduced in cases 1, 2 and 3, there is a notable increase in the binding energies of Fe and Zr in cases a and b, respectively.

$[Ni(phen)H_2O]{\cdot}V_2O_6$, designated as POMOF(V,Ni), were successfully prepared by a simple one-pot hydrothermal reaction by adjusting different solvent ratios (H_2O/EtOH). Liu et al. [40] mixed 1.0 mmol of NH_4VO_3, $Ni(CH_3COO)_2{\cdot}4H_2O$ and phenanthroline, and dissolved in three different solvent ratios (solvent 1 – 60 ml of H_2O, solvent 2 – 30 ml of H_2O and 30 ml of EtOH and solvent 3 – 60 ml of EtOH). The final compounds were labelled as POMOF(V,Ni)_1, POMOF(V,Ni)_2 and POMOF(V,Ni)_3, accordingly, based on the solvent number indicated above. The morphology of the different composites is different depending on the solvent. When synthesized in pure water, the microstructures formed are NPs, while those formed in ethanol are nanosheets (Figure 8.5), which led to different electrocatalytic

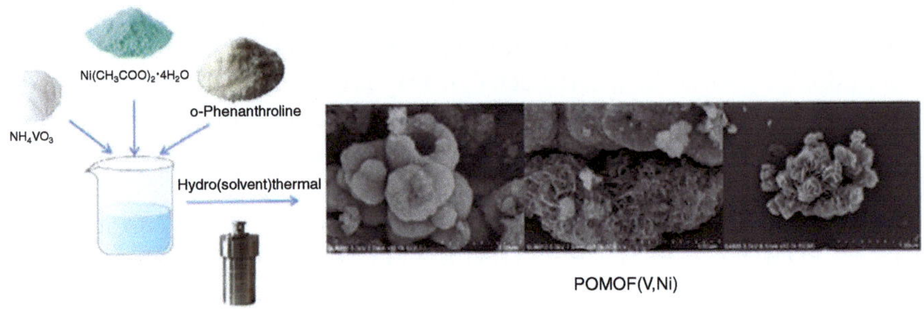

Figure 8.5 POMOF(V,Ni) composite preparation scheme. Source: Reproduced with permission of Liu et al. [40]. © 2022/Springer Nature.

results. The POMOF(V,Ni)_3 showed high catalytic performance in 1 M KOH, with an overpotential of 159 mV and a TS of 271 mV dec^{-1}. The authors attributed the better performance of POMOF(V,Ni)_3 to the fact that microsphere nanosheets can provide more exposed electroactive sites than the NPs.

8.3.3 Transition-metal/POM Composites

The combination of the electrochemical properties of metals (precious or non-precious) and POMs revealed to be extremely promising for designing new efficient HER electrocatalysts [10]. Table 8.2 provides a summary of the electrocatalytic HER activities of POM/metal composites.

Zhang et al. [20] prepared several POM-based electrodes modified with low amounts of Pt through co-deposition of the POM and Pt on a GCE, with the Pt being obtained from the anodic dissolution of the counter electrode. According to

Table 8.2 HER performance of POM/metal composites.

Catalyst	Electrolyte	Tafel slope (mV dec^{-1})	η_{10} (mV)	References
POM-$ZnFe_2O_4$	1.0 M KOH	191.6	225	[41]
NiCo-POM/Ni	1.0 M KOH	75	64	[42]
Co/Ni-POM/NF	1.0 M KOH	60.8	68	[43]
Ni-WO/NF	1.0 M KOH	86	75	[44]
POM@ZnCoS/NF	1.0 M KOH	53.5	170	[45]
$H_3[(Cp^*Rh)_4PMo_8O_{32}]$@NF	0.1 M KOH	110	190	[46]
$H_5[Na_2(Cp^*Ir)_4PMo_8O_{34}]$@NF	0.1 M KOH	84	179	[46]
Rh/MoO_2@NF	0.1 M KOH	63	125	[46]
Ir/MoO_2@NF	0.1 M KOH	42	70	[46]
$NiMo_6O_{24}$@Cu/TNA	0.5 M H_2SO_4	89.2	215	[47]
Pt-STA-CB	0.5 M H_2SO_4	27.9	33.8	[48]
PDDA:Fe_3O_4/$P5W_{30}$	0.1 M H_2SO_4	585	—	[49]

ICP-MS results, the Pt loadings were around 50–100 ng cm^{-2} by tuning the POM concentration (1–10 mM) and electrolysis time (2–18 hours). The presence of POMs was found to be determining for HER activity in acidic electrolyte, as the catalytic performance increased with increasing POM loading. Furthermore, the saturated lacunary POMs $SiW_{12}O_{40}{}^{4-}$ and $P_2W_{18}O_{62}{}^{6-}$ showed less promising results than the unsaturated SiW_{11}. Indeed, an electrode containing 10 mM of SiW_{11} displayed similar results to commercial Pt electrodes. Also using noble metals, Yuan et al. [50] reported gold nanocrystals grown on phosphotungstic acid (PTA, $PW_{12}O_{40}{}^{3-}$) with adequate HER activity in acidic medium. Here, hexagonal Au NPs with diameters of 30–50 nm were prepared in the presence of the Keggin ions, which acted as stabilizer for growth of the nanocrystals, *via* a one-pot synthesis route.

As an alternative to precious metals, TMs, such as Fe, Zn, Co and Ni, have been combined with POMs under the form of oxides [41, 49] or sulphides [45] in order to enhance the overall HER activity and stability. It is worth to mention that these metals display important roles in electrocatalysis due to their variable oxidation states, while their natural abundance, high specific capacity, low toxicity and cost effectiveness are extremely advantageous to use in energy-related applications [51, 52].

Gautam et al. [41] prepared a bifunctional electrocatalyst for HER and OER consisting of a heterostructure of a Dawson-type POM nanoplates ($(NH_4)_6P_2Mo_{18}O_{62}$) and $ZnFe_2O_4$ (denoted POM-$ZnFe_2O_4$) through a two-step hydrothermal method. The electrocatalyst obtained a comparable HER performance to commercial Pt/C in alkaline electrolyte (η of 268 mV *vs.* 213 at 20 mA cm^{-2} and 356 mV *vs.* 316 mV at 50 mA cm^{-2} and TS of 191.6 mV dec^{-1} *vs.* 159.0 mV dec^{-1}) and excellent stability over 20 hours. The great potential of this material was mainly attributed to the formation of POM/zinc-iron-oxide nanoplates heterostructures, which created additional active sites and enhanced the electrical conductivity. Besides, its hierarchical porosity also increased the number of active sites and facilitated the diffusion of reactants and products. At last, the high concentration of Fe^{3+} on the catalyst surface was also shown to favour the occurrence of HER.

Deposition of POMs on electroactive metal foam substrates is another popular route to obtain improved HER electrocatalysts. In fact, Ni foams are well established in this field due to their outstanding porosity and electrical conductivity [46]. In 2017, Singh et al. [46] immobilized organometallic POM-based clusters ($H_3[(Cp^*Rh)_4PMo_8O_{32}]$ and $H_5[Na_2(Cp^*Ir)_4PMo_8O_{34}]$) onto a Ni foam through a hydrothermal route, reporting some HER activity in alkaline medium: η_{10} of 190 and 179 mV and TSs of 110 and 84 mV dec^{-1}, respectively. However, both electrocatalysts performance improved after 48 hours of electrolysis, displaying considerably lower η_{10} (125 and 70 mV, respectively) and TSs (63 and 42 mV dec^{-1}, respectively). As Pt anodic dissolution was excluded by using a graphite counter electrode, the authors explained the improvement by the POMs reduction to Rh/MoO_2 and Ir/MnO_2, which revealed to be more catalytical active towards HER. Indeed, Ir/MnO_2 showed very close results compared to Pt/C, also deposited on Ni foam (η_{10} of 50 mV and TS of 31 mV dec^{-1}).

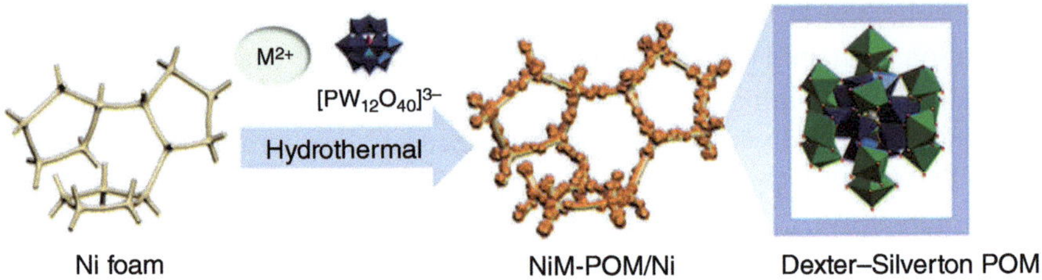

Figure 8.6 Synthesis procedure of NiM-POM/Ni (M = Co, Zn or Mn) composites. Source: Jia et al. [42]/John Wiley & Sons/CC BY 4.0.

Later, Jia et al. [42] reported several Dexter–Silverton POMs anchor on Ni foam electrodes (NiCo-POM/Ni, NiZn-POM/Ni and NiMn-POM/Ni) with excellent activity towards HER in alkaline electrolyte. The composites were prepared using Keggin-type POM clusters ($[PW_{12}O_{40}]^{3-}$) as starting material *via* a hydrothermal route in presence of commercial Ni foam and the selected TM (Ni, Co, Mn, Zn) in the form of nitrate salt (Figure 8.6). From previous studies, the authors knew that Ni^{II} ions could be incorporated into the POM lattice after their released from the Ni foam during the hydrothermal synthesis. The most promising electrocatalyst, NiCo-POM/Ni, achieved an η_{10} of 64 mV, a TS of 75 mV dec^{-1} and a faradaic efficiency around 97% for the generation of H_2. Furthermore, the composite showed outstanding stability. The authors justified these results, comparable with the ones displayed by Pt/C, with a strong linkage between both components of the composites and a high concentration of accessible active sites for HER.

A similar approach was used by Zhang et al. [43] a few years later to prepare HER electrocatalysts. A Co/Ni-POM composite with a 3D nanoflower morphology (Figure 8.7a) was prepared onto a commercial Ni foam substrate *via* a hydrothermal procedure at 180 °C. The authors studied the effect of different hydrothermal treatment times on the formation of the 3D hierarchical nanoflowers. In order to do that, hydrothermal treatments were performed for 1, 3, 6, 8 and 10 hours. The simulated Co/Ni-POM nanoflowers growth process is shown in Figure 8.7b. The results showed that the Co/Ni-POM primary grows into columns and then some

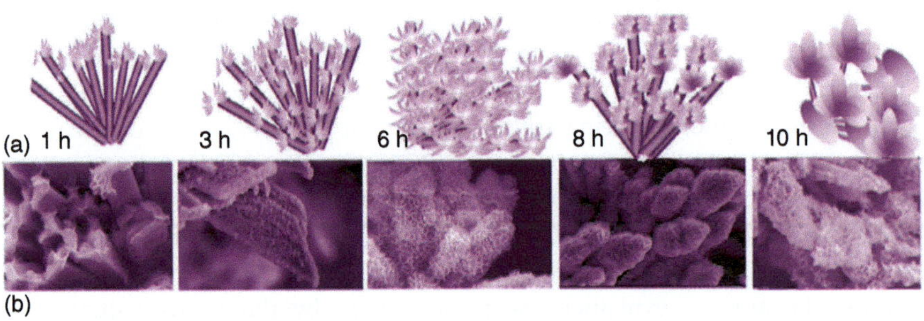

Figure 8.7 (a) Simulated growth of Co/Ni-POM nanoflowers and (b) evolution of its morphology over time by SEM. Source: Reproduced with permission of Zhang et al. [43]/Royal Society of Chemistry.

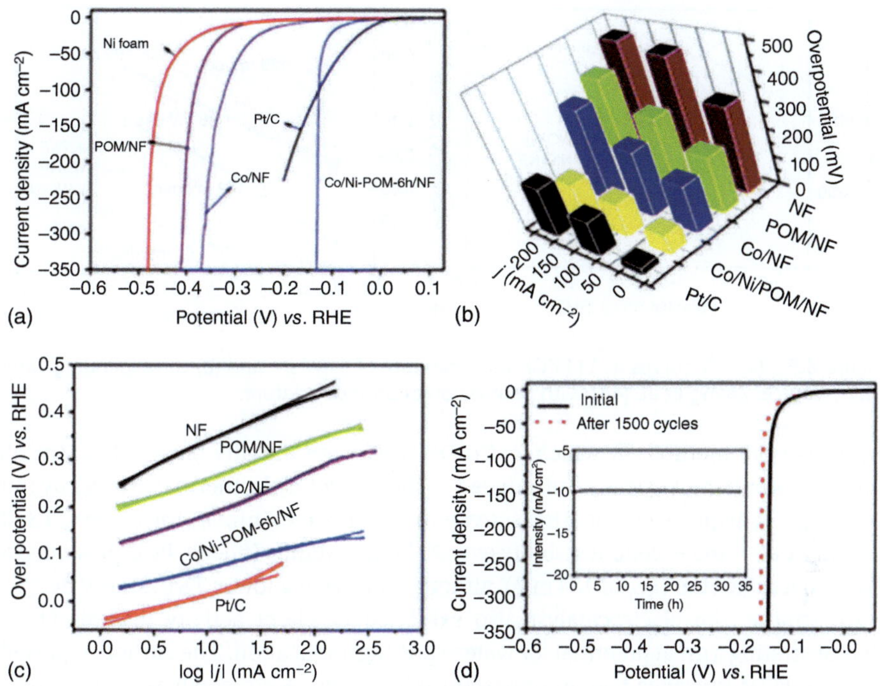

Figure 8.8 (a) Polarization curves of NF, Co/NF, POM/NF, Co/Ni-POM-6h/NF and Pt/C, (b) overpotentials required at 10, 100 and 200 mA cm^{-2} current densities, (c) Tafel plots and (d) polarization curves of Co/Ni-POM-6h/NF before and after 1500 cycles in 1 M KOH at a scan rate of 5 mV s^{-1}. Source: Zhang et al. [43]/with permission of Royal Society of Chemistry.

3D nanoflowers progressively produced on the columns over 6 hours. However, for a reaction time over 8 hours, the flowers gradually became too dense leading to a collapse of the whole structure.

The resulting material after 6 hours of reaction, Co/Ni-POM-6h/NF, showed an η_{10} of 68 mV, TS of 60.8 mV dec^{-1} and great stability over 1000 cycles in alkaline medium (Figure 8.8). Comparatively, the remaining materials with non-optimized nanoflower morphologies (Co/Ni-POM-1h/NF, Co/Ni-POM-3h/NF, Co/Ni-POM-8h/NF and Co/Ni-POM-10h/NF) achieved η_{10} of 255, 203, 163 and 177 mV and TSs of 132, 92.5, 85 and 89 mV dec^{-1}, respectively. Moreover, the electrocatalytic behaviour of Co/Ni-POM-6h/NF was also compared with the bare Ni foam, its individual components (POM/NF and Co/NF) and commercial Pt/C (Figure 8.8), demonstrating the advantage of simultaneously combining Co, Ni and POM for enhanced HER performance. Indeed, the authors attributed Co/Ni-POM-6h/NF promising HER activity to the occurrence of a synergistic effect between the mixed TMs present in the composite, its high surface area, the large number of active sites and an outstanding electron transfer over the Ni foam.

Gautam et al. [45] anchored PW_{12} NPs on ZnCoS nanowires supported over Ni foam (POM@ZnCoS/NF). The nanohybrid was prepared through a simple two-step

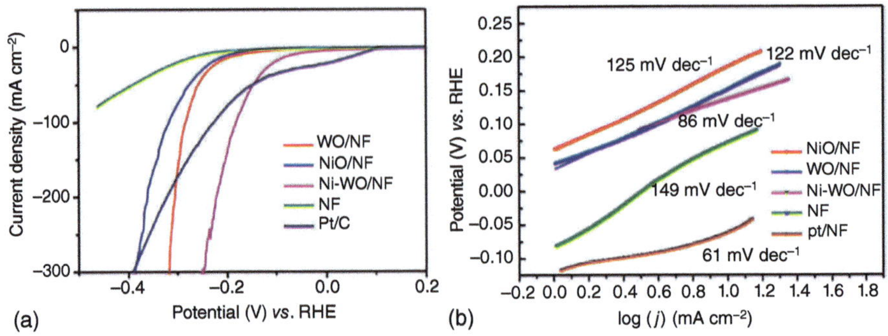

Figure 8.9 (a) LSV curves in 1 M KOH at a scan rate of 5 mV s^{-1} and (b) corresponding Tafel plots. Source: Zhang et al. [44]/with permission of Springer Nature.

hydrothermal method, in which ZnCo DHNWs were initially obtained as ZnCoS precursors and the POM was introduced in the second hydrothermal process by ion exchange. It displayed great HER activity in alkaline medium: η_{10} of 170 mV, TS of 53.5 mV dec^{-1} and excellent stability over 20 hours cycling. Indeed, Pt/C presented a larger overpotential (η_{10} of 184 mV), although showing a lower TS (29.4 mV dec^{-1}). Furthermore, the electrocatalyst also exhibited excellent activity towards OER, demonstrating a great potential for water-splitting application. The authors reported that the POM presence had an important role for the electrocatalytic performance by modulating the overall structural and electronic properties. Besides, it enhanced the ECSA of the ZnCoS nanowires and increased the number for HER active sites. As so, the HER performance was mostly explained by a synergistic effect between the POM NPs and ZnCoS nanowires.

Lastly, Zhang et al. [44] reported in 2023 the HER activity of Ni-modified PW_{12} clusters integrated *in situ* on Ni foam. The material, denoted Ni-WO/NF, was prepared *via* a hydrothermal route using Ni modified $H_3PW_{12}O_{40}$ as starting material and exhibited a hierarchical structure and a flower-like morphology by carefully adjusting the volume ratio of the solvents used (water and ethanol): the desired structure, with enhanced electrocatalytic behaviour, was obtained for a volume ratio of water:ethanol of 1 : 1. In alkaline medium, it achieved η_{10} of 75 mV and TS of 86 mV dec^{-1}, which are results comparable to Pt/C (Figure 8.9). This performance was explained by a combination of factors: (i) the catalyst unique structure, accompanied by its large specific surface area and fast charge-transfer properties, (ii) a large number of active sites and (iii) a synergistic effect between all the metals present in the material.

8.3.4 Polymer/POM

Lately, the HER activity of some POM–organic frameworks has been reported [53, 54]. In these works, POMs were combined with organic polymers in order to optimize the former properties, including their crystalline structure [53].

In 2020, Huang et al. [53] reported the HER performances of Lindqvist-type POMs functionalized with a flexible or rigid organic ligand (succinyl and phthalyl, respectively) and cyclopropylamine as terminal ligand. The authors concluded that the material with rigid ligand showed the greatest potential for HER in acidic medium, with an η_{10} of 670 mV and a TS of 83 mV dec^{-1}, due to the stronger interactions between POM clusters. Nevertheless, no comparison with Pt/C or other state-of-the-art electrocatalyst was performed in this study.

On the other hand, Sakthinathan et al. [54] developed a more promising POM/polymer nanostructure through a LbL approach, in which POM layers ($Na_{0.3}[N(C_4H_9)_4]_{7.7}[(Mo_3O_8)_4(O_3PC(O)(C_3H_6NH_2CH_2C_4H_3S)PO_3)_4]$, abbreviated to $[(TBA)Mo_{12}(AleThio)_4]$), were alternated with polymer layers of polyethylenimine (PEI); 1 to 12 bilayers were assembled and evaluated (Figure 8.10a). The material containing 8 bilayers, denoted as PEI/$[(TBA)Mo_{12}(AleThio)_4]$, obtained the best HER performance in acidic medium, with a small η_{10} of 77 mV, a TS of 80 mV dec^{-1} and excellent stability over 2000 cycles or 20 hours of use (Figure 8.10b). Indeed, an increase of the number of bilayers up to 8 was associated with an increase in the number of active sites, active surface area and electrical conductivity, thus explaining the enhanced catalytic performance. However, more than 8 bilayers led to an excessive electrocatalyst thickness, increasing the pathway for electron transfer and consequently limiting the charge transfer rate. Analysing the results, the authors observed that the presence of the polymer revealed to be fundamental to

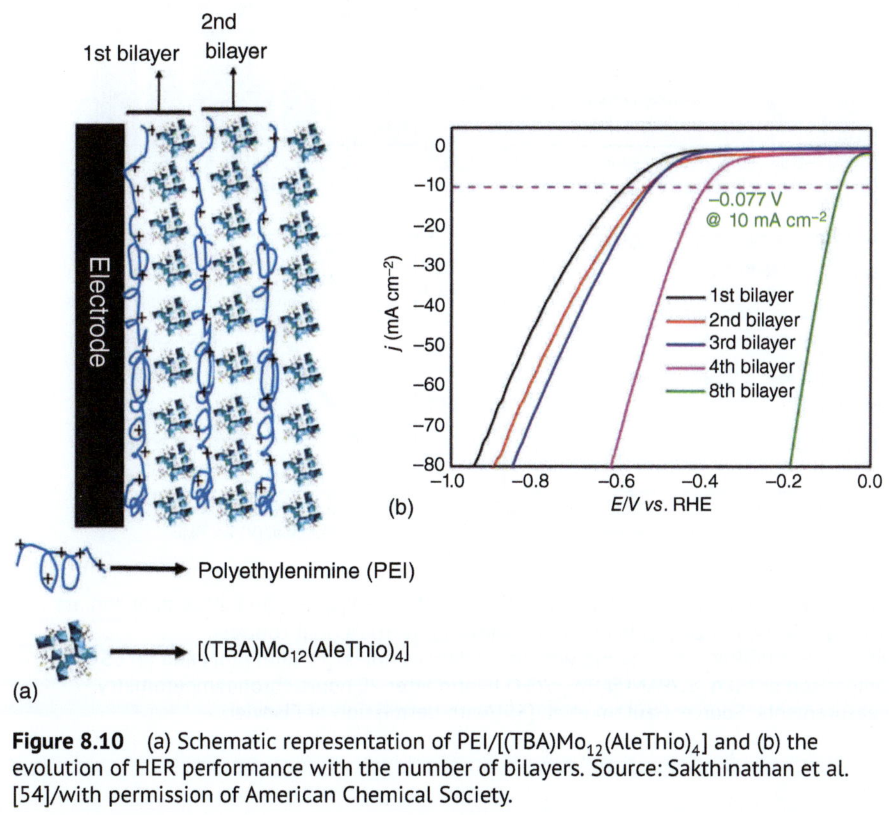

Figure 8.10 (a) Schematic representation of PEI/$[(TBA)Mo_{12}(AleThio)_4]$ and (b) the evolution of HER performance with the number of bilayers. Source: Sakthinathan et al. [54]/with permission of American Chemical Society.

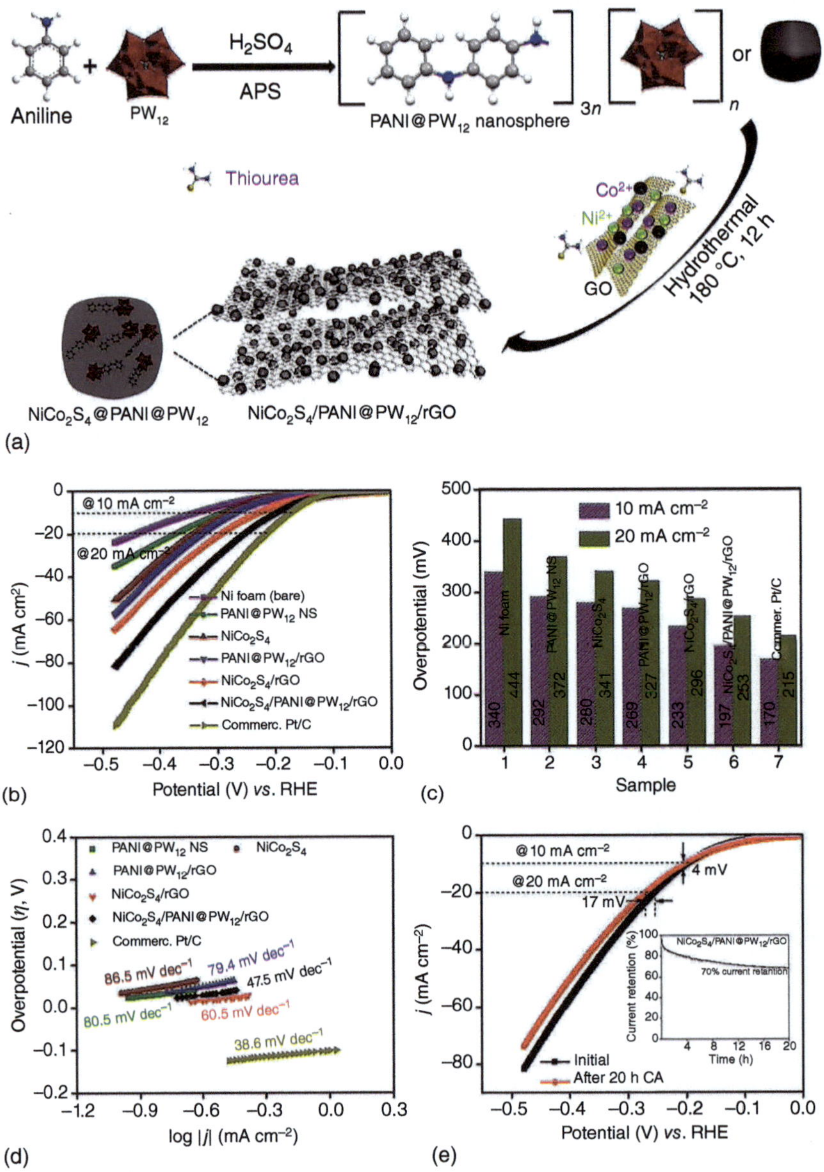

Figure 8.11 (a) Preparation route of $NiCo_2S_4$/PANI@PW_{12}/rGO, (b) LSV comparison ate 10 mV s^{-1} in 1.0 M KOH, (c) bar diagram showing overpotential contrast $NiCo_2S_4$/PANI@PW_{12}/rGO hybrid with secondary samples, (d) Tafel plots and (e) LSV comparison of $NiCo_2S_4$/PANI@PW_{12}/rGO hybrid after 20 hours chronoamperometry measurements. Source: Gautam et al. [55]/with permission of Elsevier.

increase the overall electrical conductivity and catalytic activity, besides promoting the interconnection of the nanolayers. Furthermore, the material benefited from a synergistic effect between the polymer and POM.

In a completely different strategy, Gautam et al. [55] developed recently a four-component nanohybrid containing a bimetallic sulphide ($NiCo_2S_4$), an intrinsic conductive polymer (polyaniline, PANI), a POM (PW_{12}) and a carbon nanomaterial (rGO). This 3D nano-assembly was prepared through a simple procedure, involving the *in situ* polymerization of aniline to form PANI-coated PW_{12} spheres, followed by a hydrothermal step to generate $NiCo_2S_4$ and introduce rGO (Figure 8.11a). By combining the electrochemical properties of all these components, the resulting material, denoted $NiCo_2S_4$/PANI@PW_{12}/rGO, presented outstanding behaviour in lithium storage and HER electrocatalysis. Indeed, the coating of POM with a conductive polymer not only enhanced considerably the electrical conductivity but also protected the PW_{12} from leaching. $NiCo_2S_4$ also displayed an important role in catalytic process, as it contained a favourable distribution of Ni^{3+}/Co^{3+} in octahedral sites, improving the HER kinetics. Furthermore, its interconnected porous structure resulted in high ECSA, abundant active sites and numerous ion exchange channels. Also, the nanohybrid presented an η_{10} of only 197 mV, a small TS of 47.5 mV dec^{-1} and excellent stability over 20 hours in alkaline medium (Figure 8.11b–e).

8.4 POM-derived Electrocatalysts

8.4.1 SACs

Single-atom catalysts (SACs) are a novel class of electrocatalysts in which the active metallic species are atomically dispersed into a support, as so there is no metal-metal bonds in these materials [56, 57]. Furthermore, the positively charged metal atoms modify the electronic distribution of the material, promoting enhanced catalytic activity [56, 57]. Several methodologies have been reported to prepared SACs [58], including the thermal treatment of POMs [59, 60]. Additionally, POMs can act as excellent supports for SACs, allowing a suitable coordination environment for electrocatalysis [61, 62].

Yu et al. [61] reported the HER activity of Pt SACs supported on POMs and carbon black. The four materials ({PtW_6}/C, {$PtMo_6$}/C, {PtV_9}/C and {$PtSiW_{11}$}/C) were prepared by mixing the respective POM crystals ($Na_5[H_3PtW_6O_{24}]$, $K_4[H_4\ \beta\text{-}PtMo_6O_{24}]\cdot 2H_2O$, $Na_5[H_2PtV_9O_{28}]\cdot 21H_2O$ and $[(C_3H_9)_4N]_4[SiPtW_{11}O_{40}H_2]$, respectively) and Ketjen black carbon with a small amount of Nafion in aqueous solution. This study revealed that the electron gaining properties of Pt SACs can be tune by using different POM supports. Moreover, tuning the Pt coordination environment (Figure 8.12a1–d1) also revealed to be fundamental to optimize the HER performance of these materials, as it affects the reaction pathway and associated energy barrier (Figure 8.12b). {PtW_6}/C showed the best HER activity, achieving a TS of 29.8 mV dec^{-1} and a small η_{10} of 22 mV, indicating that (i) Pt

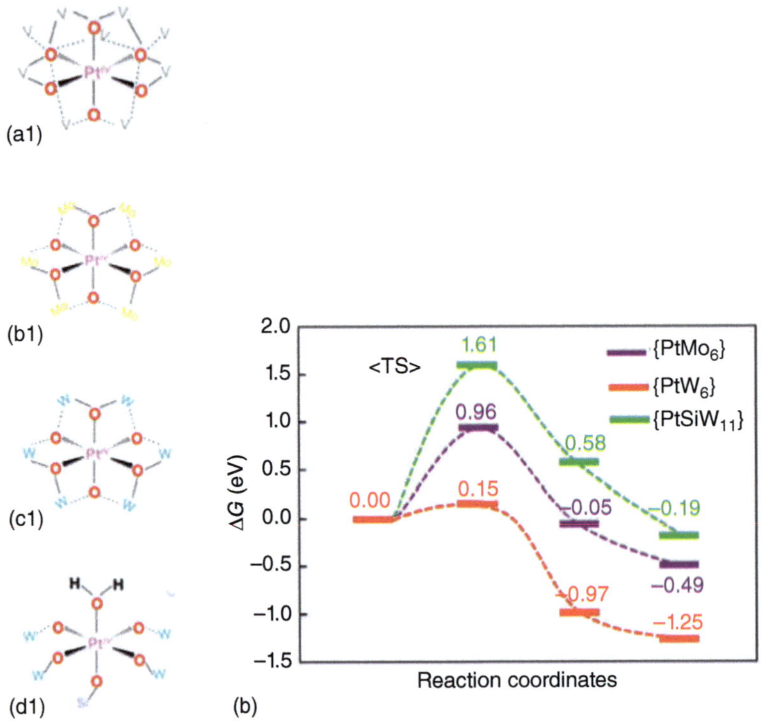

Figure 8.12 Schematic representation of Pt coordination on (a1) $\{PtV_9\}$, (b1) $\{PtMo_6\}$, (c1) $\{PtW_6\}$ and (d1) $\{PtSiW_{11}\}$ POMs. (b) Prediction of the evolution of energy barriers in $\{PtMo_6\}$, $\{PtW_6\}$ and $\{PtSiW_{11}\}$ POMs. Source: Yu et al. [61]/with permission of American Chemical Society.

coordinated to six O is advantageous and (ii) $\{PtW_6\}$ improved the electron gaining ability of Pt. At last, the excellent performance was also attributed to the presence of a carbon support which enhanced the electron transfer. Comparatively, $\{PtMo_6\}$/C and $\{PtV_9\}$/C obtained significantly higher η_{10} (139 and 310 mV, respectively) and TSs (74.3 and 86.1 mV dec^{-1}, respectively).

On the other hand, Lu et al. [63] developed atomically dispersed Ni in Mo_2C NPs encapsulated by N, P co-doped carbon, using an Anderson-type POM ($(NH_4)_4[NiMo_6O_{24}H_6]$, abbreviated to $NiMo_6$) as Ni and Mo precursor. The catalyst, denoted as Ni-Mo_2C/NPC, was prepared by pyrolysis of $NiMo_6$ in H_2/Ar atmosphere in presence of cetyltrimethylammonium bromide (CTAB), followed by doping with melamine and phytic acid to introduce N and P heteroatoms, respectively. Ni-Mo_2C/NPC achieved excellent HER activity and stability in alkaline and acidic media, obtaining η_{10} of 144 and 183 mV and TSs of 59 and 64 mV dec^{-1} in 0.5 M H_2SO_4 and 1.0 M KOH, respectively. These outstanding results were attributed to several factors: (i) the excellent electrical conductivity of the N, P co-doped carbon, resulting in enhanced charge transfer, (ii) the dispersion of Mo_2C NPs on the carbon support, limiting the particle agglomeration and consequently blockage of active sites, (iii) the synergistic effect between the N,

P co-doped carbon and Ni-Mo_2C NPs, with the generation of strong and highly active Ni—P bonds at the interface of both components and (iv) the presence of atomically dispersed Ni in Mo_2C NPs promoted an alteration on the electronic distribution, creating new active sites and leading to the activation of surrounding C atoms.

8.4.2 Transition-metal Carbides

TM carbides like molybdenum (Mo) and tungsten (W) carbides have been studied for HER due to their low price, high physical stability and electrical conductivity [64, 65]. For the preparation of different TM carbides, POMs have been used as fine precursors. Table 8.3 provides an overview of the HER activities of POM-derived metal carbides electrocatalysts.

In 2015, Wu et al. [66] developed a MOF-assisted strategy for synthesizing porous molybdenum carbide octahedral NPs (MoC_x nano-octahedrons). This strategy (Figure 8.13) is based on the confined and *in situ* carburization reaction (800 °C in N_2 atmosphere) occurring in a unique MOFs-based composite (NENU-5) consisting of a Cu-based MOFs (HKUST-1) host and Mo-based Keggin POMs ($H_3PMo_{12}O_{40}$) as porous guests, which allows the uniform formation of metal carbide nanocrystallites without coalescence and excessive growth. The as-prepared MoC_x nano-octahedrons consisted of ultrafine nanocrystallites with an unusual η-MoC phase embedded in an amorphous carbon matrix, and exhibited a uniform and highly mesoporous structure. Taking advantage of the desirable nanostructure, these porous MoC_x nano-octahedrons exhibited remarkable electrocatalytic activity for HER in both acidic (0.5 M H_2SO_4) and basic solutions (1.0 M KOH), with good stability. The resulting material exhibited an η_{10} of 142 and 151 mV and TSs of 53 and 59 mV dec^{-1} in acidic and basic solutions, respectively.

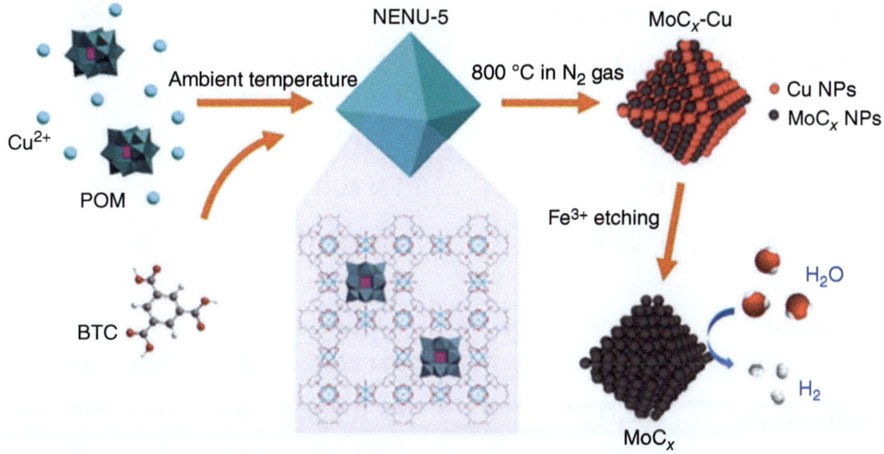

Figure 8.13 Schematic illustration of the synthesis procedure for porous MoC_x nano-octahedrons. Source: Wu et al. [66]/Springer Nature/CC BY 4.0.

Table 8.3 HER performance of POM-derived metal carbides.

Catalyst	POM	Electrolyte	Tafel slope (mV dec^{-1})	η_{10} (mV)	References
MoC_x	$H_3PMo_{12}O_{40}$	0.5 M H_2SO_4	53	142	[66]
		1 M KOH	59	151	[66]
MoP/Mo_2C@C	P_4Mo_6	0.5 M H_2SO_4	45	89	[67]
		1 M KOH	58	75	[67]
		0.1 M PBS	93	136	[67]
Co-Mo_2C@NC	PMo_{12}	0.2 M PBS	118	260	[68]
N@Mo_2C-3CFP	$(TBA)_2[Mo_6O_{19}$ −n $(NAr)_n]$	0.5 M H_2SO_4	51	56	[69]
		1 M KOH	49	66	[69]
Ni-Mo_2C@C	$(NH_4)_n[TMMo_6O_{24}H_6]\cdot 5H_2O$ (TM = Ni^{2+}, Co^{2+}, $n = 4$; TM = Fe^{3+}, Cr^{3+}, $n = 3$)	0.5 M H_2SO_4	65.6	72	[70]
Co-Mo_2C@C			80.9	122	[70]
Fe-Mo_2C@C			102.4	129	[70]
Cr-Mo_2C@C			114.2	147	[70]
P-Mo_2C@NC	$H_3[PMo_{12}O_{40}]$	0.5 M H_2SO_4	76	109	[71]
		1 M KOH	—	83	[71]
		0.1 M PBS	—	159	[71]
$Ni_3Mo_3C/Mo_2C/Ti_3C_2T_x$/NC	PMo_{12}	0.5 M H_2SO_4	58	99	[72]
		1 M KOH	69	98	[72]
Ni/N-Mo_2C@C0.6	$[H_3NiMo_6O_{21}\{(CH_2O)_3CNH_3\}]^{3-}$	1 M KOH	105.7	148	[73]
MoOPC@NC-200	$[PMo_{12}O_{40}]^{3-}$	1 M KOH	60.4	69	[74]
Co_2P/WC@NC/CNTs	$Co_4(PW_9)_2$	0.5 M H_2SO_4	65	171	[75]
		1 M KOH	76	198	[75]
		1 M PBS	204	525	[75]
W/Co_6W_6C@NC	$Co_4P_2W_{18}O_{70}$	1 M KOH	60.1	80	[76]
CO/WC@NC	$Co_9P_5W_{27}$	0.5 M H_2SO_4	93	129	[77]
		1 M KOH	91	142	[77]
CO/WC@NC	$Co_4P_4W_{30}$	0.5 M H_2SO_4	93	142	[78]
		1 M KOH	95	158	[78]
N@MOPCx-800	$[Mo_8O_{26}]^{4-}$	0.5 M H_2SO_4	69.4	108	[79]
		1 M KOH	86.6	139	[79]

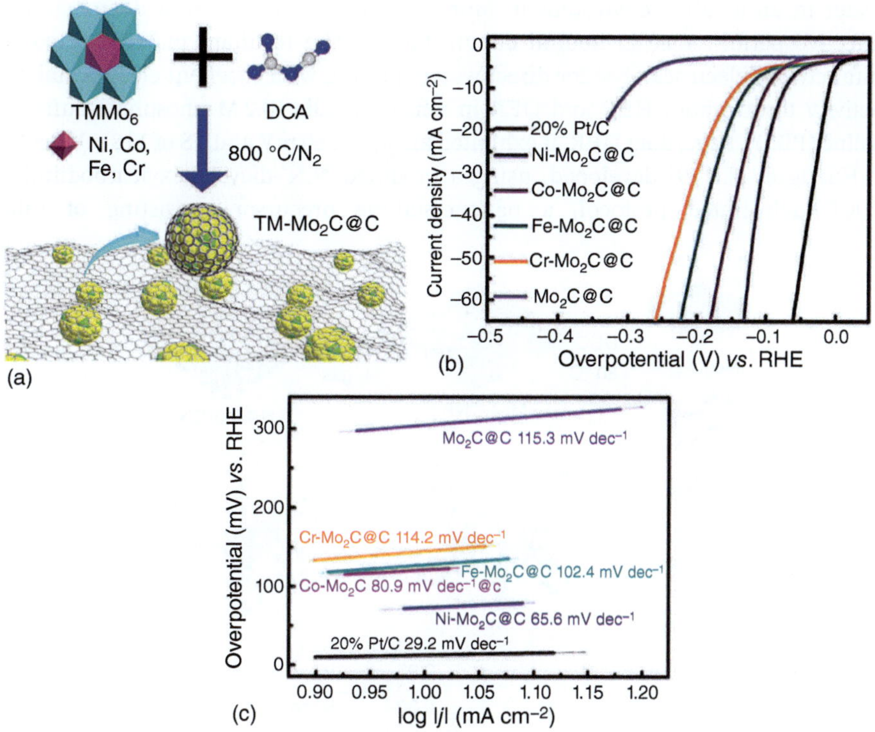

Figure 8.14 (a) Scheme of the preparation of TM-Mo_2C@C, (b) polarization curves for Ni-Mo_2C@C, Co-Mo_2C@C, Fe-Mo_2C@C, Cr-Mo_2C@C and Mo_2C@C after i_R correction and (c) Tafel slopes at 0.5 M H_2SO_4. Source: Yu et al. [70]/with permission of Royal Society of Chemistry.

The study of the electrocatalytic performance of ultra-small Mo_2C influenced by different TM dopants was defined by Yu et al. in 2018 [70] using the Anderson-type POMs $(NH_4)_n[TMMo_6O_{24}H_6]{\cdot}5H_2O$ (TM = Ni^{2+}, Co^{2+}, $n = 4$; TM = Fe^{3+}, Cr^{3+}, $n = 3$). A series of Ni-, Co-, Fe- and Cr-doped Mo_2C composite NPs covered in few-layer graphitic carbon shells (abbreviated TM-Mo_2C@C) were produced when these POMs were combined with dicyandiamide (DCA) by solid grinding and carbonized at a high temperature (Figure 8.14a). The HER catalytic activity of Mo_2C was significantly influenced by TM dopants, as demonstrated by electrocatalytic experiments in 0.5 M H_2SO_4. The electrocatalytic activity was carried out in the following order: Ni-Mo_2C > Co-Mo_2C > Fe-Mo_2C > Cr-Mo_2C (Figure 8.14b and c). Ni-Mo_2C@C had the highest activity among them, requiring a 72 mV overpotential at a current density of 10 mA cm^{-2}. Compared to other TM-doped Mo_2C and pure Mo_2C, its TS of 65.6 mV dec^{-1} was better.

Later, in 2018, Wang et al. [68] developed a new hybrid with POM (PMo_{12}), zeolitic imidazolate frameworks (ZIF-67), DCA and molybdenum carbide coated with nitrogen-doped graphite (Co-Mo_2C@NC). By combining the strengths of two metal components and adding non-metallic compositions, the electrical properties of the composites were tuned to enhance their ability to electrocatalytically split water in an alkaline environment. More significantly, the catalyst demonstrated effective performance in neutral circumstances. The resultant material acted as bifunctional electrocatalyst for direct water splitting with excellent electrocatalytic activity toward both HER and OER in neutral media (0.2 M phosphate-buffered saline (PBS)). Regarding HER, it exhibited an $\eta_{10} = 260$ mV and TS of 118 mV dec^{-1}.

Huang et al. [79] developed, using a modified N,N′-dicyclohexylcarbodiimide (DCC)-dehydrated protocol, a nanocrystalline precursor consisting of fully

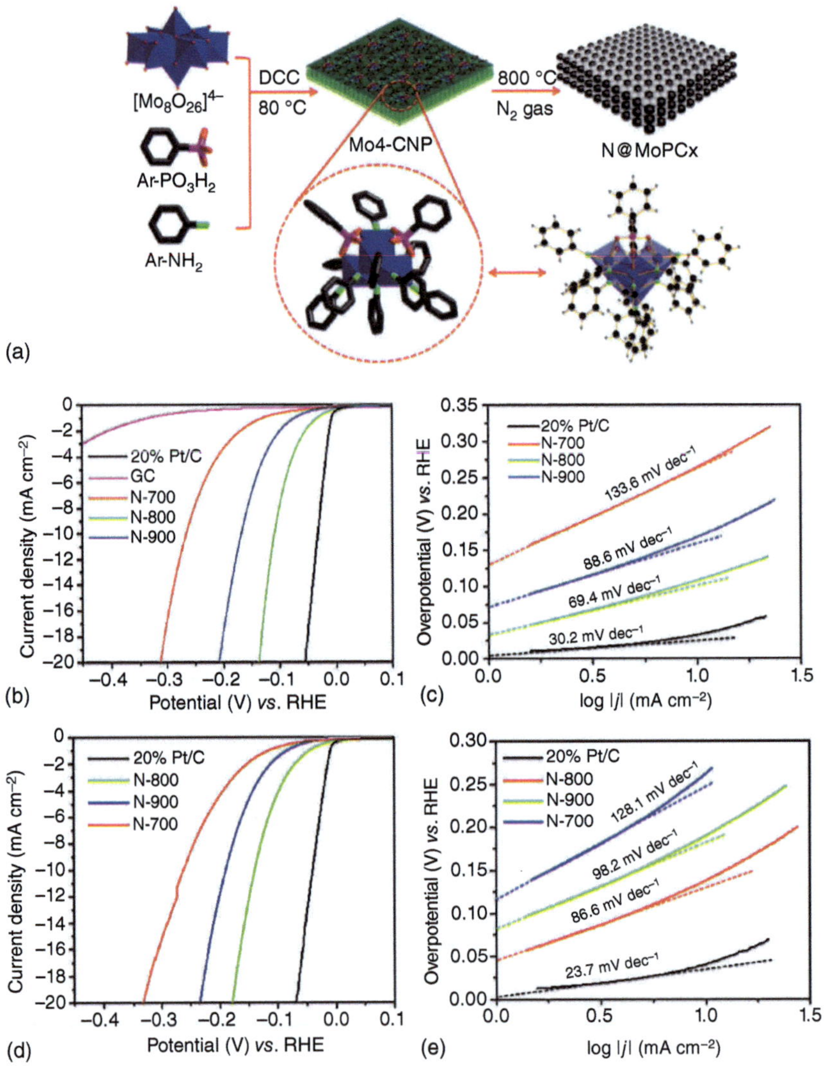

Figure 8.15 (a) Schematic illustration of the synthetic procedure for the preparation of porous nitrogen-doped molybdenum carbide and phosphide hybrid N@MoPCx, LSV curves and Tafel plots of the N@MoPCx nanosheet hybrids prepared at different temperatures, (b,c) in 0.5 M H_2SO_4 and (d,e) in 1.0 M KOH. Source: Huang et al. [79]/with permission of John Wiley and Sons.

substituted polyorganoimido-derivatized heteropolyoxometalates, $Mo_4O(ArPO_3)_2(NAr)_4(\mu_2\text{-}NAr)_5$ (Ar = phenyl, Mo_4-CNP) (Figure 8.15a). Three different carbonization temperatures (700, 800 and 900 °C) were tested and labelled as N@MOPCx-700, N@MOPCx-800 and N@MOPCx-900, respectively. Characterization results showed that the organoimido-derivatized heteropolyoxometalate Mo_4-CNP acted as N, P and C source, leading to the formation of the N-doped molybdenum carbides/phosphides nanosheet structures. The hybrids were evaluated for HER in 0.5 M H_2SO_4 and 1 M KOH. The N@MOPCx-800 exhibited excellent electrocatalytic activity with an overpotential of 108 and 139 mV in acidic and basic electrolyte (Figure 8.15b and d) and TSs of 69.4 and 86.6 mV dec^{-1} (Figure 8.15c and e), respectively. The N@MOPCx-700 and N@MOPCx-900 exhibited an overpotential of 266 and 170 mV in acidic electrolyte and 263 and 190 mV in basic electrolyte. Therefore, N@MOPCx-800 displayed the most promising electrocatalytic performance, which was attributed mainly to the N-doping, the highly porous nanosheet morphology and a synergistic effect between Mo_2C and MoP. Furthermore, the optimized synthesis procedure allowed an uniform distribution of Mo_2C and MoP particles, resulting in more exposed HER active sites.

Xu and colleagues [77] developed cobalt and tungsten-based bimetallic electrocatalysts (Co/WC@NC) using polyethylenimine as a carbon source and bimetallic POM ($[Co_9(OH)_3(H_2O)_6(HPO_4)_2(PW_9O_{34})_3]^{16-}$ (abbr. $Co_9P_5W_{27}$)) as soft templates. This was done in a single step using a straightforward high temperature calcination procedure (700 °C in nitrogen atmosphere). Co/WC@NC exhibited good HER performance in 0.5 M H_2SO_4 and 1.0 M KOH with overpotentials of 129 and 142 mV at the current density of 10 mA cm^{-2} and TSs of 93 and 91 mV dec^{-1}, respectively.

In 2023, $MoO_2/Mo_3P/Mo_2C$ triple-interface heterojunctions encapsulated into nitrogen-doped carbon (MoOPC@NC) were synthetized using ammonium phosphomolybdate and dopamine as raw material, by Xiao et al. [74]. It's interesting to note that the successful formation of the triple interface heterojunction was significantly influenced by the mass ratio of ammonium phosphomolybdate and dopamine. For this reason, the effects of various mass ratios of ammonium molybdate and dopamine were examined by adding different amounts of the latter: 50, 100, 200 and 400 mg, obtaining MoOPC@NC-50, MoOPC@NC-100, MoOPC@NC-200 and MoOPC@NC-400, respectively. Using 200 mg of dopamine led to the formation of an optimized structure, presenting $MoO_2/Mo_3P/Mo_2C$ heterojunctions encapsulated in N-doped carbon. The HER performance was tested in 1 M KOH and MoOPC@NC-200 exhibited the best results, with an overpotential of 69 mV at a current density of 10 mA cm^{-2} which is lower than those obtained for MoOP@NC-50 (155 mV), MoOPC@NC-100 (84 mV) and MoOPC@NC-400 (75 mV). The same sample presented the lowest TS (60.36 mV dec^{-1}) compared to the remaining samples, MoOP@NC-50 (68.7 mV dec^{-1}), MoOPC@NC-100 (65.5 mV dec^{-1}) and MoOPC@NC-400 (61.8 mV dec^{-1}). The authors attribute the best electrocatalytic performance of MoOPC@NC-200 to the fact that it had the largest ECSA of all the prepared heterostructures and the lowest resistance to charge transfer (R_{ct} = 8.41 Ω), which may be due to the fact that the nitrogen-doped carbon facilitates the electron transfer process at the interface.

8.4.3 Transition-metal Chalcogens

Transition-metal chalcogens (TMCs), such as MoS_2, WS_2 and $MoSe_2$, have also been studied for HER electrocatalysis due to their structure and morphology that can improve their proton reduction performance. It has been demonstrated that the conductive polymorph lowers overpotential, leading to excellent HER performance. Furthermore, their geometry can increase the potential of HER by defining more active sites through architecture and system configuration optimization. Also, interfacial engineering significantly affects the HER activity, for instance by promoting the diffusion of protons to reaction sites. In addition, TMCs have different semiconductor and metallic crystalline phases, which enable control of their functionality and also morphological flexibility to accommodate dopants [80]. Table 8.4 provides an overview of the HER activities of POM-derived chalcogens electrocatalysts.

In 2013, Huang et al. prepared a nanocomposite Co-doped WSe_2 nanosheets using Co-containing sandwich-type POM ($Bi_2W_{22}Co_2$) as metal precursor, through a chemical vapor deposition (CVD) selenization method [81]. The synthesis consisted of adding the POM precursor into a rGO suspension, followed by the addition of polyethylenimine and the pyrolysis of the resulting mixture in presence of Se powders, under an Ar/H_2 atmosphere. To improve the overall conductivity and

Table 8.4 HER performance of POM-derived metal chalcogens.

Catalyst	POM	Electrolyte	Tafel slope (mV dec^{-1})	η_{10} (mV)	References
Co-WSe_2/rGO_2	$Ni_2W_{22}Co_2$	0.5 M H_2SO_4	64	217	[81]
Co-WSe_2/rGO3			48	217	[81]
Pt/MoS_2	$H_3PW_{12}O_{40}$	0.5 M H_2SO_4	45	—	[82]
HPOM-rGO	$A_3[PMo_{12}O_{40}]$, where $A = H^+$)	0.5 M H_2SO_4	41.7	176	[83]
APOM-rGO	$A_3[PMo_{12}O_{40}]$, where $A = NH_4^+$)		44.2	217	[83]
MoS_2/C800	$H_3PW_{12}O_{40}$	0.5 M H_2SO_4	73	207	[84]
CoMoS-600	PMA, a Mo-containing POM	0.5 M H_2SO_4	65.5	235	[85]
O-CoMoS	$(NH_4)_4[Co^{II}Mo_6 O_{24} H_6]\cdot 6H_2O)$	1 M KOH	70	10	[86]
Co_5 $Mo_{10}S_x$/CC	Co_2Mo_{10}	1 M KOH	56	36	[87]
CoMoS/KB	$(NH_4)_3PMo_{12}O_{40}$	1 M KOH	64	81	[88]
300-NiMoS/CC	$(NH_4)_4[(HPO_3)_2 Mo_5O_{15}]\cdot 4H_2O$	0.5 M H_2SO_4	37	39	[89]
		1 M KOH	36	30	[89]
1T-Co-MoS_2 @HMCS	$(NH_4)_6[Co_2Mo_{10} H_4O_{38}]\cdot 4H_2O$	0.5 M H_2SO_4	78	132	[90]
		1 M KOH	63	74	[90]

facilitate charge transfer, the authors optimized the amount of rGO presented in the initial suspension (1, 3.5, 5, 10 and 20 wt%) leading to different samples labelled from Co-WSe_2/rGO_1 to Co-WSe_2/rGO_5. The electrocatalytic studies were performed in acidic electrolyte (0.5 M H_2SO_4). The results revealed that rGO loadings between 3.5 and 5 wt% resulted in better HER performance. Indeed, the hybrids containing 3.5 and 5 wt% of rGO showed the lowest overpotentials (217 mV at 10 mA cm^{-2} in both materials) and TSs of 64 and 48 mV dec^{-1}, respectively. On the other hand, the hybrids with a higher amount of rGO did not perform as well electrocatalytically, as the catalytic performance deteriorates with increasing rGO content due to blocking of the active sites of the transition-metal dichalcogenides (TMDs). Moreover, the authors also prepared an electrocatalysts without rGO (Co-WSe_2), following the same synthesis procedure described above but in absence of the carbon material, and other *via* the physical mixture of Bi_2O_3, WO_3, Co_2O_3 and 3.5 wt% rGO (labelled as physical mixture/rGO_2). Both electrocatalysts showed a significantly worse HER performance compared to Co-WSe_2/rGO_2 or Co-WSe_2/rGO_3 (η_{10} of 262 and 395 mV and TSs of 70 and 88 mV dec^{-1}, respectively), due to an insufficient electrical conductivity (in the first case) and the lack of synergistic effects between rGO and Co-WSe_2 (in the second case). As so, this work demonstrated the fundamental role of rGO and the advantages of using a POM as TM precursor instead of other inorganic salts to obtain an enhanced electrocatalytic performance.

TMDs were also grown using a POM ($H_3PW_{12}O_{40}$) as tungsten precursor and two different chalcogenide precursors: thioacetamide (for WS_2 rods) or selenium dioxide (for $MoSe_2$ hallow spheres) [82]. A solution-phase reaction was used to synthesize the hierarchical TMDs. The authors also synthesized noble metal NPs on the surface of TMDs. The epitaxial growth of NPs on TMDs led to the improvement of HER electrocatalysis in acidic medium due to the synergistic effect between Pt NPs and TMDs. The hybrid Pt/MoS_2 presented an onset potential of 35 mV and TS of 45 mV dec^{-1} while the pure MoS_2 showed a larger onset potential of 230 mV and TS of 60 mV dec^{-1}.

By using the Anderson-type $(NH_4)_4[Co^{II}Mo_6O_{24}H_6]\cdot 6H_2O$ POM as a bimetal precursor, Hou et al. developed, by *in-situ* hydrothermal growth strategy, oxygenated-CoS_2-MoS_2 (O-CoMoS) heteronanosheets grown on flexible carbon fibre cloth as bifunctional (OER and HER) electrocatalyst (Figure 8.16) [86]. The O-CoMoS exhibited an η_{10} of 97 mV and a TS of 70 mV dec^{-1} in an alkaline electrolyte. The O-CoMoS electrocatalytic performance was primarily attributed to its disordered structure, the exposure of active heterointerfaces, the 'highway' of charge transport on two-dimensional conductive channels, and the abundance of active catalytic sites resulting from the synergistic effect of the heterostructures.

Using polyoxometalates PMo_{12} as oxidants for the polymerization of pyrrole and a subsequent sulfurization process, MoS_2/C hollow nanospheres were created by Sun et al. in 2019 [84]. These MoS_2/C were annealed at different temperatures (MoS_2/C_T, where T stands for 700, 800, 900 and 1000 °C) and thoroughly investigated in 0.5 M H_2SO_4 solution as HER electrocatalysts. MoS_2/C_{800} presented a much lower overpotential at 10 mA cm^{-2} (207 mV) than MoS_2/C_{900} and MoS_2/C_{1000} (both with 277 mV) and MoS_2/C_{700} (267 mV). MoS_2/C_{800} also showed a lower TS (73 mV dec^{-1}) than

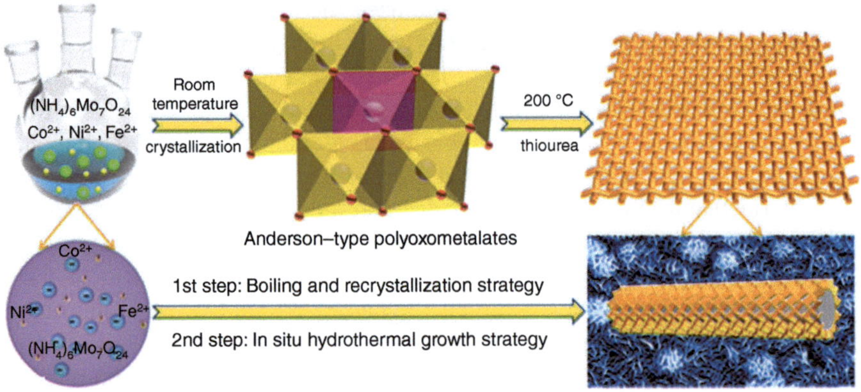

Figure 8.16 Schematic illustrations of the synthesis of as-prepared oxygenated-bimetal-sulfides heteronanosheets arrays grown on carbon fibre cloth. Source: Reproduced with permission of Hou et al. [86]/American Chemical Society.

MoS_2/C_{700} (94 mV dec^{-1}), MoS_2/C_{900} (84 mV dec^{-1}) and MoS_2/C_{1000} (91 mV dec^{-1}). The better performance of mesoporous hollow MoS_2/C_{800} nanospheres was ascribed to their significantly higher ECSA and excellent electronic transfer properties, arising from the strong interaction between MoS_2 nanosheets and the highly conductive carbon support. Moreover, this electrocatalyst displayed an ideal quantity of structural defects, which can expose active sites but also affect negatively the electrical conductivity; structural defects in MoS_2 also led to uncoordinated S atoms and, therefore, to the generation of negative charges that can act as active sites for HER. Finally, XPS results showed lower contents of N, P, O and S on the composite annealed above 900 °C, affecting its wettability and the coupling effects between MoS_2 and the N, P doped carbon.

A hydrothermal method was also used in 2020 by Lu et al. to prepare Co-promoted MoS_2 nano-crystallites, using a special Evans-Showell-type POM $(NH_4)_6[Co_2Mo_{10}O_{38}H_4]$ (denoted as Co_2Mo_{10}) as a precursor [87]. The electrocatalyst performance was attributed to a combination of factors: (i) the addition of CoMo-POMs lowers the intrinsic free energy of water dissociation and increases the intrinsic activity of CoMoS active sites; (ii) the use of carbon cloth (CC) substrate can increase the number of active sites and improve electrical conductivity from the substrate. This hybrid showed high HER (and OER) activity requiring an overpotential of 36 mV to achieve 10 mA cm^{-2} and showed a TS of 56 mV dec^{-1} in 1.0 M KOH.

Huang et al. [85] synthetized a bimetallic Co-Mo sulphide NPs supported on N,S-doped porous carbon composite through carbonization and sulfurization using POM@ZIF-67@polydopamine (Figure 8.17a) at three different temperatures (600, 800 and 1000 °C). The resulting bimetallic CoMoS-600 exhibited an overpotential of 235 mV at 10 mA cm^{-2} and a TS of 65.5 mV dec^{-1}, while CoMoS-800 and CoMoS-1000 showed an overpotential of 327 and 359 mV and TSs of 61.47 and 84.24 mV dec^{-1}, respectively (Figure 8.17b and c). The exposed active sites of the Co-Mo-S NPs and the N,S-doped porous carbons in the composites have a

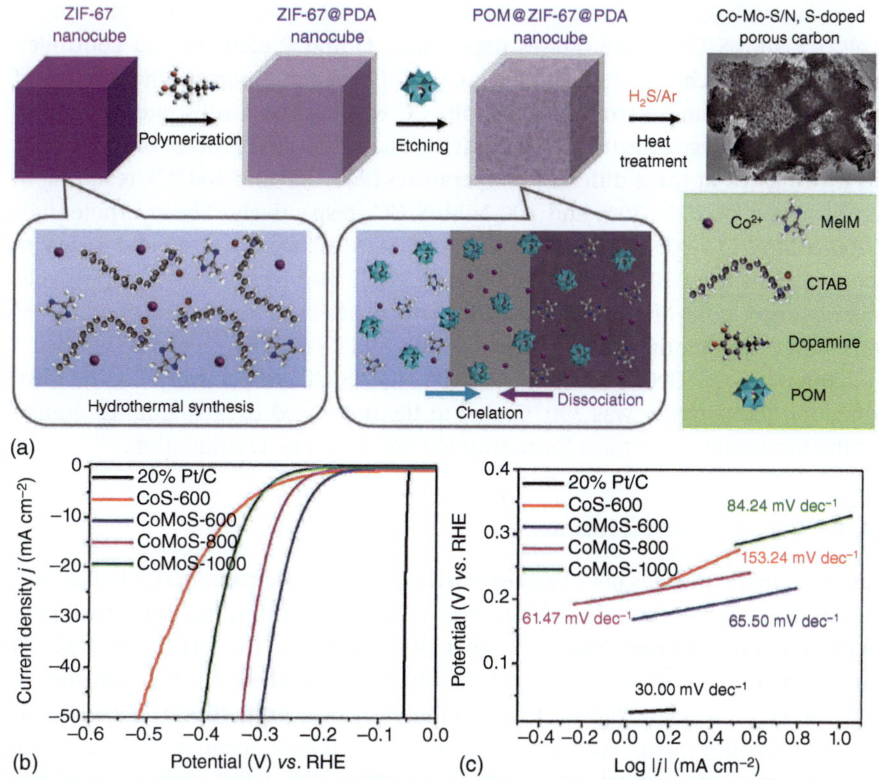

Figure 8.17 (a) Diagram of the synthesis of the POM@ZIF-67@PDA precursor and bimetallic Co-Mo-S/C composites, (b) HER polarization curves and (c) corresponding Tafel plots in 0.5 M H_2SO_4. Source: Huang et al. [85]/Royal Society of Chemistry/CC BY 3.0.

synergistic effect, according to the authors, which contributed to the electrocatalytic performance. It's interesting to note that sample CoMoS-800 had a slightly lower TS than sample CoMoS-600. This could be because the graphitized carbon matrix of CoMoS-800 had a much higher surface area and better conductivity. Nevertheless, CoMoS-600 displayed the most promising electrocatalytic behaviour of this series. Its excellent performance resulted from the optimization of structural and morphologic aspects, namely (i) the degree of graphitization of the carbon support and, consequently, its conductivity; (ii) the size of the NPs, with the formation of ultra-small NPs leading to exposed active sites; and (iii) the content of N, S and P heteroatoms.

An Strandberg-type POM $((NH_4)_4[HPO_3]_2Mo_5O_{15}]\cdot 4H_2O$ was used as pre-assembled molecular platform to construct and regulate NiMoS active sites at the atomic level. Through a straightforward ion-exchange route, Sun et al. [89] synthesize $Ni_2P_2Mo_5$. Next, O,P-doped NiMoS/CC composites were prepared by growing $Ni_2P_2Mo_5$ on CC and sulfurizing it. Three efficient methods were used to prepare these composites in order to increase the HER performance: (i) NiMo-POM was used as a precursor in the design and construction of bimetallic composite catalysts, which can modulate the active sites of NiMoS/CC at the atomic level; (ii) NiMoS/CC active phases doped with O and P elements *via* controlled sulfurization, which can change the band gap of MoS_2 and increase the number of active centres at the substrate surface; (iii) CC was used as a substrate in order to improve the catalyst's electrical conductivity. The sulfidation process of NiMoS/CC was investigated at three different temperatures (200, 300 and 400 °C), resulting in the formation of 200-, 300- and 400-NiMoS/CC, respectively. The overpotentials of 200-NiMoS/CC, 300-NiMoS/CC and 400-NiMoS/CC were 104, 39 and 86 mV at a current density of 10 mA cm^{-2} and TSs of 90, 37 and 69 mV dec^{-1}, respectively, were obtained in 0.5 M H_2SO_4. The HER performance was also investigated in 1 M KOH with overpotentials of 107, 30 and 92 mV and TSs of 103, 36 and 63 mV dec^{-1} for 200-NiMoS/CC, 300-NiMoS/CC and 400-NiMoS/CC, respectively. So, the ideal sulfidation temperature was 300 °C, due to the optimized oxygen-sulphur doping and the formation of the most S-unsaturated bonds at 300 °C sulfidation.

8.4.4 Transition-metal Nitrates

Transition-metal nitrides have been extensively used in electrocatalytic applications [91, 92], in part due to their stability over a broad pH range, high electrical conductivity and unique electronic properties [93]. Moreover, the formation of M—N bonds (M being an early TM) narrows the filled states of the *d*-band; consequently, up to the Fermi level, nitrides exhibit similar electronic properties to noble metals [91, 94]. Nitrides-based electrocatalysts can be prepare using POMs as metal precursor, with the added advantage of retaining the latter unique structural properties [73, 94].

For instance, Men et al. [73] reported recently the outstanding HER activity of a bimetallic nitride encapsulated on N-doped carbon, denoted as Co/WN@NC, in alkaline and acidic media. The electrocatalyst was prepared by the calcination of DCA, $Co_9P_5W_{27}$ (abbreviated from $[Co_9(H_2O)_6(OH)_3(HPO_4)_2(PW_9O_{34})_3]^{16-}$) and cobalt nitrate; for comparison purposes, the material was also prepared in absence of the cobalt salt (WN@NC), POM (Co@NC) or both (NC). Due to a synergistic effect between WN and Co, Co/WN@NC achieved a η_{10} of 143 mV in acidic and alkaline media (Figure 8.18a and b) and TSs of 90 and 118 mV dec^{-1}, respectively (Figure 8.18c and d). Furthermore, the material did not present stability issues in both media within 30 hours.

Huang et al. [94] developed a hybrid material consisting of single-atom Al and O co-doped Mo_2N quantum dots decorated on N-doped graphene (AlO@Mo_2N-NrGO) with excellent HER activity in alkaline medium. Here, the atomically co-doped quantum dots were obtained from Anderson-type POMs, more concretely $(NH_4)_3[AlMo_6O_{24}H_6]$, and were prepared *in situ* of GO nanosheets previously coated with PANI; the intrinsic conductive polymer was added not only to improve the electrical conductivity but also to promote a uniform dispersion of POM clusters on GO surface. The conversion of POM to Mo_2N occurred through nitridation at high temperature (Figure 8.19a). AlO@Mo_2N-NrGO showed a self-activation by multiple CV running (430–1710 cycles) (Figure 8.19b). After this activation,

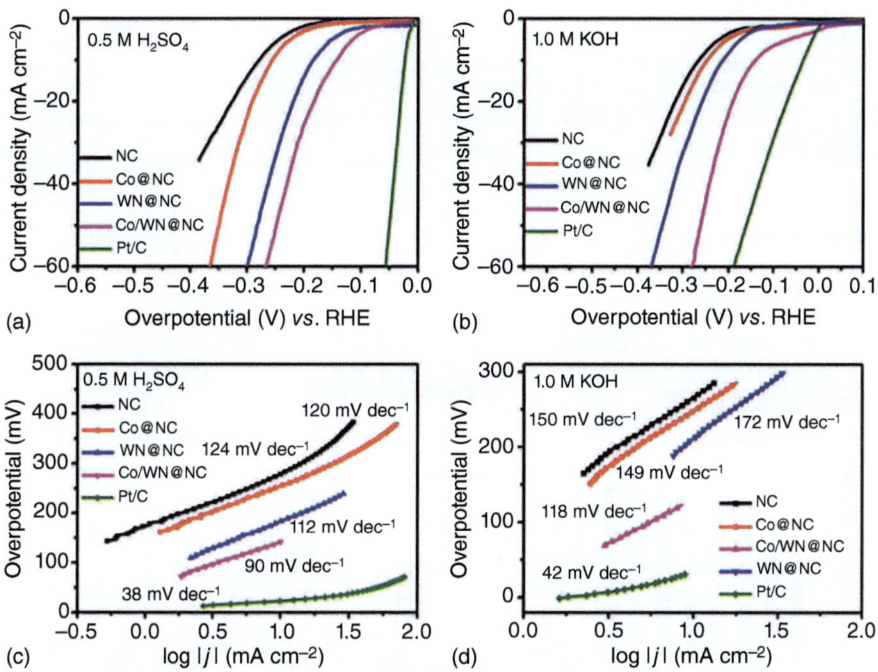

Figure 8.18 (a) HER performance of POM-derived nitrides: polarization curves in 0.5 M H_2SO_4 and (b) 1.0 M KOH electrolyte, (c) Tafel plots in 0.5 M H_2SO_4 and (d) 1.0 M KOH electrolyte. Source: Men et al. [73]/with permission of Elsevier.

the material displayed current densities at industrial level, up to 110 mA cm^{-2} at −0.3 V *vs.* RHE (Figure 8.19c). By performing i_R compensation, the current density neared 400 mA cm^{-2} at the same potential (Figure 8.19d). Moreover, it obtained a η_{10} of 111 mV, which is significantly inferior to Pt/C value (203 mV), and a TS of 69 mA cm^{-2} (Figure 8.19e). It also showed an outstanding stability, maintaining current densities around 114 mA cm^{-2} over 300 hours (Figure 8.19f). According to the authors, the presence of Al and O single atoms modulated the electronic properties of Mo_2N quantum dots and promoted the water dissociation and H_2 desorption. Additionally, the formation of Mo—N bonds between Mo_2N and N-doped graphene enhanced the electron transfer rate and simultaneously improved the stability of the material, explaining the excellent HER performance of AlO@Mo_2N-NrGO.

8.4.5 Transition-metal Phosphides

The HER activity of TM phosphides has been extensively reported [95, 96]. In the last decades, several Mo, W and Co phosphides derived from POMs were successfully applied in HER, although usually combined with a carbon support [67, 75, 97–100]. Table 8.5 gives an overview of the HER activities of the POM-derived TM phosphide electrocatalysts.

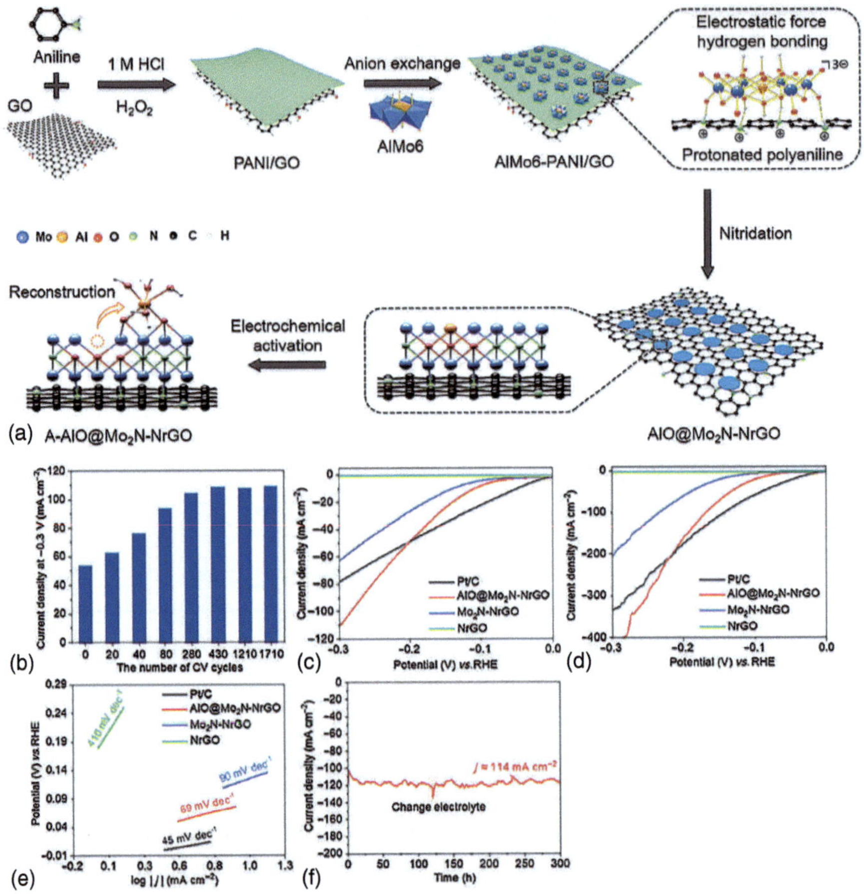

Figure 8.19 (a) Synthesis procedure of AlO@Mo$_2$N-NrGO, (b) current density at −0.3 V of AlO@Mo$_2$N-NrGO after different CV cycles, (c,d) Polarization curves of Pt/C, AlO@Mo$_2$N-NrGO, Mo$_2$N-NrGO and NrGO (c) without and (d) with i_R compensation, (e) Tafel plots and (f) long-term durability tests at η = 300 mV for 300 hours. Source: Huang et al. [94]/John Wiley & Sons/CC BY 4.0.

In 2017, Zhang et al. [67] reported a composite containing MoP and Mo_2C NPs encapsulated by N-doped carbon (MoP/Mo_2C@C) with excellent HER activity over the entire pH range (0–14). The catalyst was prepared *via* annealing of $(NH_4)_6\{Mo_2{}^{V}O_4[(Mo_2{}^{VI}O_6)NH_3CH_2CH_2CH_2C(O)(PO_3)_2]_2\}\cdot 10H_2O$ (denoted as P_4Mo_6) and DCA under N_2 atmosphere (Figure 8.20). The mass ratio between P_4Mo_6 and DCA, as well as the annealing temperature were carefully tuned in order to improve the HER performance: mass ratios of 1 : 1, 1 : 2, 1 : 3 and 1 : 4 and annealing temperatures of 700, 750, 800, 850 and 900 °C were tried. The optimized material (P_4Mo_6:DCA ratio of 1 : 2 and annealed at 800 °C) achieved small η_{10} of 89, 136 and 75 mV and TSs of 45, 93 and 58 mV dec^{-1} in acidic, neutral and alkaline media, respectively. Furthermore, it also exhibited excellent stability for 14 hours in all media. This promising electrocatalytic performance was attributed to

Table 8.5 HER performance of POM-derived transition-metal phosphides.

Catalyst	POM	Electrolyte	Tafel slope (mV dec^{-1})	η_{10} (mV)	References
$MoP/Mo_2C@C$	$(NH_4)_6\{Mo_2VO_4[(Mo_2VIO_6)NH_3CH_2CH_2CH_2C(O)(PO_3)_2]_2\}\cdot 10H_2O$ (P_4Mo_6)	0.5 M H_2SO_4	45	89	[67]
		1.0 M PBS	93	136	[67]
		1.0 M KOH	58	75	[67]
CoP/MoP@NC/CC	PMo_{12}	1.0 M KOH	40	94	[97]
Mo-W-P/CB	$H_3PW_{12}O_{40}$ and $H_3PMo_{12}O_{40}$	0.5 M H_2SO_4	62	92	[98]
CoP-WP/rGO	Co_8W_{18}	0.5 M H_2SO_4	54	130	[99]
		1.0 M PBS	80	138	[99]
		1.0 M KOH	62	96	[99]
$Co_2P/WC@NC/CNTs$	$K_{10}[Co_4(H_2O)_2(PW_9O_{34})_2]$	0.5 M H_2SO_4	65	171	[75]
		1.0 M KOH	76	198	[75]
		1.0 M PBS	204	332	[75]
MoP-QDs@PC	PMo_{12}	0.5 M H_2SO_4	59	118.8	[100]
		1.0 M KOH	66	98.8	[100]
MoOPC@NC-200	$(NH_4)_3[PMo_{12}O_{40}]$	1.0 M KOH	60.36	69	[74]

a synergistic effect between MoP and Mo_2C NPs and between MoP/Mo_2C and the carbon shells, the improved electrical conductivity and corrosion resistance caused by the graphitic carbon; and the modulation of the electronic distribution induced by the N doping.

CoP/MoP@NC/CC, an electrocatalyst consisting of cobalt and molybdenum phosphides coated on N-doped graphite carbon layers (NC) on CC, revealed promising HER and OER activities in alkaline medium [97]. The material was prepared through phosphidation of POM-based zeolitic imidazolate frameworks, previously synthesized *via* a hot-pressing route on CC. The formation of hydroxide/oxide during the catalysis and their synergistic behaviour with CoP and MoP, combined with high electrical conductivity and exposed active sites, led to a current density of 1000 mA cm^{-2} at a small HER overpotential of 475 mV. Moreover, it displayed an η_{10} of 94 mV and TS of 40 mV dec^{-1}, besides an excellent long-term stability.

On the other hand, Liu et al. [98] used $H_3PW_{12}O_{40}$ and $H_3PMo_{12}O_{40}$ to prepare Mo-W-P NPs. First, both POMs were immobilized on carbon black *via* 1-methylimidazole following an optimized Mo/W molar ratio of 1.8 : 1 and then calcinated. The resulting hybrid material, Mo-W-P/CB, presented a TS of 62 mV dec^{-1} and a η_{10} of 92 mV in acidic medium. Also, it reached 50 mA cm^{-2} at an overpotential of only 231 mV and showed excellent stability over 2000 cycles for 12 hours. The authors reported a synergistic effect between W and Mo which, combined with the high electrical conductivity of the CC and the large active surface area of the hybrid, explained the HER performance.

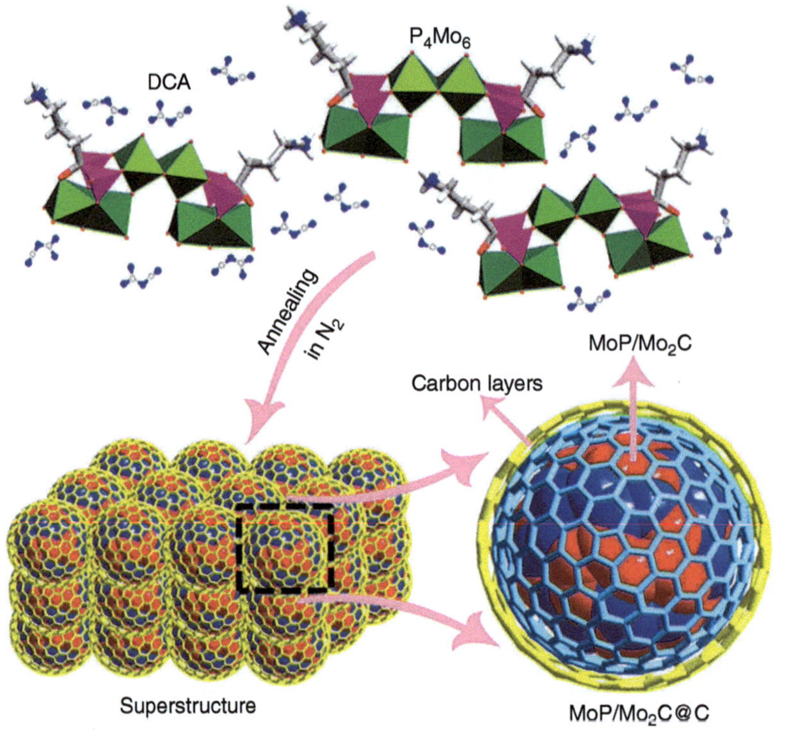

Figure 8.20 Synthesis procedure of $MoP/Mo_2C@C$. Source: Zhang et al. [67]/with permission of American Chemical Society.

Recently, Jiao et al. [99] demonstrated the excellent HER activity of Co_8W_{18}-derived CoP-WP supported on rGO (CoP-WP@rGO) in acidic, neutral and alkaline media: η_{10} of 130, 138 and 96 mV, TSs of 54, 80 and 62 mV dec^{-1}, respectively, and long-term stability in all media. Yue et al. [75] also reported outstanding catalytic performance over a wide pH range of Co_2P-WC NPs wrapped on N-doped carbon and supported on CNTs, obtained from the pyrolysis of a MOF-POM heterostructure (MOF: ZIF-8; POM: $K_{10}[Co_4(H_2O)_2(PW_9O_{34})_2]$) and CNTs; the material was denoted Co_2P/WC@NC/CNTs and presented an η_{10} of 198, 171 and 332 mV and TSs of 76, 65 and 204 mV dec^{-1} in alkaline, acidic and neutral media, respectively. In both works, the electrocatalyst promising performance was mostly attributed to (i) the formation of multiple heterojunctions (CoP-WP in the first case and Co_2P-WC in the second), leading to enhanced electron transfer and lower energy barrier for H_2 desorption and (ii) the presence of a highly conductive carbon material.

Lastly, Liu et al. [100] designed a promising POM-derived electrocatalyst for HER constituted by super-hydrophilic MoP quantum dots supported on porous carbon (MoP-QDs@PC). A POM-MOF structure (PMo_{12}@UiO-66) was used as precursor for the synthesis of MoC@PC, which was then phosphorizated to the final material; this second step also resulted in O and P doping into the carbon support, conferring additional HER active sites. This electrocatalyst outperformed Pt/C at overpotentials superior to 233 mV (acidic medium) or 193 mV (alkaline medium). Moreover, it achieved η_{10} of 118.8 and 98.8 mV and TSs of 59 and 66 mV dec^{-1} in acidic and alkaline conditions, respectively, besides an excellent stability over 50 hours or 2000 cycles. According to the authors, the formation of pores on carbon and the high wettability of the material led to enhanced mass transport and promoted the accessibility of active sites, resulting in faster HER kinetics, while the abundant active sites, high specific surface area and the unique structure of quantum dots contributed to the HER activity. Finally, the strong interaction between MoP quantum dots and the porous carbon explained its outstanding long-term stability.

8.4.6 Transition-metal Oxides

Transition-metal oxides and metal-oxide-based composites are a popular class of electrocatalysts for several reactions, including HER, especially due to their low cost and easily tuneable properties [101, 102]. Due to their low electrical conductivity, metal oxides electrocatalysts are usually combined with a highly conductive support, such as carbon materials or Ni foams [101, 103].

Xiao et al. [74] reported the synthesis of urchin-like $MoO_2/Mo_3P/Mo_2C$ heterostructure encapsulated in N-doped carbon, *via* carbonization of 400 mg of $(NH_4)_3[PMo_{12}O_{40}]$ and dopamine; the mass of dopamine (50, 100, 200 or 400 mg) and the annealing temperature (800, 900 or 1000 °C) were carefully optimized in order to obtain heterojunctions with triple interface. The most promising material, prepared with 200 mg of dopamine and annealed at 800 °C (MoOPC@NC-200), obtained a small η_{10} of 69 mV, TS of 60.4 mV dec^{-1} and excellent stability over 125 hours. The $MoO_2/Mo_3P/Mo_2C$ triple interface promoted the water dissociation and hydrogen desorption by lowering the energy barriers. Moreover, the authors reported a synergistic effect between the N-doped carbon and the heterostructure, contributing to the outstanding HER performance. Comparatively, MoOPC@NC-50, MoOPC@NC-100, MoOPC@NC-400 displayed significantly higher η_{10} of 155, 84, 75 mV and TSs of 68.47, 65.53 and 61.83 mV dec^{-1}, respectively, while the MoOPC@NC-200 annealed at 900 and 1000 °C presented even larger η_{10} (193 and 215 mV, respectively).

In other work, PMo_{12} and a Co(II) salt were used to prepare P-doped $CoMoO_4$ nanoflowers supported on Ni foam via a two-step route: first, different amounts of metal precursors (Co/Mo molar ratios of 2 : 1, 1 : 1, 1 : 2 and 3 : 1) were subjected to a hydrothermal procedure in presence of Ni foam and urea, followed by annealing in air at 500 °C [103]. P-$CoMoO_4$@Ni-1, prepared with Co/Mo ratio of 2 : 1, showed the highest specific surface area and the most promising electrocatalytic behaviour, achieving high current densities, up to 600 mA cm^{-2} at −0.6 V *vs.* RHE. Furthermore, it exhibited small overpotentials of 98 and 162 mV at 100 and 150 mA cm^{-2}, respectively, and TS of 66.24 mV dec^{-1}. In contrast, the other prepared electrocatalysts showed larger overpotentials of 292 (n_{Co}:n_{Mo} = 1 : 1), 326 (n_{Co}:n_{Mo} = 1 : 2) and 321 (n_{Co}:n_{Mo} = 3 : 1) mV at 100 mA cm^{-2}. The HER performance of P-$CoMoO_4$@Ni-1 was mainly attributed to favourable charge transfer and large surface area, which improved mass transport and H_2 bubbles release from the catalyst surface.

8.5 Concluding Remarks

This chapter demonstrates the wide application of POM-based/derived electrocatalysts towards HER. POMs in general show great potential for electrocatalysis, due to their advantageous structures containing several TMs able to participate in redox processes and abundant exposed active sites. However, their low electrical conductivity may limit the electrocatalytic performance, being the major drawback of POM electrocatalysts.

While POMs by themselves present moderate-to-good HER activity, it is usual to combine them with other catalytic active compounds (e.g., carbon materials, MOFs, TMs and polymers), in order to enhance the overall electrocatalytic performance; POM/carbon, POMOFs and POM-modified Ni foams are particularly popular combinations. POM-based hybrid materials present excellent HER activity, even in alkaline media, often overcoming Pt/C performance. The enhanced HER performance of these hybrids can mostly be explained by synergistic effects between the POM clusters and the other component. In most cases, the second component also contributes positively to the electrical conductivity and specific surface area.

On the other hand, POMs can be used as metal precursors for preparing SACs and TM carbides, sulfides, nitrides, phosphides or oxides. This synthesis route is particularly popular for HER application, obtaining POM-derived electrocatalysts with outstanding long-term stability and excellent HER performance, even in alkaline or neutral media.

Although the application of POM-based/derived electrocatalysts in HER is well studied, some points should be taken into consideration: (i) Pt counter electrodes are often used in contact with the electrolyte and the effect of platinum dissolution is not accounted for; (ii) post-catalysis characterization is rarely presented or performed, despite giving important information regarding electrocatalysts leaching, and structural and chemical alterations induced by the electrocatalytic process. Moreover, the electrocatalytic mechanism of these materials is rarely explored, hampering the design of more effective electrocatalysts. Finally, the evaluation of POM-based/derived electrocatalysts in electrolysers, besides the conventional 3-electrode cell, is a fundamental step to generate H_2 on industrial levels and should be attempted in future studies.

Acknowledgements

I.S. Marques thanks FCT for her PhD fellowship 2022.10224.BD. R. Matos acknowledges FCT for the PhD fellowship 2020.05342.BD. D.M. Fernandes thanks FCT for funding through the Individual Call to Scientific Employment Stimulus 2021.00771.CEECIND/CP1662/CT0007. This work received support and help from FCT/MCTES (LA/P/0008/2020 DOI 10.54499/LA/P/0008/2020, UIDP/50006/2020 DOI 10.54499/UIDP/50006/2020 and UIDB/50006/2020 DOI 10.54499/UIDB/50006/2020), through national funds.

List of Abbreviations

Ag/AgCl	Ag/AgCl reference electrode
C	carbon material
CB	carbon black
CC	carbon cloth
CFP	carbon fibre paper
CNT	carbon nanotube
Commerc.	commercial
CP	carbon paper
CTAB	cetyltrimethylammonium bromide
CV	cyclic voltammetry
CVD	chemical vapor deposition
DCA	dicyandiamide
Dec	decade
DFT	density functional theory
DHNW	double-hydroxide nanowires
DMF	*N*,*N*-dimethylformamide
EC	electrocatalyst
ECSA	electrochemical active surface area
FE	Faradaic efficiency
GCE	glassy carbon electrode
GO	graphene oxide
HER	hydrogen evolution reaction
HMCS	hollow mesoporous carbon sphere
ICP-MS	inductively coupled plasma mass spectrometry
KB	Ketjen black
LbL	layer-by-layer
LSV	linear sweep voltammetry
M	metal
MOF	metal–organic framework
MWCNT	multiwall carbon nanotube
NC	N-doped carbon
NF	nickel foam
NHE	normal hydrogen electrode
NP	nanoparticle
OER	oxygen evolution reaction
ORR	oxygen reduction reaction
PANI	polyaniline
PBS	phosphate-buffered saline
PEI	polyethylenimine
POM	polyoxometalate
POMOF	polyoxometalate–organic framework
PPDA	polydiallydimethylammonium
Pt/C	platinum immobilized on carbon black

PTA	phosphotungstic acid
QD	quantum dot
Ref	reference
rGF_ox	reduced oxidized graphene flakes
rGO	reduced graphene oxide
RHE	reversible hydrogen electrode
SAC	single-atom catalyst
SEM	scanning electron microscopy
TBA	tetrabutylammonium
TM	transition metal
TMC	transition-metal chalcogens
TMD	transition-metal dichalcogenides
TOF	turnover frequency
Trim	triangular 1,3,5-benzene tricarboxylate
TS	Tafel slope
ZIF	zeolitic imidazolate frameworks

List of Symbols

η	overpotential
η_{10}	overpotential at 10 mA cm^{-2}
η_{50}	overpotential at 50 mA cm^{-2}
e^-	electron
H*	adsorbed hydrogen atom
i	current intensity
j	current density
j_0	exchange current density
j_{max}	maximum current density (in module)
E	potential
R	resistance
A	ampere
V	volt
ΔG	Gibbs free energy
eV	electrovolt

References

1 Kumar, S.S. and Lim, H. (2022). An overview of water electrolysis technologies for green hydrogen production. *Energy Rep.* 8: 13793–13813.

2 Wang, S., Lu, A.L., and Zhong, C.J. (2021). Hydrogen production from water electrolysis: role of catalysts. *Nano Converg.* 8 (1): 23.

3 Shiva Kumar, S. and Himabindu, V. (2019). Hydrogen production by PEM water electrolysis – a review. *Mater. Sci. Energy Technol.* 2 (3): 442–454.

4 Xiang, C.X., Papadantonakis, K.M., and Lewis, N.S. (2016). Principles and implementations of electrolysis systems for water splitting. *Mater. Horiz.* 3 (3): 169–173.

5 Ruqia, B. and Choi, S.I. (2018). Pt and Pt-Ni(OH)(2) electrodes for the hydrogen evolution reaction in alkaline electrolytes and their nanoscaled electrocatalysts. *ChemSusChem* 11 (16): 2643–2653.

6 Kumar, S., Kaur, R., and Sharma, S. (2022). Recent reports on hydrogen evolution reactions and catalysis. *Results Chem.* 4: 11.

7 Freire, C., Freire, C., Fernandes, D.M. et al. (2018). POM & MOF-based electrocatalysts for energy-related reactions. *ChemCatChem* 10 (8): 1703–1730.

8 Zhang, L.L., Xiao, J., Wang, H. et al. (2017). Carbon-based electrocatalysts for hydrogen and oxygen evolution reactions. *ACS Catal.* 7 (11): 7855–7865.

9 Yang, H.-M. and Yuan, Z.-Y. (2023). Transition metal-based electrocatalysts for hydrogen generation and related energy carrier. In: *Transition Metal-Based Electrocatalysts: Applications in Green Hydrogen Production and Storage* (ed. P. Pathak and L. Singh), 1–20. American Chemical Society.

10 Zeb, Z., Huang, Y., Lulu, C. et al. (2023). Comprehensive overview of polyoxometalates for electrocatalytic hydrogen evolution reaction. *Coord. Chem. Rev.* 482: 53.

11 Li, C., Zhang, H., Liu, M. et al. (2023). Recent progress in metal–organic frameworks (MOFs) for electrocatalysis. *Ind. Chem. Mater.* 1 (1): 9–38.

12 Verma, J. and Goel, S. (2022). Cost-effective electrocatalysts for hydrogen evolution reactions (HER): challenges and prospects. *Int. J. Hydrogen Energy* 47 (92): 38964–38982.

13 Xu, H., Shang, H., Wang, C., and Du, Y. (2020). Surface and interface engineering of noble-metal-free electrocatalysts for efficient overall water splitting. *Coord. Chem. Rev.* 418: 20.

14 Chen, Z.J., Duan, X., Wei, W. et al. (2019). Recent advances in transition metal-based electrocatalysts for alkaline hydrogen evolution. *J. Mater. Chem. A* 7 (25): 14971–15005.

15 Xu, C., Zhang, M., Tin, X. et al. (2023). Recent advances in two-dimensional nanomaterials as bifunctional electrocatalysts for full water splitting. *J. Mater. Chem. A* 11 (35): 18502–18529.

16 Mahmood, N., Yao, Y., Zhang, J.W. et al. (2018). Electrocatalysts for hydrogen evolution in alkaline electrolytes: mechanisms, challenges, and prospective solutions. *Adv. Sci.* 5 (2): 23.

17 Zhao, J.L., Ding, Y., Wei, J. et al. (2014). A molecular Keggin polyoxometalate catalyst with high efficiency for visible-light driven hydrogen evolution. *Int. J. Hydrogen Energy* 39 (33): 18908–18918.

18 Keita, B. and Nadjo, L. (1989). New oxometalate-based materials for catalysis and electrocatalysis. *Mater. Chem. Phys.* 22 (1–2): 77–103.

19 Keita, B. and Nadjo, L. (1985). Activation of electrode surfaces - application to the electrocatalysis of the hydrogen evolution reaction. *J. Electroanal. Chem.* 191 (2): 441–448.

20 Zhang, C., Hong, Y., Dai, R. et al. (2015). Highly active hydrogen evolution electrodes via co-deposition of platinum and polyoxometalates. *ACS Appl. Mater. Interfaces* 7 (21): 11648–11653.

21 Rong, C.Y. and Pope, M.T. (1992). Lacunary polyoxometalate anions are pi-acceptor ligands - characterization of some tungstoruthenate(II,III,IV,V) heteropolyanions and their atom-transfer reactivity. *JACS* 114 (8): 2932–2938.

22 Biboum, R.N., Keita, B., Franger, S. et al. (2010). Pd-0@Polyoxometalate nanostructures as green electrocatalysts: illustrative example of hydrogen production. *Materials* 3 (1): 741–754.

23 Barsukova-Stuckart, M., Izarova, N.V., Jameson, G.B. et al. (2011). Synthesis and characterization of the dicopper(II)-containing 22-palladate(II) (Cu2Pd22P12O60)-Pd-II-P-II-O-V(OH)(8) (20-). *Angew. Chem. Int. Ed.* 50 (11): 2639–2642.

24 Wang, P., Zhang, H., Wang, P. et al. (2022). A crown ether supramolecular host-guest complex with Keggin polyoxometalate: synthesis, crystal structure and electrocatalytic performance for hydrogen evolution reaction. *Catal. Commun.* 165: 6.

25 Fernandes, D.M. and Freire, C. (2015). Carbon nanomaterial-phosphomolybdate composites for oxidative electrocatalysis. *ChemElectroChem* 2 (2): 269–279.

26 Zhang, Y.L., Luo, M., Yang, Y. et al. (2019). Advanced multifunctional electrocatalysts for energy conversion. *ACS Energy Lett.* 4 (7): 1672–1680.

27 Liu, R.J., Zhang, G., Cao, H. et al. (2016). Enhanced proton and electron reservoir abilities of polyoxometalate grafted on graphene for high-performance hydrogen evolution. *Energy Environ. Sci.* 9 (3): 1012–1023.

28 Jiang, M., Zhu, D., Cai, J. et al. (2014). Electrocatalytic hydrogen evolution and oxygen reduction on polyoxotungstates/graphene nanocomposite multilayers. *J. Phys. Chem. C* 118 (26): 14371–14378.

29 Ensafi, A.A., Heydari-Soureshjani, E., Jafari-Asl, M., and Rezaei, B. (2016). Polyoxometalate-decorated graphene nanosheets and carbon nanotubes, powerful electrocatalysts for hydrogen evolution reaction. *Carbon* 99: 398–406.

30 Ensafi, A.A., Heydari-Soureshjani, E., and Rezaei, B. (2017). Nanostructure polyoxometalates containing Co, Ni, and Cu as powerful and stable catalysts for hydrogen evolution reaction in acidic and alkaline solutions. *Int. J. Hydrogen Energy* 42 (8): 5026–5034.

31 Fernandes, D.M., Araújo, M., Haider, A. et al. (2018). Polyoxometalate-graphene electrocatalysts for the hydrogen evolution reaction. *ChemElectroChem* 5 (2): 273–283.

32 Yu, F.Y., Lang, Z., Yin, L.-Y. et al. (2020). Pt-O bond as an active site superior to Pt-0 in hydrogen evolution reaction. *Nat. Commun.* 11 (1): 7.

33 Jawale, D.V., Fossard, F., Miserque, F. et al. (2022). Carbon nanotube-polyoxometalate nanohybrids as efficient electrocatalysts for the hydrogen evolution reaction. *Carbon* 188: 523–532.

34 Keita, B., Kortz, U., Holzle, L.R.B. et al. (2007). Efficient hydrogen-evolving cathodes based on proton and electron reservoir behaviors of the phosphotungstate $H_7P_8W_{48}O_{184}$ (33-) and the Co(II)-containing silicotungstates Co-6(H_2O)(30){Co_9Cl_2(OH)(3)(H_2O)(9)(beta-SiW_8O_{31})(3)}(5-) and {Co-3(B-beta-SiW_9O_{33}(OH))(B-beta-SiW_8O_{29}OH)(2)}(2) (22-). *Langmuir* 23 (19): 9531–9534.

35 Zheng, S.T., Zhang, H., and Yang, G.Y. (2008). Designed synthesis of POM-organic frameworks from {Ni_6PW_9} building blocks under hydrothermal conditions. *Angew. Chem. Int. Ed.* 47 (21): 3909–3913.

36 Shen, Q.B., Gómez-García, C., Sun, W. et al. (2021). Improving the photocatalytic H-2 evolution activity of Keggin polyoxometalates anchoring copper-azole complexes. *Green Chem.* 23 (8): 3104–3114.

37 Nohra, B., El Moll, H., Rodriguez Albelo, L.M. et al. (2011). Polyoxometalate-based metal organic frameworks (POMOFs): structural trends, energetics, and high electrocatalytic efficiency for hydrogen evolution reaction. *JACS* 133 (34): 13363–13374.

38 Qin, J.S., Du, D.-Y., Guan, W. et al. (2015). Ultrastable polymolybdate-based metal organic frameworks as highly active electrocatalysts for hydrogen generation from water. *JACS* 137 (22): 7169–7177.

39 Shah, W.A., Ibrahim, S., Abbas, S. et al. (2021). Nickel containing polyoxometalates incorporated in two different metal-organic frameworks for hydrogen evolution reaction. *J. Environ. Chem. Eng.* 9 (5): 10.

40 Liu, M.M., Wu, X.-M., Guo, H.-X. et al. (2023). Solvent effect on the fabrication of POMs-based MOFs microspheres: dual-function in electrocatalytic hydrogen evolution and catalytic reduction of Cr^{6+}. *Catal. Lett.* 153 (9): 2803–2812.

41 Gautam, J., Kannan, K., Dahal, B. et al. (2022). Heterostructure of polyoxometalate/zinc-iron-oxide nanoplates as an outstanding bifunctional electrocatalyst for the hydrogen and oxygen evolution reaction. *J. Colloid Interface Sci.* 618: 419–430.

42 Jia, X.Y., Streb, C., and Song, Y.F. (2019). Devisable POM/Ni foam composite: precisely control synthesis toward enhanced hydrogen evolution reaction at high pH. *Chem. Eur. J.* 25 (68): 15548–15554.

43 Zhang, Z.F., Meng, Y., Su, H. et al. (2021). Controllable design of 3D hierarchical Co/Ni-POM nanoflower compounds supported on Ni foam for the hydrogen evolution reaction. *New J. Chem.* 45 (26): 11536–11543.

44 Zhang, Z.F., Ran, J., Wu, Z., and Zhao, B. (2022). Hierarchical Ni modified PW_{12} clusters in situ integrated on Ni foam for efficient alkaline hydrogen evolution. *J. Cluster Sci.* 8.

45 Gautam, J., Liu, Y., Gu, J. et al. (2021). Fabrication of polyoxometalate anchored zinc cobalt sulfide nanowires as a remarkable bifunctional electrocatalyst for overall water splitting. *Adv. Funct. Mater.* 31 (46): 13.

46 Singh, V., Ma, P., Drew, M. et al. (2017). Heterooctamolybdate-based clusters H-3(Cp*Rh)(4)PMo_8O_{32} and H-5 Na-2(Cp*Ir)(4)PMo_8O_{34} and derived hybrid nanomaterials with efficient electrocatalytic hydrogen evolution reaction activity. *Inorg. Chem.* 56 (20): 12520–12528.

47 Zang, D.J., Huang, Y., Li, Q. et al. (2019). Cu dendrites induced by the Anderson-type polyoxometalate $NiMo_6O_{24}$ as a promising electrocatalyst for enhanced hydrogen evolution. *Appl. Catal. B Environ.* 249: 163–171.

48 Zhang, T.R., Weng, S., Wang, X. et al. (2022). Platinum atomic clusters embedded in polyoxometalates-carbon black as an efficient and durable catalyst for hydrogen evolution reaction. *J. Colloid Interface Sci.* 624: 704–712.

49 Huang, M.H., Bi, L., Shen, Y. et al. (2004). Nanocomposite multilayer film of preyssler-type polyoxometalates with fine tunable electrocatalytic activities. *J. Phys. Chem. B* 108 (28): 9780–9786.

50 Yuan, J.H., Chen, Y., Han, D. et al. (2006). Synthesis of highly faceted multiply twinned gold nanocrystals stabilized by polyoxometalates. *Nanotechnology* 17 (18): 4689–4694.

51 Golmohammadi, F. and Amiri, M. (2021). Fabrication of MEA from biomass-based carbon nanofibers composited with nickel-cobalt oxides as a new electrocatalyst for oxygen reduction reaction in passive direct methanol fuel cells. *Electrocatalysis* 11 (5): 485–496.

52 Xiong, X., Jiang, R., Deng, B. et al. (2020). Bionic structural design and electrochemical manufacture of WC/N-doped carbon hybrids as efficient ORR catalyst. *J. Electrochem. Soc.* 167 (6): 7.

53 Huang, B., Ke, D., Xiong, Z. et al. (2020). Covalent hybrid materials between polyoxometalates and organic molecules for enhanced electrochemical properties. *J. Mater. Sci.* 55 (13): 5554–5570.

54 Sakthinathan, I., Köhling, J., Wagner, V. et al. (2023). Layer-by-layer construction of a nanoarchitecture by polyoxometalates and polymers: enhanced electrochemical evolution reaction. *ACS Appl. Mater. Interfaces* 15 (2): 2861–2872.

55 Gautam, J., Liu, Y., Gu, J. et al. (2022). Three-dimensional nano assembly of nickel cobalt sulphide/polyaniline@polyoxometalate/reduced graphene oxide hybrid with superior lithium storage and electrocatalytic properties for hydrogen evolution reaction. *J. Colloid Interface Sci.* 614: 642–654.

56 Zhang, L.L., Ren, Y., Liu, W. et al. (2018). Single-atom catalyst: a rising star for green synthesis of fine chemicals. *Natl. Sci. Rev.* 5 (5): 653–672.

57 Yang, X.F., Wang, A., Qiao, B. et al. (2013). Single-atom catalysts: a new frontier in heterogeneous catalysis. *Acc. Chem. Res.* 46 (8): 1740–1748.

58 Guo, J.W., Liu, H., Li, D. et al. (2022). A minireview on the synthesis of single atom catalysts. *RSC Adv.* 12 (15): 9373–9394.

59 Yang, Y., Fu, J., Zhang, Y. et al. (2021). Molecular engineering for bottom-up construction of high-performance non-precious-metal electrocatalysts with well-defined active sites. *J. Phys. Chem. C* 125 (41): 22397–22420.

60 Han, Y.H., Dai, J., Xu, R. et al. (2021). Notched-polyoxometalate strategy to fabricate atomically dispersed Ru catalysts for biomass conversion. *ACS Catal.* 11 (5): 2669–2675.

61 Yu, F.Y., Lang, Z., Zhou, Y.-J. et al. (2021). Revealing hydrogen evolution performance of single-atom platinum electrocatalyst with polyoxometalate molecular models. *ACS Energy Lett.* 6 (11): 4055–4062.

62 Liu, R.J. and Streb, C. (2021). Polyoxometalate-single atom catalysts (POM-SACs) in energy research and catalysis. *Adv. Energy Mater.* 11 (25): 13.

63 Lu, Y.K., Yue, C., Li, Y. et al. (2021). Atomically dispersed Ni on Mo_2C embedded in N, P co-doped carbon derived from polyoxometalate supramolecule for high-efficiency hydrogen evolution electrocatalysis. *Appl. Catal. B Environ.* 296: 11.

64 Chen, W.F., Muckerman, J.T., and Fujita, E. (2013). Recent developments in transition metal carbides and nitrides as hydrogen evolution electrocatalysts. *Chem. Commun.* 49 (79): 8896–8909.

65 Jun, H., Kim, S., and Lee, J. (2020). Development strategies in transition metal carbide for hydrogen evolution reaction: a review. *Korean J. Chem. Eng.* 37 (8): 1317–1330.

66 Wu, H.B., Xia, B.Y., Yu, L. et al. (2015). Porous molybdenum carbide nano-octahedrons synthesized via confined carburization in metal-organic frameworks for efficient hydrogen production. *Nat. Commun.* 6: 8.

67 Zhang, L.N., Li, S., Huaqiao, T. et al. (2017). MoP/Mo_2C@C: a new combination of electrocatalysts for highly efficient hydrogen evolution over the entire pH range. *ACS Appl. Mater. Interfaces* 9 (19): 16270–16279.

68 Wang, M., Dilpazir, S., Lu, P. et al. (2018). Synthesis of polyoxometalates derived bifunctional catalyst towards efficient overall water splitting in neutral and alkaline medium. *J. Colloid Interface Sci.* 532: 774–781.

69 Huang, Y.C., Hu, J., Xu, H. et al. (2018). Fine tuning electronic structure of catalysts through atomic engineering for enhanced hydrogen evolution. *Adv. Energy Mater.* 8 (24): 11.

70 Yu, F.Y., Gao, Y., Lang, Z. et al. (2018). Electrocatalytic performance of ultrasmall Mo_2C affected by different transition metal dopants in hydrogen evolution reaction. *Nanoscale* 10 (13): 6080–6087.

71 Yan, G., Feng, X.-J., Khan, S. et al. (2018). Polyoxometalate and resin-derived P-doped Mo_2C@N-doped carbon as a highly efficient hydrogen-evolution reaction catalyst at all pH values. *Chem. Asian J.* 13 (2): 158–163.

72 Xu, C.Y., Wu, F., Yang, X. et al. (2022). Synergistically coupling of $Ni_3Mo_3C/Mo_2C/Ti_3C_2T_x$ MXene/N-doped carbon electrocatalyst towards enhanced hydrogen evolution activity. *J. Alloys Compd.* 920: 10.

73 Men, L.L., Shi, T., Li, J. et al. (2022). Bimetallic polyoxometalate derived Co/WN composite as electrocatalyst for high-efficiency hydrogen evolution. *Int. J. Hydrogen Energy* 47 (64): 27452–27459.

74 Xiao, J.M., Zhang, S., Sun, Y. et al. (2023). Urchin-like structured $MoO_2/Mo_3P/Mo_2C$ triple-interface heterojunction encapsulated within nitrogen-doped carbon for enhanced hydrogen evolution reaction. *Small* 19 (12): 8.

75 Yue, C.L., Liu, N., Li, Y. et al. (2023). From atomic bonding to heterointerfaces: Co_2P/WC constructed by lacunary polyoxometalates induced strategy as efficient hydrogen evolution electrocatalysts at all pH values. *J. Colloid Interface Sci.* 645: 276–286.

76 Gao, J., Yang, J., Cai, X. et al. (2020). Facile preparation of $Co_6W_6C/W@NC$ electrocatalyst for hydrogen evolution reaction using transition metal substituted polyoxometalates as precursor. *Mater. Lett.* 271: 4.

77 Xu, C.Y., Li, J., Sun, D. et al. (2021). Co/WC@NC electrocatalysts derived from polyoxometalates (POM) for efficient hydrogen evolution. *Nanotechnology* 32 (37): 9.

78 Liu, Y., Li, J., Sun, D. et al. (2021). Self-assembly of bimetallic polyoxometalates and dicyandiamide to form for efficient electrochemical hydrogen generation. *New J. Chem.* 46 (1): 178–184.

79 Huang, Y.C., Ge, J., Hu, J. et al. (2018). Nitrogen-doped porous molybdenum carbide and phosphide hybrids on a carbon matrix as highly effective electrocatalysts for the hydrogen evolution reaction. *Adv. Energy Mater.* 8 (6): 9.

80 Shiraz, H.G., Crispin, X., and Berggren, M. (2021). Transition metal sulfides for electrochemical hydrogen evolution. *Int. J. Hydrogen Energy* 46 (47): 24060–24077.

81 Huang, Y.M., Ma, Z., Hu, Y. et al. (2016). An efficient $WSe_2/Co_{0.85}Se$/graphene hybrid catalyst for electrochemical hydrogen evolution reaction. *RSC Adv.* 6 (57): 51725–51731.

82 He, P.L., Li, H., Liu, H., and Wang, X. (2016). Generalized synthesis of hierarchical transition metal dichalcogenide nanosheets from polyoxometalates. *ChemNanoMat* 2 (7): 665–670.

83 Ghosh, S., Azad, U., Singh, A. et al. (2017). Facile synthesis of MoS_x and MoS_x-rGO composite: excellent electrocatalyst for hydrogen evolution reaction. *ChemistrySelect* 2 (35): 11590–11598.

84 Sun, H.H., Ji, X., Qiu, Y. et al. (2019). Poor crystalline MoS_2 with highly exposed active sites for the improved hydrogen evolution reaction performance. *J. Alloys Compd.* 777: 514–523.

85 Huang, Z., Yang, Z., Jia, Q. et al. (2022). Bimetallic Co-Mo sulfide/carbon composites derived from polyoxometalate encapsulated polydopamine-decorated ZIF nanocubes for efficient hydrogen and oxygen evolution. *Nanoscale* 14 (12): 4726–4739.

86 Hou, J.G., Zhang, B., Li, Z. et al. (2018). Vertically aligned oxygenated-CoS_2-MoS_2 heteronanosheet architecture from polyoxometalate for efficient and stable overall water splitting. *ACS Catal.* 8 (5): 4612–4621.

87 Lu, Y.K., Guo, X., Yang, L. et al. (2020). Highly efficient CoMoS heterostructure derived from vertically anchored Co_5Mo_{10} polyoxometalate for electrocatalytic overall water splitting. *Chem. Eng. J.* 394: 11.

88 Tang, J.Y., Chu, Y., Wang, K. et al. (2021). Facile in situ synthesis of an ultrafine and defect-rich Co-doped MoS_2 nanocluster catalyst for efficient hydrogen evolution reaction in alkaline medium. *ACS Appl. Energy Mater.* 4 (3): 2300–2306.

89 Sun, F.Y., Yang, L., Yue, C. et al. (2022). Strandberg-type polyoxometalate deriving O,P co-doped NiMoS/CC catalyst for highly efficient hydrogen evolution electrocatalysis. *Int. J. Hydrogen Energy* 47 (61): 25571–25582.

90 Yue, C.L., Zhou, Y., Liu, Y. et al. (2022). Achieving ultra-dispersed 1T-Co-MoS_2@HMCS via space-confined engineering for highly efficient hydrogen evolution in the universal pH range. *Inorg. Chem. Front.* 9 (11): 2617–2627.

91 Dong, S.M., Chen, X., Zhang, X., and Cui, G. (2013). Nanostructured transition metal nitrides for energy storage and fuel cells. *Coord. Chem. Rev.* 257 (13–14): 1946–1956.

92 Xie, J.F. and Xie, Y. (2016). Transition metal nitrides for electrocatalytic energy conversion: opportunities and challenges. *Chem. Eur. J.* 22 (11): 3588–3598.

93 Gao, B.A., Li, X., Ding, K. et al. (2019). Recent progress in nanostructured transition metal nitrides for advanced electrochemical energy storage. *J. Mater. Chem. A* 7 (1): 14–37.

94 Huang, Y.C., Zhou, W., Kong, W. et al. (2022). Atomically interfacial engineering on molybdenum nitride quantum dots decorated N-doped graphene for high-rate and stable alkaline hydrogen production. *Adv. Sci.* 9 (36): 12.

95 Du, H.T., Kong, R.-M., Guo, X. et al. (2018). Recent progress in transition metal phosphides with enhanced electrocatalysis for hydrogen evolution. *Nanoscale* 10 (46): 21617–21624.

96 Bhunia, K., Chandra, M., Sharma, S.K. et al. (2023). A critical review on transition metal phosphide based catalyst for electrochemical hydrogen evolution reaction: Gibbs free energy, composition, stability, and true identity of active site. *Coord. Chem. Rev.* 478: 33.

97 Tang, Y.J., Zhu, H.-J., Dong, L.-Z. et al. (2019). Solid-phase hot-pressing of POMs-ZIFs precursor and derived phosphide for overall water splitting. *Appl. Catal. B Environ.* 245: 528–535.

98 Liu, Y.H., Zhang, D., Zhang, K. et al. (2020). Facile immobilization of polyoxometalates for low-cost molybdenum/tungsten phosphide nanoparticles on carbon black for efficient electrocatalytic hydrogen evolution. *J. Coord. Chem.* 73 (17–19): 2590–2601.

99 Jiao, Y.Q., Yan, H., Wang, D. et al. (2022). Multi-touch cobalt phosphide-tungsten phosphide heterojunctions anchored on reduced graphene oxide boosting wide pH hydrogen evolution. *Sci. China Mater.* 65 (5): 1225–1236.

100 Liu, Y., Yue, C., Sun, F. et al. (2023). Superhydrophilic molybdenum phosphide quantum dots on porous carbon matrix for boosting hydrogen evolution reaction. *Chem. Eng. J.* 454: 12.

101 Xiong, W., Yin, H., Wu, T. et al. (2022). Challenges and opportunities of transition metal oxides as electrocatalysts. *Chem. Eur. J.* 8.

102 Pratama, D.S.A., Haryanto, A., and Lee, C.W. (2023). Heterostructured mixed metal oxide electrocatalyst for the hydrogen evolution reaction. *Front. Chem.* 11: 8.

103 Wen, Y., Li, C., Wang, J. et al. (2022). Hierarchically self-supporting phosphorus-doped $CoMoO_4$ nanoflowers arrays toward efficient hydrogen evolution reaction. *ACS Appl. Energy Mater.* 5 (6): 6814–6822.

9

POM-based Electrocatalysts for Oxygen Reduction Reactions

Henrique Araújo, Biljana Šljukić, and Diogo M.F. Santos

Universidade de Lisboa, Instituto Superior Técnico, Center of Physics and Engineering of Advanced Materials, Laboratory for Physics of Materials and Emerging Technologies, Chemical Engineering Department, Lisbon 1049-001, Portugal

9.1 Introduction

Oxygen reduction reaction (ORR) is essential in life processes, as well as gas sensors and electrochemical energy conversion (and electrochemistry in general). There is no need to stress out the importance of, for instance, biological respiration. The reduction of oxygen is essential for the operation of Clark cells, electrochemical detectors of molecular oxygen. Oxygen is reduced to hydroxyl ion (OH^-), releasing four electrons per O_2 molecule and thus generating a current that is directly proportional to the amount of present oxygen. The first choices for the electrode materials were noble metals such as gold or silver or noble-metal-plated materials. Still, it was observed that O_2 might be reduced to hydrogen peroxide (H_2O_2) under certain conditions, interfering with the established protocol of oxygen sensing in both liquid and gaseous samples.

As for electrochemical energy production, its importance has become more evident with the ongoing energy crises that have become more severe in the last few years. As society moves away from its global dependency on fossil fuels, which undeniably lead to climate change, greener alternatives are required, particularly by using green hydrogen as a promising energy carrier. Hydrogen has many applications in the chemical industry (e.g. ammonia production, oil refining, steelmaking) and the transportation sector, with the latter relying on the efficient energy conversion of fuel cells. Fuel cells convert the chemical energy of a fuel into electricity through a clean and green operation. The fuel oxidation (typically hydrogen, but also alcohols, borohydride, etc.) proceeds at the anode, paired with the reduction of an oxidant, i.e. oxygen, at the cathode. As for metal–air batteries, oxidation of a metal (typically aluminium, zinc, lithium, manganese, etc.) proceeds at the anode in parallel with oxygen reduction at the cathode during the battery discharge. Thus, ORR is a critical reaction for the operation of fuel cells and metal–air batteries. Still, ORR is a rather complex and kinetically slow reaction that limits the efficiency of the abovementioned devices. The reaction can take different pathways depending on electrode

Applied Polyoxometalate-based Electrocatalysis, First Edition. Edited by Diana M. Fernandes.

material and electrolyte composition. Two main ORR pathways have been identified: direct four-electron reduction and two-electron reduction. In some circumstances, oxygen reduction can also proceed as a one-electron process.

The most active ORR electrocatalyst is carbon-supported platinum (Pt/C). The complexity of ORR is well illustrated by the fact that even the use of Pt-based electrocatalysts results in a loss of ca. 400 mV at the cathode in proton-exchange membrane fuel cells. Moreover, the use of Pt/C hinders the large-scale application of fuel cells as it suffers from two main drawbacks: high price due to Pt scarcity and low stability under operating conditions. Selectivity towards ORR and electrocatalyst stability under fuel cell long-term operation are crucial issues to be tackled [1]. The problem of lack of selectivity towards ORR and low fuel-crossover tolerance is again well illustrated in the case of Pt-based electrocatalysts when the loss mentioned above becomes even more pronounced in the case of, for instance, direct methanol fuel cells. In this case, the additional loss results from the electrocatalyst's surface poisoning and so-called mixed potential caused by the crossover of methanol from the anodic to the cathodic compartment. Carbon monoxide (CO) generated during methanol oxidation at Pt-based electrocatalysts strongly adsorbs on these electrocatalysts' surface, blocking the active centres for ORR, i.e., poisoning the electrocatalysts' surface. Additionally, the positive current generated by methanol oxidation at the cathode needs to be compensated by the negative current generated by oxygen reduction. The overpotential necessary to produce this amount of reduction current is called 'mixed potential'. The durability of electrocatalysts, including those Pt-based ones, is severely jeopardized by metal dissolution to some degree and instability, particularly in acidic media.

Thus, it is imperative to design novel electrocatalysts for ORR that are efficient, stable and economically viable. The alternative groups of electrocatalysts investigated so far encompass noble metal alloys, different transition-metal compounds (macrocyclic, chalcogenides, carbides, nitrides, phosphides and sulphides) and carbon materials.

Polyoxometalates (POM) emerged only recently as potential electrocatalysts for ORR. They have unique functionality and thus represent active components of materials for applications in catalysis [2] and electrocatalysis [3] for green chemical processes and energy. Several POM characteristics make them suitable for the electrocatalysis of ORR. Synthesis of POMs is rather simple and inexpensive and thus scalable to the commercial level. POM composition and activity can be tuned via structural design and mechanistic considerations to meet the demands of a specific application. The anionic character of POMs enables linking with a particular cationic species to create a water-insoluble species while maintaining characteristics of the anionic and cationic moieties. POMs can store a high number of protons/electrons, enabling their 'burst' transfer. Finally, they have high thermal stability. However, POMs suffer from low conductivity and can suffer from solubility in aqueous media. Still, a vital shortcoming of POMs for electrocatalysis of ORR can be the lack of active sites. Thus, the use of POMs and POM-based materials for electrocatalysis of ORR is a developing area. POMs are studied solely or as composites with highly conductive, high-surface area carbon materials. Linking of a POM to a carbon material occurs via spontaneous strong electrostatic interaction or charge transfer between the two without additional linkers [4, 5]. This further leads to maximal accessibility of POM's metal active sites by the electrolyte and easy charge transfer at the POM/carbon interface, significantly boosting the ORR [6, 7].

9.2 Fundamentals of Oxygen Reduction Reaction

ORR represents a complex reaction involving several steps and intermediates and, as such, represents the rate- and efficiency-limiting step in electrochemical energy conversion and storage devices. The exact mechanism will depend on factors, including electrode material and electrolyte composition. As mentioned, two main mechanisms have been established – direct four-electron reduction (to water or OH^-, depending on the electrolyte pH) and two-electron reduction to H_2O_2, Scheme 9.1. One-electron reduction to superoxide (O_2^-) was also reported for non-aqueous aprotic solvents and/or alkaline electrolytes.

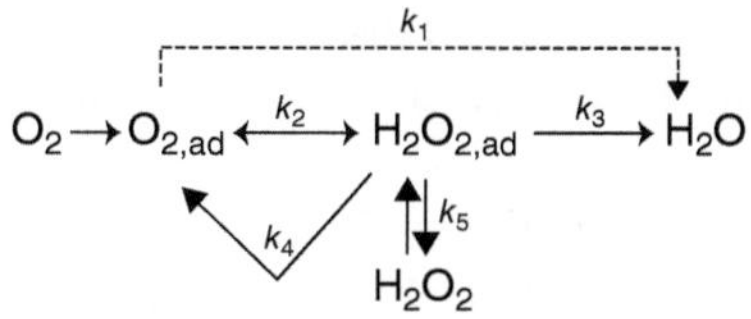

Scheme 9.1 Simplified ORR pathways. Source: Benzbiria et al. [8]/with permission of Elsevier.

The direct four-electron pathway can be described by Eqs. (9.1) and (9.2), for alkaline and acidic media, respectively.

$$O_2 + 2H_2O + 4e^- \rightarrow 4OH^- \quad E^0 = +0.401\ \text{V}\ vs.\ SHE \tag{9.1}$$

$$O_2 + 4H^+ + 4e^- \rightarrow 2H_2O \qquad E^0 = +1.229\ \text{V}\ vs.\ SHE \tag{9.2}$$

As for the two-electron or peroxide pathway, it can be described in alkaline solution by Eq. (9.3),

$$O_2 + H_2O + 2e^- \rightarrow HO_2^- + OH^- \quad E^0 = -0.065\ \text{V}\ vs.\ SHE \tag{9.3}$$

followed either by the further reduction (Eq. 9.4),

$$HO_2^- + H_2O + 2e^- \rightarrow 3OH^- \qquad E^0 = +0.867\ \text{V}\ vs.\ SHE \tag{9.4}$$

or by decomposition under a disproportionation reaction (Eq. 9.5).

$$2HO_2^- \rightarrow 2OH^- + O_2 \tag{9.5}$$

The peroxide pathway in acidic solutions is given by Eq. (9.6),

$$O_2 + 2H^+ + 2e^- \rightarrow H_2O_2 \qquad E^0 = +0.67\ \text{V}\ vs.\ SHE \tag{9.6}$$

followed either by the process described in Eq. (9.7)

$$H_2O_2 + 2H^+ + 2e^- \rightarrow 2H_2O \qquad E^0 = +1.77\,\text{V}\ \textit{vs.}\ \textit{SHE} \tag{9.7}$$

or by the H_2O_2 disproportionation (Eq. 9.8),

$$2H_2O_2 \rightarrow 2H_2O + O_2 \tag{9.8}$$

where E^0 is the standard electrode potential *vs.* the standard hydrogen electrode (SHE) at 25 °C.

Reduction of oxygen in aqueous media involves several steps (i) O_2 molecule diffuses to the electrocatalyst/electrolyte interface and is then adsorbed at the electrocatalyst's surface, (ii) electron released during the oxidation of fuel at the anode is transferred to the adsorbed O_2 molecule, (iii) O=O bond weakens and breaks and (iv) generated OH− ions are desorbed into electrolyte solution [9]. Computational charge density distribution pointed out two possible adsorption models of oxygen species on the surface of electrocatalysts [9]. The Yeager model suggests bidentate adsorption of the O_2 molecule, i.e. coordination of two oxygen atoms with the active site. The Pauling model suggests the end-on O_2 molecule adsorption, i.e. the coordination of one oxygen atom vertically to the electrocatalyst's surface. Oxygen reduction mechanisms are also classified as associative and dissociative mechanisms, depending on whether the adsorbed O_2 molecule dissociates before or after the protonation.

Intermediate species generated during the ORR include O*, OOH* and OH*, with * being an active site. Scheme 9.2 represents the ORR mechanism in alkaline media, where pathway (2) is less probable due to the high-energy barrier for O—O bond breaking. Moreover, two types of four-electron transfer reduction have been suggested: mono-site reduction and dual-site reduction. The rate-determining step (RDS) of mono-site mechanism is the removal of O* and OH*, whereas in the dual-site mechanism, it is the breaking of the O—O bond. The four-electron mechanism is desirable for maximum current efficiency, while the two-electron mechanism must be suppressed in fuel cells and metal–air batteries. This being said, it should be kept in mind that two-electron transfer reduction of O_2 is sometimes favoured to produce H_2O_2.

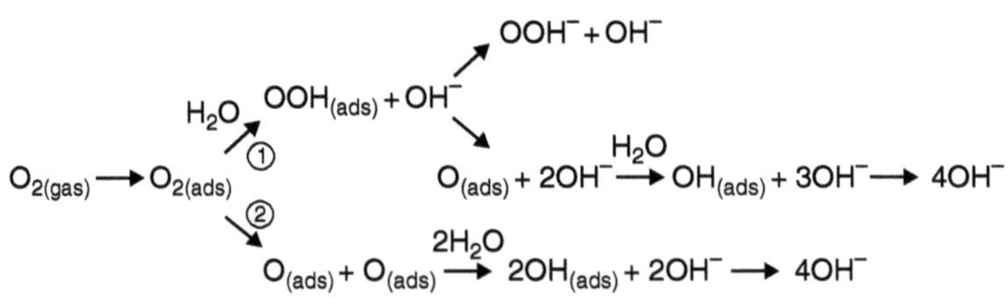

Scheme 9.2 ORR pathways in alkaline media. Source: Yu et al. [10]/with permission of Elsevier.

The performance of an electrocatalyst for the ORR is usually scrutinized by linear scan voltammetry with a rotating disc electrode (RDE) or rotating ring-disc electrode (RRDE). An example of an ORR polarization curve is presented in Figure 9.1

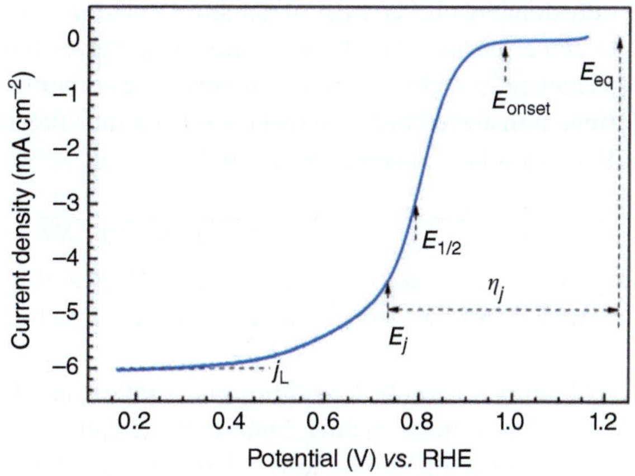

Figure 9.1 ORR polarization curve with relevant performance parameters. Source: Ge et al. [11]/with permission of American Chemical Society.

[11]. Commonly used indicators of materials' performance for ORR include onset potential (E_{onset}), half-wave potential ($E_{1/2}$), Tafel slope (b), number of electrons exchanged (n), diffusion-limited current density (j_d) and kinetic current density (j_k). There are different definitions of ORR onset potential [12], and it has been pointed out that this activity indicator needs to be taken into consideration with care [13]. E_{onset} might be defined as the potential at which the recorded current evidently diverges from the residual current. It can further be defined as the intersection of the tangent to the baseline and the rising part of the polarization curve. Finally, onset potential is defined as the potential at the current density of $-3\,\mu A\,cm^{-2}$. Overpotential at a given current density (η_j) has been suggested as a better activity indicator than onset potential [13]. Half-wave potential represents the potential at the point on the rising part of a polarization curve, where the difference between the diffusion-limited current density and the residual current density is half the limiting current density [14]. Higher onset and half-wave potentials reflect a lower overpotential necessary to reach a defined current density and thus reflect a higher ORR performance. It should be kept in mind that, depending on the employed electrocatalyst, diffusion-limited control might not be established, making determination of the half-wave potential redundant. As for diffusion-limited current density, it takes a typical value close to $6\,mA\,cm^{-2}$ at 1600 rpm in the case of Pt-based electrocatalysts in 0.1 M KOH.

The Tafel slope (b) reflects the current increment rate with the overpotential. The lower the Tafel slope value, the higher the current density increment with the given overpotential and, thus, the higher the activity of a material for the ORR. Moreover, it indicates the ORR pathway at the given electrocatalyst. For instance, typical values of Tafel slope for ORR at Pt-based electrocatalysts in alkaline media are 60 and $120\,mV\,dec^{-1}$, depending on the overpotential range. A Tafel slope of $60\,mV\,dec^{-1}$ suggests a pseudo-two-electron transfer as the RDS, while $120\,mV\,dec^{-1}$ suggests the first-electron reduction of O_2 as the RDS and suggests facilitated subsequent reduction and O—O bond breaking.

Finally, a feasible and efficient ORR electrocatalyst should catalyse a direct mechanism, i.e. direct four-electron reduction of O_2 to H_2O or OH^-. Direct reduction not only delivers higher current densities (as four electrons are transferred as opposite to two electrons), but it also avoids the generation of H_2O_2 that could damage parts of fuel cells such as carbon support or membrane separator. The number of electrons transferred per O_2 molecule during the ORR is typically determined using the Koutecký–Levich equation (Eq. 9.9),

$$\frac{1}{j} = \frac{1}{j_k} + \frac{1}{j_d} = \frac{1}{j_k} + \frac{1}{0.62nFD^{2/3}\nu^{-1/6}C}\frac{1}{\omega^{1/2}} \tag{9.9}$$

where j is the measured current density, F is the Faraday constant, D and C are oxygen diffusion coefficient and concentration, respectively, and ν is the kinematic viscosity of the electrolyte solution.

Stability under ORR conditions is usually evaluated by the accelerated ageing test, i.e. by continuous cycling. Stability tests can also be carried out using chronoamperometry at a potential lower than the potential of the peak corresponding to the oxygen reduction or by chronopotentiometry at a defined current density.

The binding energy of the oxygen intermediates represents a descriptor of the materials' activity trends towards ORR. Essentially, the so-called Sabatier principle says that a catalyst that binds oxygen intermediates slightly weaker than metals on top of the volcano plots, i.e. at an intermediate strength, will boost the ORR kinetics, Figure 9.2. A weak interaction, i.e. a weak binding, limits the adsorption of the O_2 molecule and its subsequent dissociation, generating O*. In contrast, a very strong interaction, i.e. a strong binding, hinders the removal of generated O* and OH* intermediates, thus blocking the active sites. Accordingly, theoretical considerations such as density functional theory (DFT) calculations are performed for deeper insight and a better understanding of the ORR mechanism [7, 14]. DFT calculations are used to estimate the binding energy between the reactants/reaction intermediates and the electrocatalyst's surface, as the Sabatier principle clearly indicates their interaction as crucial for the catalysis of ORR [17, 18]. The difference in the adsorption energy of two oxygen intermediates represents the reaction energy of a particular elementary step. Hence, such determined reaction energy is taken as a descriptor

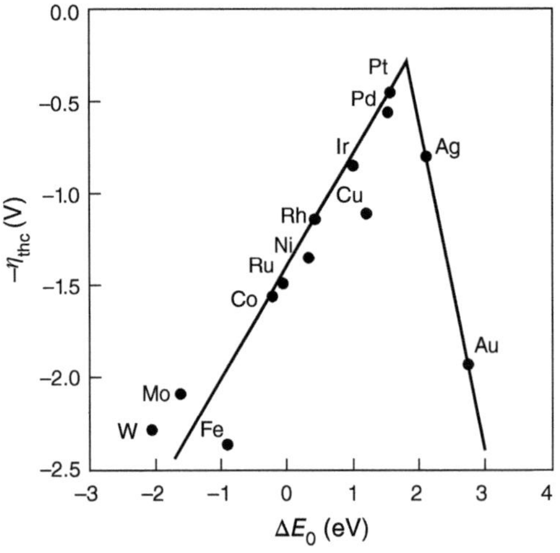

Figure 9.2 Activity trend of noble and non-noble metals towards ORR (ΔE_0 – oxygen adsorption energy). Source: Adapted from Loyola et al. [15] and Mladenović et al. [16].

related to the overpotential at a given current density for ORR. The step with the maximum free energy will give the rise to overpotential, and thus it will be the potential-determining step. The theoretical considerations guide the design of novel electrocatalysts by pointing out the direction to overpotential minimization (e.g. by tuning the oxygen intermediates-electrocatalyst's surface binding energy considering the metal and its electronic structure).

9.3 State-of-the-Art Electrocatalysts for the ORR

The ORR possesses inherently sluggish kinetics, mainly due to the reaction mechanism that involves four-electron transfers resulting in a high activation barrier, meaning efficient electrocatalysts are needed to accelerate production. The classically considered best-performing electrocatalyst is Pt/C, which shows high activity and selectivity towards the $4e^-$ mechanism, which leads to high currents. However, it is based on the noble, scarce and expensive Pt metal, increasing total production costs for the energy conversion and storage technologies that ORR is a key part of. As a result, much scientific effort is devoted to finding efficient, inexpensive and stable electrocatalysts for the cited reaction, based mostly on common transition metals such as Ni, Co, Mn and Cu. Another perspective also tackled by the scientific community in the search for alternatives to Pt/C is the lowering of noble metal content in the applied electrocatalysts via alloying with other metals to control morphology, structure and composition [19]. Other noble metals studied for ORR include palladium (Pd), gold (Au) and silver (Ag).

Various carbon materials are commonly employed as metal nanoparticle supports due to their large surface area along with a high electronic conductivity. These include carbon black, carbon nanotube (CNT), graphene and reduced graphene oxide (rGO) [20]. Another important characteristic of carbon materials for application in electrocatalysis is their porosity (micro-, meso- or macroporosity), where large pore size typically facilitates diffusion of reactants and intermediates to/from the electrode. The use of pure carbon materials as electrocatalysts for the ORR is somewhat hampered by their lower intrinsic activity and low density of active sites. However, carbon materials doped with several heteroatoms have demonstrated high activity for ORR. Typically, the most used heteroatom is nitrogen (N), with some other examples being sulphur (S), phosphorus (P), boron (B) and fluor (F). These carbon materials can overperform commercial Pt/C electrocatalysts in alkaline media in terms of onset and half-wave potential. However, their performance in acidic media is still lower than that of Pt/C.

Transition-metal-oxides (TMO)-based electrocatalysts are receiving attention as ORR electrocatalysts due to transition metals' high availability and, thus, low cost, along with good activity and improved stability [16]. The good redox properties and high electrocatalytic performance of TMO-based electrocatalysts might be related to the existence of a transition metal in different oxidation states. Among the most studied TMOs for ORR are manganese-, cobalt-, copper- and iron oxides. Titanium-

and nickel oxides are also being studied. TMOs' performance for ORR might be further boosted by combining TMOs of different transition metals due to an increment in their electrical conductivity compared to that of pure components. Alternatively, TMOs conductivity can be increased by combining them with a carbon material.

Transition metals might also be used to lower the cost and, at the same time, increase the stability of noble metal electrocatalysts. Namely, alloying Pt with transition metals enhances performance by both electronic and geometric effects of the presence of a transition metal: the former is reflected in the change of the Pt binding energy, and the geometric effect is reflected in the reduction of Pt-Pt bond length.

The use of TMOs as support for noble metal nanoparticles also overcomes stability issues observed when using Pt/C electrocatalyst. Namely, agglomeration of Pt nanoparticles due to Ostwald ripening decreases the electrocatalyst's performance. Interaction of Pt nanoparticles with TMO support prevents this.

Other transition-metal-based compounds explored for the catalysis of ORR include transition-metal nitrides (TMNs) that are typically even more stable than TMOs [21]. This higher stability is attributed to the high energy of triple bonds between TM and N atoms. Still, their performance might be lower than that of Pt/C, especially in acidic media. Transition-metal carbides (TMCs) show activity on the Pt/C level but high instability in acidic media. One of the strategies to increase their stability in acidic media is encapsulation in carbon material. Finally, transition-metal sulphides (TMSs) and phosphides (TMPs) are also considered for ORR catalysis due to rich chemical valence of S and P. These are expected to bring variable redox couples for electrocatalysis. Some TMSs and TMPs have indeed shown activity towards ORR, but very often, it is still lower compared to the Pt/C.

9.4 POM-based Electrocatalysts for the ORR

Xian et al. synthesized a ternary nanohybrid composed of Ag nanoparticles (NPs), phosphotungstic acid and reduced graphene oxide, Ag NPs@POM/rGO, via a photocatalytic reduction method where the POM functioned as both the catalyst and the bridging molecules for this process. They tested three nanohybrids with different Ag loadings of 10, 20 and 30 wt% towards the ORR in 0.1 M KOH electrolyte. The results demonstrated that a higher Ag loading led to a higher activity, and that all ternary hybrids had higher limiting-current densities and onset potentials when compared to their binary POM/rGO and Ag NPs@POM counterparts. Still, the nanohybrids could not outperform Pt/C, with only 30 wt% Ag NPs@POM/rGO having a comparable activity. Additionally, RDE measurements suggested that the ternary nanohybrids could catalyse the ORR through the $4e^-$ pathway, and the number of exchanged electrons was higher than for their binary counterparts. The observed ORR performance was attributed to the synergy between the three components of the hybrid: rGO enhanced the overall conductivity, and POMs captured the electrons transferred from the electrode to the rGO and delivered them to the Ag NPs, enhancing electron collection efficiency. Finally, chronoamperometry measurements for 14.000 s indicated that 30 wt% Ag NPs@POM/rGO was more stable than Pt/C [22].

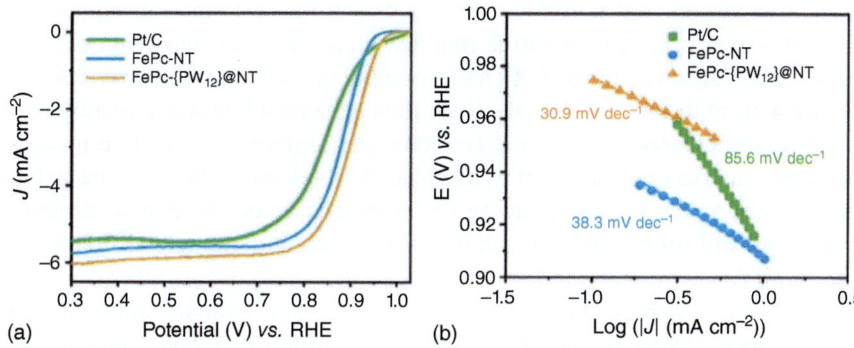

Figure 9.3 (a) ORR LSV polarization curves and (b) respective Tafel plots. Source: Adapted from Zhu et al. [24].

Ensafi et al. fabricated mono-substituted lacunary Keggin $[PW_{11}MO_{39}]^{5-}$ structures (M=Co, Ni, Cu) supported on Ru-reduced graphene oxide ($PW_{11}MO_{39}$@Ru-rGO) via a simple and inexpensive method and tested their activities towards the ORR in alkaline medium (0.1 M KOH). The catalysts presented good overall activity. Still, they were not able to surpass Pt/C, with only $PW_{11}CuO_{39}$@Ru-rGO being comparable to it. The number of exchanged electrons was between 3.9 and 4.2 for all tested POMs, suggesting oxygen reduction catalysis through the $4e^-$ pathway. Chronoamperometry tests demonstrated the good methanol (1 M) crossover tolerance and long-term durability of the catalysts, surpassing Pt/C in those characteristics and showing good ORR selectivity. The observed activity is attributed to the presence of heteroatoms in the POM structure (i.e. O and P) that produce a local spin or charge density favourable to the ORR, as well as to the activity near the edges of graphene nanosheets, where oxygen adsorption is facilitated [23].

Zhu et al. produced iron phthalocyanine (FePc) guest molecules immobilized on the outer surface of single-walled carbon nanotubes encapsulated with Keggin-type phosphotungstic acid (FePc-{PW_{12}}@NTs) to act as an electrocatalyst for the ORR in alkaline medium (0.1 M KOH). Figure 9.3 shows that the studied composite presented an activity that surpassed its non-POM-containing counterpart and even benchmark Pt/C electrocatalysts, demonstrating the impact that POM incorporation had in the catalytic activity since it lowered the Tafel slope and increased the half-wave potential to 30.9 mV dec^{-1} and 0.90 V *vs.* RHE. Additionally, the number of exchanged electrons was ca. 3.9, suggesting a $4e^-$-reduction pathway, and the catalyst presented superior long-term stability and methanol crossover tolerance. Density functional theory calculations demonstrated that PW_{12}@NTs deform the symmetric structure of FePc, forming a biaxially strained structure with a modified electronic structure favourable towards the ORR [24].

Li et al. disclosed the synthesis of a Cube-ZIF-67-derived electrocatalyst via coetching method that applied phosphomolybdic acid and potassium ferricyanide as strong and weak etchants, respectively, followed by calcination at 900 °C. The resulting material possessed a hollow nanopompoms structure active towards the ORR. The reaction was tested in 0.1 M KOH, attaining an onset potential of 0.96 V *vs.* RHE and a half-wave potential of 0.84 V *vs.* RHE, with a diffusion-limited current density comparable to that of Pt/C. Rotating ring-disc electrode tests demonstrated a low H_2O_2 production (<6%) during ORR and a number of exchanged electrons between 3.99 and 4.01, with rotating disc electrode tests agreeing (3.98–4.03). Stability and durability tests demonstrated that after 5000 cycles and 50 h of operation no activity was lost, and the electrocatalyst was resistant to methanol crossover. The activity was attributed to the synergy between the heterophase and the hollow structure, ensuring fast diffusion, sufficient catalytic active sites and enhanced ORR performance [25].

Allah et al. reported the synthesis of a polyoxomolybdate supported on nitrogen-doped carbon xerogel/diamond matrix (carbon diamond nanoparticles, CDNPs) and tested its ORR activity in 0.1 M KOH electrolyte. The POM@CDNP presented a structure of pyramidal POMs anchored in the nitrogen-doped porous carbon matrix. The good distribution of molybdenum-oxo clusters, the CDNP chemical structure, which combined both sp^2 and sp^3 carbons, and the abundant nitrogen-doping, consisting mainly of graphitic-N and pyridinic-N, were all underlined as factors that enhanced the observed ORR performance, allowing the material to achieve a limiting current density of $-7.3\ mA\ cm^{-2}$ at 0.17 V *vs.* RHE and a half-wave potential of 0.773 V *vs.* RHE, proceeding through direct $4e^-$ reduction path and presenting stability in ORR polarization conditions [26].

Ingavale et al. described the anchoring of phosphomolybdic acid (PMA) clusters on the surface of reduced graphite oxide (PMA/rGO), forming a material where Mo centres act as active sites and rGO increases the available surface area, and studied its behaviour as an ORR electrocatalyst in 0.1 M KOH solution. The optimized sample, PMA/rGO-2-200, which had a 1 : 2 PMA/rGO weight ratio and was prepared by treatment at 200 °C, presented the best overall ORR activities, attaining a 1.0 V *vs.* RHE onset potential, around $4\ mA\ cm^{-2}$ limiting current density and enabling the reaction through the $4e^-$ direct mechanism. Good stability properties and methanol crossover resistance were also verified. The problem of PMA's high solubility in aqueous media is circumvented by anchoring it in a carbon-based support, which also improves catalytic activity due to the homogeneous distribution of PMA clusters in the carbon matrix [27].

Novais et al. studied the activity of four composites prepared by the anchoring of a 48-tungsto-8-phosphate anion, $(H_7P_8W_{48}O_{184})^{33-}$ (P_8W_{48}), on four different carbon supports: graphene flakes (GF), single-walled carbon nanotubes (SWCNT), multi-walled carbon nanotubes (MWCNT) and N-doped MWCNT (N-MWCNT) for the ORR in alkaline medium (0.1 M KOH). Their studies led to the conclusion that, in general, the composites had superior performance when compared to the pristine carbon supports, especially for the N-doped samples, which presented the highest activity in both groups. The composites presented onset potentials in the range of 0.75–0.94 V *vs.* RHE, Tafel slopes between 44 and $72\ mV\ dec^{-1}$ and limiting current densities between −2.4 and $-3.7\ mA\ cm^{-2}$, with P_8W_{48}@N-MWCNT presenting the best performance values, except for the Tafel slope. The number of exchanged electrons suggests that there is a mixed mechanism between the $4e^-$ and $2e^-$ process, which is further supported by the RRDE measurements, showing significant peroxide production during the reaction. The observed activities were comparable, albeit lower, to benchmark Pt/C. Finally, the composites showed interesting stability properties and resistance to methanol crossover, a necessary feature if the catalyst is to be applied in methanol fuel cells [28].

Lee et al. tested the ORR-catalysing properties of a Ketjenblack (KB) carbon-supported Ru-based POM (RuPOM) in seawater (pH ~ 8.1) for application in seawater batteries. The ORR activity was observed to increase drastically when RuPOM was supported in KB, with the limiting current density becoming comparable to that of Pt/C. The reaction pathway was also modified, with RuPOM/KB catalysing the ORR through the $4e^-$ mechanism and pristine RuPOM catalysing it through the $2e^-$ mechanism. This is argued to be the reason for better activity in the carbon-supported POM and is thought to happen because the mechanism depends greatly on conductivity, which is greatly modified in RuPOM/KB. Despite this catalyst still being noble-metal-based, it has much higher specific activity when compared to benchmark Pt/C [29].

Novais et al. proposed the use of a Wells–Dawson POM salt, $K_7[P_2W_{17}(FeOH_2)O_{61}]\cdot 20H_2O$, supported onto graphene flakes and multiwalled carbon nanotubes doped with nitrogen, P_2W_{17}Fe@GF_N8 and P_2W_{17}Fe@MWCNT_N8, as ORR electrocatalysts in 0.1 M KOH. The pristine and N-doped carbon supports were also tested for comparison. The studies found that the composites had activities very similar to the N-doped carbon supports, be it in terms of activity or performance parameters. It is known that N-doping increases ORR activity and depends mostly on the type of nitrogen inserted, and a possible explanation for the similar behaviour between the composites and the N-doped supports is that the active sites created by POM immobilization were not sufficient to offset the eventual loss of N-induced active sites. Still, P_2W_{17}Fe@MWCNT_N8 exhibited a Tafel slope lower than benchmark Pt/C catalyst, and better stability properties, showing better current retention after 10 h of operation, which was attributed to the trapping of POM particles in the nanonetworks, and higher methanol crossover tolerance [30].

Zhang et al. designed a series of carbonaceous-supported (thermalized triazine-based frameworks (TTF), fluorine-doped TTF (TTF-F), reduced graphene oxide, or carbon Vulcan XC-72) precious-metal free POMs (based on Ni, Co and Cu) and tested their activities for ORR in alkaline medium (0.1 M KOH). The best-performing composite, Cu_6Ni_7/TTF-F (Cu_6Ni_7 structure shown in Figure 9.4), presented an onset potential of 0.91 V *vs.* RHE and had a number of exchanged electrons in the 3.92–3.94 range, while also presenting excellent stability (negligible changes after 6000 cycles). They also tested the influence of the POM-core in the electrocatalyst, concluding that the Ni- and Cu-based POMs (either individually or having both metals in the POM) induce higher activity when compared to the Co-POMs [7].

Zheng et al. produced a new Ni-POM compound, NiP_4Mo_6, and later vulcanized it, testing both materials' electrocatalytic activity towards the ORR in alkaline conditions (0.1 M KOH). The materials presented different selectivity for the ORR, with NiP_4Mo_6 being selective to the $4e^-$ path and its vulcanized counterpart being selective to the $2e^-$ path. The vulcanized material also presented higher half-wave potential and lower Tafel slope: 0.77 and 0.73 V *vs.* RHE, and 98.1 and

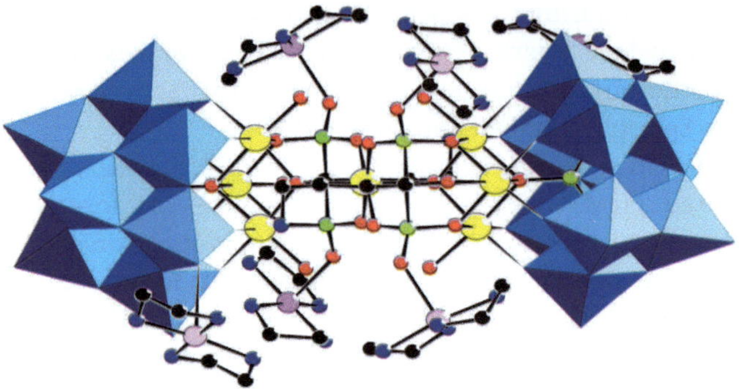

Figure 9.4 Mixed polyhedral and ball-and-stick representation of Cu_6Ni_7: light blue octahedra, WO_6; yellow spheres, Ni; green spheres, P; purple spheres, Cu; black spheres, C; dark blue spheres, N; red spheres, O. H atoms were omitted for clarity. Source: Zhang et al. [7]/with permission of American Chemical Society.

106.2 mV dec^{-1}, respectively, for vulcanized and unvulcanized samples. Both materials also presented excellent stability with virtually no change after 8000 cycles. The change in the ORR mechanism was attributed to the vulcanization process, which substituted O^{2-} atoms with S^{2-} atoms, affecting the adsorption energies, thus weakening it in such a way that H_2O_2 production would be favoured [31].

Sanji et al. reported the use of a POM-modified Pd_8Ni_2/rGO (Pd_8Ni_2/rGO-POM) as an ORR electrocatalyst in alkaline conditions (0.1 M KOH). It presented an onset potential of ~0.83 V *vs.* RHE, a Tafel slope of 59.8 mV dec^{-1} and an average of 3.9 electrons exchanged during the ORR. Furthermore, the composite exhibited excellent methanol crossover tolerance and better current retention properties after 10 h at ORR conditions, outperforming the Pt/C state-of-the-art electrocatalyst. Lastly, the performance was attributed to the Pd–Ni alloy interaction (their interactions modify their electronic structures, contributing to enhanced ORR performance), the rGO support (improves conductivity and mass transfer) and the synergistic effects between them and the POM [32].

Ammam et al. performed cyclic voltammetry tests on the Mn-substituted dissymmetrical Dawson-sandwich type and the Keggin sandwich banana-shaped POMs, $[Mn^{II}_4(H_2O)_2(H_4AsW_{15}O_{56})_2]^{18-}$ and $[((Mn^{II}OH_2)Mn^{II}_2PW_9O_{34})_2(PW_6O_{26})]^{17-}$ ($Mn_4As_2W_{30}$ and $Mn_6P_3W_{24}$). Subsequently, they evaluated their activity towards the ORR at pH 5 and pH 3, respectively. The results indicated that both catalysts demonstrated activity towards the ORR, but a deposited layer of Mn oxides (mainly MnO_2) acted as the true catalyst. This was further corroborated by performing additional CV cycles, revealing an increase in the thickness of the oxide layer and a corresponding enhancement in ORR activity [33].

Fernandes et al. disclosed the use of Co-based sandwich $(TBA)_7H_3[Co_4(H_2O)_2(PW_9O_{34})_2]$ POM, TBA-$Co_4(PW_9)_2$ supported on different carbon-based materials, such as SWCNT, GF, nitrogen-doped carbon nanotubes (N-CNT) and nitrogen-doped few-layer graphene (N-FLG), as ORR electrocatalysts in alkaline medium (0.1 M KOH). The electrocatalysts had onset potentials ranging from 0.77 to 0.90 V *vs.* RHE, Tafel slopes ranging from 68 to 96 mV dec^{-1} and number of exchanged electrons ranging from 2.7 to 4, with some of them even surpassing state-of-the-art Pt/C electrocatalyst. Additionally, the electrocatalysts presented a cyclic stability similar to that of commercial Pt/C and much higher methanol stability. Lastly, all composite materials performed better than their isolated supports, with $Co_4(PW9)_2$@N-CNT catalytic activity outperforming Pt/C [34].

Song et al. confined a cobalt-centred POM, $[CoPW_{12}]^{6-}$, in the void space of the ZIF-8 MOF and later prepared a Co-POM@MOF-derived N-doped porous carbon composite via pyrolysis treatment to test its performance towards ORR in alkaline electrolyte (0.1 M KOH). The material presented a half-wave potential of 0.84 V *vs.* RHE, a Tafel slope of 63 mV dec^{-1}, a number of exchanged electrons of 4.1 and good stability properties (retains activity after 1000 cycles) [35].

Rousseau et al. synthesized a noble-metal-free Co/Zn POM complex supported on Vulcan carbon XC-72 (Co_7Zn/C) and tested its activity towards the ORR in alkaline medium (0.1 M KOH). The material had an electron-exchange number between 3.6 and 4, signifying its selectivity towards the $4e^-$ reduction path. It presented activity comparable to commercial Pt/C, showing higher stability after 11 h at ORR conditions and much higher methanol crossover tolerance. The material also showed good activity in neutral medium, compared to commercial Pt/C, but with higher stability [6].

Liu et al. disclosed a top-down approach to deposit POM-like metal oxo clusters by using the solution-stable Keggin-type polyoxomolybdate $H_3PMo_{12}O_{40}$ (PMo_{12}) as a precursor to deposit $[\text{Mo-oxo}]_n$ ($n = 1$–20) cluster on mesoporous carbon, synthesizing three different composites Their electrocatalytic activity for the ORR in alkaline electrolyte (0.1 M KOH) was tested. The composites had an onset potential in the range of 0.75–0.93 V *vs.* RHE, Tafel slopes in the range of 44.2–109 mV dec^{-1} and the first two composites favoured the $2e^-$ reaction path, while the third composite favoured the $4e^-$ reaction pathway. Additionally, the third composite demonstrated an activation behaviour during chronoamperometric tests to determine its stability, showing an increase in current after 20 h and no further changes after a total period of 50 h, and demonstrated high methanol crossover tolerance, outperforming the commercial Pt/C electrocatalyst both in activity and stability. The activity was attributed to good dispersion of the $[\text{Mo-oxo}]_n$ clusters, high surface area and N-doping of the carbon substrate [36].

Zhao et al. combined a lacunary Keggin POM cluster, $[PW_{11}O_{39}]^{7-}$, with 3D ordered microporous graphitic carbon nitride (3DOM g-C_3N_4) to generate an electrocatalyst with favourable selectivity for the two electrons ORR pathway, testing it electrochemically on neutral medium and via light-driven reaction. The synthesized composite showed activity for the light-driven production of H_2O_2, with a production rate of 2.4 μmol h^{-1}, and electrochemical tests demonstrated that it had a number of exchanged electrons around 2.3, indicating its selectivity towards the intended ORR mechanism [37].

Marques et al. assembled sandwich-type Dawson family POMs anchored in N-doped multi-walled carbon nanotubes (MWCNT_N6) and tested their electrocatalytic properties towards ORR in alkaline medium (0.1 M KOH). The ORR results,

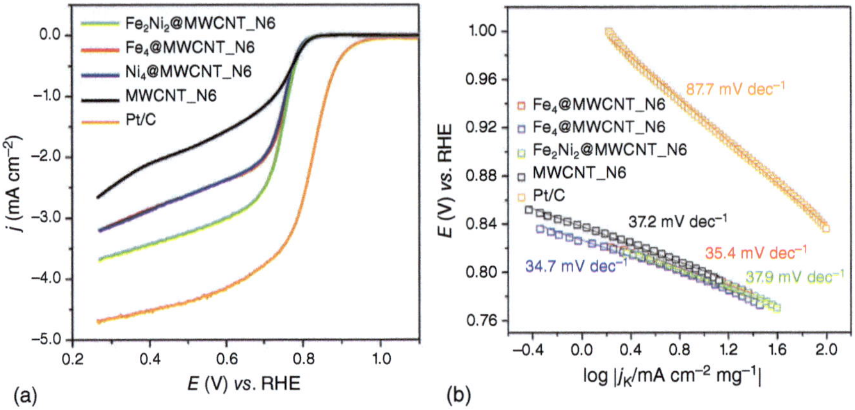

Figure 9.5 (a) ORR LSV curves and (b) Tafel plots in O_2-saturated 0.1 M KOH using MWCNT_N6, Fe_4@MWCNT_N6, Ni_4@MWCNT_N6, Fe_2Ni_2@MWCNT_N6 and Pt/C at 1600 rpm and 5 mV s^{-1}. Source: Marques et al. [38]/MDPI/CC BY 4.0.

shown in Figure 9.5, indicate that an onset potential of ~0.8 V *vs.* RHE was obtained for all materials and Tafel slopes of 35.4, 34.7 and 37.9 mV dec^{-1} were obtained for Fe_4, Ni_4 and Fe_2Ni_2, respectively. Lastly, the number of transferred electrons is around three, thus suggesting a mixed regime between the four- and two-electron pathway, and good stability for the Fe_4 and Ni_2Fe_2 (84% and 80% of current retention after 12 h) was observed at ORR conditions [38].

Limani et al. prepared a series of Co-POMs anchored on doped carbon materials (MWCNT_N8_Co4, GF_N8_Co4, GF_ND8_Co4 and GF_NS8_Co4) and tested their performance for ORR in alkaline conditions (0.1 M KOH). The ORR results demonstrated that the best-performing material (MWCNT_N8_Co4) had a performance comparable to that of benchmark Pt/C and good stability properties, along with high methanol crossover tolerance, indicating its durability [39]

The key performance indicators for the discussed POM electrocatalysts are presented in Table 9.1.

9.5 Conclusions

Fuel cells and metal–air batteries rely on the ORR as their cathodic reaction. As ORR requires high overpotentials, related to its sluggish kinetics, highly active and selective electrocatalysts (e.g. Pt) are required to decrease the reaction overpotential and promote direct $4e^-$ oxygen reduction. Currently, the prohibitive cost of noble metals has demanded the development of cost-effective, active and durable electrocatalysts based on earth-abundant transition metals (e.g. Ni, Co, Cu, Mn) that can allow the industrial implementation of these electrochemical energy conversion devices.

The present chapter started by presenting the fundamentals of ORR and state-of-the-art electrocatalysts for ORR. It then focused on a comprehensive literature review of the use of transition-metal-based POMs for ORR, as cost-effective

Table 9.1 Performance comparison of different POMs tested as electrocatalysts for ORR.

Electrocatalyst	Onset potential (V *vs.* RHE)	Tafel slope (mV dec^{-1})	Number of exchanged e$^-$	Diffusion-limited current density (mA cm^{-2})	Electrolyte	Source
30%AgNPs@POM/rGO	0.87	—	3.7–3.9	−4.4	0.1 M KOH	[22]
$PW_{11}CuO_{39}$@Ru-rGO	—	—	3.9–4.2	—	0.1 M KOH	[23]
$PW_{11}CoO_{39}$@Ru-rGO	—	—	3.9–4.2	—	0.1 M KOH	[23]
$PW_{11}NiO_{39}$@Ru-rGO	—	—	3.9–4.2	—	0.1 M KOH	[23]
FePc-{PW_{12}}@NTs	—	30.9	3.9	−6.0	0.1 M KOH	[24]
Nanopompoms	0.96	—	3.98–4.03	−5.5	1 M KOH	[25]
POM@CDNPs	0.89	—	4.1	−7.3	0.1 M KOH	[26]
PMA/rGO	1.0	—	3.4–3.8	−4.0	0.1 M KOH	[27]
P_8W_{48}@N-MWCNT	0.94	72	3.0	−3.7	0.1 M KOH	[28]
P_8W_{48}@MWCNT	0.77	47	2.2	−2.4	0.1 M KOH	[28]
P_8W_{48}@SWCNT	0.76	48	2.7	−3.0	0.1 M KOH	[28]
P_8W_{48}@GF	0.75	44	2.7	−3.0	0.1 M KOH	[28]
RuPOM/KB	—	—	~4	−4	Seawater (pH ~8.1)	[29]
P_2W_{17}Fe@MWCNT_N8	0.84	48	3.4	−3.9	0.1 M KOH	[30]
P_2W_{17}Fe@GF_N8	0.83	58	3.1	−3.3	0.1 M KOH	[30]
Cu_6Ni_7/TTF-F	0.906	—	3.92–3.94	—	0.1 M KOH	[7]
NiP_4Mo_6	—	98.1	3.86–3.88	—	0.1 M KOH	[31]
S-NiP_4Mo_6	—	106	2.02–2.08	—	0.1 M KOH	[31]
Pd_8Ni_2/rGO-POM	~0.83	59.8	3.9	—	0.1 M KOH	[32]
$Co_4(PW_9)_2$@N-CNT	0.9	92	4.0	−8.5	0.1 M KOH	[34]

(Continued)

Table 9.1 (Continued)

Electrocatalyst	Onset potential (V *vs.* RHE)	Tafel slope (mV dec^{-1})	Number of exchanged e^-	Diffusion-limited current density (mA cm^{-2})	Electrolyte	Source
$Co_4(PW_9)_2$@ SWCNT	0.77	68	3.9	−3.2	0.1 M KOH	[34]
$Co_4(PW_9)_2$@GF	0.89	71	2.7	−3.3	0.1 M KOH	[34]
$Co_4(PW_9)_2$@N-FLG	0.89	63	3.6	−4.4	0.1 M KOH	[34]
Co-W-NC	—	63	4.1	—	0.1 M KOH	[35]
Co_7Zn/C	—	—	3.6–4	−3	0.1 M KOH	[6]
MoO_x@P-C	0.75	109	~2	—	0.1 M KOH	[36]
$[Mo\text{-}O]_x$@P-C, $x = 1–20$	0.88	68.1	~2	—	0.1 M KOH	[36]
$[Mo\text{-}O]_x$@N,P-C, $x = 1–20$	0.93	44.2	3.8–3.9	—	0.1 M KOH	[36]
3DOM g-C_3N_4	—	—	2.3	—	0.1 M PBS[a)]	[37]
Ni_4@MWCNT_N6	0.80	34.7	2.7	−3.2	0.1 M KOH	[38]
Fe_4@MWCNT_N6	0.81	35.4	2.9	−3.2	0.1 M KOH	[38]
Ni_2Fe_2@MWCNT_N6	0.80	37.9	3.2	−3.7	0.1 M KOH	[38]
MWCNT_N8_Co4	0.85	41	3.5	−3.5	0.1 M KOH	[39]
GF_N8_Co4	0.83	50	3.5	−2.9	0.1 M KOH	[39]
GF_ND8_Co4	0.85	90	3.7	−3.2	0.1 M KOH	[39]
GF_NS8_Co4	0.84	40	3.2	−3.0	0.1 M KOH	[39]

a) PBS, phosphate buffer solution.

alternatives to benchmark noble metal electrocatalysts. Many different POM structures are being currently tested for ORR. In fact, it is clear that an advantage of POMs is their structures' tailorability, as they can be easily modified by straightforward changes in their synthesis procedures. Moreover, POMs performance is already comparable to or better than traditional ORR electrocatalysts, showing a promising future for energy conversion applications.

Acknowledgements

The authors would like to thank Fundação para a Ciência e a Tecnologia (FCT, Portugal) for contract IST-ID/156/2018 (B. Šljukić) and a contract in the scope of programmatic funding UIDP/04540/2020 (D.M.F. Santos).

References

1 Manthiram, A., Zhao, X., and Li, W. (2012). Developments in membranes, catalysts and membrane electrode assemblies for direct methanol fuel cells (DMFCs). *Funct. Mater. Sustain. Energy Appl.* 312–369. https://doi.org/10.1533/9780857096371.3.312.

2 Sundar, A., Maksimchuk, N.V., Ivanchikova, I.D. et al. (2023). Peroxo-Zr/Hf-containing Wells-Dawson polyoxometalates for H_2O_2-based homogeneous, heterogeneous, and biphasic oxidation catalysis. *Chem. Eur.* 0: https://doi.org/10.1002/ceur.202300066.

3 Gusmão, F.M.B., Mladenović, D., Radinović, K. et al. (2022). Polyoxometalates as electrocatalysts for electrochemical energy conversion and storage. *Energies* 15: https://doi.org/10.3390/en15239021.

4 Skunik, M., Baranowska, B., Fattakhova, D. et al. (2006). Electrochemical charging and electrocatalysis at hybrid films of polymer-interconnected polyoxometallate-stabilized carbon submicroparticles. *J. Solid State Electrochem* 10: 168–175.

5 Vidoeski, B., Svetlana, J., Holclajtner-Antunović, I. et al. (2016). Raman study of the interactions between highly ordered pyrolytic graphite (HOPG) and polyoxometalates: the effects of acid concentration. *J. Serb. Chem. Soc.* 81: 777–787. https://doi.org/10.2298/JSC160301055V.

6 Rousseau, G., Zhang, S., Oms, O. et al. (2015). Sequential synthesis of 3d–3d, 3d–4d, and 3d–5d hybrid polyoxometalates and application to the electrocatalytic oxygen reduction reactions. *Chemistry* 21: 12153–12160. https://doi.org/10.1002/chem.201501609.

7 Zhang, S., Oms, O., Hao, L. et al. (2017). High oxygen reduction reaction performances of cathode materials combining polyoxometalates, coordination complexes, and carboneous supports. *ACS Appl. Mater. Interfaces* 9: 38486–38498. https://doi.org/10.1021/acsami.7b10989.

8 Benzbiria, N., Zertoubi, M., and Azzi, M. (2021). Influence of copper surface pretreatment on the kinetics of oxygen reduction reaction in 0.5 M NaCl solution: surface characterization and electrochemical studies. *Appl. Surf. Sci. Adv.* 4: https://doi.org/10.1016/j.apsadv.2021.100069.

9 Ma, R., Lin, G., Zhou, Y. et al. (2019). A review of oxygen reduction mechanisms for metal-free-carbon-based electrocatalysts. *npj Comput. Mater.* 5: https://doi.org/10.1038/s41524-019-0210-3.

10 Yu, L., Pan, X., Cao, X. et al. (2011). Oxygen reduction reaction mechanism on nitrogen-doped graphene: a density functional theory study. *J. Catal.* 282: 183–190. https://doi.org/10.1016/j.jcat.2011.06.015.

11 Ge, X., Sumboja, A., Wuu, D. et al. (2015). Oxygen reduction in alkaline media: from mechanisms to recent advances of catalysts. *ACS Catal.* 5: 4643–4667. https://doi.org/10.1021/acscatal.5b00524.

12 Zdolšek, N., Vujković, M., Metin, O. et al. (2022). boosting electrocatalysis of oxygen reduction and evolution reactions with cost-effective cobalt and nitron-doped carbon prepared by simple carbonization of ionic liquids. *Int. J. Hydrogen Energy* 47: 14847–14858. https://doi.org/10.1016/j.ijhydene.2022.02.225.

13 Li, D., Batchelor-McAuley, C., and Compton, R.G. (2020). Some thoughts about reporting the electrocatalytic performance of nanomaterials. *Appl. Mater. Today* 18: https://doi.org/10.1016/j.apmt.2019.05.011.

14 Wang, J., Zhao, C.-X., Liu, J.-N. et al. (2021). Quantitative kinetic analysis on oxygen reduction reaction: a perspective. *Nano Mater. Sci.* 3: 313–318. https://doi.org/10.1016/j.nanoms.2021.03.006.

15 Loyola, C.Z., Ureta-Zañartu, S., Zagal, J.H., and Tasca, F. (2022). Activity volcano plots for the oxygen reduction reaction using FeN_4 complexes: from reported experimental data to the electrochemical meaning. *Curr. Opin. Electrochem.* 32.

16 Mladenović, D., Mladenović, A., Santos, D.M.F. et al. (2023). Transition metal oxides for bifunctional ORR/OER electrocatalysis in unitized regenerative fuel cells. *J. Electroanal. Chem.* 946: https://doi.org/10.1016/j.jelechem.2023.117709.

17 Milikić, J., Nikolić, N., Santos, D.M.F. et al. (2022). Platinum-dysprosium alloys as oxygen electrodes in alkaline media: an experimental and theoretical study. *Nanomaterials* 12: https://doi.org/10.3390/nano12142318.

18 Novčić, K.A., Dobrota, A.S., Petković, M. et al. (2020). Theoretical analysis of doped graphene as cathode catalyst in Li-O_2 and Na-O_2 batteries – the impact of the computaional scheme. *Electrochim. Acta* 354: https://doi.org/10.1016/j.electacta.2020.136735.

19 Wu, D., Shen, X., Pan, Y. et al. (2019). Advances, challenges, and perspectives of platinum alloy catalysts for oxygen reduction reaction. *ChemNanoMat* 6: 32–41. https://doi.org/10.1002/cnma.201900319.

20 Zhang, W., Li, J., and Wei, Z. (2023). Carbon-based catalysts of the oxygen reduction reaction: mechanistic understanding and porous structures. *Chin. J. Catal.* 48: 15–31. https://doi.org/10.1016/S1872-2067(23)64427-4.

21 Tian, X., Lu, X.F., Xia, B.Y., and Lou, X.W. (2020). Advanced electrocatalysts for the oxygen reduction reaction in energy conversion technologies. *Joule* 4: 45–68. https://doi.org/10.1016/j.joule.2019.12.014.

22 Xian, Z., Liu, R., Li, H. et al. (2016). Photocatalytic reduction synthesis of ternary Ag nanoparticles/polyoxometalate/graphene nanohybrids and its activity in the electrocatalysis of oxygen reduction. *J. Clust. Sci.* 27: 241–256. https://doi.org/10.1007/s10876-015-0926-2.

23 Ensafi, A.A., Heydari-Soureshjani, E., and Rezaei, B. (2017). $[PW_{11}MO_{39}]^{5-}$ decorated on Ru-reduced graphene oxide nanosheets, characterizations and application as a high performance storage energy and oxygen reduction reaction. *Chem. Eng. J.* 330: 1109–1118. https://doi.org/10.1016/j.cej.2017.08.062.

24 Zhu, S., Ding, L., Zhang, X. et al. (2023). Biaxially-strained phthalocyanine at polyoxometalate@carbon nanotube heterostructure boosts oxygen reduction catalysis. *Angew. Chem. Int. Ed.* 62: https://doi.org/10.1002/anie.202309545.

25 Li, S., Zhang, J., Liu, Z. et al. (2022). Hollow nanopompoms engineered via a weak-strong coetching strategy for high-performance electrocatalysts. *Chem. Eng. J.* 446: https://doi.org/10.1016/j.cej.2022.136815.

26 Allah, A.E., EL-Deeb, M.M., Farghali, A.A. et al. (2023). Growth of polyoxomolybdate with a porous pyramidal structure on carbon xerogel nanodiamond as an efficient electrocatalyst for oxygen reduction reaction. *RSC Adv.* 13: 8090–8100. https://doi.org/10.1039/D2RA07543A.

27 Ingavale, S.B., Patil, I., Parse, H. et al. (2019). Polyoxomolybdate anchored graphite oxide: noble metal-free electrocatalyst for oxygen reduction reaction. *Int. J. Hydrogen Energy* 44: 24922–24933. https://doi.org/10.1016/j.ijhydene.2019.07.180.

28 Novais, H.C., Jarrais, B., Haider, A. et al. (2023a). Electrocatalytic oxygen reduction reaction on 48-tungsto-8-phosphate wheel anchored on carbon nanomaterials. *Electrocatalysis* 14: 294–305. https://doi.org/10.1007/s12678-022-00792-w.

29 Lee, C., Jeon, D., Park, J. et al. (2020). Tetraruthenium polyoxometalate as an atom-efficient bifunctional oxygen evolution reaction/oxygen reduction reaction catalyst and its application in seawater batteries. *ACS Appl. Mater. Interfaces* 12: 32689–32697. https://doi.org/10.1021/acsami.0c07225.

30 Novais, H.C., Jarrais, B., Mbomekallé, I.-M. et al. (2023b). Hybrids composed of an Fe-containing Wells-Dawson polyoxometalate and carbon nanomaterials as promising electrocatalysts for the oxygen reduction reaction. *Inorganics* 11: https://doi.org/10.3390/inorganics11100388.

31 Zheng, Y., Xu, X., Chen, J., and Wang, Q. (2021). Surface O_2-regulation on POM electrocatalyst to achieve accurate $2e^-/4e^-$ ORR control for H_2O_2 production and Zn-air battery assemble. *Appl. Catal. B Environ.* 285. https://doi.org/10.1016/j.apcatb.2020.119788.

32 Sanji, F.D., Balakrishnan, P., Su, H. et al. (2021). Fabrication of polyoxometalate-modified palladium–nickel/reduced graphene oxide alloy catalysts for enhanced oxygen reduction reaction activity. *RSC Adv.* 11: 39118–39129. https://doi.org/10.1039/D1RA06936E.

33 Ammam, M., Keita, B., Nadjo, L. et al. (2011). Cyclic voltammetry study of the Mn-substituted polyoxoanions $[Mn^{II}{}_4(H_2O)_2(H_4AsW_{15}O_{56})_2]^{18-}$ and $[((Mn^{II}OH_2)Mn^{II}{}_2PW_9O_{34})_2(PW_6O_{26})]^{17-}$: electrodeposition of manganese oxides electrocatalysts for dioxygen reduction. *Electroanalysis* 23: 1427–1434. https://doi.org/10.1002/elan.201,000735.

34 Fernandes, D.M., Novais, H.C., Bacsa, R. et al. (2018). Polyoxotungstate@carbon nanocomposites as oxygen reduction reaction (ORR) electrocatalysts. *Langmuir* 34: 6376–6387. https://doi.org/10.1021/acs.langmuir.8b00299.

35 Song, Y., Peng, Y., Yao, S. et al. (2022). Co-POM@MOF-derivatives with trace cobalt content for highly efficient oxygen reduction. *Chin. Chem. Lett.* 33: 1047–1050. https://doi.org/10.1016/j.cclet.2021.08.045.

36 Liu, R., Cao, K., Clark, A.H. et al. (2020). Top-down synthesis of polyoxometalate-like subnanometer molybdenum-oxo clusters as high-performance electrocatalysts. *Chem. Sci.* 11: 1043–1051. https://doi.org/10.1039/C9SC05469C.

37 Zhao, S., Zhao, X., Zhang, H. et al. (2017). Covalent combination of polyoxometalate and graphitic carbon nitride for light-driven hydrogen peroxide production. *Nano Energy* 35: 405–414. https://doi.org/10.1016/j.nanoen.2017.04.017.

38 Marques, I.S., Jarrais, B., Mbomekallé, I.-M. et al. (2022). Synergetic effects of mixed-metal polyoxometalates@carbon-based composites as electrocatalysts for the oxygen reduction and the oxygen evolution reactions. *Catalysts* 12: https://doi.org/10.3390/catal12040440.

39 Limani, N., Marques, I.S., Jarrais, B. et al. (2022). Cobalt phosphotungstate-based composites as bifunctional electrocatalysts for oxygen reactions. *Catalysts* 12: https://doi.org/10.3390/catal12040357.

Part V

Polyoxometales for Batteries and Supercapacitors

10

POM-based Nanomaterials for Battery Applications

Sanaz Taghaddosi and Ulrich Stimming

Litricity GmbH, Anton-Jakob-Street 5, Rosenheim 83026, Germany

10.1 Introduction

Considering population growth, industrial development around the world and environmental pollution arising from using fossil fuels – the main source of energy production nowadays – the demand for reliable and sustainable power sources has never been more pressing than today. From powering our homes and businesses to fuelling the intricate networks of industry and vehicles, electricity serves as the so-called lifeblood of modern civilization [1].

However, the inherent intermittency and variability of renewable energy sources such as wind and solar energies underscores the critical need for effective energy storage systems. In fact, energy storage devices, including supercapacitors (SCs) with high power density, fuel cells with high energy density and rechargeable batteries in between (depicted in Figure 10.1), emerge as indispensable components in addressing the challenges of modern energy management. In fact, they play a pivotal role in bridging the gap between energy generation from wind or sun and consumption by storing surplus energy during periods of low demand and releasing it when it is needed. In doing so, they not only enhance grid stability and reliability but also unlock the full potential of renewable energy sources such as solar and wind power [2–4].

In the realm of energy storage systems, redox flow batteries (RFBs) have specific advantages in terms of safety, long lifetime, scalability, maintenance cost and flexible operation to store energy from intermittent renewable energy sources such as wind and solar [5]. While conventional batteries have dominated the market so far, RFBs offer a new and versatile design, making them suitable for different applications. In contrast to solid-state batteries (SSBs), RFBs have a very specific feature, their energy and power can be scaled independently. In other words, unlike conventional batteries such as Li-ion batteries which store the energy into the active materials bound on the electrodes themselves, positive and negative electrodes, RFBs store energy in the redox-active materials dissolved or suspended in the electrolytes, i.e. negolyte and posolyte. Therefore, by changing the tank size or cell design, RFBs can be customized

Applied Polyoxometalate-based Electrocatalysis, First Edition. Edited by Diana M. Fernandes.

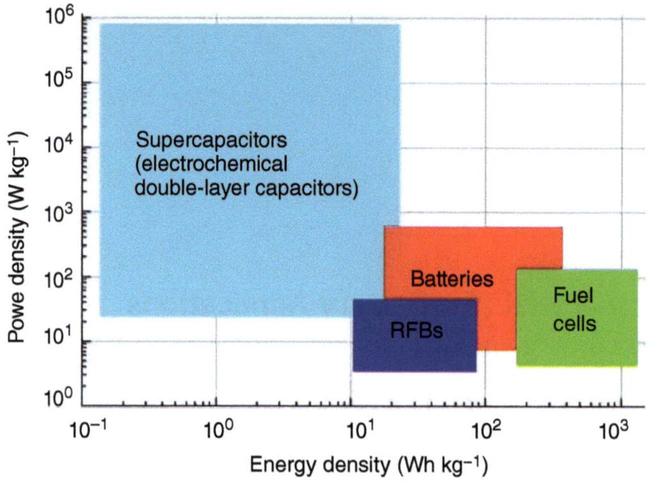

Figure 10.1 Ragone plot for energy storage devices. Source: Reproduced from Holland-Cunz et al. [2]/Springer Nature.

for different applications requiring different energy content or power capability. One other advantage of RFBs is their ability to have their electrolyte changed and refilled, allowing for easy alteration or replacement of the active materials in the reservoirs. This feature facilitates seamless troubleshooting in case of issues or enables straightforward upgrades in the future [2, 6].

10.2 Criteria for Efficient Redox Flow Batteries

Figure 10.2 shows a common design of RFBs, consisting of two tanks of active materials: negolyte and posolyte, the inert electrodes, membrane as separator, pumps and power supply/load. Electrodes in RFBs are often inert carbon electrodes on which the electrochemical electron-transfer reaction of the active materials, pumped from the reservoirs to the electrodes, takes place. Separators are typically ion-exchange membranes (anion or cation exchange) placed between two electrodes to prevent the crossover of active materials but allow the permeation of ions from supporting electrolyte to close the electric circuit [7].

RFBs are composed of two tanks containing active materials: negolyte, which is the reservoir containing active materials that will be reduced during charge process, and posolyte, which will be oxidized during charging. The redox reactions will be reversed in the discharge process as can be seen for active species A and B as an example ($U_{0B}{}^{(z+n)+}/_{B}{}^{z+} > U_{0A}{}^{(y+n)+}/_{A}{}^{y+}$) in Eq. (10.1) [5]. It should be mentioned that these equations are written for cations as an example and can be written for anions as well.

$$\begin{aligned} &\text{Posolyte:}\quad B^{z+} \xrightarrow{\text{Charge}} B^{(z+n)+} + ne^- \quad B^{(z+n)+} + ne^- \xrightarrow{\text{Discharge}} B^{z+} \\ &\text{Negolyte:}\quad A^{(y+n)+} + ne^- \longrightarrow A^{y+} \quad A^{y+} \longrightarrow A^{(y+n)+} + ne^- \end{aligned} \tag{10.1}$$

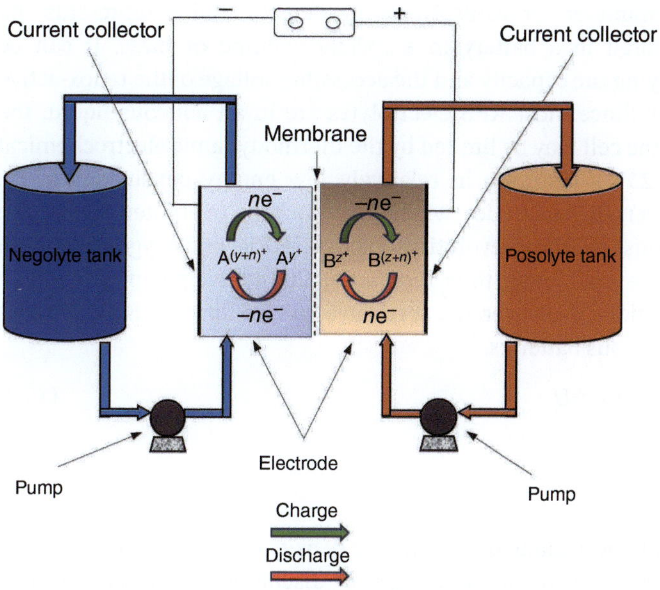

Figure 10.2 Schematic representation of a redox flow battery.

As shown in Eq. (10.1), for posolyte, which undergoes oxidation during charge, the active materials should be in the reduced form, so that they can release an electron. Conversely, the negolyte should be in the oxidized form to accept the electron(s) to be reduced and complete the electric circuit.

To evaluate the performance of an RFB, there are several important parameters to be discussed here, namely theoretical capacity, energy density, power density, coulombic efficiency (CE), voltage efficiency (VE), energy efficiency (EE) and capacity retention (CR). Theoretical capacity, a fundamental factor for assessing the performance of RFBs, is determined by the maximum amount of charge (in coulombs) per unit volume which can be stored in a battery during charge/discharge processes as shown in Eq. (10.1) and can be calculated by Eq. (10.2).

$$Q = nFcv \tag{10.2}$$

where

n is the number of electrons which can be transferred during redox reactions,
F is Faraday constant (approximately 96,485 C/mol),
c is the concentration of active materials in the electrolyte solution and
v is the volume of the electrolyte solution in a reservoir.

Theoretically, based on Eq. (10.2), the capacity of an RFB can be increased in three ways: (i) using a high concentration of negolyte or posolyte, (ii) using an electrolyte containing multi-electron-transfer reactions and (iii) increasing the size of the tanks based on the desired application. Points (i) and (ii) are related to the chemicals used as redox-active species, which will be further discussed in Section 10.3.

Another crucial parameter for RFBs is energy density, which represents the amount of energy stored in a battery in a specific volume or mass. It can be calculated by multiplying the capacity and the accessible voltage of the redox-active materials (Eq. (10.3)). Since most RFB electrolytes are in an aqueous liquid, the applicable voltage of the cell may be limited by the thermodynamic electrochemical window of water (1.23 V), resulting in relatively low energy density for RFBs. However, there are certain electrodes which have a high overpotential for the water-splitting reactions (hydrogen evolution reaction [HER] and oxygen evolution reaction [OER] as side reactions), thus increasing the actual potential window available to redox reactions. However, the accessible voltage for most SSBs remains higher than that of aqueous batteries.

$$E = n\,F\,c\,\Delta U = Q \times \Delta U \tag{10.3}$$

where

ΔU is the cell voltage.

Another parameter to be evaluated is the power density (as defined in Eq. (10.4)), which means how fast an RFB can be charged or discharged. It should be mentioned that power density can be limited by the ohmic resistance of the system. As a constant current (I) flows through the cell, because of the resistance of different parts, including electrodes, current collectors, pumps, membrane and electrolytes, an ohmic voltage drop equal to IR_{tot} can affect the cell decreasing its EE.

Equation (10.5) illustrates that R_{tot} consists of resistances of different sources. R_{elec} represents electronic resistance, related to the design and geometry of the cell and electrodes, R_{mem} is the membrane resistance, the higher the conductivity of membrane, the lower the R_{mem}, R_{elyt} is the electrolyte resistance, R_{mass} refers to the mass-transfer resistance, indicating the rate at which the active materials reach the surface of electrodes to be oxidized or reduced by convection or diffusion, and R_{ct} denotes the charge-transfer resistance, which is determined by the rate constant of the reaction. From among different sources of voltage loss, only R_{ct} together with transport limitations is related to the active materials, which will be discussed further on.

Equation (10.6) shows the formula for R_{ct} which is proportional to the inverse of electron-transfer rate constant k_0, the faster the electron transfer the lower R_{ct} and the more power density.

$$P = I \times \Delta U \tag{10.4}$$

$$R_{tot} = R_{elec} + R_{mem} + R_{elyt} + R_{mass} + R_{ct} \tag{10.5}$$

$$R_{ct} = \frac{RT}{n^2F^2Ak_0C} \tag{10.6}$$

where

R is the gas constant,
T is temperature,

F is Faraday constant,
k_0 is the electron-transfer rate constant,
c is the concentration in molar and
A is the electrochemical surface area.

There are three parameters to be used to evaluate the efficiency of an RFB: CE, VE and EE. CE (as defined in Eq. (10.7)) is the ratio of the output charge that can be extracted during discharge process to the input or stored charge during charge process for the same cycle.

$$\mathrm{CE} = \frac{Q_{\mathrm{dch}}}{Q_{\mathrm{ch}}} \times 100\% \tag{10.7}$$

An RFB with CE of 100% means that all the charge stored during the charging process can be completely retrieved during discharge cycle. However, undesirable side reactions such as HER and OER or crossover of the redox-active chemicals from one reservoir to the other side can cause a deviation of CE from 100%. The other parameter to assess is the VE of an RFB (as defined in Eq. (10.8)), which is the ratio of discharge voltage to the theoretical or charge voltage. As discussed before, ohmic resistance and charge-transfer kinetics of redox-active materials together with mass-transport limitations are the main sources of voltage loss.

The other parameter is EE, which is the overall parameter which describes the effectiveness of the energy conversion process. To put it differently, EE is the ratio of the usable energy stored in the battery to the total supplied energy during charging; alternatively, multiplying the CE and VE yields EE (Eq. (10.9)).

$$\mathrm{VE} = \frac{V_{\mathrm{dch}}}{V_{\mathrm{ch}}} \times 100\% \tag{10.8}$$

$$\mathrm{EE} = \frac{E_{\mathrm{dch}}}{E_{\mathrm{ch}}} \times 100\% = \mathrm{VE} \times \mathrm{CE} \times 100\% \tag{10.9}$$

Last but not least, CR or the inverse of capacity fade is one of the most important parameters to assess the battery performance. CR can be monitored and reported per cycle or over time, showing the gradual loss of the battery's storage capacity. Crossover of active materials from one side to the opposite side, side reactions such as HER and OER, decomposition of active materials, or precipitation of active materials in case of high concentrations can be sources of capacity fade [8]. Equation (10.10) shows the formula for calculating CR of an RFB, where Q_{t} is the capacity after a defined time, Q_n is the capacity after defined cycles and Q_{int} is the initial capacity of the battery.

$$\mathrm{CR} = \frac{Q_{\mathrm{t}} \text{ or } Q_n}{Q_{\mathrm{int}}} \times 100\% \tag{10.10}$$

10.3 Electrolyte Requirements for Redox Flow Batteries (RFBs)

As discussed earlier, active species in RFBs are dissolved or suspended in an external tank, enabling the energy and power to be scaled independently. Since energy density and power density are important parameters for RFBs, choosing the electrolyte – active materials, solvents and supporting electrolyte – is of paramount importance. Based on Eq. (10.3), energy density of an RFB is proportional to n, c and ΔU, all of which are the characteristics of electrolytes or, more precisely, active materials as such. It should be mentioned that solvents and supporting electrolytes can affect solubility and accessible potential window of active materials (more explanation can be found in Section 10.4). Also, active materials with fast electron-transfer kinetics offer high power densities since there is a proportional relationship between the electron-transfer rate constant and power density.

In this section, we will review the most important parameters to consider choosing the active materials from among the available candidates [5].

10.3.1 Wide Potential Window

It should be noted that the electrolytes of RFBs can be divided into two categories, aqueous and non-aqueous electrolytes. Based on Eq. (10.3), the energy density is proportional to the voltage of the battery. In the case of aqueous electrolytes, the voltage of the cell is limited by the electrochemical reactions of water (HER and OER). However, by using non-aqueous electrolytes, a wider potential window may be accessible. Despite this, non-aqueous electrolytes have some disadvantages including their potentially high cost, possible flammability and limited solubility of active materials in some cases. All in all, the more separated the redox peaks of posolyte and negolyte on the potential scale, the higher the energy density [5].

10.3.2 Energy Density and High Solubility

An important factor that determines the energy density of an RFB is the solubility of active materials. The higher the concentration of active materials, the higher is the energy density in the battery. Therefore, choosing a redox-active material which is highly soluble is of interest to build a high-energy-density RFB. In addition, the number of exchanged electrons determines the energy density; higher numbers of electrons exchanged mean a higher energy density.

10.3.3 Fast Electron-transfer Kinetics

The electron-transfer kinetics is one of the most important factors to consider when choosing the active materials. Fast electron transfer implies that the active material can be reduced and oxidized on the surface of the electrodes in a fast way which directly affects the power density of an RFB [2].

10.3.4 High Ionic Conductivity

The ionic conductivity of the electrolyte is an important parameter; high ionic conductivity is essential for reducing ohmic losses within the cell and ultimately improving power density.

10.3.5 Mass Transfer and Viscosity of Electrolyte

In RFBs, the mass transfer of redox-active materials from the bulk solution to the surface of the electrode is crucial for the oxidation or reduction reactions to occur. Additionally, low viscosity of the electrolyte is advantageous for RFB performance. A lower viscosity enables faster mass transfer of the redox-active species to and from the electrode surface, facilitating rapid reaction kinetics and enhancing power density [9].

10.3.6 Long-term Stability of Active Materials

The stability of active materials during charge/discharge cycles is crucial for the overall performance and longevity of RFBs. When active materials undergo oxidation and reduction reactions during charging and discharging, it is essential that they maintain their structural integrity and remain chemically stable. However, some active materials may undergo undesired transformations, leading to the formation of new species, for example the formation of different oxidation states. These newly formed species may not be soluble in the electrolyte or may undergo irreversible redox reactions, which can compromise the efficiency and cycling stability of the battery. In addition, in certain cases, the newly generated species may be redox inactive, meaning they do not participate in the charge/discharge process, decreasing the effective capacity and energy output of the battery. These results will ultimately impact the overall performance of the battery [5].

In addition to chemical stability, thermal stability is also a critical consideration for the long-term cycling of batteries. High temperatures can accelerate side reactions and promote the decomposition of active species, leading to the formation of insoluble precipitates. In addition, in the case of high concentrations of the active materials, because of the evaporation at high temperatures, the precipitation of active materials may decrease the performance of the battery. These precipitates can accumulate on the electrode surfaces or within the electrolyte, hindering ionic transport and electron-transfer processes. As a result, the battery's performance deteriorates over time, highlighting the importance of ensuring the thermal stability of active materials for reliable and durable battery operation [2].

10.3.7 Costs and Safety

Last but not least, for RFBs to be commercialized, the final price, which comprises the cost of active materials, solvents, supporting electrolyte, cell design and pumps, is of great importance. It has been said that an important factor of the total price is attributed to the cost of electrolytes. Therefore, inexpensive electrolytes are promising for RFBs as large-scale systems which can be successful in the market. In addition, the electrolytes of an RFB should be environmentally friendly and should not produce corrosive or toxic materials during production, operation and disposal [2, 10].

10.4 Classification of POMs

Polyoxometalates (POMs), inorganic clusters, are composed of coordinated transition metals with oxygen atoms, offering variable structures. There are two categories of POMs: iso-POMs, in which there is one type of transition metals known as addenda atoms, and hetero-POMs, in which there is a heteroatom apart from transition metals or addenda atoms. The general formula of POMs can be shown as $(X_xM_mO_y)^{n-}$ in the presence of heteroatom, where X represents the heteroatom, M represents the addenda atom and x, m and y represent the stoichiometry of respective elements, and in the absence of heteroatoms $(M_6O_{19})^{n-}$, where M represents the addenda atom [11]. It is worth mentioning that the addenda atoms in POMs are responsible for redox reactions and can be molybdenum, tungsten, vanadium and others at their highest oxidation states, and heteroatoms or central atoms can be silicon, phosphorus, aluminium, germanium and iodine but also lacunary structures are known.

Based on the identity of heteroatom and addenda atom and the type of metal-oxo bonding, there is a wide range of POM structures, including Keggin $(XM_{12}O_{40})^{n-}$, Dawson $(X_2M_{18}O_{62})^{n-}$, Lindqvist $(M_6O_{19})^{n-}$, Anderson, and polymeric structures, offering quite different possible applications as electrode materials, catalysts, sensors, in medicine and for energy storage devices [12, 13]. Apart from the tuneable structures and variable compositions, the usually fast redox behaviour of POMs, due to the transition metals with empty orbitals, makes them suitable materials in the field of energy storage, ranging from electrode materials and catalysts to electrolytes for RFBs.

There is another classification of POMs, substituted POMs, which consists of two different addenda atoms responsible for redox reactions. In fact, one or several of the addenda atoms in $(X_xM_mO_y)^{n-}$ or $(M_6O_{19})^{n-}$ formulas can be substituted by another metal such as Mn, Co, Ni and Cu by replacing the addenda atoms in basic media. The formula for the substituted-POMs can be shown as $(XZ_zM_{12-z}O_{40})^{n-}$ for Keggin structures and $(X_2Z_zM_{18-z}O_{62})^{n-}$ for Dawson structures. Therefore, by substituting the addenda atoms with another atom, there will be two different sites responsible for redox reactions. Hence, based on the application and need, different POMs can be tailored and tuned [14–16].

It is worth mentioning that the size of the heteroatom, the overall charge of the cluster, the basicity of the cluster, and the presence of substitute elements to the addenda atoms have an influence on the redox potentials of POMs [13]. As an example, Mbomekalle et al. investigated the effect of heteroatom size on the redox potential of the Keggin $(XSi_{12}O_{40})^{n-}$ for different elements in the same group in the periodic table as the heteroatom. From their study, it can be concluded that the smaller the heteroatom size, the more negative the redox potential is. To put it differently, the addenda atoms in POMs are responsible for redox reactions and the heteroatom can delocalize the added electron or electrons during reduction process; the smaller the size of heteroatom, the less delocalized is the added electron; as a result, the tendency of the cluster for reduction is lower, causing more negative redox potentials [17].

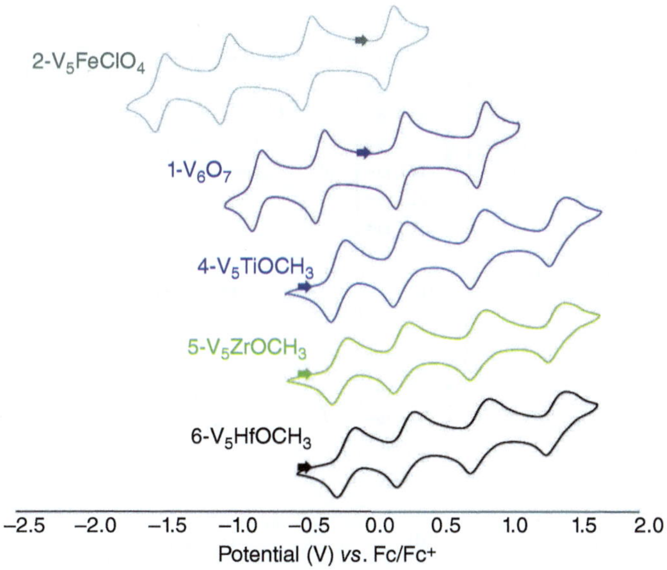

Figure 10.3 CVs of $V^{IV}V^{V}O_7(OCH_3)_{12}$, 2-$V_5FeClO_4$, 4-$V_5TiOCH_3$, 5-$V_5ZrOCH_3$ and 6-V_5HfOCH_3 in acetonitrile with 0.1 M $[Bu_4N][PF_6]$. Source: Adapted from VanGelder et al. [18].

In other work, Matson et al. investigated the effect of substituting electron-rich metals (such as Fe^{III}) and electron-deficient metals (such as titanium [Ti], zirconium [Zr] and hafnium [Hf]) within the polyoxovanadate-alkoxide cluster core ($V^{IV}V^{V}O_7(OCH_3)_{12}$), 1-$V_6O_7$. They found that, by substituting one of the vanadium atoms with iron metal, the redox peaks related to the vanadium shifted towards more negative potential (about −0.6 V), while substituting the electron-deficient metals resulted in the positive shifts of the vanadium redox peaks (Figure 10.3). Hence, depending on the desired application, the electronic configuration of POMs can be tuned [18].

In another study, Chen and Barteau investigated the effects of different counter-cations (such as H, Li, Na, K, Mg and Ca), different heteroatoms including Si, P and As, and different addenda atoms including Mo, W and V for a series of Keggin structure of POMs [19]. Cyclic voltammograms of $Li_3PMo_{12}O_{40}$ and $Na_3PMo_{12}O_{40}$ (Figure 10.4) show three redox peaks, which are quite similar for both compounds. The first two peaks from the positive side show fast kinetics, while the third peak is slower. However, all three peaks for $Mg_{1.5}PMo_{12}O_{40}$ and $Ca_{1.5}PMo_{12}O_{40}$ exhibit a slower redox kinetics, especially for $Mg_{1.5}PMo_{12}O_{40}$. These results show that the type of the counter cation can affect the rate of the electron-transfer reaction, influencing the rate of the redox reactions.

They also investigated the effect of different heteroatoms on the electrochemical behaviour of $Li_3PMo_{12}O_{40}$, $Li_3AsMo_{12}O_{40}$ and $Li_4SiMo_{12}O_{40}$ in acetonitrile with 100 mM LiTf (Figure 10.5). It can be observed that, for $Li_4SiMo_{12}O_{40}$ with a higher negative charge than that of $Li_3PMo_{12}O_{40}$ and

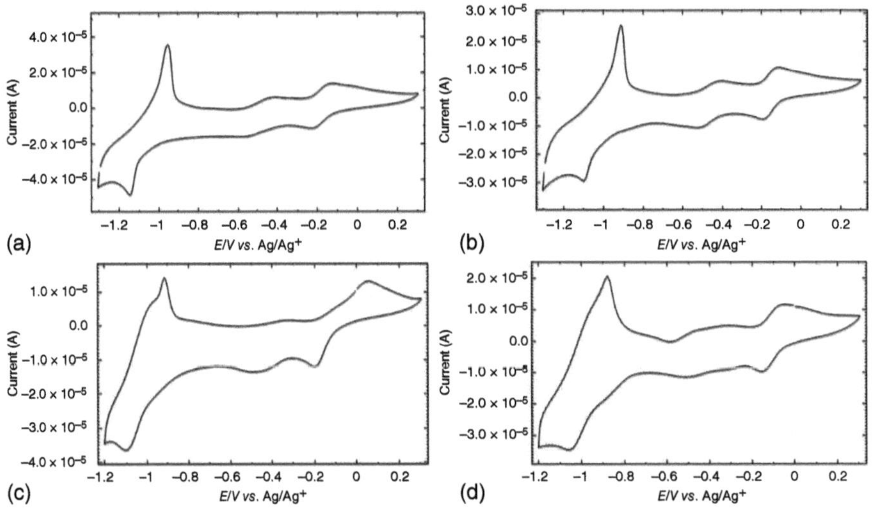

Figure 10.4 The influence of different countercations. CVs of 0.5 mM (a) $Li_3PMo_{12}O_{40}$, (b) $Na_3PMo_{12}O_{40}$, (c) $Mg_{1.5}PMo_{12}O_{40}$ and (d) $Ca_{1.5}PMo_{12}O_{40}$ in acetonitrile with 100 mM LiTf at a scan rate of 100 mV s^{-1}; GCE with a surface area of 0.071 cm^2 as working electrode. Source: Reproduced from Chen and Barteau [19]/American Chemical Society.

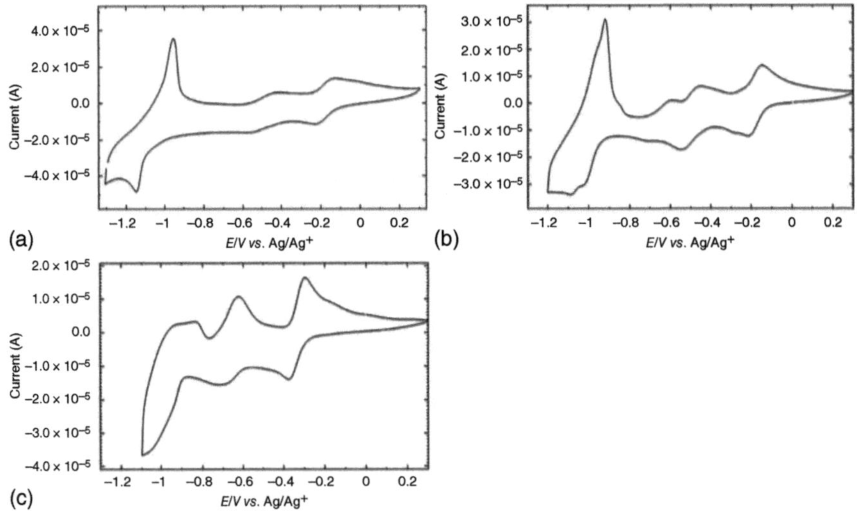

Figure 10.5 The influence of different heteroatoms. CVs of 0.5 mM (a) $Li_3PMo_{12}O_{40}$, (b) $Li_3AsMo_{12}O_{40}$ and $Li_4SiMo_{12}O_{40}$ in acetonitrile with 100 mM LiTf at a scan rate of 100 mV s^{-1}; GCE with a surface area of 0.071 cm^2 as working electrode. Source: Reproduced from Chen and Barteau [19]/American Chemical Society.

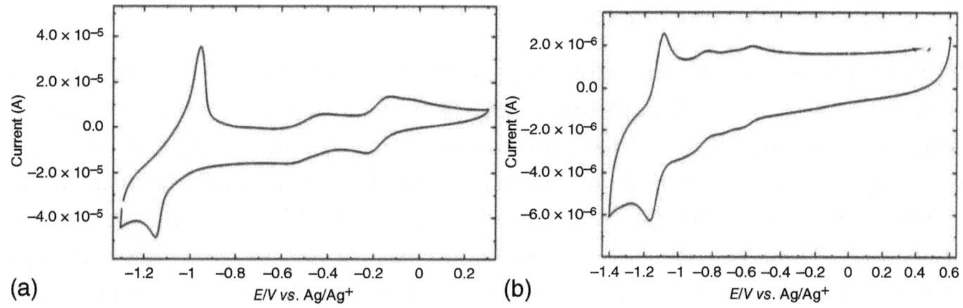

Figure 10.6 The influence of different addenda atoms. CVs of 0.5 mM (a) $Li_3PMo_{12}O_{40}$ and (b) $Li_3PW_{12}O_{40}$ in acetonitrile with 100 mM LiTf at a scan rate of 100 mV s^{-1}; GCE with a surface area of 0.071 cm^2 as working electrode. Source: Reproduced from Chen and Barteau [19]/American Chemical Society.

$Li_3AsMo_{12}O_{40}$, the peak potentials shifted to more negative potential. Similarly, the redox potentials of $Li_4SiW_{12}O_{40}$ were more negative than those of $Li_3PW_{12}O_{40}$, because of the different electronegativity of Si and P atoms. These results demonstrate that, by changing the heteroatom, POMs with different redox potentials can be tailored for an application as posolyte or negolyte. They, furthermore, showed that, by changing or substituting the addenda atoms, not only the peak potentials but also the number of redox peaks or the number of electrons transferred may be altered. These results indicate the vast potential POMs have to be modified and tailored for different purposes (Figure 10.6).

10.5 Suitability of POMs for Energy Conversion and Storage Devices

POMs with their unique structures have several advantages that make them quite suitable candidates for energy storage devices:

First, there are several transition-metal centres as active sites in POMs, making them an effective "electron sponge" materials capable of accommodating multi-electron redox reactions.

Second, the delocalization of the added electron(s) between different active sites as well as the heteroatom facilitates the electron-transfer reactions.

Third, POMs are large very stable molecules where the redox reactions do not significantly alter the structure upon charge transfer – making them suitable for long-term use.

Fourth, the electron transfer in POMs is often coupled with proton or cation transfer, avoiding highly charged clusters [20].

The focus of this section will be on the application of POMs for energy storage devices, namely SCs, Li- and Na-ion batteries and RFBs. We will discuss how POMs can increase the performance of energy storage devices.

10.5.1 POMs in Supercapacitors

Unlike traditional batteries, SC store electricity electrostatically via the electrochemical double layer rather than through electrochemical reactions. Thanks to such a mechanism, SC are endowed with advantages, such as fast charge and discharge cycles, high power density and outstanding durability. In general, SCs consist of two electrodes in contact with electrolytes and an electronically insulating separator in between [2]. The electrodes are typically composed of porous materials with high surface areas, such as activated carbon, graphene oxide, or transition-metal oxides, enabling the storage of large quantities of electric charge. The electrolyte, often an aqueous or organic solution containing ions, facilitates the flow of charge between the electrodes.

Despite the advantages of SCs, they suffer from low energy density as reflected in the Ragone plot (Figure 10.1). To address this issue, POMs, with their features such as fast and multielectron-transfer reactions, are interesting candidates for increasing the energy density of SCs while maintaining high stability. When using POMs for increasing the energy density of SCs, two parameters should be considered: first, the proper anchoring of POMs into carbon-based materials; second, maintaining the long life cycle stability. To achieve this, extensive research has been conducted to investigate stable procedures for creating POM-based hybrid materials, such as POM-conductive polymers and POM-nanocarbon, for electrodes of SCs [9, 13].

As an example, in 2018, Wang et al. synthesized a nanocomposite of POM-based metal organic framework (POMOF) covered with polypyrrole (PPy) [21]. They assessed the performance of the symmetric POMOF/PPy SC, which offered the areal capacitance of 1879 $mF\,cm^{-2}$, much higher than that of the complex without the presence of POM. In another work, Suárez-Guevara et al. incorporated PMo_{12} Keggin-type POM into the structure of graphene oxide during its reduction. The prepared composite showed the superior capacitance of 276 $F\,g^{-1}$ and good stability of 96% retention over 10,000 cycles [22]. Also, Chen et al. used $Na_6V_{10}O_{28}$ POM with a high specific capacitance of 354 $F\,g^{-1}$ as the negative electrode [23]. The asymmetric SC with activated carbon as positive electrode provided the energy density of 73 $Wh\,kg^{-1}$ and power density of 312 $W\,kg^{-1}$.

10.5.2 POMs in Li-ion Batteries

Lithium-ion (Li-ion) batteries, the most widely used rechargeable battery technology today, play an essential role in our daily life, powering everything from smartphones and laptops to electric vehicles and small-scale stationary energy storage. Their popularity stems from high energy density, long cycle life and relatively lightweight construction. They consist of a lithium-based compound as the positive electrode (e.g. lithium cobalt oxide, lithium iron phosphate, lithium manganese oxide or similar) and a carbon-based material (usually graphite) as the negative electrode. The principal operation of Li-ion batteries is based on the intercalation and de-intercalation of Li ions in electrodes during charge and discharge. Apart from cost and safety issues, the structural change of the electrode materials during intercalation and de-intercalation is one of the major issues driving research into alternative battery chemistries and electrode materials.

To accommodate these challenges, POMs with their special features, such as reversible multi-electron redox reactions, high stability without significant change in their crystal structure and structural tunability, may lead to enhanced energy storage devices. In recent years, researchers have explored the use of POMs as electrode materials in lithium-ion batteries, some of which will be discussed here.

For example, in 2016, Wang et al. synthesized a POM-based MOF, NENU-507 $(TBA)_3[PMo_8^VMo_4^{VI}O_{38}(OH)_2Zn_4(PBA)_2]{\cdot}H_2O$, by a hydrothermal method as material for the negative electrode with a discharge capacity of 640 mAh g^{-1} after 100 cycles [24]. Additionally, in 2018, Liu et al. synthesized EMI-$Mo_{72}V_{30}$@rGO (EMI: 1-ethyl-3-methylimidazolium) for the negative electrode with a discharge capacity of 1145 mAh g^{-1} at a current density of 100 mA g^{-1} with 100% CR after 500 cycles [25]. Also, Zhang et al. investigated a series of modified encapsulated ionic liquids into POM-based MOF (POMs-ILs@MOFs) as anode materials. From among different synthesized POMs-ILs@MOFs, $PMo_{10}V_2$-ILs@MIL-100 had the better performance, with a discharge capacity of 1248 mAh g^{-1} at 0.1 A g^{-1} and 75% CR after 100 cycles [26]. In addition, in 2019, Ni et al. synthesized and investigated the performance of $K_9PV_{14}O_{32}$ coated by conductive polypyrrole as cathode material with a capacity of 500 mAh g^{-1} and 90% CR after 50 cycles [27].

10.5.3 POMs in Na-ion Batteries

Sodium-ion (Na-ion) batteries have attracted increasing attention as possible alternatives to Li-ion batteries on the account of their lower cost. These batteries operate on similar principles to Li-ion batteries but use sodium ions instead of lithium ions.

While sodium-ion batteries offer the advantage of utilizing more abundant resources than lithium, they face challenges such as lower energy density and decreased cycle life compared to their lithium counterparts. Nonetheless, ongoing research and developments aim to address these limitations and unlock the full potential of sodium-ion battery technology for various applications, including grid-scale energy storage and stationary power systems. In this regard, POMs are considered as a suitable candidate to reach this goal. As an example, Lin et al. synthesized and utilized the $Na_7[H_2PV_{14}O_{42}]$ as the anode material for a Na-ion battery, offering a capacity of 322 mAh g^{-1} and a CR of 87% over 120 cycles [28]. These efforts represent a promising avenue for advancing the development of next-generation Li-ion and Na-ion batteries.

10.5.4 POMs in RFBs

As shown in the Ragone plot (Figure 10.1), RFBs usually suffer from low specific energy and power density. To solve these issues, POMs have attracted the attention of many researchers in the field of RFBs due to their beneficial features which are as follows:

(1) Since the transition metals used in POMs possess d^0 or d^1 electronic configuration, they have accessible orbitals capable of undergoing multi-electron redox processes, making them a "pool of electrons," which as such can increase the energy density of an RFB.
(2) POMs are highly soluble materials, offering high concentrations as active materials, increasing the energy density of RFBs.
(3) Because of the conjugated structures of POMs, the electron or electrons added to the cluster during reduction can be highly delocalized, bringing about fast electron-transfer kinetics without significant changes in their molecular structure. This enhances the efficiency and power density of RFBs.
(4) As POMs are usually negatively charged clusters, they are very stable materials and do not cross the membrane of RFBs which are typically cation-exchange membranes, improving the CE.
(5) Often, the electron transfer in POMs is coupled with proton or cation transfer, the total negative charge of the cluster stays constant, thus maintaining a high stability.
(6) It is possible to substitute the addenda atoms in POMs with other metal centres which are redox active, offering a versatile composition based on the desired application specifically for symmetric RFBs [13, 20].

The term symmetric in RFBs means that both negolyte and posolyte active materials are the same, while, in asymmetric RFBs, different materials are used as posolyte and negolyte. Herein, we will review some POM-based RFBs.

As early as 1997, Peiju et al. used a series of Keggin POMs – $H_3PMo_{12}O_{40}$, $H_3PW_{12}O_{40}$, $H_4SiMo_{12}O_{40}$ and $H_4SiW_{12}O_{40}$ – as posolyte and Fe^{2+}/Fe^{3+} as the negolyte in H_2SO_4 for an asymmetric RFB. They assembled the RFB with the activated porous carbon felt as the electrodes (with an active area of $70\,cm^2$), graphite plates as current collectors and a cation-exchange membrane. From among the different POMs used in this study, H_3PMo_{12}/Fe exhibited the lowest internal resistance and demonstrated the best performance with the EE of 73%, CE of 93% and VE of 78% at a current density of $40\,mA\,cm^{-2}$ for 24 cycles [29]. However, the effect of the long-term cycling remained not clear.

In 2013, Pratt and Anderson synthesized and investigated three different mixed-addenda POMs – A-α-$PV_3W_9O_{40}{}^{6-}$ (AP), B-α-$PV_3W_9O_{40}{}^{6-}$ (BP), $P_2V_3W_{15}O_{62}{}^{9-}$ (WD) – in symmetric RFBs [30]. The difference between two Keggin AP and BP structures is that, in AP, the V atoms are substituted with corner-sharing W atoms, while in BP, the substitution is from edge-sharing W atoms. To gain an insight in their kinetic behaviour and peak positions, cyclic voltammetry was used (Figure 10.7). The potential range in CVs between 0.2 and 0.9 (V vs. SHE) for AP and BP is associated to vanadium redox reactions, while the potential range between 0 and – 0.8 (V vs. SHE) is associated to the redox reactions of tungsten atoms. It is worth noting that, because of the separation of vanadium and tungsten redox peaks in terms of potential, their application in symmetric RFBs with suitable voltage is possible. Also, for WD, the potential between 0.4 and 0.9 (V vs. SHE) is related to vanadium and 0.1 to –0.6 (V vs. SHE) is related to tungsten, providing a

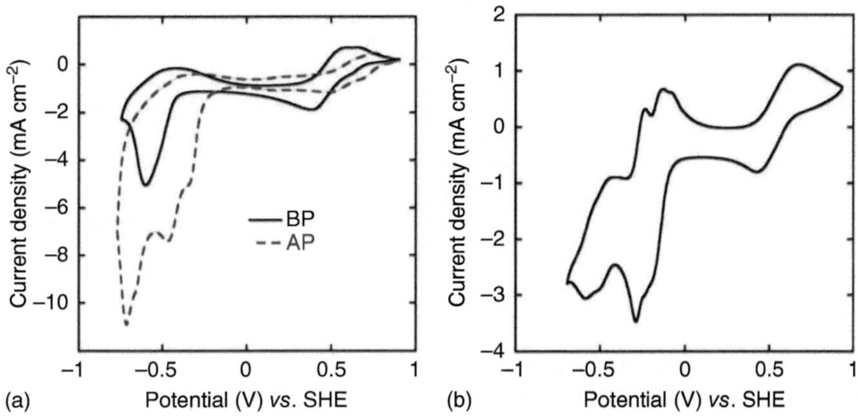

Figure 10.7 The CVs of 20 mM (a) $PV_3W_9O_{40}^{6-}$ (AP) and $PV_3W_9O_{40}^{6-}$ (BP) and (b) WD in H_2SO_4 at a scan rate of $50\,mV\,s^{-1}$. Source: Reproduced from Pratt and Anderson [30]/Royal Society of Chemistry.

suitable voltage for a symmetric cell. In the next step, they determined the number of electrons involved in the redox reactions via bulk electrolysis, which was three electrons for both AP and BP and one electron for WD. The performance of the symmetric RFBs for AP, BP and WD was investigated in H_2SO_4 during 20 cycles by carbon felt as the electrodes (with the active area of $5\,cm^2$) and Nafion 117 membrane. The CE of WD, AP and BP was 95%, 80% and between 50% and 70%, respectively. Although the CE of WD is higher than that of AP and BP, since there is one electron involved in the redox process of WD, its energy density was low. The low VE and EE can be attributed to the slower kinetics of the redox reactions of the active materials and obvious precipitation on the carbon felts [30].

In 2018, Chen et al. claimed that $[P_2W_{18}O_{62}]^{6-}$ can undergo 18 proton-coupled electron transfers in a reversible manner, enabling a high energy density RFB [31]. To investigate the electrochemical behaviour of $[P_2W_{18}O_{62}]^{6-}$, they conducted the cyclic voltammetry of $[P_2W_{18}O_{62}]^{6-}$ by a three-electrode system for solutions at different pH (Figure 10.8a) and different concentrations (Figure 10.8b). Via bulk electrolysis and UV–VIS titration, they found that 2 mM $[P_2W_{18}O_{62}]^{6-}$ at pH 7.0 undergoes four one-electron transfer processes in the range of −0.6 to 0.6 (V vs. SHE). Furthermore, when they decreased the pH to 4, they observed that the two one-electron redox peaks below 0 V (vs. SHE) changed to two-electron transfer processes (Figure 10.8a). In addition, as can be seen in Figure 10.8b, they found that, by increasing the concentration to 100 mM, there are more redox peaks in the same potential range. Based on these results, they reduced different concentrations of the cluster with 18 electrons to assess the number of electrons which can be extracted. As can be seen in Figure 10.8c, 2, 50 and 100 mM of $[P_2W_{18}O_{62}]^{6-}$ can store and release 4, 16 and 17.2 electrons, respectively. These results indicate that, by increasing the concentration, the ability of $[P_2W_{18}O_{62}]^{6-}$ to release stored electrons increases.

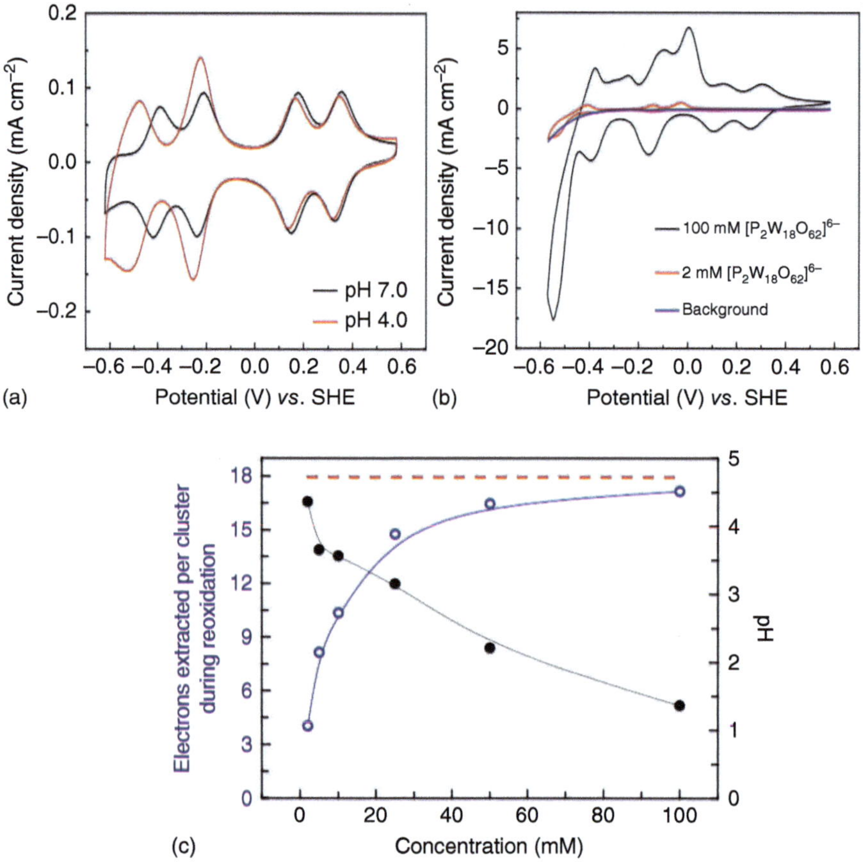

Figure 10.8 (a) CVs of 2 mM $[P_2W_{18}O_{62}]^{6-}$ in 1 M Li_2SO_4 at different pH = 7 and 4 at a scan rate of 10 mV s^{-1}. (b) The CVs of $[P_2W_{18}O_{62}]^{6-}$ with different concentrations and the blank solution of 1 M H_2SO_4 (blue line). (c) The relationship between POM concentration, solution pH and the number of electrons that can be extracted from a reduced solution by 18 electrons. Source: Reproduced from Chen et al. [31]/Springer Nature.

In the next step, they investigated the stability of 50 mM $[P_2W_{18}O_{62}]^{6-}$ in 1 M H_2SO_4 by galvanostatic charge/discharge cycles at a current density of ±50 mA cm^{-2} for 100 cycles (Figure 10.9a). The battery offered a CE > 95% and a CR of 97.3%, as depicted in Figure 10.9b.

Finally, they assembled an asymmetric RFB with $[P_2W_{18}O_{62}]^{6-}$ as negolyte and HBr/Br_2 as posolyte, and the schematic of the cell is depicted in Figure 10.10a. As can be seen in the CVs of Figure 10.10b, 100 mM $Li_6[P_2W_{18}O_{62}]$ (blue line) and 0.5 M HBr in 1 M H_2SO_4 (orange line) offered the open circuit voltage of 1.25 V. Further, they investigated the energy density of the battery for different concentrations 0.1, 0.3 and 0.5 M of $[P_2W_{18}O_{62}]^{6-}$ as the negolytes and a constant concentration of 4 M HBr and 0.1 M Br_2 in 1 M H_2SO_4 at current density 50 mA cm^{-2}, which was 42.6, 130 and 225 Wh l^{-1}, respectively. In addition, the CE and EE of each of these cells were 96% and 76%, respectively (Figure 10.10c). They postulate that, by increasing

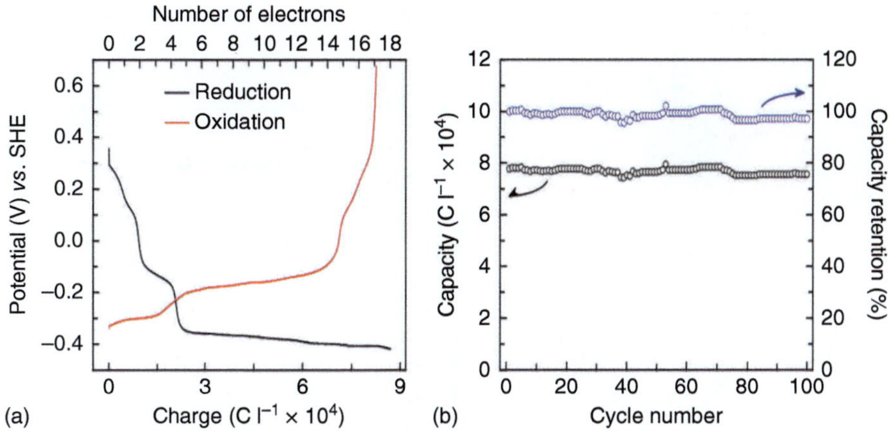

Figure 10.9 (a) Charge/discharge of 50 mM $[P_2W_{18}O_{62}]^{6-}$ in 1 M H_2SO_4 at a current density of ±50 mA cm^{-2}. (b) The capacity retention of the mentioned cell over 100 cycles. Source: Reproduced from Chen et al. [31]/Springer Nature.

the concentration to the limit solubility (1.9 M), the theoretical energy density of 1000 Wh l^{-1} will be accessible (Figure 10.10d). Also, they investigated the power density of the batteries with different concentrations of the negolyte. As can be seen in Figure 10.10e the power density for 0.3 M $[P_2W_{18}O_{62}]^{6-}$ reached the maximum amount of 0.52 W cm^{-2}. In addition, the stability of the 0.5 M $[P_2W_{18}O_{62}]^{6-}$ in 0.1 M H_2SO_4/4 M HBr + 0.1 M Br_2 battery was studied at a current density of 0.1 A cm^{-2} for 20 cycles. As shown in Figure 10.10f, the volumetric capacity of 210 Ah l^{-1} and CE of 98% was reached.

Barteau et al. have utilized $Li_3PMo_{12}O_{40}$ as a posolyte in a non-aqueous RFB (in DMF and acetonitrile) and investigated the performance of the battery as well as the effect of the two solvents [32]. The electrochemical behaviour of $Li_3PMo_{12}O_{40}$, was reflected in cyclic voltammetry. Figure 10.11 shows data of 0.5 M $Li_3PMo_{12}O_{40}$ in 0.1 M LiTf in (a) DMF and (b) acetonitrile. $Li_3PMo_{12}O_{40}$ in DMF offers three redox peaks, 2 one-electron and 1 two-electron processes, while $Li_3PMo_{12}O_{40}$ in acetonitrile has 2 one-electron redox peaks. In the next step, the authors assembled a symmetric battery by using $Li_3PMo_{12}O_{40}$ as both posolyte (with pre-reduction of the active materials) and negolyte in both DMF (Figure 10.12a) and acetonitrile (Figure 10.12b). The battery with 10 mM $Li_3PMo_{12}O_{40}$ at the state of charge 60% (providing 1.2 electrons) in DMF with 0.1 M LiTf at a current density of 0.2 mA cm^{-2} was tested. The battery provided a volumetric capacity of 0.322 Ah l^{-1} and a cell voltage of 0.7 V with a CE more than 90% and VE over 70%. However, the same condition of 10 mM $Li_3PMo_{12}O_{40}$ even at the state of charge 80% in acetonitrile provided one electron and a volumetric capacity of 0.213 Ah l^{-1} and cell voltage of 0.35 V.

To increase the energy density of the battery, they increased the concentration of active materials. As shown in Figure 10.13, a symmetric RFB with 0.1 M $Li_3PMo_{12}O_{40}$ at the same state of charge 60% in DMF provided two electrons and cell voltage of 1 V with CE more than 93%, VE more than 50% and EE more than

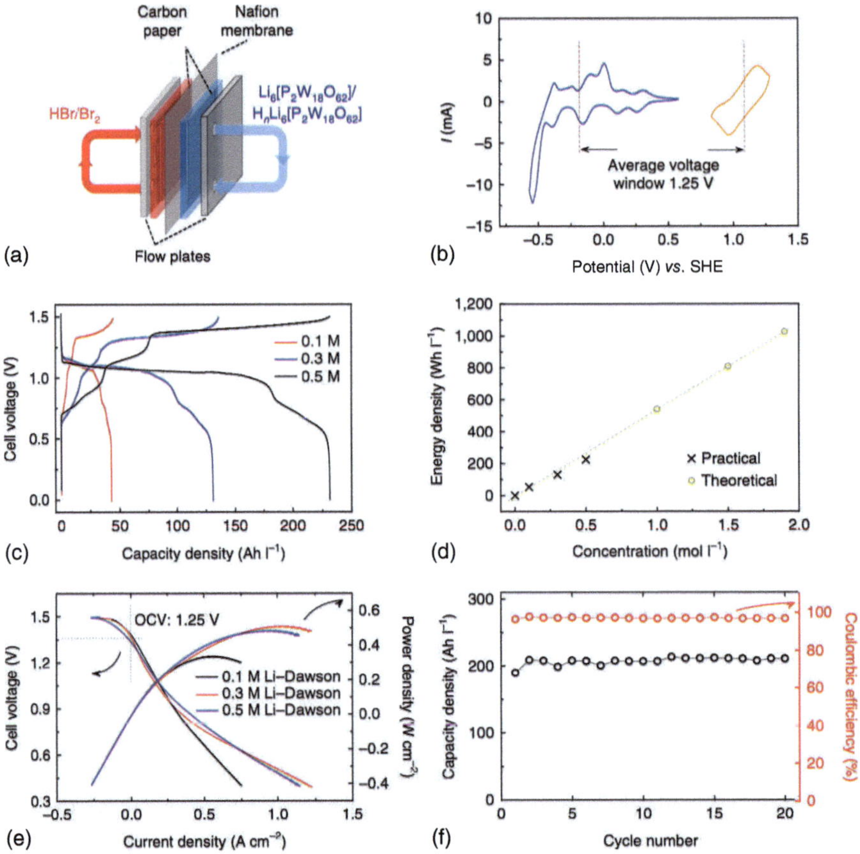

Figure 10.10 (a) Schematic of the redox flow cell. (b) CVs of 100 mM $Li_6[P_2W_{18}O_{62}]$ in 1 M H_2SO_4 (blue) and 0.5 M HBr in 1 M H_2SO_4 (orange) at a scan rate of 10 mV s^{-1}; GCE with a surface area of 0.071 cm^2 as working electrode. (c) Charge/discharge curves for different concentrations of POM at a current density of 50 mA cm^{-2} at 20 °C. (d) The relationship between energy density and concentration of $Li_6[P_2W_{18}O_{62}]$ for the mentioned cell. (e) Polarization curves representing the power density of the cell versus current density at 20 °C. (f) Battery performance of 0.5 M $Li_6[P_2W_{18}O_{62}]$ as negolyte at a current density of 0.1 A cm^{-2} with the cell voltage window 0–1.65 V. Source: Reproduced from Chen et al. [31]/Springer Nature.

50% at a current density of 2 mA cm^{-2}. The reason for low VE and EE values may be the viscosity and mass-transfer limitations. Therefore, they showed that, by increasing the concentration of active materials, they can increase not only the number of accessible electrons but also the cell voltage, bringing higher energy density and power density of the battery.

Furthermore, they assembled an asymmetric RFB with 10 mM $Li_3PMo_{12}O_{40}$ as posolyte and 10 mM $Li_6P_2W_{18}O_{62}$ in DMF with 0.1 M LiTf at the state of charge 50% to improve the performance of the battery. In this configuration, four electrons were accessible, and the cell voltage increased to 1.3 V with CE, EE and VE of 95–100%, 60% and 60%, respectively, for 24 cycles as can be seen in

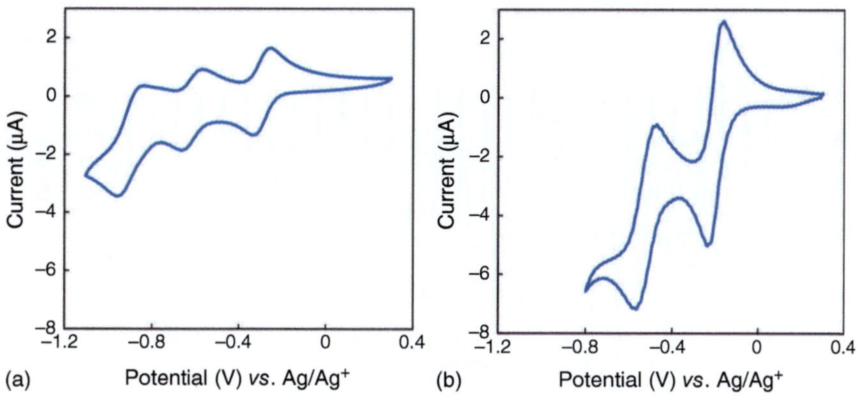

Figure 10.11 The CVs of 0.5 M $Li_3PMo_{12}O_{40}$ in 0.1 M LiTf in (a) DMF and (b) acetonitrile; GCE with a surface area of 0.071 cm^2 as working electrode. Source: Reproduced with permission from Cao et al. [32].

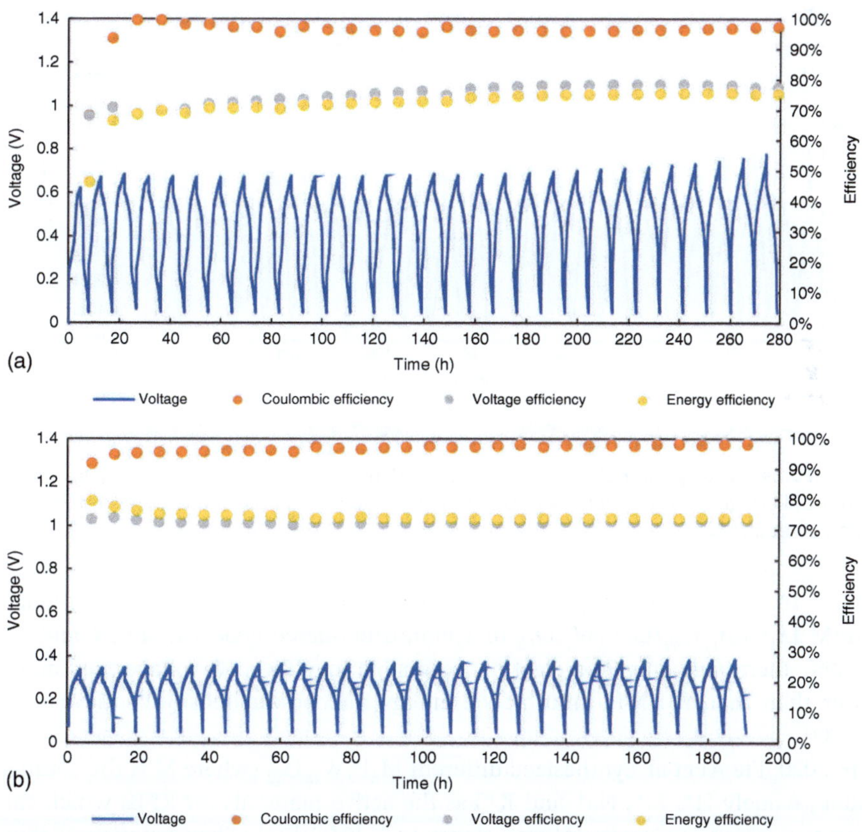

Figure 10.12 Charge/discharge profile and cycle efficiencies of 10 mM $Li_3PMo_{12}O_{40}$ in 0.1 M LiTf at a current density of 0.2 mA cm^{-2} (a) in DMF and (b) in acetonitrile. Source: Reproduced with permission from Cao et al. [32].

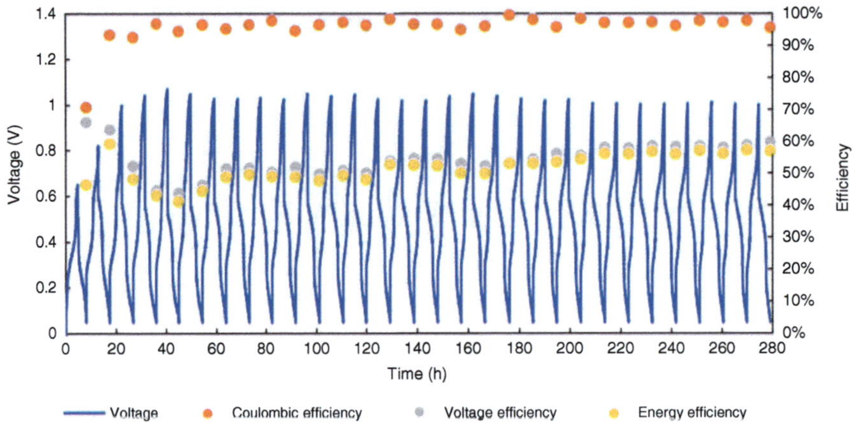

Figure 10.13 Charge/discharge profile and cycle efficiency of 0.1 M $Li_3PMo_{12}O_{40}$ in DMF with 1 M LiTf at a current density of 2 mA cm^{-2}. Source: Reproduced with permission from Cao et al. [32].

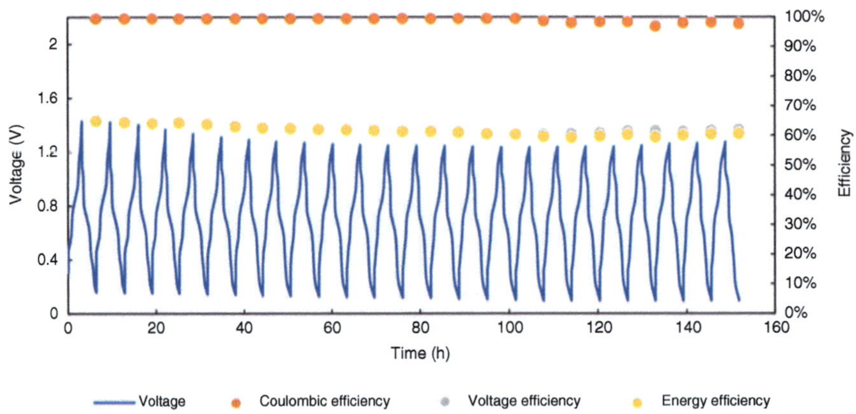

Figure 10.14 Voltage and efficiency profiles of 10 mM $Li_6P_2W_{18}O_{62}/Li_3PMo_{12}O_{40}$ in DMF with 0.1 M LiTf at a current density of 0.5 mA cm^{-2}. Source: Reproduced with permission from Cao et al. [32].

Figure 10.14. The same experiment was done in acetonitrile as well. Although 10 mM $Li_3PMo_{12}O_{40}/Li_6P_2W_{18}O_{62}$ in acetonitrile offered good CE of 95% and EE of 60%, there was only one electron accessible and the cell voltage was much lower than that of DMF, limiting the energy and power density of the battery [32, 33].

In 2020, Fie Ai et al. synthesized different $M_6P_2W_{18}O_{62}$ (where M is the counter cation, namely H^+, Li^+, Na^+ and K^+) as the active materials for RFBs which can operate even at low temperatures, down to −20 °C [34]. They mentioned that, from among different cations, $H_6P_2W_{18}O_{62}$ (herein shown as HPOM) offers higher solubility and conductivity at both room temperature and lower temperature −20 °C. Figure 10.15 shows the CVs of HPOM at (a) 25 °C and (b) −20 °C at different

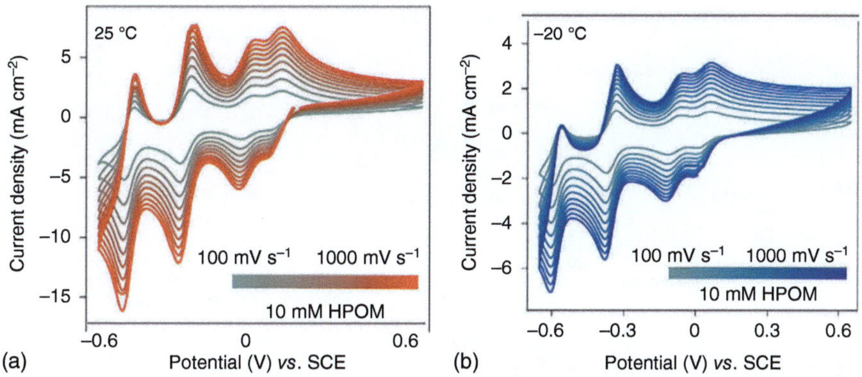

Figure 10.15 The CVs of 10 mM HPOM at different scan rates from 100 to 1000 mV s^{-1} at (a) 25 °C and (b) −20 °C. Source: Reproduced from Ai et al. [34]/Springer Nature.

scan rates, confirming the stability of HPOM at low temperatures with fast kinetics, $k_0 = 6.8\ 10^{-3}$ cm s^{-1}.

To investigate the performance of HPOM as negolyte at room temperature and −20 °C, they utilized $VOSO_4$ as posolyte which is stable up to −40 °C to build an asymmetric RFB. Figure 10.16a,b shows the rate performance of 0.5 M HPOM as negolyte and 1.5 M $VOSO_4$ in 3 M H_2SO_4 as posolyte at 25 °C with different current densities from 100 to 500 mA cm^{-2}. The battery showed the best performance at a current density of 300 mA cm^{-2} with a volumetric capacity of 78.8 Ah l^{-1} and a CE of 98%. Figure 10.16c indicates that the battery reached a power density of 397.9 mW cm^{-2} at a current density of 600 mA cm^{-2}. In addition, as can be seen in Figure 10.16d, the battery shows a very stable performance at a current density of 300 mA cm^{-2} over 450 cycles with a CE about 98%. The performance of the same battery was investigated at −20 °C (Figure 10.16e–h). The asymmetric battery at −20 °C still reached a power density of 282.4 mW cm^{-2} at a current density of 400 mA cm^{-2} and a volumetric capacity of 79.6 Ah l^{-1} at a current density of 160 mA cm^{-2} over 800 cycles with a CE of 99.06% [34].

In 2020, Wang et al. proposed an asymmetric aqueous RFB which is endowed with high power density and the advantage of cooling down the chip temperature at the same time for electronic devices by enjoying the electronic blood (E-blood) concept [35]. They assembled the battery with 0.05 M $Li_6P_2W_{18}O_{62}$ in 2 M H_2SO_4 as the negolyte, 0.04 M cerium sulphate ($Ce(SO_4)_2$) in 2 M H_2SO_4 as the posolyte and Nafion 115 as membrane with the electrode area of 5 cm^2. They claim that, although the solubility of cerium in aqueous solutions is limited to 0.05 M, the fast redox process and the high potential (about 1.45 V vs. SHE, Figure 10.17) will yield a high-performance battery. Figure 10.18a depicts the charge/discharge profiles and the efficiencies of the battery over 150 cycles at a current density 50 mA cm^{-2} resulting in a CE of 98%, EE of 80% and capacity fade of 0.003% per cycle. To investigate the rate performance of the battery, they tested the full cell at different current densities from 40 to 200 mA cm^{-2}. They observed an increase of CE from 98% to 100% but a reduction of EE from 82% to 44% upon increasing the current density. The reduction

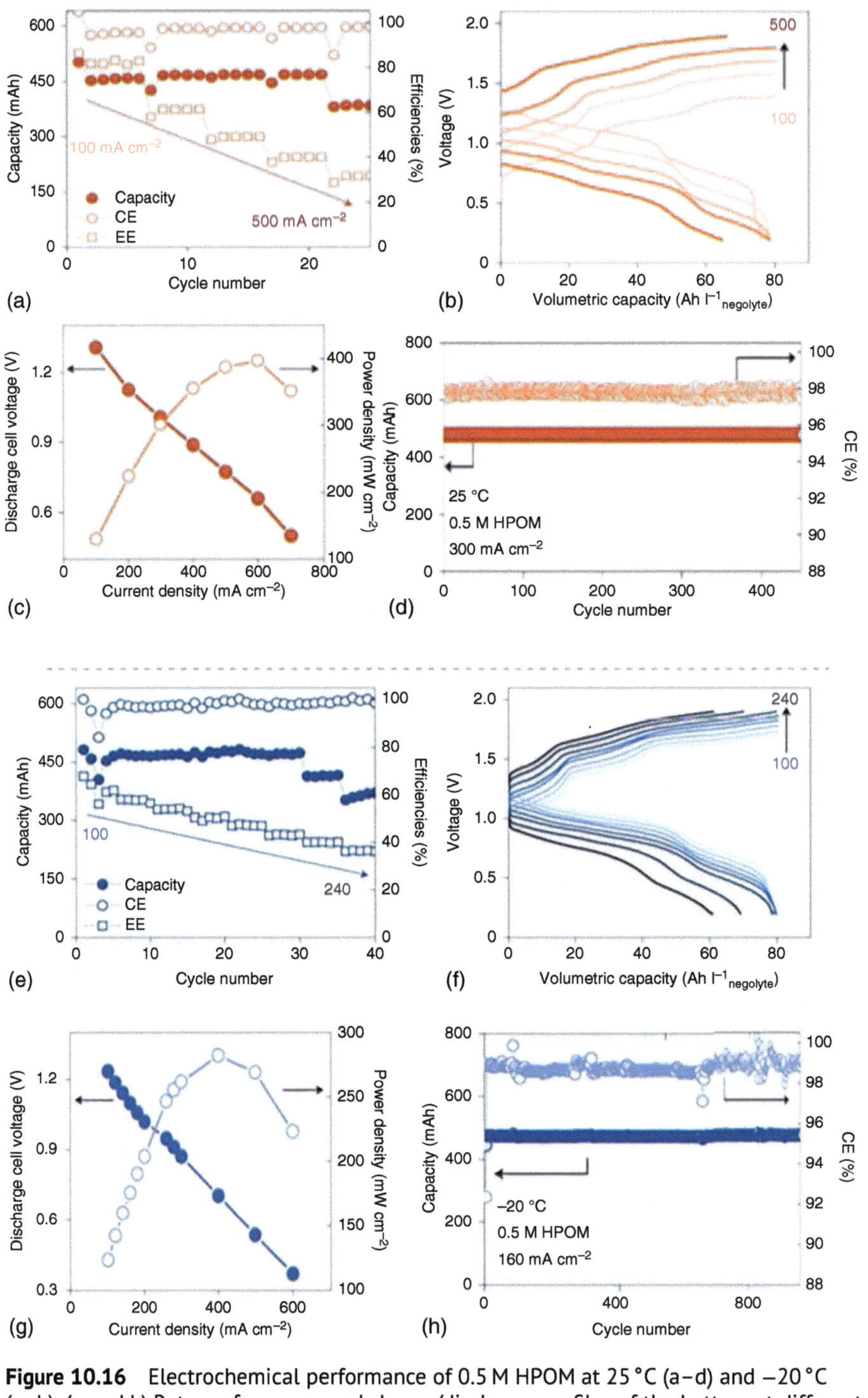

Figure 10.16 Electrochemical performance of 0.5 M HPOM at 25 °C (a–d) and −20 °C (e–h). (a and b) Rate performance and charge/discharge profiles of the battery at different current densities from 100 to 500 mA cm^{-2}. (c) Discharge polarization curves. (d) Cycling capacity and CE at 300 mA cm^{-2}. (e and f) Rate performance and charge/discharge profiles of the battery at −20 °C from 100 to 240 mA cm^{-2}. (g) Discharge polarization curves. (h) Cycling capacity and CE at 160 mA cm^{-2}. Source: Reproduced from Ai et al. [34]/Springer Nature.

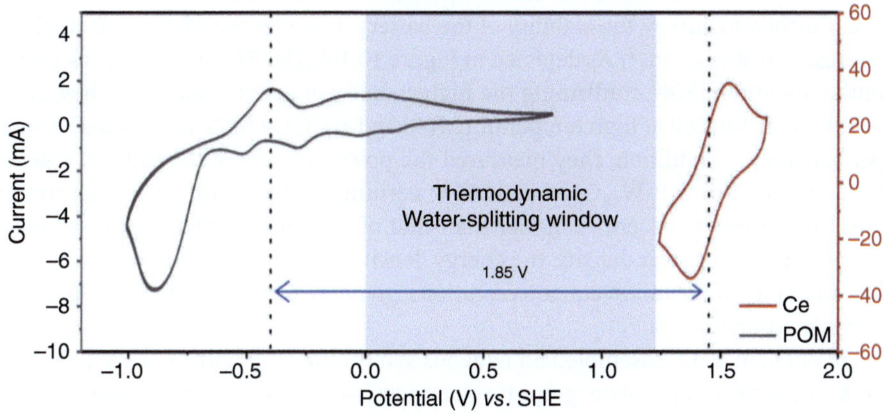

Figure 10.17 The CVs of $Li_6P_2W_{18}O_{62}$ and Ce showing a full-cell voltage of 1.8 V; graphite plate with 6 mm diameter as working electrode. Source: Reproduced with permission from Wang et al. [35].

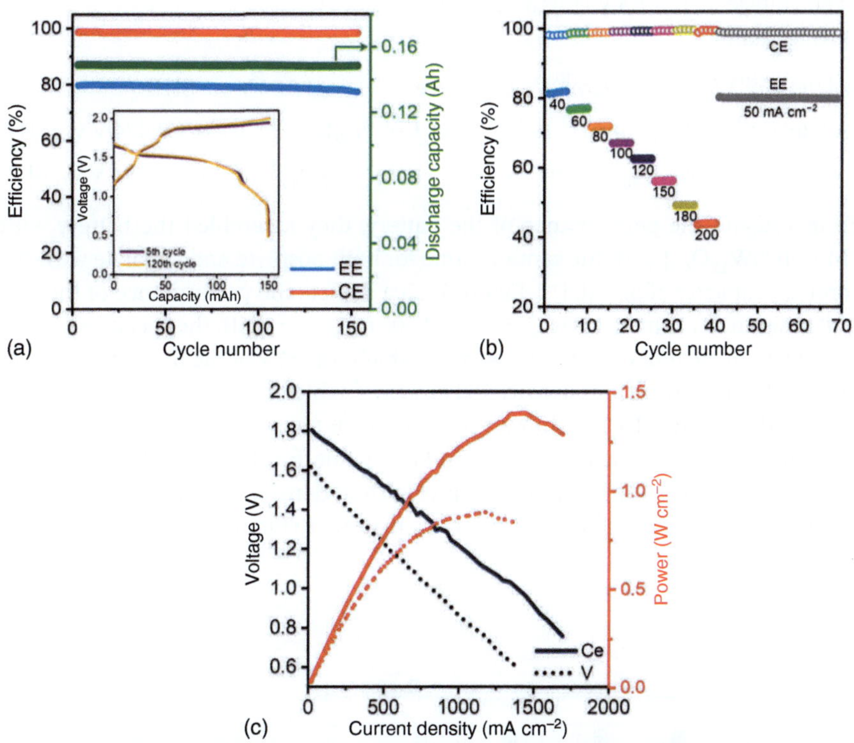

Figure 10.18 (a) Cycling test of the POM-Ce flow battery at 50 mA cm^{-2}, inset: charge/discharge profiles of the cell over cycling. (b) Rate performance test at different current densities. (c) Polarization and power density plots of the POM-Ce flow battery and its comparison with a vanadium redox flow battery. Source: Reproduced with permission from Wang et al. [35].

in EE may be attributed to the ohmic resistance rather than the kinetics which is fast. Further, to survey the stability of the battery, they decreased the current density again to 50 mA cm^{-2}. As depicted in Figure 10.18b, the EE recovered back to the initial amount of 80%, confirming the high stability of active materials. They, also, operated the full cell at high temperatures (40 and 50 °C), which also showed a high performance. In addition, they measured the power density of the full cell assembled with 0.07 M $Li_6P_2W_{18}O_{62}$ and 0.065 M cerium sulphate with a 1 cm^2 electrode area, which was 1.4 W cm^{-2}, higher than that of vanadium RFBs (Figure 10.18c). Despite the high power density, the energy density of the cell is restricted due to the low solubility of cerium in aqueous solutions, giving rise to further exploration for a suitable posolyte.

Liu et al., in 2017, assembled an aqueous symmetric RFB with $H_6[CoW_{12}O_{40}]$ as both, negolyte and posolyte, graphite electrodes with a surface area of 4 cm^2 and a Nafion membrane [36]. In such a system, based on the CVs depicted in Figure 10.19, the Co centres will be oxidized by a one-electron reaction during charge at potential about 1.103 V vs. SHE, and W centres will be reduced during charge by two two-electron processes at potentials about −0.074 and −0.19 V vs. SHE, providing the cell voltage of 1.2–1.4 V (Figure 10.19).

During charge:

$$\text{Posolyte: } Co^{II}W_{12}O_{40}{}^{6-} \quad Co^{III}W_{12}O_{40}{}^{5-} + e^- \quad U_0 = 1.103\ \text{V vs. SHE}$$

$$\text{Negolyte: } Co^{II}W_{12}O_{40}{}^{6-} + 2H^+ + 2e^- \quad H_2Co^{II}W_{12}O_{40}{}^{6-} \quad U_0 = -0.074\ \text{V vs. SHE}$$

$$H_2Co^{II}W_{12}O_{40}{}^{6-} + 2H^+ + 2e^- \quad H_4Co^{II}W_{12}O_{40}{}^{6-} \quad U_0 = -0.191\ \text{V vs. SHE}$$

To investigate the performance of the battery, they assembled the battery with 0.4 M $H_6[CoW_{12}O_{40}]$ with the same volume for both posolyte and negolyte with the theoretical capacity of 5.3 Ah l^{-1}. Figure 10.20a depicts the performance of the battery at different current densities, 25, 50 and 100 mA cm^{-2}, with the specific capacity of 4.61 Ah l^{-1}at current density 25 mA cm^{-2} which was 87% of the theoretical capacity and 3.71 Ah l^{-1} at current density 100 mA cm^{-2} which was 70% of the theoretical capacity. Also, Figure 10.20b shows the CE and EE at these current densities; the CE for these current densities was about 98%, while EE decreased from 86.7% to 54.2%. However, by going back to the initial current density, the initial amount was recovered, showing the stability of the active materials (Figure 10.20b).

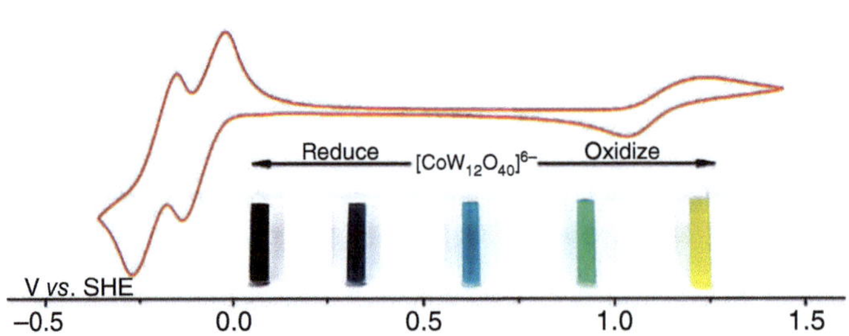

Figure 10.19 CV of $H_6[CoW_{12}O_{40}]$ at a wide potential window, showing the accessible cell voltage for a symmetric RFB. Source: Reproduced from Liu et al. [36]/John Wiley & Sons.

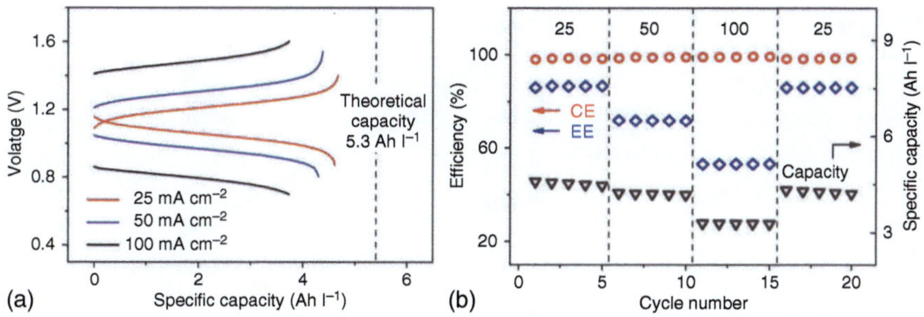

Figure 10.20 (a) Voltage profile of 0.4 M $H_6[CoW_{12}O_{40}]$ at different current densities, and (b) cycle performance of $H_6[CoW_{12}O_{40}]$ at these current densities and back to the initial amount. Source: Reproduced from Liu et al. [36]/John Wiley & Sons.

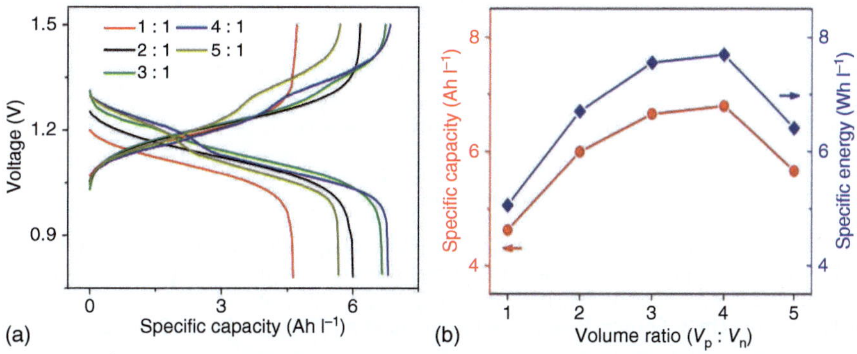

Figure 10.21 (a) The cell performance of 0.4 M $H_6[CoW_{12}O_{40}]$ with different volume ratios of posolyte (V_P) and negolyte (V_N) at a current density of 25 mA cm^{-2} and (b) the relationship between specific capacity, energy density and volume ratio of posolyte to negolyte ($V_P : V_N$). Source: Reproduced from Liu et al. [36]/John Wiley & Sons.

In the next step, since the redox reaction in posolyte involves one electron and in negolyte involves four electrons, they increased the volume of posolyte (V_P) to see whether they can improve the specific capacity of the battery by balancing the charge or not. To do so, they investigated the performance of the battery with the volume ratio $V_P : V_N$ 1 : 1, 2 : 1, 3 : 1, 4 : 1 and 5 : 1. As it can be seen in Figure 10.21a,b, the specific capacity and energy density increased from the ratio of 1 : 1 to 4 : 1, with a decrease for the ratio of 5 : 1. These results confirm that the electron-transfer process in the negolyte is a four-electron process and by balancing the charge of the negolyte and posolyte, the performance of the battery will be improved.

To further improve the performance of the battery, they used an elevated concentration of active materials, 0.8 M $H_6[CoW_{12}O_{40}]$, with the volume ratio of 4 : 1 ($V_P : V_N$) at a current density of 25 mA cm^{-2} (Figure 10.22). As can be seen in Figure 10.22a, this battery provided the specific capacity of 13.5 Ah l^{-1} and energy density of 15.4 Wh l^{-1} which was higher than that of 0.4 M active materials. In addition, Figure 10.22b shows the EE and CE of this cell at different current densities, 25, 50 and 100 mA cm^{-2}; CE for these current densities was constant about 99% and

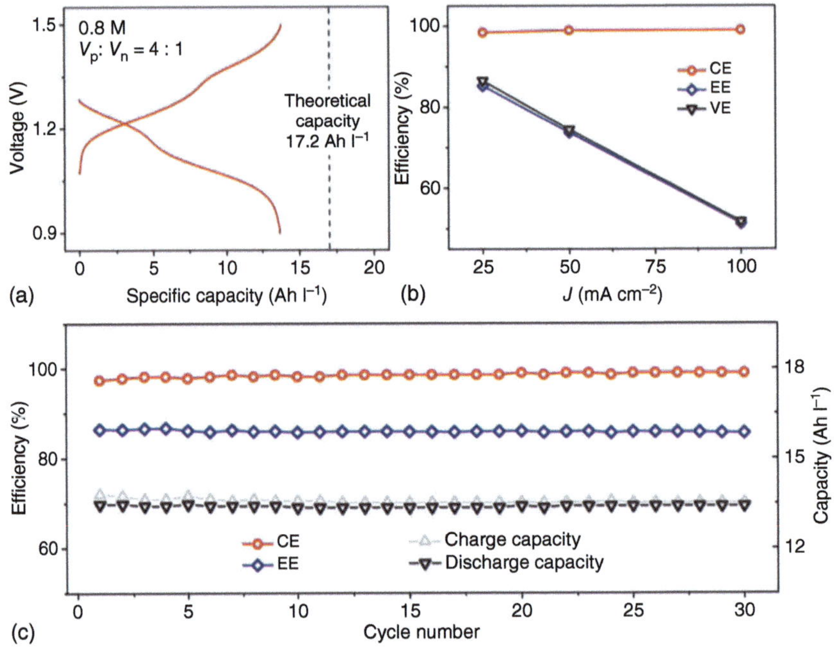

Figure 10.22 The cell performance of 0.8 M $H_6[CoW_{12}O_{40}]$ with the volume ratio of 4 : 1 ($V_P : V_N$). (a) Voltage profile of the cell at a current density of 25 mA cm^{-2}, (b) the CE and EE of the cell at different current densities and (c) cycling stability of the cell at a current density of 25 mA cm^{-2}. Source: Reproduced from Liu et al. [36]/John Wiley & Sons.

EE was 85.4%, 73.8% and 54.9% for the current densities of 25, 50 and 100 mA cm^{-2}, respectively. Moreover, the cycling stability of this high-concentration battery was tested and showed the CE of 99% and EE of 86% at a current density of 25 mA cm^{-2} for 30 cycles (Figure 10.22c).

As we discussed in Section 10.2, an important factor of the total cost of the battery is the cost of the active materials. In this regard, Barros et al. proposed a simple, facile, affordable procedure to synthesize $[CoW_{12}O_{40}]^{6-}$ with the same electrochemical behaviour which is not only time-saving but also of high yield [37]. Therefore, $[CoW_{12}O_{40}]^{6-}$ is an interesting compound because of its high solubility, fast kinetics, high cell voltage for symmetric RFBs and the size of the clusters which do not permeate the membrane.

In 2018, Matson et al. investigated the RFB performance of the functionalized Lindqvist POMs $[TiV_5(OCH_3)_{13}]^-$ (herein known as 1-TiV_5) and $[Ti_2V_4(OCH_3)_{14}]$ (herein known as 2-Ti_2V_4) for symmetric batteries [38]. In a follow-up work [18], they found that, by substituting one or two of the addenda atoms with heteroatoms such as Ti, they can modify not only the redox potentials of the cluster but also the number of electrons involved in the reactions. By looking at the CVs, it is apparent that, because of the enough separation of oxidation and reductio peaks of two redox centres, the positive peaks related to the oxidation of vanadium atoms and the negative potentials related to the reduction of titanium atoms, the RFB can utilize the

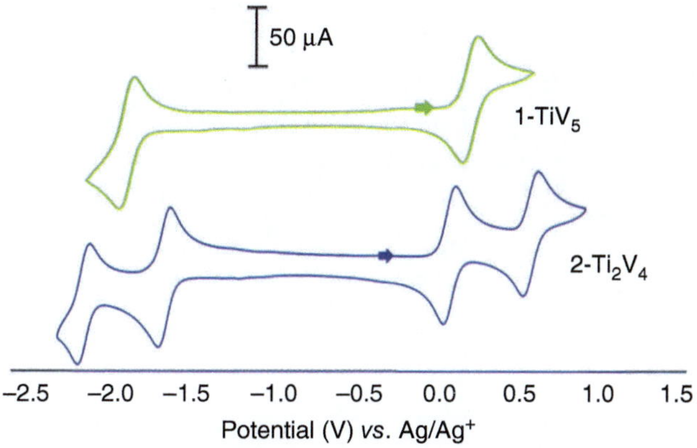

Figure 10.23 The CVs of 1-TiV_5 and 2-Ti_2V_4 in acetonitrile and 0.1 M $[NBu_4][PF_6]$ at a scan rate of 20 mV s^{-1}; GCE with a surface area of 0.071 cm^2 as working electrode. Source: Reproduced from VanGelder and Matson [38]/Royal Society of Chemistry.

1-TiV_5 (with the cell voltage of 2.3 V) or 2-Ti_2V_4 (with the cell voltage of 2.74 V) for both electrolytes-negolyte and posolyte (Figure 10.23).

In the next step, they evaluated the electron-transfer kinetics of these two materials by conducting CVs at different scan rates, which resulted in approx. 3.9×10^{-3} and 3×10^{-2} cm s^{-1} for Ti and V redox peaks, respectively. Further, they determined the solubility of 1-TiV_5 and 2-Ti_2V_4 in 0.1 M MeCN in tetrabutylammonium hexafluorophosphate ($[NBu_4][PF_6]$) with UV–VIS spectroscopy, which was 0.513 M for 1-TiV_5 and 0.193 M for 2-Ti_2V_4. Considering these solubilities and the practical cell voltages of 2.3 and 1.73 V, the theoretical energy density for 1-TiV_5 and 2-Ti_2V_4 is calculated to 15.83 and 4.472 Wh l^{-1}, respectively. For long-term stability tests, they used an H-cell with 5 mM 1-TiV_5 in acetonitrile with 0.1 M $[NBu_4][PF_6]$ for both negolyte and posolyte, anion-exchange membrane and carbon mesh electrodes ($2 \times 1 \times 0.5$ cm). The full cell was charged at 0.15 mA cm^{-2} and discharged at 0.015 mA cm^{-2} (to provide enough time to be fully discharged) for 20 cycles. As depicted in Figure 10.24a,b, the battery provided a CE of 91%. To confirm the stability of the solution, they conducted CVs after charge–discharge cycles. As it can be seen in Figure 10.24c, the voltammograms of the solutions before and after battery cycling are practically identical, confirming the stability of the clusters. For 2-Ti_2V_4, also, the same charge–discharge experiment was done with the restricted cell voltage of 1.9 V (one electron involved) due to the instability of the cluster at more extended potentials. Figure 10.24d,e shows the charge–discharge performance of the symmetric battery with 5 mM 2-Ti_2V_4 for 20 cycles with a CE of 89%. Also, post-cycling voltammograms of 2-Ti_2V_4 shows good stability of the cluster (Figure 10.24f).

Friedl et al. utilized two POMs, $[SiW_{12}O_{40}]^{4-}$ and $[PV_{14}O_{42}]^{9-}$, as negolyte and posolyte to build asymmetric RFBs [39]. Both POMs are of Keggin structure, but $[PV_{14}O_{42}]^{9-}$ has two more metal-oxygen caps. Figure 10.25a depicts the CVs of $[SiW_{12}O_{40}]^{4-}$, exhibiting two one-electron and one two-electron processes at equilibrium potentials: 0.01, −0.21 and −0.37 V vs. SHE with peak-to-peak separations

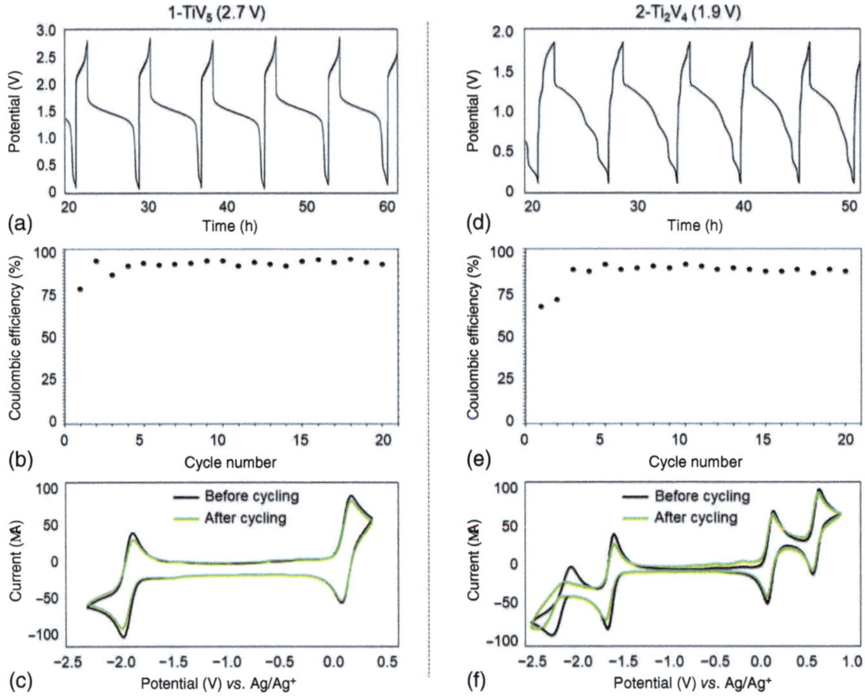

Figure 10.24 (a) Charge–discharge cycles for 5 mM 1-TiV_5 in acetonitrile with 0.1 M $[NBu_4][PF_6]$ at a charge current density of 0.15 mA cm^{-2} and discharge current density of 0.015 mA cm^{-2}. (b) Coulombic efficiency of the cell over 20 cycles and (c) the CVs of the electrolytes before and after cycling; GCE with a surface area of 0.071 cm^2 as working electrode. (d)–(f) The same characterizations for 2-Ti_2V_4. Source: Reproduced from VanGelder and Matson [38]/Royal Society of Chemistry.

of 57, 56 and 29 mV, confirming high reaction rates of the redox reaction. For further investigations, the authors used only the first two reduction peaks of $[SiW_{12}O_{40}]^{4-}$, since the third one is in the vicinity of HER. In addition, by electrochemical impedance spectroscopy, they measured the electron-transfer constant rate (k_0) of 1.1×10^{-2} and 1.8×10^{-2} cm s^{-1} for the first and second redox peaks of $[SiW_{12}O_{40}]^{4-}$, respectively. Figure 10.25b shows the CVs of $[PV_{14}O_{42}]^{9-}$, consisting of the overlap of multi-electron-transfer processes which made the determination of the exact potentials for each redox reaction difficult. By a symmetric cell of $[PV_{14}O_{42}]^{9-}$, they were able to measure the maximum number of electrons involved in the redox reactions which was $n = 7$.

To investigate the performance of the battery, they assembled the RFB with 60 ml of 0.2 M $[SiW_{12}O_{40}]^{4-}$ in 1 M LiCl at pH 1.6 as negolyte and 60 ml of 0.1 M $[PV_{14}O_{42}]^{9-}$ in 1 M LiCl at pH 2 as posolyte. The reason for using different concentrations of POMs is to balance the charge in both sides, since they used four electrons for $[PV_{14}O_{42}]^{9-}$ and two electrons for $[SiW_{12}O_{40}]^{4-}$. Additionally, prior to the experiment, they added hydrazine to posolyte in order to reduce it with four electrons to be ready for oxidation process. The theoretical capacity of this

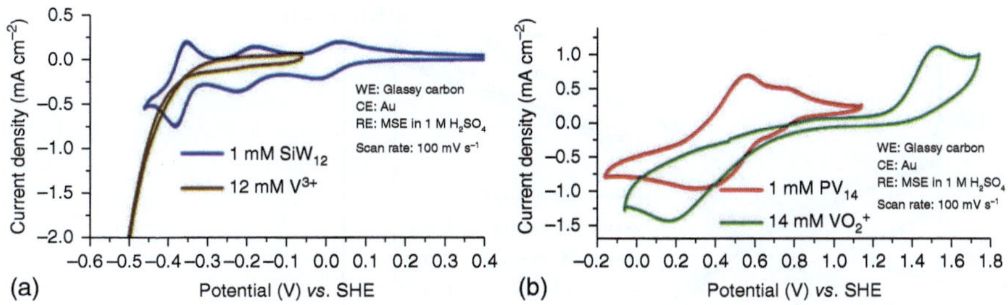

Figure 10.25 (a) CVs of 1 mM $[SiW_{12}O_{40}]^{4-}$ with comparison to 12 mM V^{3+} in 1 M H_2SO_4 and (b) 1 mM $[PV_{14}O_{42}]^{9-}$ in 1 M LiCl at pH 2.3 with comparison to 14 mM VO_2^+ in 1 M H_2SO_4. Source: Reproduced from Friedl et al. [39]/Royal Society of Chemistry.

battery would be 640 mAh for both negolyte and posolyte. The cell performance was investigated at three different current densities as shown in Figure 10.26a. At a current density of 30 mA cm^{-2}, the capacity (Q = 612 mAh) was close to the theoretical value with the CE of 94%. Also, the VE at this current density was 65% which stems from the ohmic losses.

The long-term stability of this cell was investigated over 14 days (155 cycles) at different current densities as shown in Figure 10.26b, 50 cycles at a current density of 30 mA cm^{-2}, 10 cycles at a current density of 40 mA cm^{-2} and 10 cycles at a current density of 60 mA cm^{-2}. From cycles 20 to 71, the capacity fade was calculated to be 0.16% per cycle, attributed to the presence of oxygen. To confirm this, from cycles 85 to 97, the N_2 gas was stopped, causing the capacity fade of 0.56% per cycle. After that, during cycle 98, hydrazine was added to the posolyte solution to re-reduce it and N_2 gas was resumed, resulting in increasing the capacity to the initial amount.

In a subsequent study, Friedl et al. investigated the performance of this battery-$[SiW_{12}O_{40}]^{4-}$ as negolyte and $[PV_{14}O_{42}]^{9-}$ as posolyte-for a large system with the active area of 1400 cm^2 [40]. Interestingly, the upscale cell showed better performance with CE of about 100% and EE of 86% over 88 days (1400 cycles). These results imply that the POMs used in this battery are promising candidates for RFBs to be commercialized.

10.6 Further Possibilities

As we discussed earlier, RFBs store active materials in external tanks. Therefore, one of the possibilities to increase the energy density of them, a quite simple one, is to increase the volume of the tanks, based on Eqs. (10.2) and (10.3). However, until now, the usage of RFBs is limited to the stationary energy storage devices because of the large size of their tanks. One of the possibilities to make the RFBs more portable and compact is to hybridize the POM-based RFBs with air or hydrogen gas. In other words, hybrid POM-based RFBs utilize POM active materials in one side, posolyte or negolyte, and air/oxygen or hydrogen in another side reducing the size of the considerably, probably down to 50–60%. As these gases are lighter than

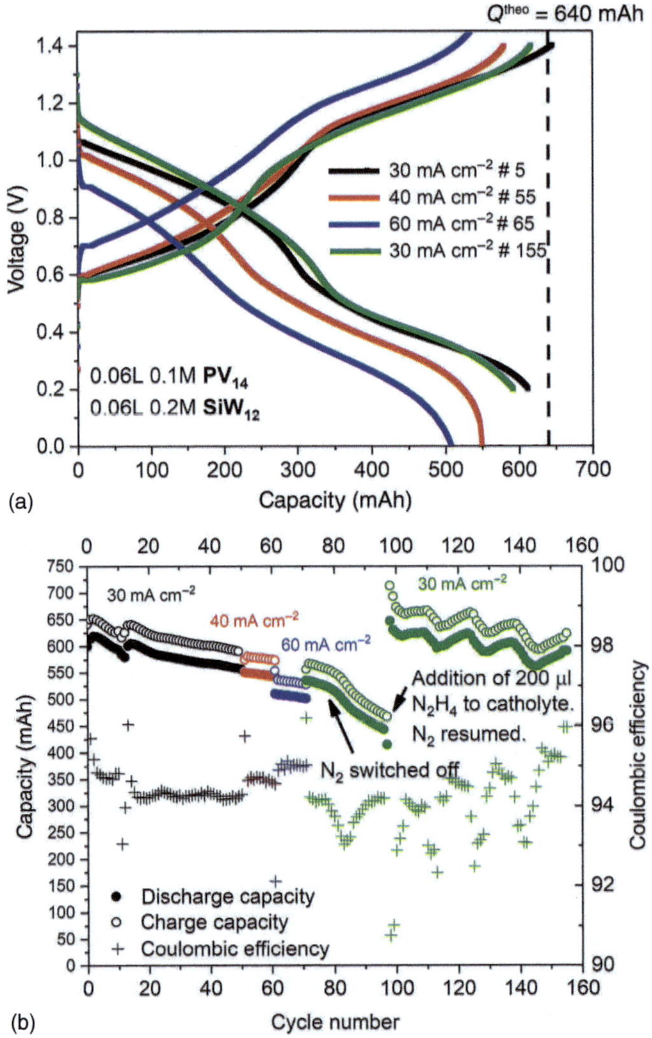

Figure 10.26 (a) Charge–discharge profiles of 0.2 M $[SiW_{12}O_{40}]^{4-}$ and 0.1 M $[PV_{14}O_{42}]^{9-}$ at different current densities. (b) Charge (open circle) and discharge (full circle) capacity over 155 cycles with current density indicated. Source: Reproduced from Friedl et al. [39]/Royal Society of Chemistry.

liquid electrolytes, they will be more suitable for smaller RFBs. For example, in the case of using air as the positive electrode, H_2O will be oxidized to oxygen and POM will be reduced in the negolyte upon charging [41]. Therefore, by using a POM with redox peaks at very negative potentials, the cell voltage will be increased, resulting in high energy density. However, there are some problems, such as the need of a bi-catalyst for the oxygen electrode since this reaction is of slow kinetics. Also, the possible crossover of oxygen will oxidize the active materials on the negative side, thus decreasing the capacity of the battery. Additionally, hydrogen gas can be used for the negative side, to be oxidized to protons and POMs can be used as posolyte, but that requires an additional tank for hydrogen [42].

10.7 POM-based RFBs in Comparison with Other RFBs

To evaluate the effectiveness of a battery chemistry, several criteria must be considered, including solubility of the active materials, chemical and thermal stability, long cycle life, electron-transfer kinetics, cost-effectiveness and safety. Here, we will compare some other chemistries used in RFBs with those used in POM-based RFBs.

10.7.1 Iron/Chromium RFBs

Fe/Cr RFBs are a type of RFBs, in which an iron solution is used as the posolyte and a chromium solution as the negolyte. During charge, Fe^{2+} is oxidized to Fe^{3+} and Cr^{3+} is reduced to Cr^{2+} (Eq. 10.11), providing a cell voltage of 1.18 V [7].

$$\begin{aligned} &\text{Posolyte}: \quad Fe^{2+} \rightarrow Fe^{3+} + e^- \quad U^\theta = 0.77\,\text{V vs. SHE} \\ &\text{Negolyte}: \quad Cr^{3+} + e^- \rightarrow Cr^{2+} \quad U^\theta = -0.41\,\text{V vs. SHE} \end{aligned} \tag{10.11}$$

Although the electron-transfer reaction of Fe^{2+}/Fe^{3+} is of intermediate reaction rate, the kinetic of the Cr^{3+}/Cr^{2+} reaction is very slow, decreasing the power density of the battery considerably. In addition, the potential of Cr^{3+}/Cr^{2+} reaction is in the vicinity of the HER, decreasing the efficiencies of the battery [7].

10.7.2 All-vanadium RFBs

All-vanadium RFBs are the most common RFBs at the moment. In all-vanadium RFBs, four different oxidation states of vanadium (V^{2+}/V^{3+} and VO_2^+/VO^{2+}) in sulfuric acid are used; the redox reactions of posolyte and negolyte during charge are shown in Eq. (10.12). The advantage of such a configuration of the battery which uses only one type of metals is that the crossover of materials through the membrane does not cross-contaminate the respective other electrolyte.

The currently used concentration of vanadium electrolytes is limited to 1.6–1.8 M in H_2SO_4. This solubility of vanadium species in sulfuric acid media (about 2 M) limits the energy density of the battery to 25 Wh kg^{-1} [43]. To increase the solubility of vanadium, a great amount of research has been done. For example, Li et al. demonstrated that, by using a mixture of electrolytes containing SO_4^{2-} and Cl^-, the solubility of vanadium can be increased to 2.5 M. with this concentration, they were able to assemble an RFB with an energy density more than 36 Wh l^{-1} with the EE of 87% over 20 days. In addition, the maximum solubility of vanadium ions in sulfuric acid electrolyte is restricted, for example at temperatures lower than 10 °C, V^{2+}/V^{3+} and at temperatures higher than 40 °C, VO_2^+ will precipitate. Nonetheless, using a mixture of SO_4^{2-} and Cl^- increased the thermal stability of species from −5 to 50 °C [42, 44].

In addition, as discussed, the other parameter affecting the energy density is the cell voltage. Inspired by this concept, Liu et al. investigated non-aqueous vanadium RFBs, since the voltage of aqueous RFBs is limited by the electrochemical window of water [45]. The non-aqueous RFB offered the accessible potential window of 4 V in acetonitrile, thereby increasing the energy density of the battery.

However, all-vanadium RFBs face some issues, including the sluggish kinetic of vanadium ions, their high cost, toxicity and corrosiveness of the electrolytes [2].

$$\begin{aligned} &\text{Posolyte}: \quad VO^{2+} + H_2O \rightarrow VO_2{}^{+} + 2H^{+} + e^{-} \quad U^{\theta} = 1.00\,\text{V vs. SHE} \\ &\text{Negolyte}: \quad V^{3+} \rightarrow V^{2+} + e^{-} \quad U^{\theta} = -0.26\,\text{V vs. SHE} \end{aligned} \tag{10.12}$$

10.7.3 Zn/Br$_2$ RFBs

The Zn/Br battery is another type of RFBs, in which Br^- is oxidized to Br_2 during charging while Zn^{2+} is deposited on the surface of the electrode as metallic Zn (Eq. 10.13). With such a configuration where one of active materials is part of the electrode, a high energy density can be accessible. However, the formation of dendrite during deposition and dissolution of the metal may cause short circuits. Additionally, the formation of Br_2 gas during the battery operation poses safety concerns [7].

$$\begin{aligned} &\text{Posolyte}: \quad 2Br^{-} \rightarrow Br_2 + 2e^{-} \quad U^{\theta} = 1.09\,\text{V vs. SHE} \\ &\text{Negolyte}: \quad Zn^{2+} + 2e^{-} \rightarrow Zn \quad U^{\theta} = -0.76\,\text{V vs. SHE} \end{aligned} \tag{10.13}$$

From among all the proposed RFBs, POM-based RFBs with their unique features offer a promising system in terms of safety, cost-effectiveness and high power and energy density. Table 10.1 shows the performance of RFBs using POMs as active materials.

10.8 Conclusions

This review examines the potential of POMs for the advancement of RFBs focusing on their chemistry, performance, characteristics and also challenges. The performance of the POM-based RFBs discussed in this chapter is summarized in Table 10.1. There are many advantages associated with POM-based RFBs as compared with traditional vanadium-based systems. The advantages include multiple electron-transfer reactions, improved redox kinetics, high solubility, as well as chemical and electrochemical stability. These attributes address critical shortcomings of other technologies such as low power density, energy density and electrolyte penetration. Various design schemes for POM-based RFBs have been summarized, highlighting their impact on battery performance. Asymmetric and symmetric configurations, along with aqueous and organic solvent options, offer different trade-offs in terms of efficiency, power density and cost-effectiveness. While symmetric RFBs offer advantages such as simplified electrolyte composition and reduced cross-contamination, challenges arise in achieving applicable electrode

Table 10.1 The performance of POM-based RFBs.

Active materials											
Negolyte	Posolyte	$n_N : n_P$ [a]	Supporting electrolyte	Cell voltage (V)	CE (%)	EE (%)	Power density ($W\,cm^{-2}$)	Energy density ($Wh\,l^{-1}$)	Number of cycles	CR (%)	References
1 M $FeSO_4$	Saturated $H_3[PMo_{12}O_{40}]$	4 : 1	0.5 M H_2SO_4	N/A	93	73	N/A	N/A	24	N/A	[28]
0.02 M K_6 [A-α-$PV_3W_9O_{40}$]		3 : 3	0.5 M H_2SO_4	≈1	80	N/A	N/A	N/A	10	N/A	[29]
0.5 M Li_6 $[P_2W_{18}O_{62}]$	4 M HBr + 0.1 M Br_2	18 : 1	N: 0.1 M H_2SO_4 P: none	1.25	98	76	0.5	225	20	N/A	[30]
0.1 M Li_3 $[PMo_{12}O_{40}]$		2 : 2	0.1 M $Li(CF_3SO_3)$ in DMF	1	93	50	N/A	1.6	30	N/A	[31]
0.01 M Li_3 $[PMo_{12}O_{40}]$		1 : 1	0.1 M $Li(CF_3SO_3)$ in acetonitrile	0.35	>95	>70	N/A	N/A	30	N/A	[31]
0.01 M Li_6 $[P_2W_{18}O_{62}]$	0.01 M Li_3 $[PMo_{12}O_{40}]$	4 : 4	0.1 M $Li(CF_3SO_3)$ in DMF	1.3	>95	≈60	N/A	0.27	24	N/A	[31]
0.5 M H_6 $[P_2W_{18}O_{62}]$	1.5 M $VOSO_4$	6:1	P: 3M H_2SO_4 N: none	N/A	98	>80	0.397	N/A	450	N/A	[33]
0.05 M Li_6 $[P_2W_{18}O_{62}]$	0.04 M $Ce(SO_4)_2$	10 : 1	2M H_2SO_4	N/A	98	80	N/A	N/A	150	99.997	[34]
0.8 M H_6 $[CoW_{12}O_{40}]$		4 : 1	None	≈1.3	99.4	88.6	0.56	15.4	100	99.98	[35]
0.005 M $[TiV_5O_6(OCH_3)_{13}]^-$		1 : 1	0.1 M $[NBu_4][PF_6]$ in acetonitrile	2.3	91	N/A	N/A	N/A	4	N/A	[37]
0.005 M $[Ti_2V_4(OCH_3)_{14}]$		1 : 1	0.1 M $[NBu_4][PF_6]$ in acetonitrile	1.9	89	N/A	N/A	N/A	20	N/A	[37]
0.2 M $H_4[SiW_{12}O_{40}]$	0.1M Na_2H_7 $[PV_{14}O_{42}]$	2 : 4	1 M LiCI	0.7	94	65	0.15	3.1	155	99.84	[38]

a) The number of electrons transferred for negolyte and posolyte.

potential and maximizing energy density. Asymmetric RFBs, on the other hand, face issues related to cross-contamination through the membrane.

The control of POM characteristics presents further opportunities and challenges. Adjusting acid strengths, electrode potential and element composition can optimize solubility, stability and electrochemical properties. However, deeper understanding of the structure–activity relationship and the impact of modified ligands and solvent environments is needed to fully exploit the potential of POMs in RFB applications.

Overall, this review underscores the importance of rational design and optimization in harnessing the benefits of POMs for next-generation RFBs. Continued research and innovation in POM chemistry, battery design and materials science will be crucial in overcoming challenges and unlocking the full potential of POM-based RFB systems for scalable and sustainable energy storage solutions.

Abbreviations and Symbols

CE	coulombic efficiency
CR	capacity retention
CV	cyclic voltammetry
ED	energy density
EE	energy efficiency
GCE	glassy carbon electrode
HER	hydrogen evolution reaction
OER	oxygen evolution reaction
PD	power density
POMs	polyoxometalates
RFBs	redox flow batteries
SHE	standard hydrogen electrode
SSBs	solid-state batteries
VE	voltage efficiency
R_{tot}	total resistance
R_{mass}	mass-transfer resistance
R_{ct}	charge-transfer resistance
R_{elec}	electronic resistance
R_{elyt}	electrolyte resistance
K_0	electron-transfer rate constant
Q	theoretical capacity
Q_{int}	initial capacity
Q_t	capacity after a defined time
Q_n	capacity after a defined number of cycles
Q_{dch}	discharge capacity
Q_{ch}	charge capacity
c	concentration

ΔU	cell voltage
U^{θ}	standard potential
U_0	equilibrium potential
v	volume of the electrolyte
F	Faraday constant

References

1 Dinesh, A., Olivera, S., Venkatesh, K. et al. (2018). Iron-based flow batteries to store renewable energies. *Environ. Chem. Lett.* 16: 683–694. https://doi.org/10.1007/s10311-018-0709-8.

2 Holland-Cunz, M.V., Cording, F., Friedl, J., and Stimming, U. (2018). Redox flow batteries—concepts and chemistries for cost-effective energy storage. *Front. Energy* 12: 198–224. https://doi.org/10.1007/s11708-018-0552-4.

3 Zhang, J., Jiang, G., Xu, P. et al. (2018). An all-aqueous redox flow battery with unprecedented energy density. *Energy Environ. Sci.* 11: 2010–2015. https://doi.org/10.1039/C8EE00686E.

4 Weber, A.Z., Mench, M.M., Meyers, J.P. et al. (2011). Redox flow batteries: a review. *J. Appl. Electrochem.* 41: 1137–1164. https://doi.org/10.1007/s10800-011-0348-2.

5 Arévalo-Cid, P., Dias, P., Mendes, A., and Azevedo, J. (2021). Redox flow batteries: a new frontier on energy storage. *Sustainable Energy Fuels* 5: 5366–5419. https://doi.org/10.1039/D1SE00839K.

6 Leung, P., Li, X., Ponce de León, C. et al. (2012). Progress in redox flow batteries, remaining challenges and their applications in energy storage. *RSC Adv.* 2: 10125–10156. https://doi.org/10.1039/C2RA21342G.

7 Pan, F. and Wang, Q. (2015). Redox species of redox flow batteries: a review. *Molecules* 20: 20499–20517. https://doi.org/10.3390/molecules201119711.

8 Yao, Y., Lei, J., Shi, Y. et al. (2021). Assessment methods and performance metrics for redox flow batteries. *Nat. Energy* 6: 582–588. https://doi.org/10.1038/s41560-020-00772-8.

9 Horn, M.R., Singh, A., Alomari, S. et al. (2021). Polyoxometalates (POMs): from electroactive clusters to energy materials. *Energy Environ. Sci.* 14: 1652–1700. https://doi.org/10.1039/D0EE03407J.

10 Ding, Y., Zhang, C., Zhang, L. et al. (2018). Molecular engineering of organic electroactive materials for redox flow batteries. *Chem. Soc. Rev.* 47: 69–103. https://doi.org/10.1039/C7CS00569E.

11 Wang, C., Wang, B., Yang, H. et al. (2024). A review on polyoxometalates-based materials in addressing challenges faced by electrochemical energy storage systems. *Chem. Eng. J.* 483: 149143. https://doi.org/10.1016/j.cej.2024.149143.

12 López, X., Carbó, J.J., Bo, C., and Poblet, J.M. (2012). Structure, properties and reactivity of polyoxometalates: a theoretical perspective. *Chem. Soc. Rev.* 41: 7537–7571. https://doi.org/10.1039/C2CS35168D.

13 Zhang, Y., Liu, J., Li, S.-L. et al. (2019). Polyoxometalate-based materials for sustainable and clean energy conversion and storage. *EnergyChem* 1: 100021. https://doi.org/10.1016/j.enchem.2019.100021.

14 Huskinson, B., Marshak, M.P., Suh, C. et al. (2014). A metal-free organic–inorganic aqueous flow battery. *Nature* 505: 195–198. https://doi.org/10.1038/nature12909.

15 Huskinson, B., Marshak, M.P., Gerhardt, M.R., and Aziz, M.J. (2014). Cycling of a quinone-bromide flow battery for large-scale electrochemical energy storage. *ECS Trans.* 61: 27–30. https://doi.org/10.1149/06137.0027ecst.

16 Lin, K., Chen, Q., Gerhardt, M.R. et al. (2015). Alkaline quinone flow battery. *Science* 1979 (349): 1529–1532. https://doi.org/10.1126/science.aab3033.

17 Mbomekallé, I.-M., López, X., Poblet, J.M. et al. (2010). Influence of the heteroatom size on the redox potentials of selected polyoxoanions. *Inorg. Chem.* 49: 7001–7006. https://doi.org/10.1021/ic100675h.

18 VanGelder, L.E., Brennessel, W.W., and Matson, E.M. (2018). Tuning the redox profiles of polyoxovanadate-alkoxide clusters via heterometal installation: toward designer redox reagents. *Dalton Trans.* 47: 3698–3704. https://doi.org/10.1039/C7DT04455K.

19 Chen, J.-J.J. and Barteau, M.A. (2016). Electrochemical properties of Keggin-structure polyoxometalates in acetonitrile: effects of countercation, heteroatom, and framework metal exchange. *Ind. Eng. Chem. Res.* 55: 9857–9864. https://doi.org/10.1021/acs.iecr.6b02316.

20 Liu, Y., Zhang, J., Lu, S., and Xiang, Y. (2022). Polyoxometalate-based electrolyte materials in redox flow batteries: current trends and emerging opportunities. *Mater. Rep.: Energy* 2: 100094. https://doi.org/10.1016/j.matre.2022.100094.

21 Wang, H.-N., Zhang, M., Zhang, A.-M. et al. (2018). Polyoxometalate-based metal–organic frameworks with conductive polypyrrole for supercapacitors. *ACS Appl. Mater. Interfaces* 10: 32265–32270. https://doi.org/10.1021/acsami.8b12194.

22 Suárez-Guevara, J., Ruiz, V., and Gómez-Romero, P. (2014). Stable graphene–polyoxometalate nanomaterials for application in hybrid supercapacitors. *Phys. Chem. Chem. Phys.* 16: 20411–20414. https://doi.org/10.1039/C4CP03321C.

23 Chen, H., Wee, G., Al-Oweini, R. et al. (2014). A polyoxovanadate as an advanced electrode material for supercapacitors. *ChemPhysChem* 15: 2162–2169. https://doi.org/10.1002/cphc.201400091.

24 Wang, Y.-Y., Zhang, M., Li, S.-L. et al. (2017). Diamondoid-structured polymolybdate-based metal–organic frameworks as high-capacity anodes for lithium-ion batteries. *Chem. Commun.* 53: 5204–5207. https://doi.org/10.1039/C6CC10208E.

25 Liu, W., Yu, G., Zhang, M. et al. (2018). Investigation of the enhanced lithium battery storage in a polyoxometalate model: from solid spheres to hollow balls. *Small Methods* 2: https://doi.org/10.1002/smtd.201800154.

26 Zhang, M., Zhang, A.-M., Wang, X.-X. et al. (2018). Encapsulating ionic liquids into POM-based MOFs to improve their conductivity for superior lithium storage. *J. Mater. Chem. A Mater.* 6: 8735–8741. https://doi.org/10.1039/C8TA01062E.

27 Ni, E., Tsukada, T., Wen, Q., and Sonoyama, N. (2019). Improved performance of nano-sized polyoxometalate as lithium-battery cathode by conductive polymer coating. *J. Electrochem. Soc.* 166: A5226. https://doi.org/10.1149/2.0341903jes.

28 Lin, C., Lin, W., Huang, S. et al. (2018). Mechanism of sodium ion storage in Na_7 $[H_2$ PV_{14} $O_{42}]$ anode for sodium-ion batteries. *Adv. Mater. Interfaces* 5: https://doi.org/10.1002/admi.201800491.

29 Sun, P.G.Y., You, W., and Tan, Z. (1997). Studies on an iron heteropolyacid redox flow battery. *J. Electrochem.* 3: https://doi.org/10.61558/2993-074X.2650.

30 Pratt, H.D. III, and Anderson, T.M. (2013). Mixed addenda polyoxometalate "solutions" for stationary energy storage. *Dalton Trans.* 42: 15650–15655. https://doi.org/10.1039/C3DT51653A.

31 Chen, J.-J., Symes, M.D., and Cronin, L. (2018). Highly reduced and protonated aqueous solutions of $[P_2W_{18}O_{62}]^{6-}$ for on-demand hydrogen generation and energy storage. *Nat. Chem.* 10: 1042–1047. https://doi.org/10.1038/s41557-018-0109-5.

32 Cao, Y., Chen, J.-J.J., and Barteau, M.A. (2020). Systematic approaches to improving the performance of polyoxometalates in non-aqueous redox flow batteries. *J. Energy Chem.* 50: 115–124. https://doi.org/10.1016/j.jechem.2020.03.009.

33 Chen, J.-J.J. and Barteau, M.A. (2017). Molybdenum polyoxometalates as active species for energy storage in non-aqueous media. *J. Energy Storage* 13: 255–261. https://doi.org/10.1016/j.est.2017.07.017.

34 Ai, F., Wang, Z., Lai, N.-C. et al. (2022). Heteropoly acid negolytes for high-power-density aqueous redox flow batteries at low temperatures. *Nat. Energy* 7: 417–426. https://doi.org/10.1038/s41560-022-01011-y.

35 Wang, X., Gao, M., Lee, Y.M. et al. (2022). E-blood: high power aqueous redox flow cell for concurrent powering and cooling of electronic devices. *Nano Energy* 93: 106864. https://doi.org/10.1016/j.nanoen.2021.106864.

36 Liu, Y., Lu, S., Wang, H. et al. (2017). An aqueous redox flow battery with a tungsten–cobalt heteropolyacid as the electrolyte for both the anode and cathode. *Adv. Energy Mater.* 7: https://doi.org/10.1002/aenm.201601224.

37 Barros, Á., Artetxe, B., Eletxigerra, U. et al. (2023). Systematic approach to the synthesis of cobalt-containing polyoxometalates for their application as energy storage materials. *Materials* 16: 5054. https://doi.org/10.3390/ma16145054.

38 VanGelder, L.E. and Matson, E.M. (2018). Heterometal functionalization yields improved energy density for charge carriers in nonaqueous redox flow batteries. *J. Mater. Chem. A Mater.* 6: 13874–13882. https://doi.org/10.1039/C8TA03312A.

39 Friedl, J., Holland-Cunz, M.V., Cording, F. et al. (2018). Asymmetric polyoxometalate electrolytes for advanced redox flow batteries. *Energy Environ. Sci.* 11: 3010–3018. https://doi.org/10.1039/C8EE00422F.

40 Friedl, J., Pfanschilling, F.L., Holland-Cunz, M.V. et al. (2019). A polyoxometalate redox flow battery: functionality and upscale. *Clean Energy* 3: 278–287. https://doi.org/10.1093/ce/zkz019.

41 Hosseiny, S.S., Saakes, M., and Wessling, M. (2011). A polyelectrolyte membrane-based vanadium/air redox flow battery. *Electrochem. Commun.* 13: 751–754. https://doi.org/10.1016/j.elecom.2010.11.025.

42 Tucker, M.C., Srinivasan, V., Ross, P.N., and Weber, A.Z. (2013). Performance and cycling of the iron-ion/hydrogen redox flow cell with various catholyte salts. *J. Appl. Electrochem.* 43: 637–644. https://doi.org/10.1007/s10800-013-0553-2.

43 Skyllas-Kazacos, M., Menictas, C., and Kazacos, M. (1996). Thermal stability of concentrated V(V) electrolytes in the vanadium redox cell. *J. Electrochem. Soc.* 143: L86. https://doi.org/10.1149/1.1836609.

44 Roe, S., Menictas, C., and Skyllas-Kazacos, M. (2016). A high energy density vanadium redox flow battery with 3 M vanadium electrolyte. *J. Electrochem. Soc.* 163: A5023. https://doi.org/10.1149/2.0041601jes.

45 Liu, Q., Sleightholme, A.E.S., Shinkle, A.A. et al. (2009). Non-aqueous vanadium acetylacetonate electrolyte for redox flow batteries. *Electrochem. Commun.* 11: 2312–2315. https://doi.org/10.1016/j.elecom.2009.10.006.

11

POM-based Nanomaterials for Supercapacitors

Ana C. Alves[1,2] *and Marta M. Alves*[1,2]

[1] *Universidade de Lisboa, Instituto Superior Técnico, Institute of Molecular Sciences, Centro de Química Estrutural, Avenida Rovisco Pais, Lisbon 1049-001, Portugal*
[2] *Universidade de Lisboa, Instituto Superior Técnico, Departamento de Engenharia Química, Avenida Rovisco Pais, Lisbon 1049-001, Portugal*

11.1 Introduction to Energy-storage Devices

Energy-storage devices are required to answer the growing demand of storing energy collected from intermittent sources, such as sun and wind, as well as delivering energy for daily life applications. The Ragone plot, Figure 11.1, shows the status of energy *vs.* power densities, as benchmarking metrics of different energy-storage devices. Supercapacitors and batteries are high-performance energy-storage devices that store charge electrochemically. The architectural feature of supercapacitors and batteries consists in two electrodes (anode and cathode) placed on current collectors, a separator and a conducting electrolyte. However, these storage devices differ on how the charge is stored, resulting in different and very complementary metrics.

Currently, rechargeable secondary batteries, in particular Li-ion batteries (LIBs), are at the forefront of high-energy devices (100–300 Wh kg^{-1}), reaching an operation time ranging from several minutes to hours, which are key metrics for the portable electronics, electric mobility and stationary storage markets. The high-energy density provided by LIBs results from a faradaic reaction and the intercalation of Li^+ on the bulk of the electrode material, e.g. graphite and a lithium metal oxide, that sometimes leads to crystalline phase transitions, Figure 11.2a. In addition to lithium, different metal oxides are nowadays used as cathode materials such as nickel (Ni), manganese (Mn), cobalt (Co) or aluminium (Al). However, modern batteries provide only modest power density ($\sim 10^3\ \text{W kg}^{-1}$) and limited lifetime (<3000–4000 cycles) is their major bottleneck, owing to diffusion rate limitations, volume changes on electrodes and phase transformations during operation.

Supercapacitors are another family of energy-storage devices, whose electrical performance complements very well with that of batteries. These devices harvest high-power density ($\sim 10^7\ \text{W kg}^{-1}$) that is delivered in a few minutes to just seconds but provide very modest energy density. Electroactive materials in supercapacitors

Applied Polyoxometalate-based Electrocatalysis, First Edition. Edited by Diana M. Fernandes.

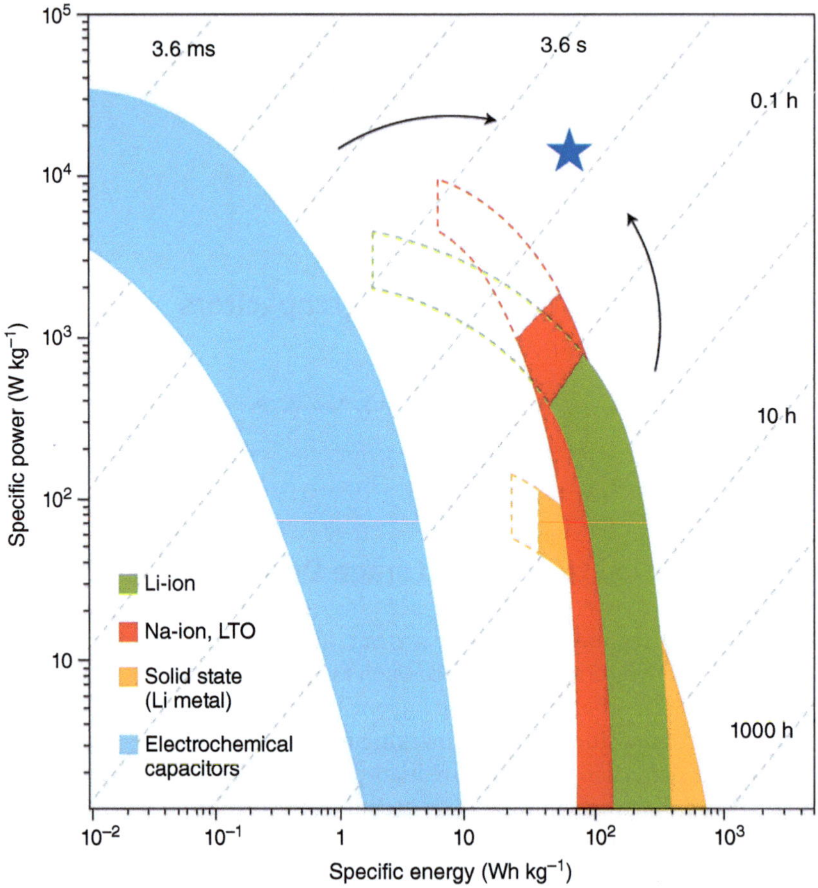

Figure 11.1 Ragone plot. Dashed lines represent zones where the cyclability of the device is altered in the case of symmetric cycling. The diagonal dotted lines and timescales represent characteristic operation timescales, obtained by dividing energy by power. The region marked with a star icon represents the current challenge regarding energy-storage systems, where high-energy-density devices meet high-power devices. Source: Simon and Gogotsi [1]/with permission of Springer Nature.

can be carbon-based materials or pseudocapacitive metal compounds. In the former case, the dominant charge-storage mechanism is based on the electrochemical double layer and, therefore, known as electrical double-layer capacitors (EDLCs). In the latter case, the charge-storage mechanism is assigned to fast and reversible surface (or near surface) processes that include surface electrostatic (non-faradaic) and faradaic processes. Thanks to these surface processes, negligible crystallographic and morphological changes occur in the electrode material, which leads to a long cyclability.

EDLCs, Figure 11.2b, follow non-faradaic processes at the surface of the electrode. In this case, high-surface area electrode materials absorb/desorb charged species from the electrolyte and, therefore, the process is essentially physical in nature.

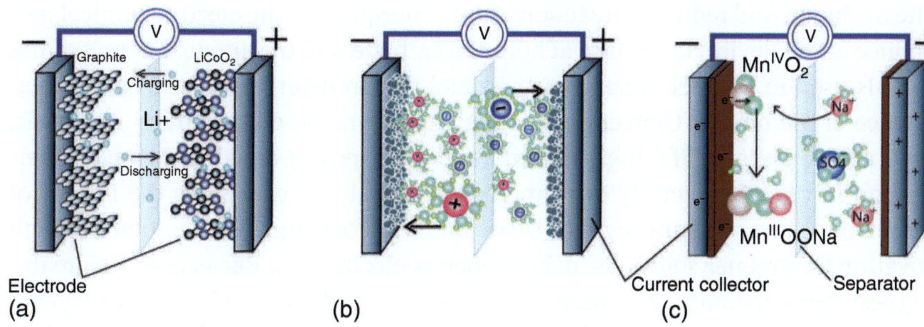

Figure 11.2 Schematics of different electrochemical energy-storage systems: (a) battery; (b) electrochemical double-layer capacitor; and (c) pseudocapacitor. Source: Jost et al. [2]/with permission of Royal Society of Chemistry.

Commercially available carbon materials are state-of-the-art technologies with high surface area (>1000 $m^2\ g^{-1}$ [3]), excellent power density, fast charge/discharge rate of typically below one second and long cycle of life. To improve their energy density, organic electrolytes are used to extend the working voltage window. The use of these electrolytes, often toxic and volatile, has been a limitation to the application of supercapacitors on emerging new technologies.

Pseudocapacitors, Figure 11.2c, on the other hand, introduce a new concept of energy storage and can operate very safely in aqueous-based electrolytes. As the 'pseudo' suffix suggests, the charge-storage mechanism of pseudocapacitors follows 'something that appears to be but is not actually'. In fact, this process arises from a mixture of storage mechanisms. It exhibits a faradaic behaviour resulting from reversible redox reactions, resembling the charge/discharge process in rechargeable batteries, but also non-faradaic electrostatic interactions similar to EDLCs. As a result, these pseudocapacitive materials show an enhanced charge-storage capability while maintaining the fast response typical from carbon materials.

To evaluate the capacitance response of an electrode material, standard tests can be carried out, namely cyclic voltammetry, galvanostatic charge/discharge cycling and electrochemical impedance spectroscopy. From this result, other important metrics such as energy (E) (Eq. 11.1) and power (P) (Eq. 11.2) densities can be determined to evaluate the supercapacitor performance, according to:

$$E = \frac{1}{2} C V^2 \tag{11.1}$$

$$P_{\max} = \frac{V^2}{4\text{ESR}} \tag{11.2}$$

where C is the cell capacitance, V is the cell voltage and ESR is the equivalent series resistance. The cycle stability overtime, potential window and operating voltage of the supercapacitor are also important parameters that affect the charge-storage capability of the device. As the performance of supercapacitors is determined by the active material employed into the electrode, it is of major importance in developing and employing materials with high surface area, charge capability and cycling stability.

The oxidation and reduction reactions, and consequently, the electrochemical performance of the device, take place at both the cathode and the anode. Therefore, the materials used in the electrodes play a crucial role in determining the overall performance of the device. Conventional supercapacitors, based on carbon materials, exhibit a long cycle of life, high stability and great power density. However, their operation in organic electrolytes is an ecological liability as well as the import of carbon from oversea countries that carries a large footprint. Hence, there is high motivation for greener solutions, using aqueous electrolytes. Because of the greater charge-storage capability, transition-metal oxides (e.g. RuO_2, MnO_2 and Nb_2O_5) are known families of electroactive materials that emerged as promising alternatives to conventional carbon electrodes. Among others, RuO_2 electrode materials show great performance for pseudocapacitors with a significantly high specific capacitance >700 $F g^{-1}$ in H_2SO_4 [4]. Despite that, the use of RuO_2-based electrodes has been declining owing to its high cost (~14,600 € kg^{-1} of Ru [5]), which motivates the development of less costly electrochemically active materials. MnO_2 are also extremely promising however care must be taken when producing and operating these materials to avoid unwanted phase changes and formation of parasitic species [6]. Additionally, the energy density for these transition-metal-based supercapacitors can be further optimized by using additives in the electrolytes to increase the potential window (Eq. 11.1).

Inserted in the metal oxide family, polyoxometalates (POMs) are a class of metal-oxygen nanoclusters that display an extraordinary range of physical, chemical and electric properties. POMs have multiple unique redox states participating in charge-storage processes. This peculiar ability to act as electron-proton reservoirs is triggering the investigation on their electrochemical response as electroactive materials or electrolyte additives. Results discussed in this book chapter overview the use of POMs as alternative electrodes in aqueous energy-storage devices and as an electrolyte additive in solid-state supercapacitors.

11.2 Properties of POMs for Supercapacitors

POMs are a class of one- or multi-transition metal oxygen nanocluster compounds (~1–5 nm) with variable functional properties with great interest for energy storage, including the reversible redox chemistry and protonic conductivity [7–9]. POM complexes exhibit an oxide nanocluster of corner- and edge-sharing octahedral $[MO_6]$ unit with different addenda atoms, M = Mo, W, V, Nb, etc., that form an ionic metallic core. POMs are considered isopolyanions, $[M_zO_y]^{n-}$ Figure 11.3a, or heteropolyanions, $[X_mM_zO_y]^{n-}$, Figure 11.3b and c, according to the absence (only one metallic element) or the presence of additional heteroelement X = P, Si, B, Ge, Sb, etc. [10, 11], respectively. Among other existing POM structures, the most commonly studied heteropolyanion is Keggin-type POM, $[XM_{12}O_{40}]^{n-}$, whose structure is represented in Figure 11.3b.

The application of POMs as electrodes for supercapacitor was firstly (and successfully) analysed by Goodenough et al. [12] in 1998. Although the energy

Figure 11.3 Different types of POMs: (a) Isopolyanions, including isopolymolybdates and isotungstates; (b) Keggin-type heteropolyanions; (c) Dawson-type heteropolyanions. MO_6 octahedral units are represented as blue polyhedral, X heteroatoms as yellow spheres and oxygen atoms as red spheres. Source: Gumerova and Rompel [10]/with permission of Springer Nature.

density of a capacitor POM-containing system was ~40% of the RuO_2 electrode material, the work developed by Goodenough et al. stimulated further investigation of the (promising) faradaic nature of POMs. On account of the multiple and reversible redox processes and tailored structure along with composition, POMs and their hybrids have emerged as electroactive materials for aqueous supercapacitors.

The versatility over POM composition during its synthesis led to a wide variety of multi-metallic oxide-based materials. The redox properties of POMs are greatly determined by the addenda metal ions on the nanocluster. Molybdate (Mo)-, tungsten (W)- and vanadium (V)-containing POMs have been electrochemically studied as addenda atoms [10, 13–15]. Polyoxovanadate undergoes reduction at a more positive potential than polyoxomolybdate that, in turn, is reduced more easily than polyoxotungstate, following the oxidizing ability of each metal: V (V) > Mo (VI) > W (VI) [10, 13]. For example, Mo-containing POMs have a redox potential that are ~400 mV more positive than those for the corresponding W-based POM [10]. Substituting W metal as an addenda atom in the Keggin-type POM for Mo atom was shown to significantly enhance the capacitance performance of the electrode owing to the higher oxidation capacity of Mo compared to that of W [14, 15].

Keggin-type POM have 12 addenda metal atoms and, accordingly, different number of electrons are transferred. Mo addenda atom in the POM accepts between 1 and 12 electrons, while W atom accepts only up to 6 electrons [10]. In the case of V, the polyoxovanadate accepts even more electrons than the Mo sites in polyoxomolybdate. The oxidation and reduction reactions of POMs are effortlessly analysed using electrochemical techniques. During redox reactions, POMs are capable of undergoing fast reversible multi-electron redox reactions with continuous intercalation of ions with negligible physical structural change, behaving like an 'electron/ion sponge' [16, 17]. Therefore, POMs are commonly reported to exhibit a pseudocapacitive behaviour. Typically, cyclic voltammograms of Keggin-type $[PMo_{12}O_{40}]^{n-}$ POMs in acidic electrolyte show relatively narrow redox peaks, assigned to successive and reversible faradaic processes occurring in a mixed-valence species, Figure 11.4. The three pairs of anodic (I–III) and cathodic (I′–III′) peaks of $[PMo_{12}O_{40}]^{n-}$ are assigned to three steps of two-electron transfer processes with a total of 6 electrons transferred

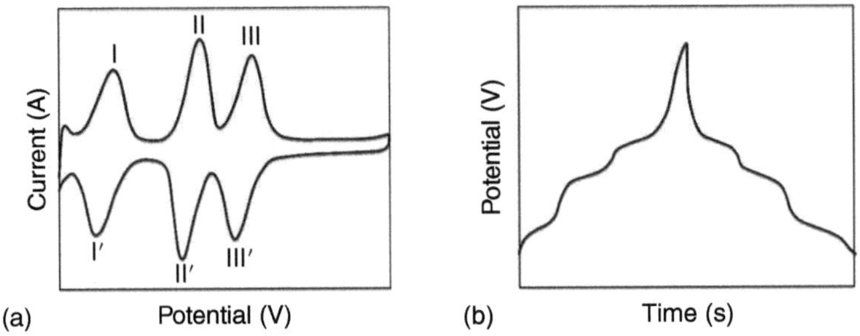

Figure 11.4 Schematics of the cyclic voltammogram of the successive and reversible redox reaction of $[PMo_{12}O_{40}]^{4-}$ polyoxometalate in aqueous H_2SO_4 solution.

in Mo^{VI}/Mo^{V} centres:

$$PMo_{12}O_{40}{}^{3-} + 2e^- + 2H^+ \rightleftharpoons H_2PMo_2{}^{V}Mo_{10}{}^{VI}O_{40}{}^{3-} \tag{11.3}$$

$$H_2PMo_2{}^{V}Mo_{10}{}^{VI}O_{40}{}^{3-} + 2e^- + 2H^+ \rightleftharpoons H_4PMo_4{}^{V}Mo_8{}^{VI}O_{40}{}^{3-} \tag{11.4}$$

$$H_4PMo_4{}^{V}Mo_8{}^{VI}O_{40}{}^{3-} + 2e^- + 2H^+ \rightleftharpoons H_6PMo_6{}^{V}Mo_6{}^{VI}O_{40}{}^{3-} \tag{11.5}$$

One must also consider the electrochemical features of the heteroatom that can contribute for enhancing the electrochemical performance of the material through synergistic effect with the multiple redox active sites of the metal [10, 15, 18].

As mentioned before, the (electro)chemical stability during the charge/discharge process ensures the efficiency and reliability of the charge-storage process overtime. Unfortunately, a crucial feature of POMs that negatively affects these parameters' stability is their solubility in liquid electrolytes that hinders their practical application in aqueous-based supercapacitors [13]. POMs are usually salts or solid acids with high solubility in water and organic solvents, becoming unsuitable for direct use as electrode materials for supercapacitors. Having this in mind, to take advantage of the great redox activity of POMs, two different strategies have been adopted, Figure 11.5. On one hand, POMs must be immobilized on highly conductive supports to prevent their dissolution into the electrolyte. This approach not only ensures the redox activity of the POM, but also boosts the overall electrochemical capacitance if the substrate actively participates on the charge storage through synergistic processes. Additionally, this combination can contribute for enhanced surface area and improved electronic conductivity. Typically, nanostructured carbon materials, conductive organic polymers and supramolecular structures are used as conductive supports for the development of POM-based materials for supercapacitors. Once stabilized, POM-based electrodes are highly active in aqueous acidic electrolyte, usually sulphuric acid solutions, because it facilitates the counterion H^+ flux and promotes the proton-induced reactions that leads to higher capacitance [19–21]. On the other hand, the capability of POMs to undergo multiple electron transfer reactions is also regarded as an interesting alternative to develop redox mediator electrolytes for solid-state supercapacitors.

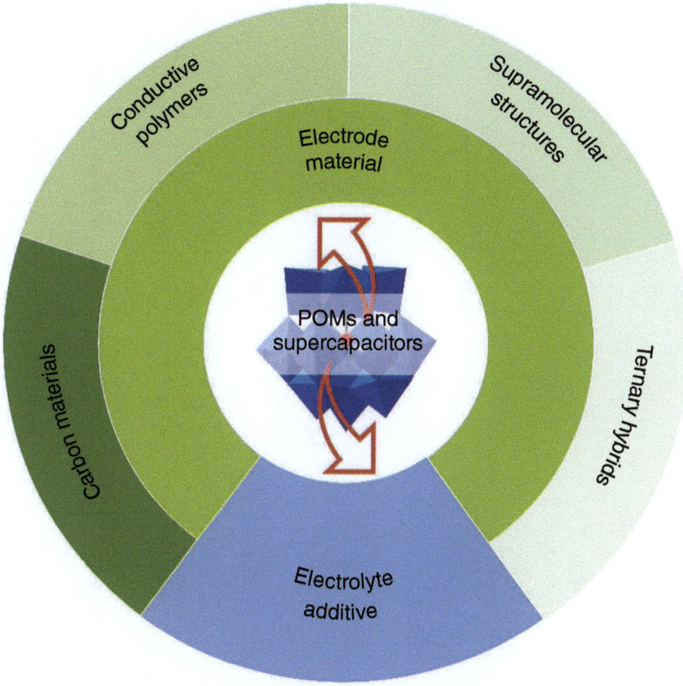

Figure 11.5 Schematics of the applications of POMs in supercapacitors, highlighting their employment as POM-based electrode materials, and electrolyte additives.

11.2.1 POMs as Electrode Materials

Wiring POMs into conductive substrates enhances their electrochemical features. Therefore, various supramolecular self-assembly strategies have been proposed for the design of hybrid POM electrode materials. Further in this sub-section, the focus will be on the specific discussion on the use of carbon-based materials, conducting polymers and metal organic frameworks (MOFs) to immobilize POMs through covalent or non-covalent bonding.

11.2.1.1 POM/Carbon Composites

Carbon-based materials have been widely used as substrates for POM structures because of their high specific surface area, diverse morphology, good electrical conductivity and exceptional physical and chemical stability [22]. Additionally, the aromatic system present in these carbon systems allows the π–π stacking interactions to attach the POM, promoting its immobilization. A wide variety of different carbon substrates were used for 'wiring' POMs into their surface, including carbon nanotubes (CNTs), activated carbon (AC) and graphene (or reduced graphene oxide) materials [15, 19, 21, 23–40] (Table 11.1), which will be individually discussed in this sub-section.

CNTs are allotropic structures of carbon in the shape of a tube with one or more layers, enriched with sp^2 hybridization that eases their chemical modification [23].

Table 11.1 Electrochemical properties of POM-based modified carbon materials. Specific capacitance from the material in a three-electrode system, unless stated otherwise. Energy and power density from the two-electrode device.

POM-based modified carbon materials	Electrolyte	Working window (*)	Specific capacitance	Energy density/scan rate	Energy density	Power density	Capacitance retention	Reference
CNT/Cs-PMo_{12}	Activated Nafion membrane with a drop of a 1 M H_2SO_4 solution	−1.2 to 1.0 V (does not present reference)	Cell: 285 $F\,g^{-1}$	200 $mA\,g^{-1}$	57 $Wh\,kg^{-1}$	—	—	[32]
MWCNT/PMo_{12}	1 M H_2SO_4	0 to 1 V	1.43 $F\,cm^{-2}$	50 $mV\,s^{-1}$	—	—	—	[23, 24]
MWCNT/$SiMo_{12}$			2.38 $F\,cm^{-2}$					
2 L $SiMo_{12}$			1.43 $F\,cm^{-2}$					
S1: MWCNT/PMo_{12}/ $SiMo_{12}$			2.38 $F\,cm^{-2}$					
AC/PMo_{12}	1 M H_2SO_4	0.305 to 0.831 V	Cell_+: 160 $F\,g^{-1}$ Cell_−: 183 $F\,g^{-1}$	2 Ag	—	—	91% after 8000 cycles @ 1600 $mA\,g^{-1}$	[25]
MWCNT/PMo_{12}	N. A.	−0.1 to 0.7 V	84 $F\,cm^{-3}$	50 $mV\,s^{-1}$	—	—	—	[15]
MWCNT/$PMo_{10}V_2$			88 $F\,cm^{-3}$					
MWCNT/PMo_{12}/$PMo_{10}V_2$		−0.1 to 0.65	158 $F\,cm^{-3}$					
AC/PW_{12}	1 M H_2SO_4	−0.6 to 1 V	254 $F\,g^{-1}$	10 $mV\,s^{-1}$	4.96 $Wh\,kg^{-1}$	115 $kW\,kg^{-1}$	98% after 30,000 cycles @ 6 $A\,g^{-1}$	[21]
MWCNT/PPMo/PW_{12}	1 M H_2SO_4	−0.15 to 0.6 V	92.7 $F\,cm^{-3}$	25 $mV\,s^{-1}$	—	—	92% after 2000 cycles @ 1 $V\,s^{-1}$	[26]
MMWCNT/PMo/SiW_{12}			NA	—			NA	
HT-rGO/PMo_{12}	1 M H_2SO_4	−0.2 to 0.8 V	276 $F\,g^{-1}$	10 $mV\,s^{-1}$	4.3 $Wh\,kg^{-1}$	30 $kW\,kg^{-1}$	Cell: 96% after 10,000 cycles @ 1 $A\,g^{-1}$	[27]
SWCNT/TBA-PV_2Mo_{10}	1 M H_2SO_4	−0.2 to 0.8 V	317 $mF\,cm^{-2}$	0.1 $mA\,cm^{-2}$	15.4 $Wh\,kg^{-1}$	15.7 $kW\,kg^{-1}$	Cell: 95% after 6500 cycles @ 1 $A\,g^{-1}$	[35]
AC/P_2Mo_{18}	1 M H_2SO_4	−0.2 to 1.0 V	275 $F\,g^{-1}$	6 $A\,g^{-1}$	—	—	Electrode: 70% after 2000 cycles @ 6 $A\,g^{-1}$	[36]

MWCNT/PMo_{12}-PW_{12}	1 M H_2SO_4	−0.1 to 0.65 V	—	—	—	—	—	[33]
MWCNT/$SiMo_{12}$-SiW_{12}								
PDDA-rGO/PMoW	1 M H_2SO_4	−0.2 to 0.8 V	279.1 F g^{-1}	1 A g^{-1}	6.15 Wh kg^{-1}	250 W kg^{-1}	Cell: 94.6% after 1700 cycles @ 5 A g^{-1}	[34]
rGO/PMo_{12}	1 M H_2SO_4	−0.2 to 1.0 V	299 F g^{-1}	5 mA cm^{-2}	39.45 Wh kg^{-1}	6583 W kg^{-1}	95% after 2000 cycles @ 10 mA cm^{-2}	[28]
rGO/PW_{12}		−0.6 to 1.0 V	370 F g^{-1}					
rGO/Mo_{132}	1 M Li_2SO_4	−1.0 to 0.2 V	617.3 F g^{-1}	5 A g^{-1}	31.6 Wh kg^{-1}	4893.3 W kg^{-1}	66.7% after 3000 cycles @ 5 A g^{-1}	[37]
rGO/PMo	0.5 M H_2SO_4	0.3 to 0.95 V	207 F g^{-1}	10 mV s^{-1}	—	—	—	[29]
DTAB-EEG/Mo_{132}	1 M H_2SO_4	0.0 to 1.0 V	65 F g^{-1}	1 mV s^{-1}	—	—	Electrode: 99% after 5000 cycles @ 100 mV s^{-1}	[38]
GO/metallacalixarene/$SiMo_{12}$	0.5 M H_2SO_4	−0.1 to 0.7 V	230.2 F g^{-1}	0.5 A g^{-1}	—	—	Electrode: 92.7% after 1000 cycles @ 10.0 A g^{-1}	[39]
AC/TEAPW_{12}	1 M TEABF4 in MeCN	−2.02 to 0.5 V *vs.* Ag/Ag^+	66 F cm^{-3}	5 mV s^{-1}	16 mWh cm^{-3}	1.35 W cm^{-3}	93% after 10,000 cycles	[19]
rGO/modified PMo_{12} (PAHB)	0.5 M H_2SO_4	−0.30 to 0.7 V *vs.* SCE	395 F g^{-1}	1 A g^{-1}	26.2 Wh kg^{-1}	600 W kg^{-1}	74% after 10,000 cycles @ 5 A g^{-1}	[30]
rGO/PMo_{12}	1 M H_2SO_4	−0.2 to 0.7 V	631 F g^{-1}	5 A g^{-1}	50.46 Wh kg^{-1}	2800 W kg^{-1}	87.12% after 10,000 cycles @ 15 A g^{-1}	[31]
HrGO/$SiVW_{11}$	0.5 M H_2SO_4	0 to 1 V (does not present reference)	377.4 mF cm^{-2}	1.5 mA cm^{-2}	52.41 μWh cm^{-2}	1500 μW cm^{-2}	—	[40]
CrGO/$SiVW_{11}$			277.2 mF cm^{-2}		38.5 μWh cm^{-2}	1500 μW cm^{-2}		
AC/PV4W8	0.5 M H_2SO_4	0 to 1 V (does not present reference)	62 F g^{-1}	0.6 A g^{-1}	8 Wh kg^{-1}	1280 W kg^{-1}	—	[41]

PDDA, poly(dimethyl diallyl) ammonium; DTAB, dodecyltrimethylammonium bromide; EEG, electrochemically exfoliated graphene. PAHB: phosphomolybdic acid heteropoly blue (*) *vs.* Ag/AgCl unless stated otherwise.

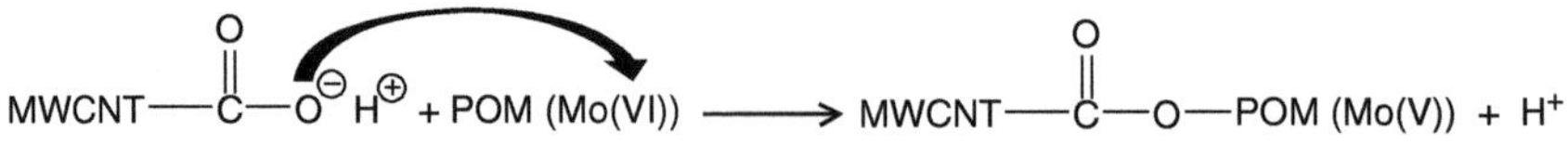

Figure 11.6 Scheme of the POM-chemisorption reaction. Source: Cuentas-Gallegos et al. [42]/with permission of Elsevier.

This process is important to ensure an adequate POM chemisorption onto the substrate, as it has been reported to provide strong and irreversible bonds into the POM/CNT hybrid systems [17]. The chemisorption strategy follows firstly the oxidation of CNT, with a strong acid (e.g. H_2SO_4, HNO_3), that creates an oxygen-rich surface capable of interacting with POM structure [42], as demonstrated by the following scheme (Figure 11.6) [42, 43]. In particular, 'layer-by-layer' (LbL) assembly is a simple chemisorption process that deposits, in an alternating way, layers of positive and negative ions (in this case POMs) to form a stable film [44]. The use of positively charged species, for instance a polymer, ensures POM anchoring through electrostatic interactions with the opposing charged layer.

The group of Gomez-Romero conducted, for the first time, a study on LbL POM/CNTs composites being used as an electrode in a symmetric solid-state supercapacitor [32]. The synergistic effect of $[PMo_{12}O_{40}]^{4-}$ (denoted PMo_{12}) and the CNTs resulted in high capacitance values (285 $F\,g^{-1}$ at 0.2 $A\,g^{-1}$), and good stability during 500 charge/discharge cycles. Although the capacitance decreased at higher current densities, the performance of the hybrid material was significantly larger compared to the one from bare CNTs, revealing the pseudocapacitive contribution of PMo_{12}. This groundbreaking study encouraged further investigations on the electrochemical properties of POM-modified CNTs for supercapacitors.

Lian et al. [23, 24] reported a LbL deposition of $[SiMo_{12}O_{40}]^{4-}$ (denoted $SiMo_{12}$) and PMo_{12} on multi-walled carbon nanotubes (MWCNTs) using polydiallyldimethylammonium chloride (PDDA) as positively charged species in the assembly, Figure 11.7a. Single-layer CNT modification with PMo_{12} or $SiMo_{12}$ showed a significant increase in the charged stored (1.66 and 1.23 $F\,cm^{-2}$, respectively, at 50 mV) [23, 24] in relation to bare CNT with 0.62 $F\,cm^{-2}$. This increase was due to the reversible multiple electron transfer of POMs. A similar shaped voltammogram was reported for PMo_{12} and $SiMo_{12}$ hybrids with three pairs of redox peaks (labelled 'ox' and 'red') assigned to the reactions Eqs. (11.3–11.5). The difference reported in peak potentials resulted from the effect of changing heteroatoms in POM structure. By the dual-layer modification of MWCNTs, a mixed contribution of PMo_{12} and $SiMo_{12}$ POMs was observed. The S1 layer, composed of PMo_{12} at the bottom, followed by $SiMo_{12}$ at the top, revealed greater charged stored values, reaching 2.68 $F\,cm^{-2}$ (at 50 $mV\,s^{-1}$), when compared with the dual-layer 2L of PMo_{12} (2.38 $F\,cm^{-2}$) [23]. The superimposition of two POMs on MWCNTs exhibited synergistic effect, not only between two similar layers of PMo_{12}, but also between PMo_{12} and $SiMo_{12}$. These results support these materials as promising for high-power supercapacitor applications.

The group of Lian et al. [15] further studied the effect of substituting two addenda atoms of Mo by V, forming a $[PV_2Mo_{10}]^{5-}$ (denoted PV_2Mo_{10}) POM material into

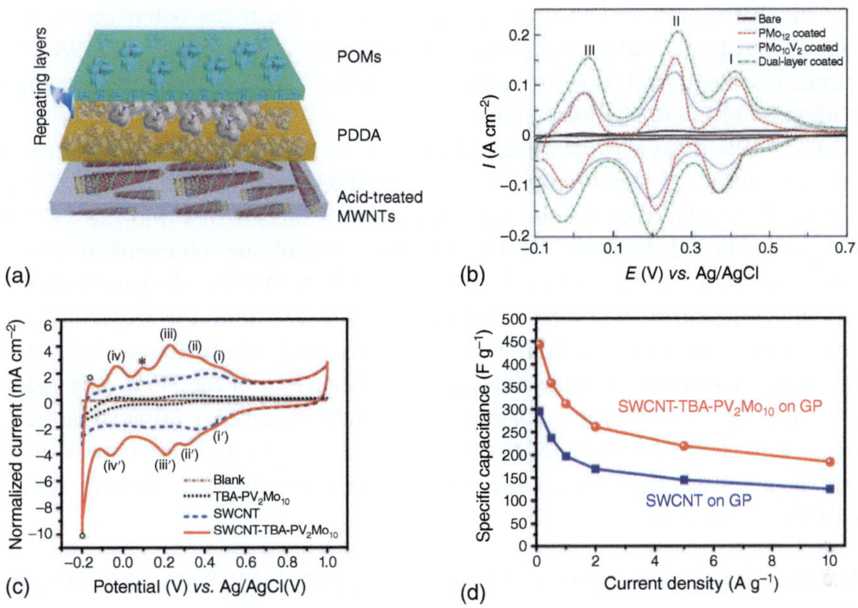

Figure 11.7 POM-based CNTs hybrids: (a) schematics of the dual-layer modification of MWCNTs using PDDA and POMs. Source: Ji et al. [44]/Royal Society of Chemistry/CC BY 3.0; Cyclic voltammetry of Mo-based POMs within carbon composites (b) single-layer PMo_{12}-coated MWCNT and single-layer $PMo_{10}V_2$-coated MWCNT compared to dual-layer coated MWCNT. Source: Bajwa et al. [15]/with permission of IOP Publishing; (c) blank (glassy carbon), TBA-PV_2Mo_{10}, SWCNT and SWCNT-TBA-PV_2Mo_{10} at $10\,mV\,s^{-1}$ in 1 M H_2SO_4. Source: Chen et al. [35]/with permission of Royal Society of Chemistry; and (d) the specific capacitance of SWCNT and SWCNT-TBA-PV_2Mo_{10} symmetric cell at various current densities of 0.1 to $10\,A\,g^{-1}$ in 1 M H_2SO_4. Source: Chen et al. [35]/with permission of Royal Society of Chemistry.

MWCNTs in single or double layer by LbL method. In a single-layer electrode, both PMo_{12}- and PV_2Mo_{10}-coated MWCNTs showed similar electrochemical behaviour with three pairs of redox peaks, Figure 11.7b. In spite of this, a more capacitive electrochemical feature was observed in PV_2Mo_{10}-coated MWCNT, as well as a larger potential window towards more positive values (~150 mV) due to the higher redox potential of V. This resulted in capacitances of 84 and $88\,F\,cm^{-3}$ (at $50\,mV\,s^{-1}$) for PMo_{12}- and PV_2Mo_{10}-coated MWCNTs, respectively (compared to $15\,F\,cm^{-3}$ from MWCNTs) [15]. The multi-layer POM-coated MWCNT exhibited a greater charge-storage capability ($158\,F\,cm^{-3}$), value that is five times larger than the single-coated, and 10 times higher than bare MWCNT [15]. The group subsequently extended their research [32] by investigating the combination of similar Keggin-type POM systems: XMo_{12} and XPW_{12}, X = Si and P, in a multi-layer system in MWCNTs, demonstrating the sensitivity of Keggin ions through the change of the addenda atom. Owing to the great electrochemical behaviour of Mo and W for energy storage, the mixture of PMo_{12} with W-based $SiW_{12}O_{40}{}^{4-}$ (denoted SiW_{12}) or $PW_{12}O_{40}{}^{4-}$ (denoted PW_{12}) was also investigated [26]. Although mixing SiW_{12} with PMo_{12} showed no effect on the shape of the voltammogram, mixing

PW_{12} and PMo_{12} resulted in a complete change of the redox behaviour of the material towards an ideally capacitive profile. The PMo_{12}/PW_{12}-modified MWCNTs electrode revealed a better performance of 92.7 F cm^{-3}, although slightly less than that of pure PMo_{12} (106.9 F cm^{-3}) [26].

Instead of PDDA as organic linker, Chen et al. [35] reported a PV_2Mo_{10}-derived POM structure using tetra-*n*-butyl ammonium (TBA) in SWCNTs, SWCNT-TBA-PV_2Mo_{10}. Figure 11.7c shows the voltammogram of the TBA-PV_2Mo_{10}, SWCNTs and the single-layer hybrid SWCNT-TBA-PV_2Mo_{10}, resembling voltametric response reported by Bajwa et al. [15]. The voltammogram of TBA-PV_2Mo_{10} deviated from the rectangular shape, showing faradaic processes. However, upon the preparation of the hybrid material, a larger current density was observed and also four well-defined redox peaks, associated to the pseudocapacitive behaviour of the POM. The redox peak couple, labelled as (i) and (i′), that is not usually observed in other voltammograms (e.g. by [15]), was attributed to the redox processes of V atoms at more positive potential values at ~0.48 V [35]. The redox reaction of PV_2Mo_{10} was summarized by the following equation:

$$PV_2{}^{V}Mo_{10}{}^{VI}O_{40}{}^{5-} + (6+n)\,e^- + (6+n)\,H^+ \rightleftharpoons H_{6+n}PV_{2-n}{}^{V}V_n{}^{IV}Mo_6{}^{VI}Mo_4{}^{V}O_{40}{}^{5-} \tag{11.6}$$

When applied in a symmetric supercapacitor with two identical SWCNT-TBA-PV_2 Mo_{10} electrodes on graphite paper, the capacitance of the cell was 444 F g^{-1} (at 0.1 A g^{-1}), while the SWCNT symmetric device revealed only 236 F g^{-1}, Figure 11.7d. Moreover, SWCNT-TBA-PV_2Mo_{10} device exhibited 95% capacitance retention after 6500 charge/discharge cycles [35].

In an effort to explore the versatility of POMs as redox active materials in carbon substrates, researchers have investigated the electrochemical behaviour of POM-modified AC. AC is a microporous carbon material also used for anchoring POMs to a conducting matrix. The first of these studies was performed by the group of Gomez-Romero et al. [25] that reported the preparation of hybrid electrode materials of PMo_{12} immobilized on a commercial AC. As expected, the voltammogram of AC showed the typical rectangular capacitive behaviour, while the AC/PMo_{12} hybrid revealed the coexistence of both double layer and redox activity, Figure 11.8a. In a two-electrode cell, the AC/PMo_{12} revealed an increase of 27% of capacitance (at 2 A g^{-1}) in relation to the bare AC, and good cyclability with 91% capacitance retention after 8000 cycles (at 1.6 A g^{-1}) [25]. In a subsequent work, the combination of PW_{12} POM with AC resulted in a capacitance of 254 F g^{-1}, in an expanded voltage range up to 1.6 V compared to only 1 V for bare AC, Figure 11.8b. The symmetric two-electrode cell cycled up to 1.6 V imposed a limitation of the charge/discharge process, owing to the limit for O_2/H_2 generation, Figure 11.8c. Increasing the operating voltage window from 1.5 to 1.7 V resulted in more pronounced curvature of the charge/discharge branches, associated to parasitic reactions. For instance, in the case of cells cycled between 0 and 1.7 V, a loss of capacitance of 65% after 30000 cycles was noticed, compared to the capacitance cells cycled up to 1.5 V, that retained ~100% capacitance, Figure 11.8d.

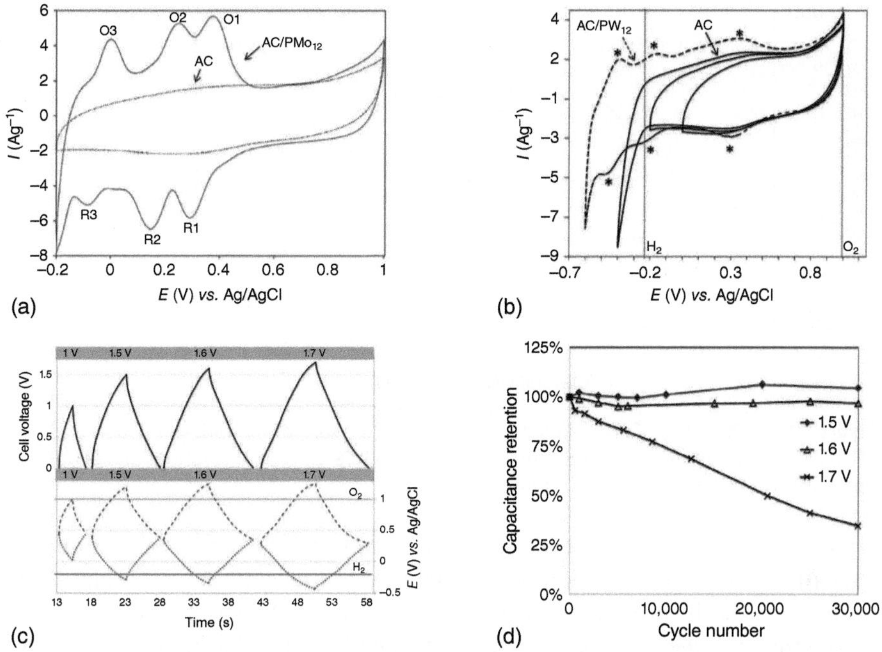

Figure 11.8 POM-modified AC materials: cyclic voltammogram of (a) AC and AC/PMo_{12} hybrid material. Source: Ruiz et al. [25]/with permission of Elsevier; and (b) AC and AC/PW_{12} hybrid material at 10 mV s^{-1} in 1 M H_2SO_4. Source: Suárez-Guevara et al. [21]/with permission of Royal Society of Chemistry; (c) galvanostatic charge/discharge cycles and the corresponding positive and negative electrode profiles for AC/PW_{12} in the voltage windows: 0–1, 0–1.5, 0–1.6 and 0–1.7 V at 6 A g^{-1} in 1 M H_2SO_4. Source: Suárez-Guevara et al. [21]/with permission of Royal Society of Chemistry; and (d) capacitance retention upon cycling for AC/PW_{12} at increasing voltage windows of 1.5, 1.6 and 1.7 V, at 6 A g^{-1} 1 M H_2SO_4. Source: Suárez-Guevara et al. [21]/with permission of Royal Society of Chemistry.

Owing to the limitation of water-splitting reactions, Gomez-Romero et al. [19] used an organic media (1 M TEABF4 in CH_3CN) for analysing the electrochemical behaviour of organic/inorganic POM, prepared with tetraethylammonium (TEA), denoted as AC/TEAPW_{12} hybrid. A nearly rectangular voltammogram was obtained in a potential window of ~2.50 V, highlighting the almost ideal capacitive behaviour, with an increase of 36% of capacitance in respect to AC. Moreover, four pairs of redox peaks were distinguished in the voltammogram of AC/TEAPW_{12}, contrary to what was observed for the same hybrids in acidic media, related to pH-dependent reactions [19]. Based on this, authors also postulated that the absence of protons in the organic electrolyte resulted in the splitting of the most negative pH-dependent redox peak [19].

Besides CNTs and AC carbon materials, graphene appears as a class of two-dimensional (2D) carbon structures that has great potential for the development of electrodes for high-power density-storage devices. In particular, POM-modified graphene-based materials have been developed for supercapacitor applications [27–31, 34, 38–40, 45]. The group of Gomez-Romero et al. [27]

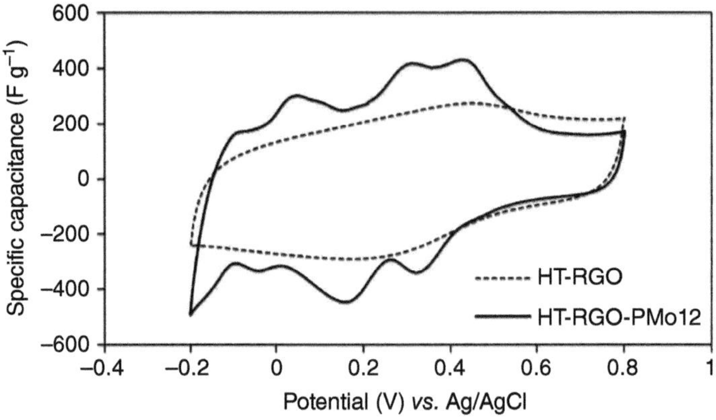

Figure 11.9 POM-modified graphene materials: cyclic voltammogram of HT-rGO and HR-rGO/PMo_{12} hybrid at 10 mV s^{-1} in 1 M H_2SO_4. Source: Suarez-Guevara et al. [27]/with permission of Royal Society of Chemistry.

reported the preparation of POM-decorated graphene layers through the simultaneous hydrothermal reduction of graphene oxide (denoted HT-rGO) and PMo_{12} anchoring. The incorporation of PMo_{12} in the HT-rGO structure resulted in the appearance of three redox peaks assigned to the PMo_{12} faradaic nature, Figure 11.9. From the voltammogram, the capacitance of the HT-rGO/PMo_{12} hybrid was 276 F g^{-1} (at 10 mV s^{-1}, in 1 M H_2SO_4) that, in relation to bare HT-rGO, revealed an increase of ~30% due to the pseudocapacitive behaviour [27]. A symmetric cell containing HT-rGO/PMo_{12} electrode showed a discharge capacitance of 123 F g^{-1} (at 1 A g^{-1}), in a 1 V potential window and 96% capacitance retention after 10000 cycles of charge/discharge. In a subsequent work, the authors extended their work towards a H_2SO_4/hydroquinone polyvinyl alcohol (PVA) gel-electrolyte all solid-state symmetric cell using rGO/PMo_{12} hybrid as electrodes [45]. In this case, the rGO/PMo_{12}-based cell provided a capacitance of 419 mF cm^{-2} (at 5 mV s^{-1}), value that is considerably higher than the one obtained without hydroquinone in the gel-electrolyte (278 mF cm^{-2} in H_2SO_4/PVA electrolyte) [45].

Biradar et al. [40] compared the effect of the deposition of the V-substituted $SiVW_{11}$ POM in hydrothermally (HT-rGO) and chemically reduced (CH-rGO) graphene oxide. In the case of HT-rGO, the POM deposited via chemisorption, HT-rGO/$SiVW_{11}$, while in CH-rGO, $SiVW_{11}$ attached to the surface through physisorption, CH-rGO/$SiVW_{11}$. Owing to the greater interaction between the POM and functional groups of graphene surface, the symmetric cell containing HT-rGO/$SiVW_{11}$ electrode showed a higher capacitance, of 377.4 mF cm^{-2} (at 1.5 mA cm^{-2}), compared to CH-rGO/$SiVW_{11}$ (277.2 mF cm^{-2}) at the same current density [40].

Considering an asymmetric supercapacitor, Dubal et al. [28] developed two POM/rGO-based electrode materials: PMo_{12} and PW_{12} POMs, working as positive and negative electrodes, respectively. In the three-electrode system, the voltammograms of both rGO/PMo_{12} and rGO/PW_{12} hybrid materials were nearly rectangular with faradaic features, assigned to a mixed-storage mechanism. In the case of

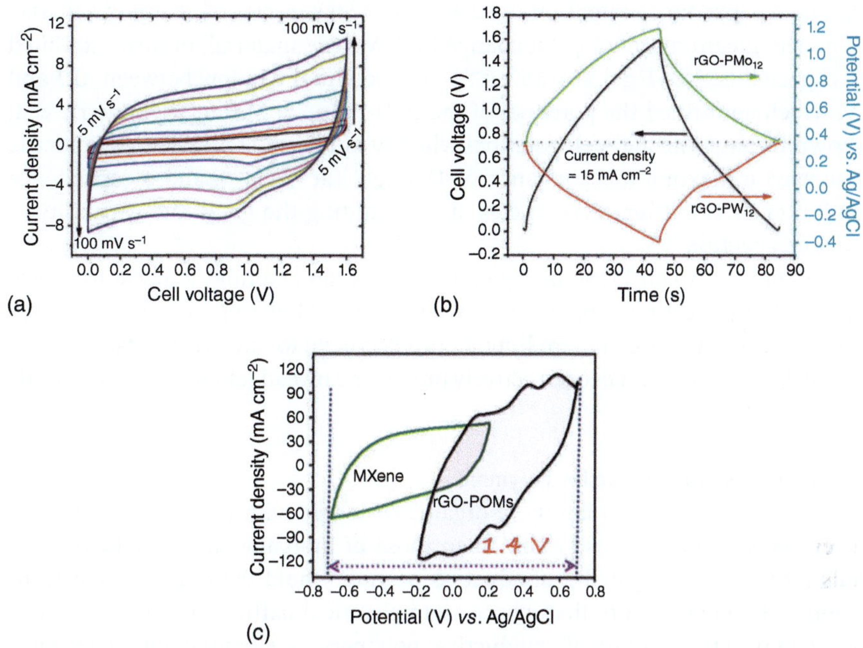

Figure 11.10 Asymmetric cells using POM-modified graphene materials: (a) cyclic voltammogram of the rGO/PMo_{12} || rGO/PW_{12} cell at different scan rates, ranging from 5 to 100 mV s^{-1} in 1 M H_2SO_4. Source: Dubal et al. [28]/with permission of John Wiley and Sons; (b) representative charge/discharge curves of the rGO/PMo_{12} || rGO/PW_{12} cell, with the corresponding potential distribution across the positive and negative electrodes at 15 mA cm^{-2} in 1 M H_2SO_4. Source: Dubal et al. [28]/with permission of John Wiley and Sons; and (c) cyclic voltammogram of MXene and rGO-PMo_{12} electrodes in a three-electrode cell at 100 mV s^{-1} in 1 M H_2SO_4. Source: Hwang et al. [31]/with permission of Elsevier.

rGO/PW_{12}, a larger potential window up to −0.6 V was obtained, compared to the one of rGO/PMo_{12} that was restrained at −0.2 V and, therefore a greater capacitance value for rGO/PW_{12} was calculated (299 *vs.* 370 F g^{-1}, at 5 mA cm^{-2}) [28]. Therefore, the adsorption of PW_{12} onto the rGO surface provides additional overpotential for the storage process in the two-electrode cell. The voltammogram of the asymmetric rGO/PMo_{12} || rGO/PW_{12} cell, collected in a voltage range of 1.6 V, exhibited a rectangular shape with redox waves, covering the electrochemical response of both PMo_{12} and PW_{12} POMs, Fig 11.10a. A representative charge/discharge curve for the rGO/PMo_{12} || rGO/PW_{12} cell shown in Figure 11.10b confirms the pseudocapacitive behaviour of the electrodes in the cell, with a nearly linear triangular-shaped curve. Moreover, it proves that both positive and negative electrodes contributed equivalent potential for the cell, creating a higher capacitance value (110.9 F g^{-1} at 15 mA cm^{-2}) in relation to the rGO/PMo_{12} symmetric cell (71 F g^{-1}), and 95% capacitance retention after 2000 charge/discharge cycles [28]. Hwang et al. [31] also developed a POM-based rGO/PMo_{12} material that worked as cathode in an asymmetric supercapacitor cell using a MXene as anode. In a three-electrode cell, the rGO/PMo_{12} hybrid revealed a capacitance of 631 F g^{-1} (at 5 A g^{-1}), between −0.2 and

0.7 V, and a capacitance retention of 93.45% after 5000 cycles of charge/discharge, with no significant morphological change [31]. MXene material, in turn, exhibited high capacitance (406 $F g^{-1}$ at 1.66 $A g^{-1}$), in a potential window between -0.7 and 0.2 V, which confirmed the potential of the rGO-PMo_{12} || MXene asymmetric cell, as shown through the half-cell potential window in Figure 11.10c. The asymmetric cell showed a maximum capacitance of 185 $F g^{-1}$ (at 4 $A g^{-1}$), and a capacitance retention of 87.12% after 10000 cycles, demonstrating the great rate capability of this supercapacitor.

The synergy between POM and graphene holds the potential to boost electrical conductivity by creating supplementary active sites. However, effectively managing the correlation between these interactions is crucial for ensuring a stable hybrid material. Hence, there is a need to actively pursue the investigation of these electrode materials.

11.2.1.2 POMs into Conductive Polymers

Conductive polymers are a family of organic materials with peculiar conductivity and reversible redox chemistry similar to those of inorganic semiconductors and metals [46]. The conjugated π conductive polymers hold delocalized π-electrons in their backbone structure that works as an electrical pathway for mobile charge carriers [46]. This feature of conductive polymers is essential for ensuring a reversible redox chemistry and ionic transport in hybrid nanocomposites. Hybrid organic/inorganic materials-based on POMs and conductive polymers led to an enhanced overall electric conductivity and charge-storage capability [47]. Usually, the conductive polymers as support for POM-based materials have several other advantages, including low cost, simple synthesis, thermal stability and, more importantly, redox properties [48]. Examples of conductive polymers used in these hybrids materials are polypyrrole (PPy), poly[3,4-(ethylenedioxy)thiophene] (PEDOT) and polyaniline (PANi) (Table 11.2).

The group of Gómez-Romero was one of the first to look beyond hybrid materials and develop organic–inorganic POM-based conductive polymer materials for energy-storage applications. In an early work, Gómez-Romero et al. [20] reported the use of electrochemically grown PANi/PMo_{12} hybrid nanocomposite as electrodes for solid-state symmetrical supercapacitors. Preliminary tests on a symmetrical cell revealed a small loss of capacitance (~10–15%) after 2000 charge/discharge cycles. In a subsequent study, the same authors showed an evolution of the capacitance of PANi/PMo_{12} hybrid over 1000 cycles, Fig 11.11a. This increase in capacitance (up to ~120 $F g^{-1}$) was assigned to an activation of the electrode due to the progressive penetration of the electrolyte into the electrode structure [54, 55]. These results [20, 54] were regarded as a promising proof-of-concept because, although authors did not attempt to optimize the cell systems, a good performance was obtained, encouraging further research. More recently, Vannathan et al. [41] developed a hybrid material based on $PV_4W_8O_{40}{}^{7-}$ POM material (denoted PV_4W_8) in AC (AC/PV_4W_8) that was compared with a PANi/PV_4W_8 hybrid electrode. The PANi/PV_4W_8 hybrid material revealed greater capacitance, of 584 $F g^{-1}$ (at 0.8 $A g^{-1}$), along with energy and power densities, compared to the AC/PV_4W_8 material, Figure 11.11b. Authors

Table 11.2 Electrochemical properties of POM-based modified conductive polymer materials. Specific capacitance from the material in a three-electrode system. Energy and power density from the two-electrode device.

POM-based modified conductive polymer materials	Electrolyte	Working window (*)	Specific capacitance	Current density/scan rate	Energy density	Power density	Capacitance retention	Reference
PANi/PMo_{12}	0.5 M H_2SO_4	−0.1 to 0.75 V	Cell: 195 mF cm^{-2}	0.125 mA cm^{-2}	24.4 mJ cm^{-2}	—	—	[20]
PPy/PMo_{12}	0.5 M H_2SO_4	−0.1 to 0.8 V	Cell: 31 F g^{-1}	1 mA	4.0 Wh kg^{-1}	0.103 kW kg^{-1}	—	[49]
PPy/PMo_{12}	0.05 M H_2SO_4	−0.1 to 0.8 V	77.0 mF cm^{-2}	1 mA cm^{-2}	—	—	100% over 3500 cycles @ 1 mA cm^{-2}	[50]
PPy/PMo_{12}	0.1 M H_2SO_4	−0.5 to 0.5 V (does not present reference)	Cell: 561.1 F g^{-1}	0.5 A g^{-1}	19.48 Wh kg^{-1}	0.078 kW kg^{-1}	95% after 4500 cycles	[51]
PPy/Mo_6O_{18}	0.1 M Bu_4NBF_4 in Ch_3CN	−0.2 to 1.8 V	54 F g^{-1}		—	—	95% over 1200 cycles	[52]
PAM/PMo_{12}	1 M H_2SO_4	0 to 0.9 V	134.6 mF cm^{-2}	2 mA cm^{-2}	33.65 μWh cm^{-2}	2000 W cm^{-2}	87.9% after 1000 cycles @ 10 mA	[53]
PANi/PV_4W_8	0.5 M H_2SO_4	0 to 1 V	584 F g^{-1}	0.8 A g^{-1}	81 Wh kg^{-1}	1598 W kg^{-1}	93.13% after 4500 cycles @ 500 mV s^{-1}	[41]

PANi, polyaniline; PPy, polypyrrole; PAM, polyacrylamide. (*) *vs.* Ag/AgCl unless stated otherwise.

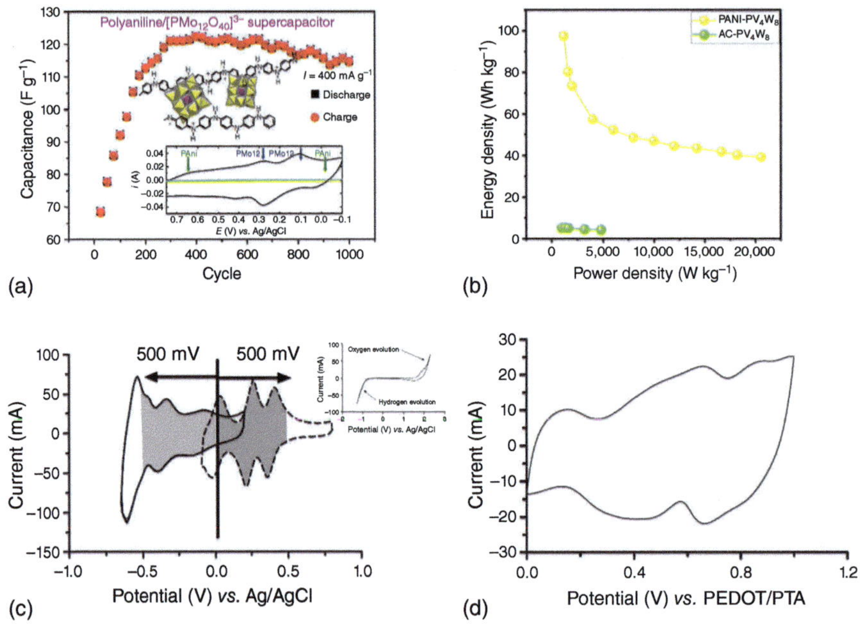

Figure 11.11 POM-modified conductive polymer materials: (a) capacitance as a function of cycle number of a symmetric supercapacitor with PANi/PMo_{12} hybrid nanocomposite as electrode material in 0.5 M Na_2SO_4. Inset image shows the cyclic voltammogram of the composite in relation to the plain carbon. Source: Gómez-Romero et al. [54]/with permission of Springer Nature; (b) Ragone plot of AC/PV_4W_8 and PANi/PV_4W_8 hybrid materials; Source: Vannathan et al. [41]/with permission of Springer Nature. (c) Superimposed cyclic voltammogram at 100 mV s^{-1}. Source: Suppes et al. [49]/with permission of IOP Publishing; and (d) cyclic voltammogram of the symmetric cell of PPy/PMo_{12} as a positive electrode and PEDOT/PW_{12} as a negative electrode at 100 mV s^{-1} in 0.5 M Na_2SO_4. Source: Suppes et al. [49]/with permission of IOP Publishing.

hypothesized that the large difference between hybrids was due to (i) a uniform integration of the POM on the PANi matrix and (ii) the synergistic effect between the π-bonded conjugated PANi backbone and the POM [41]. Both of these factors led to an increase in the number of active sites related to faradaic reactions and, consequently, to an enhanced capacitance of the electrode.

Besides PANi, PPy and PEDOT are other widely studied conductive polymers for supercapacitor applications that have been regarded as a promising support material for POMs. For instance, Suppes et al. [49] synthesized two composite materials: PPy/PMo_{12} and PEDOT/PW_{12} through chemical polymerization as positive and negative electrodes, respectively. The voltammogram of these hybrids in a three-electrode cell showed the typical redox peaks observed for POMs-based material, Figure 11.11c. Moreover, as the potential windows were superimposed, the possibility of constructing a two-electrode cell was disclosed. The shaded area in the voltammogram, Figure 11.11c, represents 1 V potential range used in the supercapacitor cell, that achieved a capacitance of 31 F g^{-1} at 1 A g^{-1}. The obtained voltammogram for this system is represented in Figure 11.11d. In this device, the decrease of capacitance reported over 200 charge/discharge cycles was attributed to the degradation of the PPy/PMo_{12} electrode, as the drift of the potential towards more positive values led to the overoxidation of PPy [49]. Working in a similar potential window, Yang et al. [50] reported a PMo_{12} PPy nanopillar arrays that exhibited stable redox reaction over 3500 cycles of charge/discharge with ~100% retention of capacitance (~77.0 mF cm^{-2} @ 1 mA cm^{-2}).

Using polyacrylamide (PAM) as support conductive material, Travas-Sejdic et al. [53] developed a POM-based hydrogel hybrid composite with PAM/sodium alginate (SA)/polyluminol (PLum) in carbon cloth. PLum is a conductive polymer that is capable of immobilizing PMo_{12}, creating a stable and electrochemically active material [53]. Anchoring PMo_{12} on PAM/SA/PLum hybrid hydrogel prevented the dissolution of the POM into the electrolyte (1 M Na_2SO_4), retaining 87.9% of its initial capacitance (253.76 mF cm^{-2} @ 10 mV s^{-1}) after 1000 cycles of charge/discharge. Owing to the hydrogel nature of the electrode, a flexible symmetric supercapacitor was fabricated with a maximum energy density of 33.65 μWh cm^{-2}, which was an improvement over the electrochemical metrics of other POM/CPs hydrogel electrodes reported [53].

While encasing POMs within polymeric matrices is suggested as a viable solution to address past limitations, this approach inherently restricts the accessibility of the redox active sites in POMs, thereby imposing constraints on their intrinsic capacitances. Hence, further research is required to explore innovative formulations of POM/polymer during its design, as well as introduce new types of POMs and polymers into the hybrid material.

11.2.1.3 POM-based Ternary Nanohybrids (TNH)

As previously mentioned, the high solubility of POMs requires combining with other conductive materials to make a feasible and direct application as electrodes for supercapacitors. Vast efforts have been made to combine POMs with carbon and conducting polymers individually [56]. While combining POM and graphene can enhance electrical conductivity, through the creation of additional active sites, the correlation between these interactions still requires proper management. Entrapping POMs into polymeric matrices is reported as a possible solution to overcome the previous limitations; however, this strategy necessarily limits the availability of the redox active sites of POMs, constraining their intrinsic capacitances [56]. Therefore, POM-modified material with carbon and conductive polymers to create ternary nanohybrids (TNHs) stood as a strategy to address the aforementioned limitations, Table 11.3. This is an approach that aims to further enhance the applications of POMs in supercapacitor devices by re-designing electrodes with improved capacitive behaviour [41, 56–59, 61, 62]. So far, few combinations of TNHs have been reported, mainly based on PMo_{12} POM combined with polymers as PANi or PPy along with different carbon materials like CNT, CD and rGO [43, 57–59]. Combinations using TiN as alternatives to carbon-based materials have also been described [60]. For other TNH structures, polymeric ionic liquid (PIL) and coordination polymers have also been reported along with graphene [56, 61].

Table 11.3 Electrochemical properties of ternary POM/conductive polymer/carbon-based hybrid materials. Specific capacitance from the material in a three-electrode system. Energy and power density from the two-electrode device.

Ternary POM/conductive polymer/carbon-based hybrid	Electrolyte	Working window/V (*)	Specific capacitance	Current density/scan rate	Energy density	Power density	Capacitance retention	Reference
CNT/PMo_{12}/PANi	1 M H_2SO_4	0 to 0.8 V *vs.* SCE	825 F g^{-1}	5 A g^{-1}	18.7 Wh kg^{-1}	2 kW kg^{-1}	72.4% after 10,000 cycles	[57]
PMo_{12}/rGO/PIL	0.5 M H_2SO_4	−0.1 to 0.9 V	408 F g^{-1}	0.5 A g^{-1}	56 Wh kg^{-1}	52 kW kg^{-1}	98% during 2000 cycles @ 10 A g^{-1}	[56]
PMo_{12}/PPy/CNT hydrogel	0.5 M H_2SO_4	−0.4 to 0.8 V	1.17 F cm^{-2}	1 A g^{-1}	67.5 μWh cm^{-2}	700 μW cm^{-2}	85.7%	[58]
PPy/PMo_{12}/rGO	0.5 M H_2SO_4	−0.2 to 0.8 V	360 F g^{-1}	0.5 A g^{-1}	—	—	—	[59]
CDs/PMo_{12}/PANi	1 M H_2SO_4	−0.2 to 1.0 V	479 F g^{-1}	1 A g^{-1}	37.1 Wh kg^{-1}	0.750 kW kg^{-1}	Electrode: 68.1% after 1000 cycles at 5 A g^{-1}	[43]
PMo_{12}/PANi/TiN NWA	1 M H_2SO_4	0 to 0.7 V *vs.* SCE	469 F g^{-1}	1 A g^{-1}	21.6 Wh kg^{-1}	0.375 kW kg^{-1}	—	[60]
$Co(itmb)_4[H_2VW_{12}O_{40}]\cdot 4H_2O$/graphene	2 M H_2SO_4	−0.6 to 0 V	Cell: 405.3 F g^{-1}	5 A g^{-1}	—	—	89% after 1000 cycles at 15 A g^{-1}	[61]
$Zn(itmb)_4[H_2VW_{12}O_{40}]\cdot 4H_2O$/graphene			Cell: 339.2 F g^{-1}				88% after 1000 cycles at 15 A g^{-1}	

PIL, polymeric ionic liquid; PANi, polyaniline; PPy, polypyrrole; (*) *vs.* Ag/AgCl unless stated otherwise.

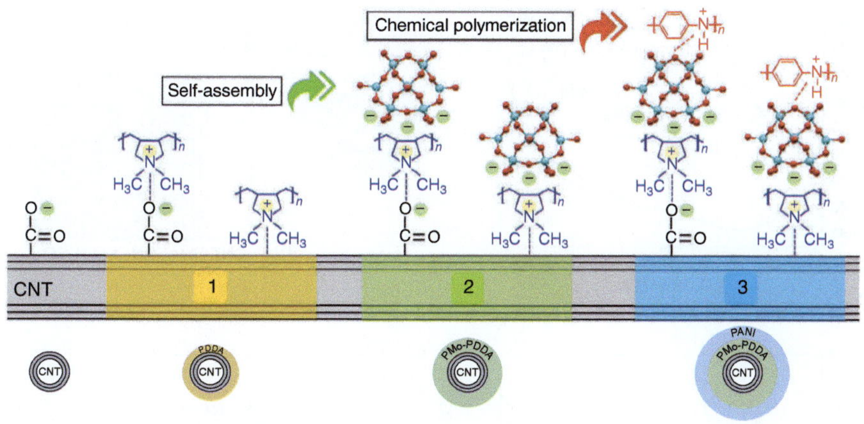

Figure 11.12 Schematics of the synthesis mechanism for the coaxial cable-like CNT/PMo_{12}/PANi by self-assembly and chemical polymerization. Source: Yang et al. [57]/with permission of John Wiley and Sons.

Huang et al. [57] prepared coaxial cable-like hybrid materials combining CNT, PANi fibres and PMo_{12} POM, denoted as CNT/PMo_{12}/PANi. Figure 11.12 schematically depicts the synthesis of the TNH through self-assembly of the POM material, followed by the chemical polymerization of PANi [57]. In this hybrid material, the oxidative properties of PMo_{12} mediated the chemical polymerization of aniline and eased building a regular connection between PANi and CNT.

A control hybrid material is prepared the same way, CNT/PANi/PMo_{12}; however, the PANi was polymerized previous to the assembly of the POM structure. The effect of this synthesis step on the electrochemical behaviour is represented by the voltammogram, Figure 11.13a. The cyclic voltammogram of hybrid materials exhibited combined redox peaks of the different components: PMo_{12}, CNT and PANi. Whitin the CNT/PMo_{12}/PANi hybrid, PANi served as a 'shield' layer for PMo_{12} integrity, Figure 11.13b, ensuring a good stability of this nanohybrid, which resulted in a higher specific capacitance of $825\,F\,g^{-1}$ (at $5\,A\,g^{-1}$), and cycling stability of ~85% retention after 6000 charge/discharge cycles [57]. The superior capacitance of CNT/PMo_{12}/PANi was attributed to the synergistic effect between PANi and PMo_{12}. However, when the POM is used as an external layer, in the CNT/PANi/PMo_{12} setup, the contact of the PMo_{12} with the electrolyte solution leads to its dissolution. This behaviour may be behind the poor cyclic stability reported for CNT/PANi/PMo_{12} fibres [57].

In ternary coupling material based on carbon dots (CDs)/PMo_{12}/PANi TNH, the PMo_{12} cluster bonded with CDs, via chemisorption interaction, kept a bridge between CDs and PANi chains that were further polymerized [43]. Comparing the storage capability of the components individually, Figure 11.13c, one could observe that each component is responsible for enhancing the overall capacitance of the TNH. For instance, PMo_{12}, as expected, contributed to the pseudocapacitive behaviour, but also improves the proton-doping level of PANi, promoting its electroactivity; the CDs promoted higher electrical conductivity for an eased electron

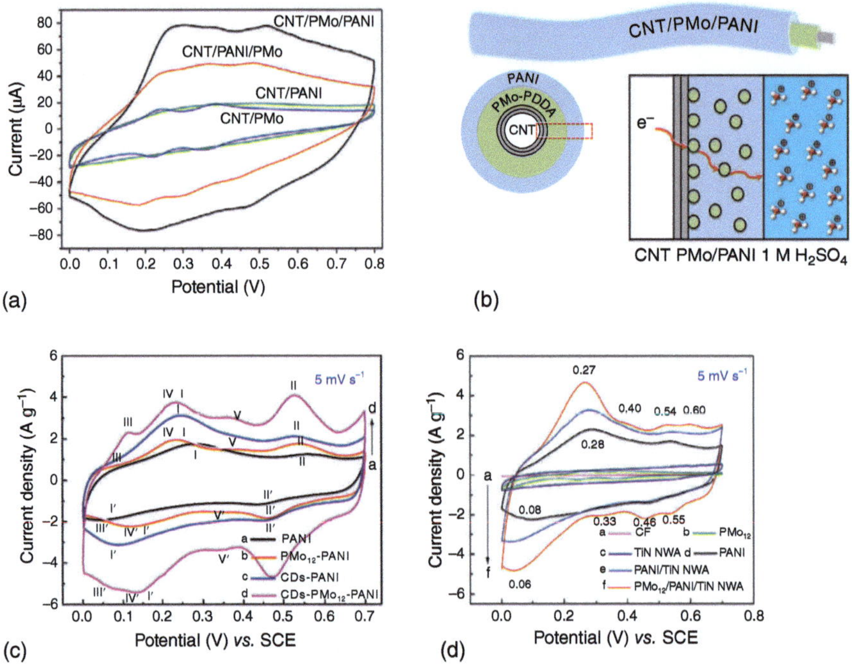

Figure 11.13 Ternary PANi-PMo_{12}-based nanohybrids: (a) cyclic voltammogram of coaxial cable-like CNT/PMo_{12}/PANi hybrid material at 100 mV s^{-1} in 1 M H_2SO_4. Source: Yang et al. [57]/with permission of John Wiley and Sons; (b) schematics of electron conductivity of CNT/PMo_{12}/PANi. Source: Yang et al. [57]/with permission of John Wiley and Sons; Cyclic voltammograms of (c) PANi, PMo_{12}/PANi, CDs/PANi and CDs/PMo_{12}/PANi. Source: Lu and Xie [43]/with permission of Springer Nature; and (d) CF (carbon fibres), PMo_{12}, TiN NWA, PANi, PANi/TiN NWA and PMo_{12}/PANi/TiN NWA at 5 mV s^{-1} in 1 M H_2SO_4. Source: Lu and Xie [60]/with permission of Royal Society of Chemistry.

transfer of the CDs/PMo_{12}/PANi [43]. As a result, the TNH CDs/PMo_{12}/PANi revealed the typical electrochemical behaviour of PMo_{12}, with a strong capacitive contribution, reaching a capacitance of 479 F g^{-1} (at 1 A g^{-1}), reasonable rate capacitance retention of 69.3% (from 0.5 to 10 A g^{-1}) and good cycling capacitance retention of 68.1% after 1000 cycles at 5 A g^{-1} [43].

Due to its superior electrical conductivity and mechanical stability, titanium nitride (TiN) holds as an alternative promise as an electrode material for supercapacitors [63]. As such Lu and Xie [60] studied the combination of PMo_{12}/PANi/TiN nanowire array (NWA) by hydrothermally growing TiO_2 on a carbon fibre, followed by annealing in NH_3 and lastly the electrodeposition of PANi and PMo_{12}. In relation to others, the voltammogram of the PMo_{12}/PANi/TiN NWA, shown in Figure 11.13d, exhibited four pairs of redox peaks, in which the pair of peaks ~ 0.27/0.06 V corresponded to the simultaneous redox reaction of PMo12 (0.26/0.23 V) and PANi (0.28/0.08 V) [60], which justified the greater current density compared to other redox peaks. This PMo_{12}/PANi/TiN NWA electrode exhibited an improved capacitance of 469 F g^{-1} (at 1 A g^{-1}) owing to: (i) the redox properties of PMo_{12};

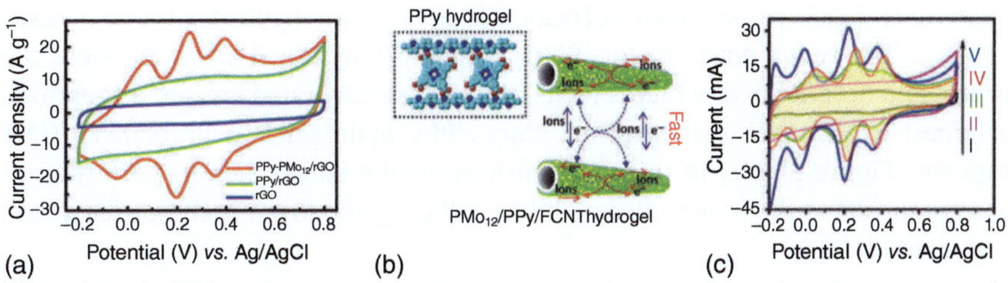

Figure 11.14 Ternary PPy-PMo_{12}-based nanohybrids: (a) cyclic voltammogram for PPy-PMo_{12}/rGO TNHs, PPy/rGO and rGO at 50 mV s^{-1} in 0.5 M H_2SO_4. Source: Chen et al. [59]/with permission of Royal Society of Chemistry; (b) schematic of the conduction mechanisms for PMo_{12}/PPy/CNT hydrogel. Source: Wang et al. [58]/with permission of Royal Society of Chemistry; and (c) cyclic voltammogram for CNT (I), PPy/CNT hydrogel (II), PMo_{12}/PPy hydrogel (III), PMo_{12}/PPy/CNT composite (IV) and PMo_{12}/PPy/CNT hybrid hydrogel (V) at 50 mV s^{-1} in 0.5 M H_2SO_4. Source: Wang et al. [58]/with permission of Royal Society of Chemistry.

(ii) the 'superhighway' of electrons provided by TiN NWA; and (iii) the core–shell nanowire structure that supplied a large surface area as active sites that shortened diffusion paths and effective electron transfer [60]. All of these reactions promoted fast and reversible redox reactions.

Chen et al. [59] synthesized PPy-POM/rGO TNH materials via a one-pot redox relay strategy, obtaining PPy-PMo_{12} nanoparticles, with sizes of about 65 nm, that were anchored on rGO nanosheets. The role of these nanoparticles was not only to enhance the pseudocapacitive behaviour of the TNH, but also to prevent the re-stacking of rGO, preserving the mesopore structure for efficient ion diffusion, Figure 11.14a. As a result of this special 'highway' for ion diffusion, the TNH exhibited excellent electrochemical performance in a 0.5 M H_2SO_4 solution, reaching a capacitance of 360 F g^{-1} at 0.5 A g^{-1} [59]. In an all-solid-state symmetric supercapacitor cell, using PVA/H_2SO_4 gel electrolyte, the PPy-POM/rGO electrode material exhibited excellent energy-storage performance even at high current densities (2.61 mF cm^{-2} at 150 mA m^{-2}) [59].

In the build-up of the PMo_{12}/PPy/CNT hybrid, reported by [58], the positively charged PPy hydrogel was doped with TCPP (5,10,15,20-tetrakis(4-carboxyphenyl) porphyrin). This TCPP cross-linker uniformly wraps into the outer surface of negatively charged CNT, simultaneously creating a cage structure for the easy storage of PMo_{12}. Density functional theory calculations revealed cage structures with a suitable size for single PMo_{12} molecules (diameter = 1.05 nm), resulting in monodispersed PMo_{12} single molecules on the hybrid hydrogel. This 'caging' effect was described to activate PMo_{12} molecules and enhance the charging/discharging performance of the TNH by introducing new active sites for proton transfer, Figure 11.14b. The electrochemical properties of the PMo_{12}/PPy/CNT systems were evaluated using a three-electrode system. Both CNT (curve I) and PPy/CNT hydrogel (curve II) exhibited dominant double-layer capacitance which is translated by the rectangle-like shape of their CV curves, Figure 11.14b. Differently, CV curves of PMo_{12}-containing materials (curves III–V) exhibited the typical pairs of reversible redox peaks, each corresponding to a two-electron process caused by PMo_{12}. The redox peaks of the PMo_{12}/PPy/CNT hydrogel electrode were the sharpest and best defined, which showed the largest integrated area, and therefore, the largest capacitance compared to PMo_{12}/PPy hydrogel and PMo_{12}/PPy/CNT composite, Figure 11.14c. In this case, authors attributed the weaker performance of the composite to the poorer dispersibility of PMo_{12} and lower specific area due to its aggregation [58].

Other polymers than PANi and PPy have been reported on the build-up of TNH structures. Yang et al. [56] adapted a PIL in the construction of POM nanohybrids. Their strategy enabled the uniform distribution of PMo_{12} molecules along the large surface area of graphene sheets, where PIL was used as a linker, Figure 11.15a. First-principles calculations and impedance spectroscopy analysis reveal that PILs enhanced the redox reactions of POMs by providing efficient ion transfer channels and facilitating the charge transfer in these TNH. The electrochemical properties, tested in a two-electrode system, revealed a high specific capacitance of $408\,F\,g^{-1}$ at $0.5\,A\,g^{-1}$, rapid rate capability, of 92% retention at $10\,A\,g^{-1}$, a long cycling life (98% retention during 2000 cycles), Figure 11.15b. Sun et al. [61] explored coordination polymers to construct hybrids of $Co(itmb)_4[H_2VW_{12}O_{40}]\cdot 4H_2O$ and $Zn\text{-}(itmb)_4[H_2VW_{12}O_{40}]\cdot 3H_2O$ (itmb = 1-(imidazo-1-ly)-4-(1,2,4-triazol-1-ylmethyl)benzene) assembled with graphene, Figure 11.15c. The highest performance was observed when the mass ratio of POM–polymer and graphene material was of 1 : 1. The specific capacitance of $Co(itmb)_4[H_2VW_{12}O_{40}]\cdot 4H_2O$/graphene was of $405.3\,F\,g^{-1}$ and that of $Zn\text{-}(itmb)_4[H_2VW_{12}O_{40}]\cdot 3H_2O$/graphene of $339.2\,F\,g^{-1}$ Figure 11.15d [61], although in their infancy, these results prove the huge potential of TNHs as electroactive materials for supercapacitor applications.

11.2.1.4 POMs Within Supramolecular Structures

At a completely different level, POMs can be assembled within supramolecular structures to create new multifunctional POMs. This is an area where the combination of POM as electroactive materials has considerably evolved. A brief overview over this combination will be provided in this subsection. The preponderance of supramolecular structures is their ability to accommodate various functional units at the molecular level, and therefore design materials with tailored energy metrics, flexibility and redox properties. Recently, supramolecular structures, including MOFs and MXene materials, have been regarded as promising supports for POM-based electrodes for supercapacitors. This is owing to the synergistic effect between the components of both structures that led to superior charge-storage capabilities. Besides enhancing the electrochemical response of POMs, the supramolecular assembly can be regarded as a key strategy to improve the dispersion of these molecules on a substrate and, therefore, optimize the exposure of electrochemical active sites at the molecular level [22].

MOFs are a class of porous crystalline materials composed of metal ions or clusters linked with organic ligands [64]. Owing to their tunability, MOFs are known as 'materials by design', meaning that one can design a synthesis method to

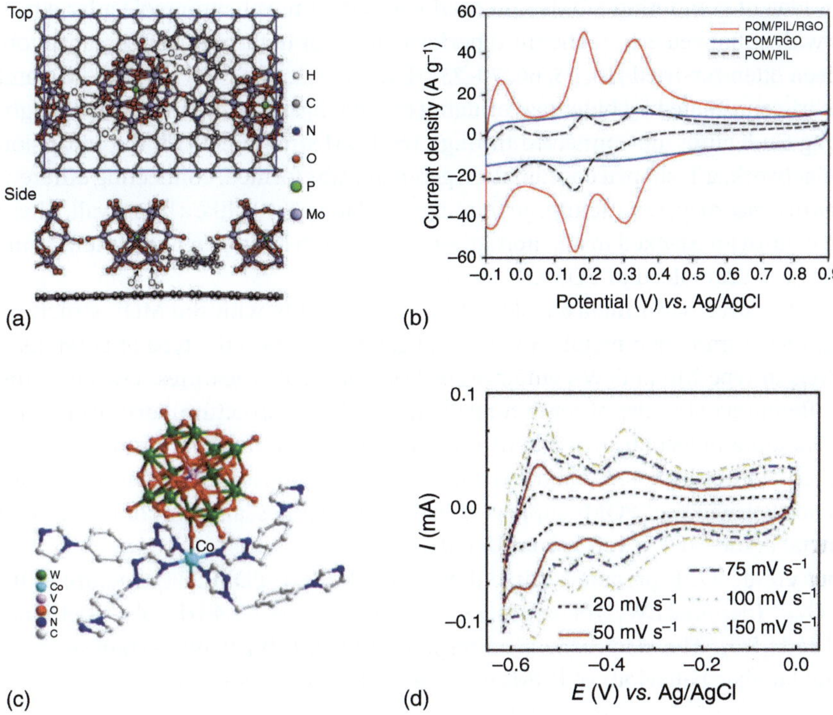

Figure 11.15 Other ternary nanohybrids: (a) PIL/POM ($2[PMo_{12}][C_{14}H_{24}N_4]$) structure on graphene. Source: Yang et al. [56]/with permission of John Wiley & Sons; (b) cyclic voltammogram for POM/PIL/rGO, POM/rGO and POM/PIL at 50 mV s^{-1} in 0.5 M H_2SO_4. Source: Yang et al. [56]/with permission of John Wiley & Sons; (c) unit cell structure of compound $(Co(itmb)_4[H_2VW_{12}O_{40}]\cdot 4H_2O$. Source: Sun et al. [61]/with permission of John Wiley and Sons; and (d) corresponding cyclic voltammogram at different scan rates in 2 M H_2SO_4. itmb, 1-(imidazo-1-ly)-4-(1,2,4-triazol-1-ylmethyl)benzene. Source: Sun et al. [61]/with permission of John Wiley and Sons.

prepare a tailored material with the desired features, for instance good electronic and fast ionic conductivity. Thanks to this, it is possible to introduce different redox-metal-containing active sites in order to enhance the storage capability of the material. Additionally, the large pore structure of MOFs is also a key feature because it facilitates the ion diffusion during charge/discharge cycling. Owing to these features, MOFs have been regarded as promising materials for supercapacitors since 2012 [65]. Although MOFs present a metallic feature on their structure, pristine MOFs used as electrode materials are electrically insulators and typically exhibit low specific capacitance value [66]. Several approaches were proposed to overcome these disadvantages [67–69], for instance by immobilizing POMs in MOFs (POM@MOFs, or POMOFs). This is a promising strategy to develop high surface area material, with adjustable porosity and enhanced electrochemistry. Integrating POMs within MOFs using a host–guest approach to synthesize a stable POMOF is expected to solve the high water solubility and increase the mechanical strength of POMs to be used as electrode materials.

The idea of combining POMs and MOFs to obtain novel supramolecular structures with enhanced electrochemical performances for high-power supercapacitors has been often reported [14, 65, 66, 70–73]. The metal in the MOF, e.g. Cu, Mn and Ag, is tri-coordinated by bidentate bridging organic ligand and, in turn, two neighbouring Cu-ligands are connected through the POM structure, With the expansion of the network, a T-shaped coordination geometry was formed, connecting different chains in a macrocycle structure, generating a 2D lattice layer, like a brick wall. These 2D structures are stacked in an alternating way through hydrogen bond interactions forming a supramolecular POMOF assembly [14].

The capacitive performance of POMOFs varies greatly with the MOF structure, including the transition metal and organic ligands; and with the type of POM, usually Keggin-type Mo and W-containing POMs and related features. Owing to the vast different possibilities of combining POMs and MOFs structure, very diverse and complex supramolecular structure with specific electrochemical performance can be prepared. It must be noted that the electrochemical response of POMOFs resembles the redox behaviour of POMs supported in other substrates, and the contribution of the metal of the MOF is rarely considered.

Hou et al. [72] prepared three-dimensional (3D) POMOF structure with $BW_{12}O_{40}{}^{5-}$ (denoted BW_{12}) of $[Cu_x(bty)_y][BW_{12}]$ (bty = bis(1,2,4-triazol-1-yl)ethane) in 0.1 M H_2SO_4. The electrochemical response of $[Cu_x(bty)_y][BW_{12}]$ translated the general faradaic behaviour of POMs with two redox peaks assigned to:

$$BW_{12}O_{40}{}^{5-} + 6e^- + 5H^+ \rightleftharpoons HBW_{12}O_{40}{}^{6-} \tag{11.7}$$

$$HBW_{12}O_{40}{}^{6-} + 6e^- + 5H^+ \rightleftharpoons H_2Bw_{12}O_{40}{}^{7-} \tag{11.8}$$

Two 3D POMOF structures, $[Cu_3(bty)_3][BW_{12}]$ **(1)** and $[Cu_{2.5}(bty)_5][BW_{12}]$ **(2)**, exhibited capacitance values of 214.59 and 186.17 $F\,g^{-1}$ at 0.48 $A\,g^{-1}$ and 91.55% and 90.2% capacitance retention after 1000 cycles, respectively. Using different Keggin-types BW_{12}, SiW_{12} and PMo_{12} POMs, Du et al. [70] synthesized $[(Ag_5(pz)_7)(BW_{12}]$ **(3)**, $[(Ag_5(pz)_7)(SiW_{12}]\text{-}(OH)\cdot H_2O$ **(4)**, $(HPpy)[(Ag(pz))_2(PMo_{12})]$ **(5)** (pz = pyrazine, Ppy = pyrrole), revealing the greater capacitance value of compound **(5)**, 1611 $F\,g^{-1}$ at 2.16 $A\,g^{-1}$, compared to **(3)** and **(4)**. Chai et al. [65] reported two Mo-based POMOFs, $[Cu^I H_2(C_{12}H_{12}N_6)(PMo_{12})]\cdot[(C_6H_{15}N)(H_2O)_2]$ **(6)** and $[Cu^{II}{}_2(C_{12}H_{12}N_6)_4(PMo^{VI}{}_9Mo^V{}_3O_{39})]$ **(7)** ($C_{12}H_{12}N_6$ = 1,4-bis(triazol-1-ylmethyl) benzene), as electrode materials for supercapacitors. Compound **(7)** showed much higher specific capacitance (249.0 $F\,g^{-1}$ at 3 $A\,g^{-1}$) than the **(6)**-based compound, attributed to the mixed valence reaching higher redox capacity [65]. Moreover, compound **(6)** revealed greater cycling stability of 93.5% after 1000 cycles at 10 $A\,g^{-1}$. These examples on the use of POMOFs as electrode materials for supercapacitors show, on one hand, the variety of electrochemical behaviour of Keggin-type POMs based MOF, and on the other hand, promising results for boosting the synthesis of new POMOFs with improved redox chemistry. The combination of POMs with MOFs opens up numerous possibilities that will settle in the intrinsic nature of each of these two families.

Likewise, combining POMs with MXene materials opens many other opportunities to leverage POMs as electroactive materials. MXene $M_{n+1}X_nT_x$ (M = transition metal, X = carbon/nitrogen, n = integer between 1 and 3 and T_x = surface functional groups), are a family of 2D structures that possess high conductivity and redox activity [74]. Due to their open structure, MXene provide an electronically conducting matrix capable of integrating POMs. The major bottleneck for the preparation of MXene/POM hybrid materials is the electrostatic repulsion occurring between negatively charged functional groups in MXene monolayers and the polyanion nature of POMs. To overcome this limitation, Gomes-Romero et al. [74] developed a strategy to intercalate PW_{12} POM clusters into MXene layers by using the interactions between POM anions and organic cations (cetyltrimethylammonium cations, CTA^+) pre-intercalated into the MXene. The CTA^+-delaminated TiC_2T_x/PW_{12} revealed improved stability over MXene/PW_{12} material, overcoming the aggregation of PW_{12} clusters. Other transition-metal complexes have been employed as supports for POMs, forming coordination hybrids for supercapacitors [75–79].

The potential for designing POM-based supramolecular structures is vast, constituting an expansive research field that encompasses both the intricate preparation processes and their diverse applications as supercapacitor electrodes.

11.2.2 POMs as Electrolyte Additives

The progress in supercapacitor systems requires, as well, the development of efficient electrolytic solutions. A good electrolyte solution for a supercapacitor should exhibit high ionic conductivity, a wide electrochemical stability window and be environmentally friendly. Usually, organic solvents offer a wider operation voltage window in relation to traditional aqueous-based solutions owing to the water-splitting reactions that constrain the cathodic and anodic limits and, therefore, a lower energy density is achieved (following Eq. 11.1). However, organic electrolytes show low ionic conductivity, high toxicity and flammability, features that constrain their sustainable use in supercapacitors. Matching the requirements for achieving high-energy density devices, redox mediator electrolytes is an approach to reduce the negative features of traditional water/organic-based electrolytes and, at the same time, maximize the performance of the electrode materials.

In solid-state electrolytes, POMs are regarded as promising additives to the electrolyte for supercapacitor applications, owing to their redox behaviour and high ionic conductivity. Compared to traditional electrolytes (e.g. H_2SO_4), POMs exhibit redox activity capable of improving the charge-storage capability of the device, and also provide high ion mobility, being considered electron–proton reservoirs. Usually, POMs are dispersed in a polymer matrix to form a quasi-solid or solid electrolyte in order to obtain an ion conductive network. Different solid polymer electrolyte-based with POMs as proton conductors have been reported, including PVA [80–85], PAM [86], PANi [87] and a polyacrylate elastomer [88]. In this section,

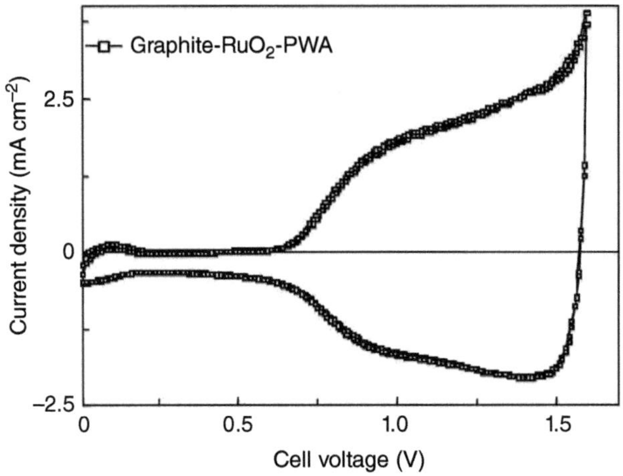

Figure 11.16 POM-based materials as additives for electrolytes: Cyclic voltammogram of all solid graphite–PVA–PWA–RuO_2 asymmetric capacitor cell at 100 mV s^{-1}. Source: Reproduced with permission of Lian and Li [82]/Elsevier.

we will discuss briefly POM/PVA electrolyte systems for solid-state supercapacitors [80–85].

In 2009, Lian et al. [82] compared the performance of the symmetrical capacitor RuO_2-SiW_{12}/PVA-RuO_2 totally immersed in aqueous electrolyte and the solid polymer electrolyte composed of SiW_{12}/PVA. A rectangular-shaped voltammogram of both aqueous-based and solid polymer-based electrolytes suggested a capacitive behaviour in both cases. Although the latter electrolyte showed slightly lower capacitance (50 mF cm^{-2} *vs.* 70 mF cm^{-2}), the SiW_{12}/PVA solid electrolyte revealed a good ionic conductivity of 0.01 S cm^{-1} at room temperature and cyclability after 20000 cycles (at 1 V s^{-1}). Applying the PVA–PWA solid electrolyte into an asymmetric all solid graphite–PVA–PWA–RuO_2 capacitor cell, two different capacitive regions were observed, supported by two different slopes in the charge/discharge curves, Figure 11.16: from 0 to 0.7 V, a double-layer predominant behaviour with a capacitance of 1–2 mF cm^{-2}; and between 0.7 and 1.55, a capacitive behaviour with 22 mF cm^{-2}. Further optimization of the POM/PVA electrolyte system revealed that the combination of PW_{12} and SiW_{12} POMs improved the proton conductivity of 0.013 S cm^{-1} at room temperature when compared to the value of the individual components, assigned to synergistic effect between structures [80]. These studies encouraged the development of others POM/PVA-based solid electrolytes not only because of the good proton conductivity, but also because of the additional capacitance resulting from the redox reactions of POMs.

In a subsequent work, Lian et al. [83, 84] continuously developed a BW_{12} (BWA)/PVA solid electrolyte that exhibited a wider working potential window than SiWA/PVA in a EDLC cell with two CNT–graphite electrodes. The presence of B heteroatom in place of Si in the POM led to an increase of the potential window from 1.4 to 1.6 V. A capacitance of ~7 mF cm^{-2} (at 1 V s^{-1}) in a cell voltage of

1.6 V was determined for BWA/PVA-based solid supercapacitor, indicating good accessibility of ions in the polymer electrolyte [83].

Despite the current application of proton-conducting POMs to supercapacitors, some research is still missing for overcome this problem, including the loss of proton conductivity over time, and the capacitance retention of POMs in solid state (whether it is the same as in solution) [9].

11.3 Conclusions and Future Perspectives

POMs have demonstrated their potential as electrode and electrolyte additives in supercapacitors. This success stems from addressing their high solubility in both aqueous and organic electrolytes by combining POMs with conductive substrates. While extensive research has focused on combining POMs with AC, limited information exists regarding the combination with conductive polymers. In the realm of electrolyte additives, polymers like PVA and PAM have also been reported, with efforts to tune POM chemistry to expand the potential window of the electrolyte's redox activity. The emerging approach involves designing TNH structures to leverage the advantages of conductive carbon's large surface area and polymers as linkers. Numerous possibilities for TNHs and combining POMs with supramolecular structures remain unexplored for electroactive materials and may be thought of as electrolyte additives.

In conclusion, the synergistic effect of POM combined with various conductive structures relies on the interactions POM establishes. The electrochemical performance of the composite material is influenced not only by the chosen materials but also by proportions and assembly design. This encompasses POM dispersion and structure. In electrodes, a structure is composed of composites or layered materials, matters. Layered structures require careful consideration of layer order; for instance, placing POMs on the outermost layer may harm stability but enhance charge transfer capability.

The prominent POM for supercapacitor electrodes is the Keggin-type PMo12 POM, exhibiting strong faradaic features with notable redox peaks attributed to reversible two-electrode transfer in Mo active sites. POMs with W and V have also been identified as electroactive materials. On the support side, CNT and rGO have been widely reported, while PANi and PPy are common polymers. Despite a wealth of data on POMs as electrode materials, comparing performances proves to be challenging due to varied formulations, preparation methods and electrochemical conditions. Standardized measurements are necessary to facilitate meaningful comparisons and pave the way for POMs as electroactive materials in supercapacitors.

Future challenges include exploring additional combinations, understanding POM-support interactions for new synthesis methodologies, investigating complex POM structures and testing diverse conductive supports. Expanding beyond aqueous electrolytes to ionic liquids or deep eutectic solvents, deeper electrochemical analysis and standardizing methods for POM electrochemical analysis is crucial for advancing POM applications in supercapacitors. This knowledge not only benefits energy storage but also extends to other areas like humidity sensing (from PANI) and self-healing, broadening the scope of POMs in multifunctional electronic and energy devices.

Acknowledgements

Centro de Química Estrutural Research Unit was funded by Fundação para a Ciência e Tecnologia (FCT) through projects UIDB/00100/2020 (DOI: 10.54499/UIDP/00100/2020) and UIDP/00100/2020 (DOI: 10.54499/UIDB/00100/2020). Institute of Molecular Sciences Associate Laboratory was funded by FCT through project LA/P/0056/2020 (DOI: 10.54499/LA/P/0056/2020). The authors thank the FCT for the financial support under the projects PTDC/QUI-ELT/2075/2020 (DOI: 10.54499/PTDC/QUI-ELT/2075/2020). A.C. Alves acknowledges the PhD fellowship 2020.05407.BD funded by FCT.

References

1 Simon, P. and Gogotsi, Y. (2020). Perspectives for electrochemical capacitors and related devices. *Nat. Mater.* 19 (11): 1151–1163.

2 Jost, K., Dion, G., and Gogotsi, Y. (2014). Textile energy storage in perspective. *J. Mater. Chem. A* 2 (28): 10776–10787.

3 Zhang, L.L. and Zhao, X.S. (2009). Carbon-based materials as supercapacitor electrodes. *Chem. Soc. Rev.* 38 (9): 2520–2531.

4 Yoshida, N., Yamada, Y., Nishimura, S.-i. et al. (2013). Unveiling the origin of unusual pseudocapacitance of $RuO_2 \cdot nH_2O$ from its hierarchical nanostructure by small-angle X-ray scattering. *J. Phys. Chem. C* 117 (23): 12003–12009.

5 umicore (2024). Ruthenium. https://pmm.umicore.com/en/prices/ruthenium/ (accessed 22 April 2024).

6 Alves, A.C., Chiavassa, L., Martins, T.D. et al. (2023). Retaining the specific capacitance under electrochemical stress: a pH-induced self-protection mechanism for manganese dioxide pseudocapacitive electrodes. *Electrochem. Commun.* 156: 107596.

7 Zhang, Y., Liu, J., Li, S.-L. et al. (2019). Polyoxometalate-based materials for sustainable and clean energy conversion and storage. *EnergyChem* 1 (3): 100021.

8 Horn, M.R., Singh, A., Alomari, S. et al. (2021). Polyoxometalates (POMs): from electroactive clusters to energy materials. *Energy Environ. Sci.* 14 (4): 1652–1700.

9 Cheng, D., Li, K., Zang, H., and Chen, J. (2022). Recent advances on polyoxometalate-based ion-conducting electrolytes for energy-related devices. *Energy Environ. Mater.* 6 (2): 12341.

10 Gumerova, N.I. and Rompel, A. (2018). Synthesis, structures and applications of electron-rich polyoxometalates. *Nat. Rev. Chem.* 2 (2): 0112.

11 Long, D.L., Tsunashima, R., and Cronin, L. (2010). Polyoxometalates: building blocks for functional nanoscale systems. *Angew. Chem. Int. Ed. Engl.* 49 (10): 1736–1758.

12 Yamada, A. and Goodenough, J.B. (1998). Keggin-type heteropolyacids as electrode materials for electrochemical supercapacitors. *J. Electrochem. Soc.* 145 (3): 737–743.

13 Genovese, M. and Lian, K. (2015). Polyoxometalate modified inorganic–organic nanocomposite materials for energy storage applications: a review. *Curr. Opin. Solid State Mater. Sci.* 19 (2): 126–137.

14 Chai, D., Gómez-García, C.J., Li, B. et al. (2019). Polyoxometalate-based metal-organic frameworks for boosting electrochemical capacitor performance. *Chem. Eng. J.* 373: 587–597.

15 Bajwa, G., Genovese, M., and Lian, K. (2013). Multilayer polyoxometalates-carbon nanotube composites for electrochemical capacitors. *ECS J. Solid State Sci. Technol.* 2 (10): M3046–M3050.

16 Wang, H., Hamanaka, S., Nishimoto, Y. et al. (2012). In operando X-ray absorption fine structure studies of polyoxometalate molecular cluster batteries: polyoxometalates as electron sponges. *JACS* 134 (10): 4918–4924.

17 Wang, C., Song, Y., Cong, W. et al. (2023). From surface loading to precise confinement of polyoxometalates for electrochemical energy storage. *Chin. Chem. Lett.* 34 (12): 108194.

18 Naskar, P., Chakraborty, P., Kundu, D. et al. (2021). Envisaging future energy storage materials for supercapacitors: an ensemble of preliminary attempts. *ChemistrySelect* 6 (5): 1127–1161.

19 Zhu, J.-J., Benages-Vilau, R., and Gomez-Romero, P. (2020). Can polyoxometalates enhance the capacitance and energy density of activated carbon in organic electrolyte supercapacitors? *Electrochim. Acta* 362: 137007.

20 Gómez-Romero, P., Chojak, M., Cuentas-Gallegos, K. et al. (2003). Hybrid organic–inorganic nanocomposite materials for application in solid state electrochemical supercapacitors. *Electrochem. Commun.* 5 (2): 149–153.

21 Suárez-Guevara, J., Ruiz, V., and Gomez-Romero, P. (2014). Hybrid energy storage: high voltage aqueous supercapacitors based on activated carbon–phosphotungstate hybrid materials. *J. Mater. Chem. A* 2 (4): 1014–1021.

22 Cheng, H., Liu, R., Zhang, R. et al. (2023). Recent advances in supramolecular self-assembly derived materials for high-performance supercapacitors. *Nanoscale Adv.* 5 (9): 2394–2412.

23 Akter, T., Hu, K., and Lian, K. (2011). Investigations of multilayer polyoxometalates-modified carbon nanotubes for electrochemical capacitors. *Electrochim. Acta* 56 (14): 4966–4971.

24 Bajwa, G., Akter, T., and Lian, K. (2011). Polyoxometalates modified carbon nanotubes for electrochemical capacitors. *ECS Trans.* 35 (25): 31–37.

25 Ruiz, V., Suárez-Guevara, J., and Gomez-Romero, P. (2012). Hybrid electrodes based on polyoxometalate–carbon materials for electrochemical supercapacitors. *Electrochem. Commun.* 24: 35–38.

26 Genovese, M. and Lian, K. (2014). Pseudocapacitive behavior of Keggin type polyoxometalate mixtures. *Electrochem. Commun.* 43: 60–62.

27 Suarez-Guevara, J., Ruiz, V., and Gomez-Romero, P. (2014). Stable graphene-polyoxometalate nanomaterials for application in hybrid supercapacitors. *Phys. Chem. Chem. Phys.* 16 (38): 20411–20414.

28 Dubal, D.P., Chodankar, N.R., Vinu, A. et al. (2017). Asymmetric supercapacitors based on reduced graphene oxide with different polyoxometalates as positive and negative electrodes. *ChemSusChem* 10 (13): 2742–2750.

29 Wang, J., Wang, X., Gu, D. et al. (2018). Synthesis and characterization of polyoxometalate/graphene oxide nanocomposites for supercapacitor. *Ceram. Int.* 44 (14): 17492–17498.

30 Fan, G., Deng, L., and Gu, Y. (2021). Preparation and capacitive properties of a free-standing phosphomolybdic acid heteropoly blue modified reduced graphite oxide composite. *Sustainable Energy Fuels* 5 (6): 1892–1903.

31 Hwang, S.-K., Patil, S.J., Chodankar, N.R. et al. (2022). An aqueous high-performance hybrid supercapacitor with MXene and polyoxometalates electrodes. *Chem. Eng. J.* 427: 131854.

32 Cuentas-Gallegos, A.K., Martínez-Rosales, R., Baibarac, M. et al. (2007). Electrochemical supercapacitors based on novel hybrid materials made of carbon nanotubes and polyoxometalates. *Electrochem. Commun.* 9 (8): 2088–2092.

33 Genovese, M., Foong, Y.W., and Lian, K. (2016). The unique properties of aqueous polyoxometalate (POM) mixtures and their role in the design of molecular coatings for electrochemical energy storage. *Electrochim. Acta* 199: 261–269.

34 Ding, Y.-H., Peng, J., Lu, H.-Y. et al. (2016). Tungsten addenda mixed heteropolymolybdates supported on functionalized graphene for high-performance aqueous supercapacitors. *RSC Adv.* 6 (84): 81085–81091.

35 Chen, H.Y., Al-Oweini, R., Friedl, J. et al. (2015). A novel SWCNT-polyoxometalate nanohybrid material as an electrode for electrochemical supercapacitors. *Nanoscale* 7 (17): 7934–7941.

36 Mu, A., Li, J., Chen, W. et al. (2015). The composite material based on Dawson-type polyoxometalate and activated carbon as the supercapacitor electrode. *Inorg. Chem. Commun.* 55: 149–152.

37 Dong, Y., Chen, L., Chen, W. et al. (2018). rGO Functionalized with a highly electronegative keplerate-type polyoxometalate for high-energy-density aqueous asymmetric supercapacitors. *Chem. Asian J.* 13 (21): 3304–3313.

38 Pakulski, D., Gorczyński, A., Czepa, W. et al. (2019). Novel Keplerate type polyoxometalate-surfactant-graphene hybrids as advanced electrode materials for supercapacitors. *Energy Storage Mater.* 17: 186–193.

39 Hou, Y., Chai, D., Li, B. et al. (2019). Polyoxometalate-incorporated metallacalixarene@graphene composite electrodes for high-performance supercapacitors. *ACS Appl. Mater. Interfaces* 11 (23): 20845–20853.

40 Biradar, B.R., Maity, S., Chandewar, P.R. et al. (2023). High areal capacitance polyoxotungstate-reduced graphene oxide-based supercapacitors. *Inorg. Chem. Commun.* 155: 110987.

41 Vannathan, A.A., Kella, T., Shee, D., and Mal, S.S. (2023). High-performance electrochemical supercapacitors based on polyoxometalate integrated into polyaniline and activated carbon nanohybrid. *Ionics* 29: 4227–4241.

42 Cuentas-Gallegos, A.K., Martínez-Rosales, R., Rincón, M.E. et al. (2006). Design of hybrid materials based on carbon nanotubes and polyoxometalates. *Opt. Mater.* 29 (1): 126–133.

43 Lu, L. and Xie, Y. (2018). Phosphomolybdic acid cluster bridging carbon dots and polyaniline nanofibers for effective electrochemical energy storage. *J. Mater. Sci.* 54 (6): 4842–4858.

44 Ji, Y., Huang, L., Hu, J. et al. (2015). Polyoxometalate-functionalized nanocarbon materials for energy conversion, energy storage and sensor systems. *Energy Environ. Sci.* 8 (3): 776–789.

45 Dubal, D.P., Suarez-Guevara, J., Tonti, D. et al. (2015). A high voltage solid state symmetric supercapacitor based on graphene–polyoxometalate hybrid electrodes with a hydroquinone doped hybrid gel-electrolyte. *J. Mater. Chem. A* 3 (46): 23483–23492.

46 Nezakati, T., Seifalian, A., Tan, A., and Seifalian, A.M. (2018). Conductive polymers: opportunities and challenges in biomedical applications. *Chem. Rev.* 118 (14): 6766–6843.

47 Gómez-Romero, P., Cuentas-Gallegos, K., Lira-Cantú, M., and N. C-P. (2005). Hybrid nanocomposite materials for energy storage and conversion applications. *J. Mater. Sci.* 40: 1423–1428.

48 Ray, S.S., Gusain, R., and Kumar, N. (2020). Conducting polymer-functionalized carbon nanomaterials-based adsorbents. In: *Carbon Nanomaterial-Based Adsorbents for Water Purification*, 327–340. Elsevier.

49 Suppes, G.M., Cameron, C.G., and Freund, M.S. (2010). A polypyrrole/phosphomolybdic acid|poly(3,4-ethylenedioxythiophene)/phosphotungstic acid asymmetric supercapacitor. *J. Electrochem. Soc.* 157 (9): A1030–A1034.

50 Yang, M., Hong, S.B., Yoon, J.H. et al. (2016). Fabrication of flexible, redoxable, and conductive nanopillar arrays with enhanced electrochemical performance. *ACS Appl. Mater. Interfaces* 8 (34): 22220–22226.

51 Vannathan, A.A., Maity, S., Kella, T. et al. (2020). In situ vanadophosphomolybdate impregnated into conducting polypyrrole for supercapacitor. *Electrochim. Acta* 364: 137286.

52 Alshehri, S.A., Al-Yasari, A., Marken, F., and Fielden, J. (2020). Covalently linked polyoxometalate–polypyrrole hybrids: electropolymer materials with dual-mode enhanced capacitive energy storage. *Macromolecules* 53 (24): 11120–11129.

53 Al-Ghaus, Z., Akbarinejad, A., Zhu, B., and Travas-Sejdic, J. (2021). Polyluminol-polyoxometalate hybrid hydrogels as flexible and soft supercapacitor electrodes. *J. Mater. Chem. A* 9 (36): 20783–20793.

54 Gómez-Romero, P., Ayyad, O., Suárez-Guevara, J., and Muñoz-Rojas, D. (2010). Hybrid organic–inorganic materials: from child's play to energy applications. *J. Solid State Electrochem.* 14 (11): 1939–1945.

55 Cuentas-Gallegos, A.K., Lira-Cantú, M., Casañ-Pastor, N., and Gómez-Romero, P. (2005). Nanocomposite hybrid molecular materials for application in solid-state electrochemical supercapacitors. *Adv. Funct. Mater.* 15 (7): 1125–1133.

56 Yang, M., Choi, B.G., Jung, S.C. et al. (2014). Polyoxometalate-coupled graphene via polymeric ionic liquid linker for supercapacitors. *Adv. Funct. Mater.* 24 (46): 7301–7309.

57 Yang, P., Xie, J., Wang, L. et al. (2020). Coaxial cable-like carbon nanotubes-based active fibers for highly capacitive and stable supercapacitor. *Adv. Mater. Interfaces* 7 (18): 2000949.

58 Wang, M., Zhang, Y., Zhang, T. et al. (2020). Confinement of single polyoxometalate clusters in molecular-scale cages for improved flexible solid-state supercapacitors. *Nanoscale* 12 (22): 11887–11898.

59 Chen, Y., Han, M., Tang, Y. et al. (2015). Polypyrrole-polyoxometalate/reduced graphene oxide ternary nanohybrids for flexible, all-solid-state supercapacitors. *Chem. Commun. (Camb.)* 51 (62): 12377–12380.

60 Lu, L. and Xie, Y. (2017). Fabrication and supercapacitor behavior of phosphomolybdic acid/polyaniline/titanium nitride core–shell nanowire array. *New J. Chem.* 41 (1): 335–346.

61 Sun, X., Qu, Y., Wang, G. et al. (2022). Two coordination polymers@graphene hybrid electrodes for high-performance supercapacitors with enhanced rate capability and specific capacitance. *ChemistrySelect* 7 (6): 202103660.

62 Wang, H.N., Zhang, M., Zhang, A.M. et al. (2018). Polyoxometalate-based metal-organic frameworks with conductive polypyrrole for supercapacitors. *ACS Appl. Mater. Interfaces* 10 (38): 32265–32270.

63 Zhou, X., Shang, C., Gu, L. et al. (2011). Mesoporous coaxial titanium nitride-vanadium nitride fibers of core-shell structures for high-performance supercapacitors. *ACS Appl. Mater. Interfaces* 3 (8): 3058–3063.

64 Liu, J., Huang, M., Hua, Z. et al. (2022). Polyoxometalate-based metal organic frameworks: recent advances and challenges. *ChemistrySelect* 7 (18): 202200546.

65 Chai, D., Xin, J., Li, B. et al. (2019). Mo-based crystal POMOFs with a high electrochemical capacitor performance. *Dalton Trans.* 48 (34): 13026–13033.

66 Liu, Y.Z., Yao, W., Gan, H.M. et al. (2019). Polyoxometalates-based metal-organic frameworks made by electrodeposition and carbonization methods as cathodes and anodes for asymmetric supercapacitors. *Chemistry* 25 (72): 16617–16624.

67 Liu, Y., Xu, N., Chen, W. et al. (2018). Supercapacitor with high cycling stability through electrochemical deposition of metal-organic frameworks/polypyrrole positive electrode. *Dalton Trans.* 47 (38): 13472–13478.

68 Wang, L., Feng, X., Ren, L. et al. (2015). Flexible solid-state supercapacitor based on a metal-organic framework interwoven by electrochemically-deposited PANI. *JACS* 137 (15): 4920–4923.

69 Choi, K.M., Jeong, H.M., Park, J.H. et al. (2014). Supercapacitors of nanocrystalline metal-organic frameworks. *ACS Nano* 8 (7): 7451–7457.

70 Du, N., Gong, L., Fan, L. et al. (2019). Nanocomposites containing Keggin anions anchored on pyrazine-based frameworks for use as supercapacitors and photocatalysts. *ACS Appl. Nano Mater.* 2 (5): 3039–3049.

71 Zheng, Z., Zhao, X., Gong, L. et al. (2020). Coral-like $\{SiW_{10}Mn_2\}$-based Mn-MOFs: facile fabrication with high electrochemical capacitor performance. *J. Solid State Chem.* 288: 121409.

72 Hou, Y., Han, P., Zhang, L. et al. (2023). pH-controlled assembling of POM-based metal–organic frameworks for use as supercapacitors and efficient oxidation catalysts for various sulfides. *Inorg. Chem. Front.* 10 (1): 148–157.

73 Ma, L., Lu, B., Li, S. et al. (2023). Two new 2D POMOFs based on octamolybdate/copper substituted keggin polymolybdate for enhancing electrochemical capacitor performance. *J. Mol. Struct.* 1281: 135133.

74 Zhu, J.J. and Gomez-Romero, P. (2022). Polyoxometalate intercalated MXene with enhanced electrochemical stability. *Nanoscale* 14 (40): 14921–14934.

75 Wang, C., Rong, S., Zhao, Y. et al. (2021). Three-dimensional supramolecular crystalline materials based on Keggin-based polyoxometalates and 1,2-bis(4-pyridyl) ethylene for supercapacitor electrodes. *Transition Met. Chem.* 46 (4): 335–343.

76 Hu, S., Li, K., Yu, X. et al. (2022). Enhancing the electrochemical capacitor performance of Keggin polyoxometalates by anchoring cobalt-triazole complexes. *J. Mol. Struct.* 1250: 131753.

77 He, L.-L., Cui, L.-P., Yu, K. et al. (2022). The pseudocapacitance and sensing materials constructed by Dawson/basket-like phosphomolybdate. *J. Solid State Chem.* 316: 123578.

78 Chen, Y., Liang, J., Chang, Z., and Wang, X. (2023). A $\{PMo_{12}\}$-based 2D sandwich-like supramolecular network constructed from a new semi-rigid amide-derived ligand with enhanced capacitive activity and electrochemical sensing performances. *Inorg. Chim. Acta* 552: 121490.

79 Zhou, C.-W., Wang, X.-Y., Duan, Z.-P. et al. (2023). Construction of Sb-capped Dawson-type POM derivatives for high-performance asymmetric supercapacitors. *Electrochim. Acta* 442: 141823.

80 Gao, H., Tian, Q., and Lian, K. (2010). Polyvinyl alcohol-heteropoly acid polymer electrolytes and their applications in electrochemical capacitors. *Solid State Ionics* 181: 874–876.

81 Lian, K. and Tian, Q. (2010). Solid asymmetric electrochemical capacitors using proton-conducting polymer electrolytes. *Electrochem. Commun.* 12: 517–519.

82 Lian, K. and Li, C.M. (2009). Solid polymer electrochemical capacitors using heteropoly acid electrolytes. *Electrochem. Commun.* 11: 22–24.

83 Gao, H., Virya, A., and Lian, K. (2015). Proton conducting $H_5BW_{12}O_{40}$ electrolyte for solid supercapacitors. *J. Mater. Chem. A* 3: 21511.

84 Gao, H. and Lian, K. (2016). A $H_5BW_{12}O_{40}$–polyvinyl alcohol polymer electrolyte and its application in solid supercapacitors. *J. Mater. Chem. A* 4: 9585.

85 Cheng, D., Li, B., Sun, S. et al. (2020). Proton-conducting polyoxometalates as redox electrolytes synergistically boosting the performance of self-healing solid-state supercapacitors with polyaniline. *CCS Chem.* 2: 1649–58a.

86 Foong, Y.W., Lian, K., Kirk, D., and Thorpe, S. (2018). Borotungstic acid – polyacrylamide solid electrolytes for electrochemical capacitors with H_3PO_4 plasticizer. *Mater. Sci. Eng., B* 229: 96–104.
87 Cheng, D., Gao, Z., Wang, W. et al. (2023). Zwitterion-dissociated polyoxometalate electrolytes for solid-state supercapacitors. *Polyoxometalates* 2: 9140019.
88 Guo, H., Zeng, M., Li, X. et al. (2021). Multifunctional enhancement of proton-conductive, stretchable, and adhesive performance in hybrid polymer electrolytes by polyoxometalate nanoclusters. *ACS Appl. Mater. Interfaces* 13: 30039–30050.

Index

Applied Polyoxometalate-based Electrocatalysis, First Edition. Edited by Diana M. Fernandes.

d

e

p

r